METHODS IN MOLECULAR BIOLOGY

For further volumes:
http://www.springer.com/series/7651

Pharmaceutical Nanotechnology

Basic Protocols

Edited by

Volkmar Weissig and Tamer Elbayoumi

Department of Pharmaceutical Sciences, College of Pharmacy, Midwestern University, Glendale, AZ, USA

Nanomedicine Center of Excellence in Translational Cancer Research, Midwestern University, Glendale, AZ, USA

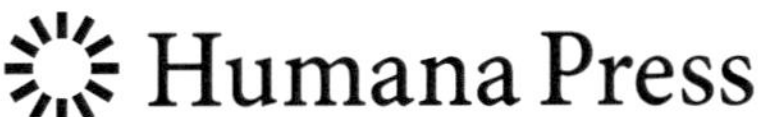

Editors
Volkmar Weissig
Department of Pharmaceutical Sciences
College of Pharmacy, Midwestern University
Glendale, AZ, USA

Nanomedicine Center of Excellence
in Translational Cancer Research
Midwestern University
Glendale, AZ, USA

Tamer Elbayoumi
Department of Pharmaceutical Sciences
College of Pharmacy, Midwestern University
Glendale, AZ, USA

Nanomedicine Center of Excellence
in Translational Cancer Research
Midwestern University
Glendale, AZ, USA

ISSN 1064-3745 ISSN 1940-6029 (electronic)
Methods in Molecular Biology
ISBN 978-1-4939-9518-9 ISBN 978-1-4939-9516-5 (eBook)
https://doi.org/10.1007/978-1-4939-9516-5

This Humana imprint is published by the registered company Springer Science+Business Media, LLC, part of Springer Nature.
The registered company address is: 233 Spring Street, New York, NY 10013, U.S.A.

Preface

This book is the continuation of our efforts to translate the progress being made in nanotechnology, as applied to Biomedical Science, into easy-to-follow protocols within the MiMB Springer Protocols series. In *Methods in Molecular Biology* volumes 605 [1] and 606 [2], we focused exclusively on liposomes, the very first FDA-approved pharmaceutical nanocarrier platform. In MiMB volume 991 [3], we expanded the scope of step-by-step protocols to a large variety of nanotechnologies used for probing, imaging, and manipulating metabolic functions on cellular and subcellular levels. In this present volume, we introduce basic protocols needed for the formulation, surface modification, characterization, and application of a variety of pharmaceutical nanocarriers such as micelles, nanoparticles, dendrimers, carbon dots, polymersomes, and others. All protocols should be of particular interest to investigators working in academic and industrial laboratories conducting research in the wide field of pharmaceutical sciences with an emphasis on drug delivery.

All chapters were written by accomplished experts in their specific fields, and we are very grateful to all authors for having spent parts of their valuable time to contribute to this book.

In an opening chapter, Sachin Kumar Singh reviews theories and practices of a large variety of nanovesicles. Vandana B. Patravale and her colleagues provide in two chapters detailed protocols for the preparation and characterization of micelles and of solid lipid nanoparticles, respectively. Amphiphilic mixed micelles—composed of polymeric phospholipid conjugates and PEG-succinate ester of tocopherol for improving berberine (an isoquinoline alkaloid) delivery into tumors—are described by Mingyi Yao and Tamer Elbayoumi, while Hayat Onyuksel and her colleagues give detailed protocols for the application of phospholipid micelles for peptide drug delivery. The design, preparation, and characterization of peptide-based nanocarriers for gene delivery are described by Saman Hosseinkhani and Mohsen Alipour. Protocols for the preparation and characterization of gelatin nanoparticles are provided by Rania M. Hathout and Abdelkader A. Metwally. Two protocol chapters were written by Ibrahim M. El-Sherbiny and co-workers; the first one discusses the green synthesis of chitosan-silver/chitosan-gold hybrid nanoparticles and the second one a method for the fabrication of chitosan-based nano-in-microparticles. Rakesh Tekade and his colleagues describe in one chapter the preparation of mucoadhesive dendrimers as solid dosage forms and, in another chapter, methods for the surface modification of nanoparticles employing polymer coating. Plant viral nanoparticles for the in situ vaccination of tumors are presented by Nicole Steinmetz and her group. A detailed and very comprehensive review chapter, by Mohamed Ismail Nounou, discusses the current state of the art in the area of bioconjugation as applied to targeted nanoscale drug delivery. Diana Guzman-Villanueva and Volkmar Weissig give a detailed protocol for the preparation of triphenylphosphonium-based mitochondria-targeted nanocarriers. Ildiko Badea and colleagues discuss the preparation and characterization of peptide-modified gemini surfactants for gene delivery. The formulation of responsive carbon dots for anticancer drug delivery is presented by Yanli Zhao and Tao Feng. The application of click chemistry for the surface modification of nanocarriers is outlined in detail by Mark Helm and his coauthors. Another review chapter in our book is dedicated to the preparation and characterization of polymersomes, written by Liyan Qiu and Yumiao Hu. Partially polymerized phospholipid vesicles for the efficient delivery of macromolecules are described by Tamer Elbayoumi and coauthors.

Medha D. Joshi and colleagues provide protocols for the fabrication of nanostructured lipid carrier-based gels from microemulsion templates. Finally, novel approaches in studying nanoparticle-biological interactions are presented. The behavior of gold nanoparticles in aqueous solutions and their interaction with lipid membranes are discussed in a review chapter by Sohail Murad and his team. Hongda Wang and Mingjun Cai describe the use of atomic force microscopy for cell membrane investigations. Protocols for the physico-chemical characterization of phthalocyanine-functionalized quantum dots via capillary electrophoresis are provided by Anne Varenne and her colleagues, and the in vitro testing of nanotherapeutics utilizing tumor spheroids is described in detail by Gerard G.M. D'Souza and his group.

It is our hope that our book will become an important source of know-how, as well as a source of inspiration to all investigators, who are as fascinated as we are about the potential of the merger of nanotechnology with pharmaceutical sciences. We would like to thank Patrick Marton and Monica Suchy for inviting us to edit this book; John Walker, the series editor of MiMB, for his unlimited guidance and help throughout the whole process; and David C. Casey for his help in getting this book into production.

Glendale, AZ, USA ***Volkmar Weissig***
Tamer Elbayoumi

References

1. Weissig V (ed) (2010) Liposomes: methods and protocols, vol 1: pharmaceutical nanocarriers. Springer protocols, MiMB 605. Humana Press, New York. pp 1–563
2. Weissig V (ed) (2010) Liposomes: methods and protocols, volume 2: biological membrane models. Springer protocols, MiMB 606. Humana Press, New York. pp 1–548
3. Weissig V, Elbayoumi T, Olsen M (eds) (2013) Cellular and subcellular nanotechnology: methods and protocols. Springer protocols, MiMB 991. Humana Press, New York, pp 1–370

Contents

Contributors

MAYS AL-DULAYMI • *College of Pharmacy and Nutrition, University of Saskatchewan, Saskatoon, SK, Canada*

MOHSEN ALIPOUR • *Department of Nanobiotechnology, Faculty of Biological Sciences, Tarbiat Modares University, Tehran, Iran*

MOHAMMED A. ALMUTERI • *School of Pharmacy-Boston, MCPHS University, Boston, MA, USA*

ILDIKO BADEA • *College of Pharmacy and Nutrition, University of Saskatchewan, Saskatoon, SK, Canada*

FETHI BEDIOUI • *Chimie ParisTech, PSL Research University, Unité de Technologies Chimiques et Biologiques pour la Santé UTCBS, Paris, France; INSERM, UTCBS U 1022, Paris, France; CNRS, UTCBS UMR 8258, Paris, France; Université Paris Descartes, Sorbonne Paris Cité, UTCBS, Paris, France*

AHMED S. F. BELAL • *Department of Pharmaceutical Chemistry, Faculty of Pharmacy, Alexandria University, Alexandria, Egypt*

MINGJUN CAI • *Changchun Institute of Applied Chemistry, Chinese Academy of Sciences, Changchun, Jilin, People's Republic of China*

FANNY D'ORLYÉ • *Chimie ParisTech, PSL Research University, Unité de Technologies Chimiques et Biologiques pour la Santé UTCBS, Paris, France; INSERM, UTCBS U 1022, Paris, France; CNRS, UTCBS UMR 8258, Paris, France; Université Paris Descartes, Sorbonne Paris Cité, UTCBS, Paris, France*

GERARD G. M. D'SOUZA • *School of Pharmacy-Boston, MCPHS University, Boston, MA, USA*

FATEMA ELAMRAWY • *Department of Pharmaceutics, Faculty of Pharmacy, Alexandria University, Alexandria, Egypt*

ANAS EL-ANEED • *College of Pharmacy and Nutrition, University of Saskatchewan, Saskatoon, SK, Canada*

TAMER ELBAYOUMI • *Department of Pharmaceutical Sciences, College of Pharmacy, Midwestern University, Glendale, AZ, USA; Nanomedicine Center of Excellence in Translational Cancer Research, Midwestern University, Glendale, AZ, USA*

IBRAHIM M. EL-SHERBINY • *Center for Materials Science, Zewail City of Science and Technology, University of Science and Technology (UST), Giza, Egypt*

PERIHAN ELZAHHAR • *Department of Pharmaceutical Chemistry, Faculty of Pharmacy, Alexandria University, Alexandria, Egypt*

KARINA ESPARZA • *Department of Biopharmaceutical Sciences, University of Illinois at Chicago, Chicago, IL, USA*

TAO FENG • *Division of Chemistry and Biological Chemistry, School of Physical and Mathematical Sciences, Nanyang Technological University, Singapore, Singapore*

STEVEN N. FIERING • *Department of Microbiology and Immunology, Geisel School of Medicine at Dartmouth Lebanon, Norris Cotton Cancer Center, Lebanon, NH, USA; Department of Genetics, Geisel School of Medicine at Dartmouth Lebanon, Norris Cotton Cancer Center, Lebanon, NH, USA*

HOLGER FREY • *Institute of Organic Chemistry, Johannes Gutenberg-University, Mainz, Germany*

THOMAS FRITZ • *Institute of Pharmacy and Biochemistry, Johannes Gutenberg-University, Mainz, Germany*

AVANTI GANPULE • *School of Pharmacy-Boston, MCPHS University, Boston, MA, USA*

VARUN GARG • *School of Pharmaceutical Sciences, Lovely Professional University, Phagwara, Punjab, India*

MEGHA GOSHI • *Arizona College of Osteopathic Medicine, Midwestern University, Glendale, AZ, USA*

ZISHU GUI • *School of Pharmacy-Boston, MCPHS University, Boston, MA, USA*

MONICA GULATI • *School of Pharmaceutical Sciences, Lovely Professional University, Phagwara, Punjab, India*

DIANA GUZMAN-VILLANUEVA • *Department of Pharmaceutical Sciences, College of Pharmacy, Midwestern University, Glendale, AZ, USA; Nanomedicine Center of Excellence in Translational Cancer Research, Midwestern University, Glendale, AZ, USA*

RANIA M. HATHOUT • *Department of Pharmaceutics and Industrial Pharmacy, Faculty of Pharmacy, Ain Shams University, Cairo, Egypt*

AMR HEFNAWY • *Center for Materials Science, Zewail City of Science and Technology, University of Science and Technology (UST), Giza, Egypt*

NADA A. HELAL • *Department of Pharmaceutics, Faculty of Pharmacy, Alexandria University, Alexandria, Egypt*

MARK HELM • *Institute of Pharmacy and Biochemistry, Johannes Gutenberg-University, Mainz, Germany*

SAMAN HOSSEINKHANI • *Department of Nanobiotechnology, Faculty of Biological Sciences, Tarbiat Modares University, Tehran, Iran; Department of Biochemistry, Faculty of Biological Sciences, Tarbiat Modares University, Tehran, Iran*

YUMIAO HU • *Ministry of Educational (MOE) Key Laboratory of Macromolecular Synthesis and Functionalization, Zhejiang University, Hangzhou, China; Department of Polymer Science and Engineering, Zhejiang University, Hangzhou, China*

RATNESH D. JAIN • *Department of Pharmaceutical Sciences and Technology, Institute of Chemical Technology, Mumbai, India*

CYNTHIA J. JAMESON • *Department of Chemistry, University of Illinois at Chicago, Chicago, IL, USA*

DULARI JAYAWARDENA • *Department of Biopharmaceutical Sciences, University of Illinois at Chicago, Chicago, IL, USA*

MEDHA D. JOSHI • *Department of Pharmaceutical Sciences, Chicago College of Pharmacy, Midwestern University, Downers Grove, IL, USA*

KIRAN KALIA • *Department of Pharmaceuticals, Ministry of Chemicals and Fertilizers, National Institute of Pharmaceutical Education and Research (NIPER)–Ahmedabad, An Institute of National Importance, Government of India, Gandhinagar, Gujarat, India*

BHUPINDER KAPOOR • *School of Pharmaceutical Sciences, Lovely Professional University, Phagwara, Punjab, India*

RAHUL MAHESHWARI • *Department of Pharmaceuticals, Ministry of Chemicals and Fertilizers, National Institute of Pharmaceutical Education and Research (NIPER)–Ahmedabad, An Institute of National Importance, Government of India, Gandhinagar, Gujarat, India*

MARK R. MENDIOLA • *Department of Pharmaceutical Sciences, College of Pharmacy, Midwestern University, Glendale, AZ, USA*

Abdelkader A. Metwally • *Department of Pharmaceutics and Industrial Pharmacy, Faculty of Pharmacy, Ain Shams University, Cairo, Egypt; Department of Pharmaceutics, Health Sciences Center, Kuwait University, Kuwait City, Kuwait*

Amit G. Mirani • *Department of Pharmaceutical Sciences and Technology, Institute of Chemical Technology, Mumbai, India*

Neeraj Mittal • *Department of Pharmaceutical Sciences and Drug Research, Punjabi University, Patiala, Punjab, India*

Waleed Mohammed-Saeid • *College of Pharmacy and Nutrition, University of Saskatchewan, Saskatoon, SK, Canada; College of Pharmacy, Taibah University, Madina, Saudi Arabia*

Sohail Murad • *Department of Chemical Engineering, University of Illinois at Chicago, Chicago, IL, USA; Department of Chemical Engineering, Illinois Institute of Technology, Chicago, IL, USA*

Abner A. Murray • *Department of Molecular Biology and Microbiology, Case Western Reserve University School of Medicine, Cleveland, OH, USA*

Huy X. Nguyen • *Department of Pharmaceutical Sciences, College of Pharmacy, Midwestern University, Glendale, AZ, USA*

Mohamed Ismail Nounou • *Department of Pharmaceutical Sciences, School of Pharmacy & Physician Assistant Studies (SOPPAS), University of Saint Joseph (USJ), Hartford, CT, USA*

Tebello Nyokong • *Department of Chemistry, Rhodes University, Grahamstown, South Africa*

Hayat Onyuksel • *Department of Biopharmaceutical Sciences, University of Illinois at Chicago, Chicago, IL, USA*

Priyanka A. Oroskar • *Department of Chemical Engineering, University of Illinois at Chicago, Chicago, IL, USA*

Vandana B. Patravale • *Department of Pharmaceutical Sciences and Technology, Institute of Chemical Technology, Mumbai, India*

Rashmi H. Prabhu • *Department of Pharmaceutical Sciences and Technology, Institute of Chemical Technology, Mumbai, Maharashtra, India*

Nicholas Pytel • *Arizona College of Osteopathic Medicine, Midwestern University, Glendale, AZ, USA*

Liyan Qiu • *Ministry of Educational (MOE) Key Laboratory of Macromolecular Synthesis and Functionalization, Zhejiang University, Hangzhou, China; Department of Polymer Science and Engineering, Zhejiang University, Hangzhou, China*

Gonzalo Ramírez-García • *Chimie ParisTech, PSL Research University, Unité de Technologies Chimiques et Biologiques pour la Santé UTCBS, Paris, France; INSERM, UTCBS U 1022, Paris, France; CNRS, UTCBS UMR 8258, Paris, France; Université Paris Descartes, Sorbonne Paris Cité, UTCBS, Paris, France; Department of Chemistry, Rhodes University, Grahamstown, South Africa*

Nidhi Raval • *Department of Pharmaceuticals, Ministry of Chemicals and Fertilizers, National Institute of Pharmaceutical Education and Research (NIPER)–Ahmedabad, An Institute of National Importance, Government of India, Gandhinagar, Gujarat, India*

Mohammed Sedki • *Center for Materials Science, Zewail City of Science and Technology, University of Science and Technology (UST), Giza, Egypt*

Mee Rie Sheen • *Department of Microbiology and Immunology, Geisel School of Medicine at Dartmouth Lebanon, Norris Cotton Cancer Center, Lebanon, NH, USA; Department of*

Genetics, Geisel School of Medicine at Dartmouth Lebanon, Norris Cotton Cancer Center, Lebanon, NH, USA; Division of Hematology and Oncology, Beth Israel Deaconess Medical Center, Harvard Medical School, Boston, MA, USA

SACHIN KUMAR SINGH • *School of Pharmaceutical Sciences, Lovely Professional University, Phagwara, Punjab, India*

NICOLE F. STEINMETZ • *Department of Biomedical Engineering, Case Western Reserve University School of Medicine, Cleveland, OH, USA; Department of NanoEngineering, University of California-San Diego, La Jolla, CA, USA; Department of Materials Science and Engineering, Case Western Reserve University School of Engineering, Cleveland, OH, USA; Department of Macromolecular Science and Engineering, Case Western Reserve University School of Engineering, Cleveland, OH, USA; Division of General Medical Sciences-Oncology, Case Western Reserve University, Cleveland, OH, USA*

RAKESH KUMAR TEKADE • *Department of Pharmaceuticals, Ministry of Chemicals and Fertilizers, National Institute of Pharmaceutical Education and Research (NIPER)–Ahmedabad, An Institute of National Importance, Government of India, Gandhinagar, Gujarat, India*

PRASHANT G. UPADHAYA • *Department of Pharmaceutical Sciences and Technology, Institute of Chemical Technology, Mumbai, India*

ANNE VARENNE • *Chimie ParisTech, PSL Research University, Unité de Technologies Chimiques et Biologiques pour la Santé UTCBS, Paris, France; INSERM, UTCBS U 1022, Paris, France; CNRS, UTCBS UMR 8258, Paris, France; Université Paris Descartes, Sorbonne Paris Cité, UTCBS, Paris, France*

FRANK A. VELIZ • *Department of Biomedical Engineering, Case Western Reserve University School of Medicine, Cleveland, OH, USA*

MATTHIAS VOIGT • *Institute of Pharmacy and Biochemistry, Johannes Gutenberg-University, Mainz, Germany*

SHEETU WADHWA • *School of Pharmaceutical Sciences, Lovely Professional University, Phagwara, Punjab, India*

HONGDA WANG • *Changchun Institute of Applied Chemistry, Chinese Academy of Sciences, Changchun, Jilin, People's Republic of China*

VOLKMAR WEISSIG • *Department of Pharmaceutical Sciences, College of Pharmacy, Midwestern University, Glendale, AZ, USA; Nanomedicine Center of Excellence in Translational Cancer Research, Midwestern University, Glendale, AZ, USA*

MATTHIAS WORM • *Institute of Organic Chemistry, Johannes Gutenberg-University, Mainz, Germany*

FRANCIS YAMBAO • *Department of Pharmaceutical Sciences, College of Pharmacy, Midwestern University, Glendale, AZ, USA*

MINGYI YAO • *Department of Pharmaceutical Sciences, College of Pharmacy-Glendale, Midwestern University, Glendale, AZ, USA; Nanomedicine Center of Excellence in Translational Cancer Research (Nanomedicine COE-TCR), College of Pharmacy-Glendale, Midwestern University, Glendale, AZ, USA*

NUSEM YU • *Department of Pharmaceutical Sciences, College of Pharmacy, Midwestern University, Glendale, AZ, USA*

YANLI ZHAO • *Division of Chemistry and Biological Chemistry, School of Physical and Mathematical Sciences, Nanyang Technological University, Singapore, Singapore; School of Materials Science and Engineering, Nanyang Technological University, Singapore, Singapore*

Chapter 1

Nanovesicles for Nanomedicine: Theory and Practices

Sheetu Wadhwa, Varun Garg, Monica Gulati, Bhupinder Kapoor, Sachin Kumar Singh, and Neeraj Mittal

Abstract

Lipid-based nanovesicles such as liposomes, niosomes, and ethosomes are now well recognized as potential candidates for drug delivery and theranostic applications. Some of them have already stepped forward from laboratory to market. The property to entrap lipophilic drugs in their bilayers and hydrophilic drugs in the aqueous milieu makes them a unique carrier for drug delivery. Delivery of drugs/diagnostics to various organs/tissues/cells via nanovesicles is considered to be a topic of long-standing interest with new challenges being posed to formulation scientists with new developments. The key challenge in this context is the physiological and pathological conditions, which make the delivery of drugs extremely difficult at the disease locus and makes their precise delivery ineffective. This chapter gives an insight into the role of novel nanovesicles in the field of drug delivery. We present an overview of the formulation and characterization and role of diverse nanovesicles. A comprehensive update about their application and current as well as potential challenges have also been discussed.

Key words Nanovesicles, Topical, Transdermal, Lipid-based, Drug delivery

1 Introduction

In 1965 citation classic, the late Alec Bangham and colleagues published the first description of swollen phospholipid systems that established the basis for model membrane systems [1–3]. Within a few years, a variety of enclosed phospholipid bilayer structures consisting of single bilayers, initially termed "Banghosomes" and then rechristened as liposomes were described. Gregory Gregoriadis, for the first time, established the concept that liposomes could entrap drugs and be used as drug delivery systems [2, 4–11].

Liposomes can load hydrophilic drugs in aqueous core while lipophilic drugs are contained inside the lipid bilayer. Liposomes demonstrate protection of entrapped moieties from degradation caused by enzymes. As they are prepared from natural materials or

Volkmar Weissig and Tamer Elbayoumi (eds.), *Pharmaceutical Nanotechnology: Basic Protocols*, Methods in Molecular Biology, vol. 2000, https://doi.org/10.1007/978-1-4939-9516-5_1,

their synthetic derivatives, liposomes are biocompatible and biodegradable [12, 13].

Conventional liposomes have some drawbacks. They are complex to prepare, have low inherent stability, and, therefore, cannot be stored for a long time. Liposomes show rapid uptake by the reticuloendothelial system thus decreasing their circulation half-life. Further, leakage of loaded drugs from liposomes results in less drug loading efficiency. Moreover, they are difficult to sterilize. New materials employed in formulation of vesicles may be toxic and nonbiocompatible, hence, alarm for regulatory clearance [12–20].

2 Composition of Nanovesicles

A number of vesicular delivery systems such as liposomes, niosomes, ethosomes, transfersomes, and pharmacosomes were developed.

Liposomes are simple microscopic vesicles in which lipid bilayer structures are present with an aqueous volume entirely enclosed by a membrane, composed of lipid molecule. Liposome composition includes natural and/or synthetic phospholipids (phosphatidylethanolamine, phosphatidylglycerol, phosphatidylcholine, phosphatidylserine, phosphatidylinositol) Phosphatidylcholine (also known as lecithin) and phosphatidylethanolamine constitutes the two major structural components of most biological membranes. Liposome bilayers may also contain other constituents such as cholesterol, hydrophilic polymer conjugated lipids, and water. Cholesterol has been largely used to improve the bilayer characteristics of the liposomes. It improves the membrane fluidity and bilayer stability and reduces the permeability of water-soluble molecules through the membrane [12, 13, 21].

Niosomes are vesicular delivery system based on nonionic surfactants. Niosomes have structural similarity with liposomes. They mainly consist of nonionic surfactants (generally from alkyl or di-alkyl polyglycerol ether class) and cholesterol. Selection of surfactant is usually done on the basis of HLB value. The surfactant molecules tend to orient themselves in such a way that the hydrophilic ends of the nonionic surfactant point outwards, while the hydrophobic ends face each other to form the bilayer [12, 13, 17].

Transfersomes are defined as bilayer vesicles mainly composed of phospholipids similar to that of liposomes and edge activator-like surfactants for flexibility. In these bilayer vesicles, aqueous volume is entirely enclosed in aqueous cavity while hydrophobic drugs are enclosed in between hydrophobic area of phospholipids [22–24]. Transfersomes are vesicles in which cholesterol is replaced by edge activator or ethanol or any other membrane fluidizer so as to make them squeeze through the narrow pores of skin tissue. They are more flexible as compared to liposome. These self-optimized ultra-deformable bilayer lipid vesicles can tolerate ambient stress as they are composed of edge activator and phospholipid [25, 26]. Edge activators used in the preparation of transfersomes are mainly single chain surfactants which are responsible for deformability of vesicles by destabilizing them. Phospholipids are main membrane forming components of the transfersomes which provide stability to vesicles. Thus, for deformable elastic vesicles (transfersomes) a membrane-stabilizing agent like phospholipid and a destabilizing agent, i.e., edge activator is necessary. Various edge activators like span 40, span 60 span 80, span 85, tween 20, tween 60, tween 80, sodium oleate, sodium cholate, sodium deoxycholate, dicetylphosphate (DCP), and KG (dipotassium glycyrrhizinate) are used in the preparation of transfersomes [26–35]. Use of different edge activators may affect different physiochemical properties of vesicles such as size, entrapment efficiency, and zeta potential [32, 35, 36]. Mainly soya phospholipids like soya phosphatidylcholine and soya hydrogenated phosphatidylcholine are used as membrane-stabilizing agents in the preparation of transfersomes [29, 37, 38].

Ethosomes are soft or sophisticated, malleable, and novel vesicular carrier systems tailored for enhanced drug delivery system. These vesicles are modulated from tens of nanometers to microns. They are modified lipid carrier systems with relatively high concentration of ethanol for efficient permeation enhancement in terms of quantity and depth. They are widely used for increasing lipid fluidity and cell membrane permeability by interacting with polar head of lipid molecules resulting in lowering of melting point of the stratum corneum's lipid. Ethosomes are mainly composed of phospholipids, e.g., soya phosphatidylcholine as a vesicle-forming component and high concentration of ethanol, i.e., 20–45%. Propylene glycol is also used which acts as a penetration enhancer [39–41]. A brief description of various nanovesicles is shown in Table 1 and the composition is shown in Table 2. A brief sketch of composition of nanovesicles is shown in Fig. 1.

3 Development of Nanovesicles

Nanovesicles are manufactured by a variety of methods based upon the type of vesicle, type of drug to be loaded, scalability, and lamellarity (Table 2).

3.1 Thin Film Hydration

In this method, lipid mixtures (phosphatidylcholine and cholesterol) are dissolved in organic solvents such as chloroform/methanol mixture. This dissolved lipid solution is added to the round-bottom flask which is attached to the rotary evaporator which is immersed in water bath and rotated at optimum rotations. It should be taken into consideration that the temperature should

Table 1
Description of existing and emerging nanovesicular systems

S. No.	Vesicular systems	Description
1	Liposomes	Liposomes are simple vesicles consisting of lipid bilayer structures with an aqueous volume entirely enclosed by a membrane
2	Niosomes	Niosomes enclose an aqueous solution in a highly ordered bilayer made up of nonionic surfactants
3	Transfersomes	Transfersomes consist of an inner aqueous compartment surrounded by a flexible lipid bilayer due to the incorporation of "edge activators" into it
4	Ethosomes	Ethosomes are vesicles consisting of phospholipids, water, and a high quantity of ethanol
5	Transethosomes	Transethosomes are lipid vesicles based on the properties of both transfersomes and ethosomes
6	Phytosomes	Phytosomes are lipid-compatible molecular complexes of plant-based extracts
7	Discomes	Discomes are large disk-shaped structures formed by addition of specific amount of surfactant to mixed vesicular dispersions
8	Pharmacosomes	Pharmacosomes are colloidal dispersions of drugs covalently bound to lipids
9	Virosomes	Virosomes are liposomes (lipid bilayer of retrovirus-based lipids) spiked with virus glycoproteins
10	Sphingosomes	Sphingosomes are vesicles in which an aqueous core is entirely enclosed by lipid bilayer membrane consisting of natural or synthetic sphingolipid
11	Enzymosomes	Enzymosomes are liposomal units in which an enzyme is covalently immobilized to the surface of liposomes
12	Ufasomes	Ufasomes (unsaturated fatty acid liposomes) are vesicles with long chain fatty acids and soap mixture
13	Bilosomes	Bilosomes are bile-salts-based delivery systems which provide the oral delivery of vaccines
14	Aquasomes	Aquasomes are nano-sized spherical vesicles that enable drug and antigen delivery
15	Emulsomes	Emulsomes are nano-sized lipid-based particles consisting of lipid assembly and a polar group
16	Cubosomes	Cubosomes are bicontinuous cubic phases consisting of two separate, continuous, but non-intersecting hydrophilic regions divided by a lipid layer
17	Cryptosomes	Cryptosomes are lipid-based vesicles with a surface coat composed of phosphatidylcholine and phosphatidyl ethanolamine
18	Colloidosomes	Colloidosomes are hollow shells whose permeability and elasticity are controlled during preparation
19	Genosomes	Genosomes are artificial macromolecular complexes used for the transfer of functional genes

(continued)

Table 1
(continued)

S. No.	Vesicular systems	Description
20	Photosome	Photosomes are photolyase-loaded liposomes which release the entrapped contents by photo-triggered charges
21	Erythrosomes	Erythrosomes are lipid-based systems coated with chemically cross-linked human erythrocyte cytoskeletons
22	Vesosomes	Vesosomes are nested compartments in the form of "interdigitated" bilayer phase formed by addition of ethanol and saturated phospholipids

Table 2
Method of preparation and composition of nanovesicles

Nanovesicular systems	Composition	General method of preparation
Liposomes	• Phospholipids • Cholesterol • Charge inducers (if required)	• Hand shaking • Thin film hydration • Reverse-phase evaporation • Membrane extrusion • Sonication (Probe and Bath) • Ethanol injection • Ether injection • Micro-emulsification • Double emulsion
Niosomes	• Nonionic surfactant • Cholesterol • Charge inducers (if required)	• Hand shaking • Thin film hydration • Reverse-phase evaporation • Pro-niosome method • Dehydration–Rehydration • Freeze and thaw • Bubble method • Ether injection • Sonication
Transfersomes	• Phospholipids • Surfactant/edge activator • Alcohol (as a solvent) • Buffering agent	• Thin film hydration
Ethosomes	• Phospholipids • Ethanol • Water • Propylene glycol (if required)	• Thin film hydration • Ethanol injection
Transethosomes	• Phospholipids • Ethanol • Surfactant/edge activator • Water • Charge inducer (if required)	• Thin film hydration • Ethanol injection
Phytosomes	• Phospholipids • Phytoconstituent(s)	• Thin film hydration

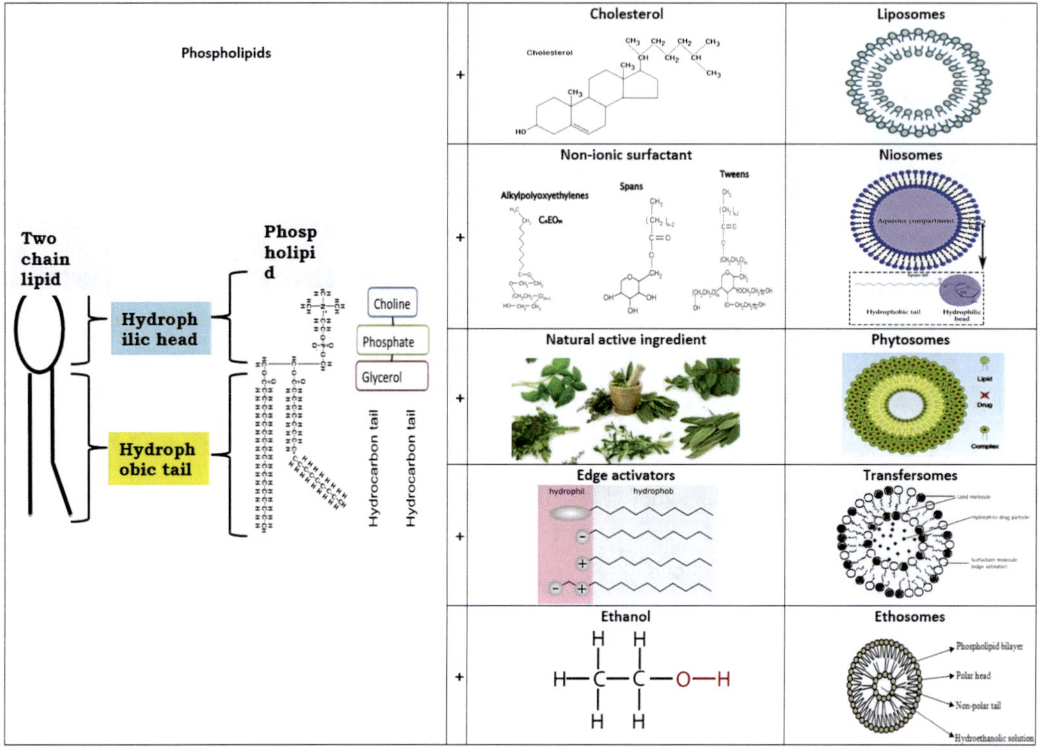

Fig. 1 Schematic representation of composition of nanovesicles

be kept above the transition temperature of lipids. The rotation is continued till dry lipid layer is formed in the flask. The flask is detached from the evaporator. Upon hydration, the lipids swell and peel off from the flask. The energy required for swelling of lipids and vesicle formation is provided by hand shaking or by rotating in a rota evaporator. Generally, multilamellar vesicles (MLV) are formed using this technique [42, 43].

3.2 Nonshaken Method

In this method, lipids are dried in a similar way as mentioned in thin film hydration method. After drying the film, hydration is done by exposing the lipid film to stream of water saturated with nitrogen, followed by swelling in aqueous medium without shaking. The vesicles formed by this method are large and unilamellar [42, 43].

3.3 Proliposomes

In this method, hydration is done later on at the time of consumption of these vesicles. In this method, lipids are dried on a finally divided particulate support like sodium chloride and sorbitol. At the time of its usage, water is added to this lipid-coated powder (proliposome) to give a suspension of MLVs [42, 43].

3.4 Freeze-Drying Method

It is based on the formation of a homogenous dispersion of lipids in water-soluble carrier materials. Liposome-forming lipids and water-soluble carrier materials such as sucrose are dissolved in tert-butyl alcohol/water co-solvent systems in appropriate ratios to form a clear isotropic monophase solution. Then the monophase solution was sterilized by filtration and filled into freeze-drying vials. In recent study, a laboratory freeze-drier was used during the drying process. On addition of water, the lyophilized product spontaneously forms homogenous liposome preparation. After investigation of the various parameters associated with this method it is found that the lipid/carrier ratio is the key factor affecting the size and the polydispersity of the liposome preparation [42, 43].

3.5 Ethanol Injection Method

This method is basically used for preparation of ethosomes. In this method, soya phosphatidylcholine (SPC) and drug are dissolved in ethanol in a beaker which is sealed to minimize the ethanol loss due to evaporation. The solution is then stirred using a magnetic stirrer and distilled water is added in a streamlined manner using syringe [44].

3.6 Ether Injection Method

This method is used for preparation of niosomes. The ether injection method differs from the ethanol injection method. Since ether is immiscible with the aqueous phase, it is heated so that the solvent is removed from the liposomal product. The method involves injection of ether–lipid solutions into warmed aqueous phases above the boiling point of the ether. The ether vaporizes upon contacting the aqueous phase, and the dispersed lipid forms primarily unilamellar liposomes. An advantage of the ether injection method compared to the ethanol injection method is the removal of the solvent from the product, enabling the process to be run for extended periods forming concentrated liposomal product with high entrapment efficiencies.

3.7 Hot Method

This method is used for preparation of ethosomes. The drug is dissolved in a mixture of ethanol and propylene glycol and the mixture is added to the phospholipid dispersion in water at 40 °C. After homogeneous mixing the preparation is then probe sonicated at 4 °C for three cycles of 5 min, with a gap of 5 min between each cycle. The formulation is then homogenized using a high-pressure homogenizer to get nano-sized ethosomes [44].

3.8 Cold Method

This is the most common and widely used method for preparation of ethosomes. The phospholipids, drug, and other lipid materials are dissolved in ethanol, in a covered vessel, at room temperature, with vigorous stirring. The mixture is heated up to 30 °C in a water bath. Further, water is heated to 30 °C in separate vessel, and added to the above mixture and stirred for 5 min in a covered vessel. The vesicle size of the ethosome-based formulation can be decreased if desired, by its further sonication or extrusion. The prepared formulation must be properly stored under refrigeration [44].

3.9 Reverse-Phase Evaporation (REV) Technique

A lipidic film is prepared by evaporating organic solvent under reduced pressure. The system is purged with nitrogen and the lipids are redissolved in a second organic phase which is usually constituted by diethyl ether and/oriso propyl ether. Large unilamellar and oligolamellar vesicles are formed when an aqueous buffer is introduced into this mixture. The organic solvent is subsequently removed and the system is maintained under continuous nitrogen. These vesicles have aqueous volume to lipid ratios that are 30 times higher than sonicated preparations and four times higher than multilamellar vesicles. Most importantly, a substantial fraction of the aqueous phase (up to 62% at low salt concentrations) is entrapped within the vesicles, encapsulating even large macromolecular assemblies with high efficiency [42, 43].

3.10 Ultrasonication

An aqueous phase is added to the surfactant/cholesterol mixture in a glass vial. The mixture is then sonicated using probe for a certain time period. The resultant vesicles are small, uniform, and unilamellar. In the case of niosomes the resulting vesicle sizes are generally larger than liposomes. Usually niosomes prepared by this technique are larger than 100 nm in diameter [42, 43].

4 Characterization of Nanovesicles

4.1 Entrapment Efficiency

The entrapment efficiency is expressed as the percentage entrapment of the drug added. Entrapment efficiency is determined by separation of the unentrapped drug by the use of dialysis, ultracentrifugation, or mini-column centrifugation method. After separation, the vesicles are disrupted using Triton X-100 or n-propanol. Alternatively, the amount of free drug is measured and subtracted from the total amount of drug added [42, 45]. The ultracentrifugation technique is reported as a simple and fast method for the separation of drug-loaded liposomes from their medium. However, prior to the centrifugation, vesicles should be aggregated in order to enable their sedimentation by adding an equal volume of protamine solution to the sample. Once drug-loaded vesicles are separated from their medium, the lipidic bilayer is disrupted with methanol or Triton X-100 and the released material is then quantified. Techniques used for this quantification depend on the nature of the encapsulant and include spectrophotometry, fluorescence spectroscopy, enzyme-based methods, and electrochemical techniques [36, 42, 45].

4.2 Morphological Visualization

Visualization of nanovesicle can be done using transmission electron microscopy (TEM) and by scanning electron microscopy (SEM). This is generally done to check the shape and other morphological characters of vesicles [26, 33, 36, 42, 46, 47].

4.3 Vesicle Size and Zeta Potential

The average size and size distribution of vesicles are important parameters especially when these are intended for therapeutic use by inhalation or parenteral route. Several techniques are available for assessing sub-micrometric liposome size and size distribution, which include microscopy techniques, size-exclusion chromatography (SEC), field-flow fractionation, and static or dynamic light scattering. Particle size and zeta potential can be determined by dynamic light scattering (DLS) using a computerized inspection system and photon correlation spectroscopy (PCS). Samples were prepared in distilled water, filtered through a 0.2 mm membrane filter and diluted with filtered saline and then size measurement done by using photon correlation spectroscopy or dynamic light scattering measurements. The zeta potential of a particle is the overall charge that a particle acquires in a particular medium. It is a physical property which is exhibited by any particle in suspension [42, 43, 48–51].

It has long been recognized that the zeta potential is a very good index of the interaction magnitude between colloidal particles. Measurements of zeta potential are commonly used to predict the stability of colloidal systems. If all the particles in suspension have a large negative or positive zeta potential then they will tend to repel each other and there will be no tendency to aggregation. However, if the particles have low zeta potential values then there will be no force to prevent the particles flocculating. To measure the zeta potential, a laser is used to provide a light source illuminating particle within the samples [42, 43, 48–51].

4.4 Transition Temperature

The transition temperature of the vesicular lipid systems can be determined by using differential scanning calorimetry.

4.5 Surface Tension Activity Measurement

The surface tension activity of the drug in aqueous solution can be measured by the ring method in a Du Nouy ring tensiometer [42, 52].

4.6 Vesicle Stability

The stability of vesicles can be determined by assessing the size and structure of the vesicles over time. Mean size is measured by DLS and structural changes are observed by TEM.

4.7 Drug Content

Drugs can be quantified by analytical methods such as high-performance liquid chromatography (HPLC) and liquid chromatography coupled with mass spectrometry (LC/MS).

4.8 Drug Release Study

In vitro drug release study is conducted using artificial membrane like cellophane membrane. While in ex vivo studies animal skin like that of porcine, mice, rat, etc., can be used. Franz diffusion cell was used as an apparatus on which drug release study was conducted. It has a donor compartment in which the formulation was kept and a receptor compartment which contains the dissolution media. With help of skin deposition study, we are able to know the amount of drug and vesicles retained in the skin [53].

4.9 Penetration and Permeation Studies

Depth of penetration from ethosomes can be visualized by confocal laser scanning microscopy (CLSM). CSLM study is done using animal or cadaver skin by using dyes like rhodamine 123 in vesicles. By this study we are able to know up to which extent and in which proportion vesicles penetrate into intact skin. This technique can be used to study the mechanism of penetration of vesicles across skin and to determine the histological organization of skin [25].

4.10 Elasticity or Deformability Study

Elasticity is the main property of transfersomes. It is checked through extrusion method, i.e., by-passing vesicles through polycarbonate membranes of known size at constant pressure and time followed by measuring the size of vesicle passing each membrane [37, 54].

Various techniques that are used to characterize these nanovesicles are deciphered in Table 3.

5 Challenges of Nanovesicles

In the pharmaceutical field, liposomes have long been of great interest by offering a promising way for both systemic and locally acting drugs used for therapeutic applications in humans and animals. As a result of the great potential of liposomes in the area of drug delivery, several companies have been actively engaged in expansion and evaluation of liposome products. Most of them concern anticancer and antifungal drugs that, administered in their free form, are toxic or exhibit serious side-effects, and their encapsulation into liposomal vesicles significantly diminishes these unwanted properties. However, there are few commercially available pharmaceutical products based on drug-in-liposome formulations. Nanovesicle-based formulation has not entered the market in great numbers because of some problems limiting their development. Even that batch-to-batch reproducibility, low drug entrapment, particle size control, and short circulation half-life of vesicles seem to have been resolved, some other problems are still limiting the widespread use of liposomes, among them the stability issues, sterilization method, and production of large batch sizes.

Another challenge is the identification of a suitable method for sterilization of vesicle formulations as phospholipids are thermolabile and are substances sensitive to procedures involving the use of heat, radiation, and/or chemical sterilizing agents. The alternative technique of nanovesicle sterilization is filtration through sterile membranes.

Table 3
Characterization techniques that are used to characterize nanovesicles

Nanovesicular systems	Parameters	Characterization techniques
Physical characterization		
L, N, T, E, TE, P	Vesicle morphology	Transmission electron microscopy, freeze fracture electron microscopy
L, N, T, E, TE, P	Vesicle size and distribution	Zeta sizer, dynamic light scattering, photon correlation spectroscopy, laser light scattering, gel permeation and exclusion
L, N, T, E, TE, P	Vesicle surface charge/zeta potential	Zeta sizer, free-flow electrophoresis
L, N, T, E, TE, P	Surface pH	pH meter
L	Lamellarity	31 P-NMR, small angle X-ray scattering
L, N, T, E, TE, P	Percent drug entrapment, percent free drug, percent drug loading	Mini-column centrifugation technique, ion-exchange chromatography
L, N, T, E, TE, P	Percent drug release	Franz diffusion cell, dialysis membrane
L, N, T, E, TE, P	Phase behavior	Differential scanning calorimeter, freeze fracture electron microscope
Chemical characterization		
L, T, E, TE, P	Phospholipid concentration	Stewart assay, Barlett assay, HPLC
L, T, E, TE, P	Phospholipid hydrolysis	HPLC, TLC
L, N, T, E, TE, P	Cholesterol concentration	Cholesterol oxidase assay, HPLC
L, T, E, TE, P	Phospholipid peroxidation	UV spectrophotometer
L, N, T, E, TE, P	Osmolarity	Osmometer
E, TE	Alcohol content	Gas chromatography
Biological characterization		
L, N, T, E, TE, P	Sterility	Aerobic and anaerobic cultures
L, N, T, E, TE, P	Pyrogenicity	Limulus amebocyte lysate (LAL) test
L, N, T, E, TE, P	Animal toxicity	Histological and pathology studies, survival rate monitoring

L liposomes, *N* niosomes, *T* transfersomes, *E* ethosomes, *TE* transethosomes, *P* phytosomes

One more challenge for liposome is the large-scale production method. Pharmaceutically acceptable procedures are those that can be easily scaled to larger batch sizes and economically feasible. However, unlike the classical pharmaceutical dosage forms (tablets, capsules, suppository, etc.) which are produced in large batch sizes, liposome-based drugs even those already in the market are produced in small size batches and thus are costly for the manufacturers. Scale-up process to larger size batches is often a monumental task for the process development scientists. Cost and large-scale production is another barrier in the commercialization of vesicular delivery system. Lack of competent researchers and infrastructure to develop new nanovesicle at many research institutes is another hindrance. Cost involved in nanovesicle research and cost of analysis and instrument involved also poses a challenge. Lack of skilled persons in industry and academia is also a challenge. Various merits and limitations of nanovesicles are given in Table 4.

6 Application of Nanovesicles

These nanovesicles have been extensively reported for oral delivery, topical, and transdermal delivery of proteins and peptides. Moreover, they are also used for targeting the drug at the diseased site,

Table 4
Advantages and limitations of nanovesicles

S. No.	Formulation	Benefits	Limitations
1	Liposomes [55]	• Controlled drug release • Biodegradable • Biocompatible • Reduced dose frequency • Improved bioavailability • Improved reduction in toxicity • Entrapped both hydrophilic and lipophilic drugs	• Stability issues • Rapid clearance • Drug leakage
2	Niosomes	• Stable as compared to liposomes • Biodegradable • Biocompatible • Controlled drug release • High drug entrapment efficiency • Reduced dose frequency • Osmotically active • Entrapped hydrophilic, lipophilic and amphiphilic drugs	• Less bioavailability • Time-consuming process • Limited shelf-life due to hydrolysis, aggregation, and leakage
3	Transfersomes	• Biocompatible • Biodegradable • High entrapment efficiency • Accommodate both hydrophilic and hydrophobic drugs	• Chemically unstable
4	Ethosomes	• Small size as compared to liposomes • Superior skin penetration • High entrapment efficiency • Encapsulate both hydrophilic and hydrophobic drugs • Squeeze through intracellular pathways of skin • Biocompatible • Biodegradable • Chemically stable	
5	Transethosomes [56]	• Transethosomes may contain advantages of both transfersomes and ethosomes • Higher entrapment efficiency than ethosomes • Higher skin penetration efficiency than ethosomes	• No particular trend on stability was determined

(continued)

Table 4 (continued)

S. No.	Formulation	Benefits	Limitations
6	Phytosomes	• Absorption enhancement • Bioavailability enhancement • Reduces dose requirement • Improve percutaneous absorption and solubility of phytoconstituents	• Phytoconstituent is rapidly eliminated • Short duration of action
7	Pharmacosomes	• Method of preparation is less time-consuming as compared to liposomes • Volume of inclusion has no effect on entrapment efficiency • Wider stability profile than liposomes • Greater shelf-life than liposomes • No drug leakage when covalent type of bonds are formed between drug and carrier	• On storage, pharmacosomes undergo fusion, hydrolysis, and aggregation
8	Virosomes [57, 58]	• Biodegradable and biocompatible • Nontoxic • Protects drug from degradation • Enable and allow drug to remain intact when they reach the cytoplasm • Higher safety profile • Extended the uptake, distribution, and elimination behavior of drug in the body	• Expensive
9	Sphingosomes [59]	• High drug stability against acid hydrolysis as compared to liposomes and niosomes • Improved drug retention properties • Better target at specific site • Better pharmacokinetic effect	• Expensive process • Low entrapment efficiency
10	Ufasomes [60]	• Less expensive than liposomes • Ready availability of fatty acids	• Sensitive to pH and ionic strength of medium • Instability
11	Bilosomes [61]	• Chemically more stable as compared to liposomes and niosomes • Do not require special condition for storage and handling • Patient-compliant	
12	Emulsomes [62]	• Increase solubility and bioavailability of poorly soluble drugs • Reduce dosing frequency of drug • Reduce toxicity • Protect drug from harsh gastric environment	

Table 5
Applications of nanovesicles

Nanovesicular systems	Applications
Liposomes	• Oral drug delivery [63] • Topical drug delivery [60] • Carriers for protein delivery [64] • Carriers for gene delivery [64] • Diagnostic imaging of tumors [65] • Cosmetics [66]
Niosomes	• As an adjuvant in immunological studies [67] • Topical drug delivery [68] • Transdermal drug delivery [69]
Transfersomes	• Carriers for protein and peptide delivery [70] • Topical drug delivery [71] • Transdermal drug delivery [72]
Ethosomes	• Topical drug delivery [73] • Transdermal drug delivery [73, 74]
Transethosomes	• Transdermal drug delivery [55, 56]
Sphingosomes	• Carriers for gene delivery [75] • Tumor targeting [76] • Immunological studies [77]

particularly to treat tumors. As theranostics, they are also used for diagnosis of tumors. Various applications of nanovesicles are shown in Table 5.

7 Conclusion and Future Prospects

Nanovesicles possess immense potential to find their application in both systemic and topical drug delivery. The major advantage of these delivery systems is their noninvasiveness and painless delivery. Development of these vesicular systems is still in infancy and more research is required in this field. Development of alternative novel drug delivery is a continuous process, and with the advancement of science and technology the proliferation of nanovesicles as delivery systems is expected in future. Despite the success of nanovesicles discussed so far, attention should be provided towards development of successful nanovesicles that can be commercialized and reach behind the bedsides of patients.

References

1. Cevc G (2004) Lipid vesicles and other colloids as drug carriers on the skin. Adv Drug Deliv Rev 56:675–711
2. Bangham AD, Standish MM, Watkins JC (1965) Diffusion of univalent ions across the lamellae of swollen phospholipids. J Mol Biol 13:238–252

3. Barry BW (2002) Drug delivery routes in skin: a novel approach. Adv Drug Deliv Rev 54: S31–S40
4. Bangham AD, Standish MM, Watkins JC et al (1967) The diffusion of ions from a phospholipid model membrane system. Protoplasma 63:183–187
5. Papahadjopoulos D, Watkins JC (1967) Phospholipid model membranes. II. Permeability properties of hydrated liquid crystals. Biochim Biophys Acta 135:639–652
6. Deamer DW (2010) From "Banghasomes" to liposomes: a memoir of Alec Bangham, 1921–2010. FASEB J 24:1308–1310
7. Batzri S, Korn ED (1973) Single bilayer liposomes prepared without sonication. Biochim Biophys Acta 298:1015–1019
8. Gregoriadis G, Ryman BE (1971) Liposomes as carriers of enzymes or drugs: a new approach to the treatment of storage diseases. Biochem J 124:58P
9. Gregoriadis G (1973) Drug entrapment in liposomes. FEBS Lett 36:292–296
10. Gregoriadis G (1976) The carrier potential of liposomes in biology and medicine. Part 1. N Engl J Med 295:704–710
11. Gregoriadis G (1976) The carrier potential of liposomes in biology and medicine. Part 2. N Engl J Med 295:765–770
12. Biju SS, Talegaonkar S, Mishra PR et al (2006) Vesicular systems: an overview. Indian J Pharm Sci 68:141–153
13. Torchilin VP (2005) Recent advances with liposomes as pharmaceutical carriers. Nat Rev Drug Discov 4:145 160
14. Kajimoto K, Yamamoto M, Watanabe M et al (2011) Noninvasive and persistent transfollicular drug delivery system using a combination of liposomes and iontophoresis. Int J Pharm 403:57–65
15. Honeywell-Nguyen PL, Bouwstra JA (2005) Vesicles as a tool for transdermal and dermal delivery. Drug Discov Today Technol 2:67–74
16. Redziniak G (2003) Liposomes et peau: passé, présent, futur. Pathol Biol 51:279–281
17. Uchegbu IF, Vyas SP (1998) Non-ionic surfactant based vesicles (niosomes) in drug delivery. Int J Pharm 172:33–70
18. Uchegbu I (1998) The biodistribution of novel 200-nm palmitoyl muramic acid vesicles. Int J Pharm 162:19–27
19. Cevc G, Blume G (2004) Hydrocortisone and dexamethasone in very deformable drug carriers have increased biological potency, prolonged effect, and reduced therapeutic dosage. Biochim Biophys Acta 1663:61–73
20. Cevc G, Mazgareanu S, Rother M (2008) Preclinical characterisation of NSAIDs in ultradeformable carriers or conventional topical gels. Int J Pharm 360:29–39
21. Rahman YE, Rosenthal MW, Cerny EA et al (1974) Preparation and prolonged tissue retention of liposome-encapsulated chelating agents. J Lab Clin Med 83:640–647
22. Jain S, Jain NK (2008) Liposomes as drug carriers. In: Jain NK (ed) Controlled and novel drug delivery, 1st edn. CBS Publisher and Distributors, New Delhi, pp 304–352
23. Patel R, Singh SK, Singh S et al (2009) Development and characterization of curcumin loaded transfersome for transdermal delivery. J Pharm Sci Res 1:71–80
24. Vyas SP, Khar RK (2008) Targeted and controlled drug delivery, 1st edn. CBS Publishers and Distributors, New Delhi
25. Cevc G, Schätzlein A, Richardsen H (2002) Ultradeformable lipid vesicles can penetrate the skin and other semi-permeable barriers unfragmented. Evidence from double label CLSM experiments and direct size measurements. Biochim Biophys Acta 1564:21–30
26. Trotta M, Peira E, Debernardi F et al (2002) Elastic liposomes for skin delivery of dipotassium glycyrrhizinate. Int J Pharm 241:319–327
27. Barry BW (2001) Novel mechanisms and devices to enable successful transdermal drug delivery. Eur J Pharm Sci 14:101–114
28. van den Bergh BA, Vroom J, Gerritsen H et al (1999) Interactions of elastic and rigid vesicles with human skin in vitro: electron microscopy and two-photon excitation microscopy. Biochim Biophys Acta 1461:155–173
29. Cevc G, Gebauer D, Stieber J et al (1998a) Ultraflexible vesicles, transfersomes, have an extremely low pore penetration resistance and transport therapeutic amounts of insulin across the intact mammalian skin. Biochim Biophys Acta 1368:201–215
30. Goosen C, Du Plessis J, Müller DG et al (1998b) Correlation between physicochemical characteristics, pharmacokinetic properties and transdermal absorption of NSAID's. Int J Pharm 163:203–209
31. Oh EK, Jin SE, Kim JK et al (2011) Retained topical delivery of 5-aminolevulinic acid using cationic ultradeformable liposomes for photodynamic therapy. Eur J Pharm Sci 44:149–157
32. Sheo DM, Shweta A, Ram CD et al (2010) Transfersomes-A novel vesicular carrier for enhanced transdermal delivery of stavudine: development, characterization and performance evaluation. J Sci Speculat Res 1:30–36

33. Trotta M, Peira E, Carlotti ME (2004) Deformable liposomes for dermal administration of methotrexate. Int J Pharm 270:119–125
34. El Zaafarany GM, Awad GA, Holayel SM et al (2010) Role of edge activators and surface charge in developing ultradeformable vesicles with enhanced skin delivery. Int J Pharm 397:164–172
35. Kim A, Lee EH, Choi SH et al (2004) In vitro and in vivo transfection efficiency of a novel ultradeformable cationic liposome. Biomaterials 25:305–313
36. Lau KG, Hattori Y, Chopra S et al (2005) Ultra-deformable liposomes containing bleomycin: in vitro stability and toxicity on human cutaneous keratinocyte cell lines. Int J Pharm 300:4–12
37. Cevc G, Blume G (1992) Lipid vesicles penetrate into intact skin owing to the transdermal osmotic gradients and hydration force. Biochim Biophys Acta 1104:226–232
38. Cevc G, Gebauer D (2003) Hydration-driven transport of deformable lipid vesicles through fine pores and the skin barrier. Biophys J 84:1010–1024
39. Bendas ER, Tadros MI (2007) Enhanced transdermal delivery of salbutamol sulfate via ethosomes. AAPS PharmSciTech 8:213–220
40. Touitou E, Dayan N, Bergelson L et al (2000) Ethosomes—novel vesicular carriers for enhanced delivery: characterization and skin penetration properties. J Control Release 65:403–418
41. Upadhyay N, Mandal S, Bhatia L et al (2011) A review on ethosomes: an emerging approach for drug delivery through the skin. Rec Res Sci Tech 3:19–24
42. New RRC (1999) Liposomes a practical approach, 1st edn. Oxford University Press, New York
43. Laouini A, Jaafar-Maalej C, Limayem-Blouza I et al (2012) Preparation, characterization and applications of liposomes: state of the art. J Colloid Sci Biotechnol 1(2):147–168
44. Verma P, Pathak K (2012) Nanosized ethanolic vesicles loaded with econazole nitrate for the treatment of deep fungal infections through topical gel formulation. Nanomedicine 8:489–496
45. Chen Y, Lu Y, Chen J et al (2009) Enhanced bioavailability of the poorly water-soluble drug fenofibrate by using liposomes containing a bile salt. Int J Pharm 376:153–160
46. Agronskia A, Valentijn J, Driel L et al (2008) Integrated fluoroscense and transmission electron microscopy. J Struct Biol 164:183–189
47. Parry K (2000) Scanning electron microscopy: an introduction. Ill-Vs Rev 13:40–44
48. Dragovic R, Gardiner C, Brooks A et al (2011) Sizing and phenotyping of cellular vesicles using nanoparticle tracking analysis. Nanomedicine 7:780–788
49. Kato H, Suzuki M, Fuzita K (2009) Reliable size determination of nanoparticles using dynamic light scattering method for in vitro toxicology accessment. Toxicol In Vitro 23:927–934
50. Fan H, Nazari M, Raval G (2014) Utilizing zeta potential to study the effective charge, membrane partitioning and membrane permeation of lipopeptide surfactine. Biochim Biophys Acta 1838:2306–2312
51. Marsalek R (2014) Particle size and zeta potential of ZnO. APCBEE Procedia 9:13–17
52. Demetzos C (2008) Differential scanning calorimetry (DSC): a tool to study the thermal behavior of lipid bilayers and liposomal stability. J Liposome Res 18(3):159–173
53. Song YK, Kim CK (2006) Topical delivery of low-molecular-weight heparin with surface-charged flexible liposomes. Biomaterials 27:271–280
54. Gillet A, Lecomte F, Hubert P et al (2011) Skin penetration behaviour of liposomes as a function of their composition. Eur J Pharm Biopharm 79:43–53
55. Ascenso A, Raposo S, Batista C et al (2015) Development, characterization, and skin delivery studies of related ultradeformable vesicles: transfersomes, ethosomes, and transethosomes. Int J Nanomedicine 10:5837–5851
56. Garg V, Singh H, Bimbrawh S et al (2017) Ethosomes and transfersomes: principles, perspectives and practices. Curr Drug Deliv 14:613–633. https://doi.org/10.2174/1567201813666160520114436
57. Sharma R, Yasir M (2010) Virosomes: a novel carrier for drug delivery. Int J Pharm Tech Res 2:2327–2339
58. Saroja CH, Lakshmi PK, Bhaskaran S (2011) Recent trends in vaccine delivery systems: a review. Int J Pharm Invest 1:64–74
59. Biju SS, Sushama T, Mishra PR, Khar RK (2006) Vesicular systems: An overview. Indian J Pharm Sci. 68:141–153
60. Patel RP, Patel H, Baria AH (2009) Formulation and evaluation of liposomes of ketoconazole. Int J Drug Deliv Technol 1:16–23

61. Ahmad J, Singhal M, Amin S, Rizwanullah M, Akhter S, Kamal MA, Haider N, Midoux P, Pichon C (2017) Bile salt stabilized vesicles (Bilosomes): a novel nano-pharmaceutical design for oral delivery of proteins and peptides. Curr Pharm Des 23:1575–1588
62. Paliwal R, Paliwal SR, Mishra N, Mehta A, Vyas SP (2009) Engineered chylomicron mimicking carrier emulsome for lymph targeted oral delivery of methotrexate. Int J Pharm 380:181–188
63. Shivhare UD, Ambulkar DU, Mathur VB et al (2009) Formulation and evaluation of pentoxifylline liposome formulation. Dig J Nanomater Biostruct 4:857–862
64. Lasic DD, Papahadjopoulos D (eds) (1998) Applications of liposomes. Elsevier, Amsterdam
65. Kirpotin DB, Lasic DD, Papahadjopoulos D (1998) Medical applications of liposomes. Elsevier, Amsterdam
66. Posner R (2002) Liposomes. J Drugs Dermatol 1:161–164
67. Conacher M, Alexander J, Brewer JM (2000) Niosomes as immunological adjuvants. In: Uchegbu IF (ed) Synthetic surfactant vesicles. International Publishers Distributors Ltd, Singapore, pp 185–205
68. Malhotra M, Jain NK (1994) Niosomes as drug carriers. Indian Drugs 31:81–86
69. Kazi KM, Mandal AS, Biswas N et al (2010) Niosome: a future of targeted drug delivery systems. J Adv Pharm Technol Res 1: 374–380
70. Hafer C, Goble R, Deering P et al (1999) Formulation of interleukin-2 and interferon-alpha containing ultra deformable carriers for potential transdermal application. Anticancer Res 19:1505–1507
71. Duangjit S, Opanasopit P, Rojanarata T et al (2011) Characterization and in vitro skin permeation of meloxicam-loaded liposomes versus transfersomes. J Drug Deliv 2011:418316. https://doi.org/10.1155/2011/418316
72. Cevc G (1996) Transferosomes, liposomes and other lipid suspensions on the skin: permeation enhancement, vesicle penetration, and transdermal drug delivery. Crit Rev Ther Drug Carrier Syst 13:257–388
73. Dkeidek I, Touitou E (1999) Ethosomes: a recent approach in transdermal/topical delivery. AAPS Pharm Sci 1:202
74. Ehab R, Bendas L, Mina I (2007) Enhanced transdermal delivery of salbutamol sulfate via ethosomes. AAPS PharmSciTech 8:213–220
75. Vyas SP, Khar RK (2002) Targeted and controlled drug delivery. CBS publisher, New Delhi
76. Lankalapalli S, Damuluri M (2012) Sphingosomes: applications in targeted drug delivery. Int J Pharm Chem Biol Sci 2:507–516
77. Saraf S, Gupta D, Kaur CD et al (2011) Sphingosomes a novel approach to vesicular drug delivery. Int J Curr Sci Res 1:63–68

Chapter 2

Preparation and Characterization of Micelles

Vandana B. Patravale, Prashant G. Upadhaya, and Ratnesh D. Jain

Abstract

Nanoformulations in the past few decades have gained tremendous attention owing to their affirmative applications in increasing the bioavailability of poorly soluble drugs. Micelles in particular are favored due to their varied advantages which include thermodynamic stability, simple formulating steps, Newtonian flow, and enhanced biological barrier penetration. Owing to these advantages micellar nanosystems find extensive applications in oral, transdermal, and parenteral administration, and are now being explored for ocular and other noninvasive novel pathways of drug delivery such as nose to brain. In this chapter, we have discussed the protocol for the preparation of sumatriptan loaded micelles for the therapy of migraine. The inner core of these micelles comprises hydrophobic region of diblock polymer which holds the drug, while the hydrophilic region of the same provides conformational stability in the aqueous environment.

Key words Polymeric micelles, Di-block copolymer, Sumatriptan, Nanocarrier, Colloid

1 Introduction

Search for pioneering medicines in the management of diseases without any compromise on the safety and efficacy front has been a major requirement of the health care sector [1, 2]. As solubility has been one of the major hurdles in the drug delivery process, significant success has been witnessed in the arena of solubilization of drugs [3]. Of the various techniques available today for increasing drug solubility, nanonization appears to be a promising one and hence trending [4].

Micelles refer to the supramolecular assembly of surfactant molecules/amphiphilic block copolymers dispersed in a liquid colloid. These surfactants or amphiphilic copolymers consist of a hydrophilic head and a hydrophobic tail [5, 6]. When the concentration of the aforesaid block copolymer or surfactant increases above a certain concentration namely critical aggregation concentration (CAC) or critical micelle concentration (CMC), they start to associate in order to experience minimal water contact. This association strives the hydrophilic head regions in contact with

Volkmar Weissig and Tamer Elbayoumi (eds.), *Pharmaceutical Nanotechnology: Basic Protocols*, Methods in Molecular Biology, vol. 2000, https://doi.org/10.1007/978-1-4939-9516-5_2,

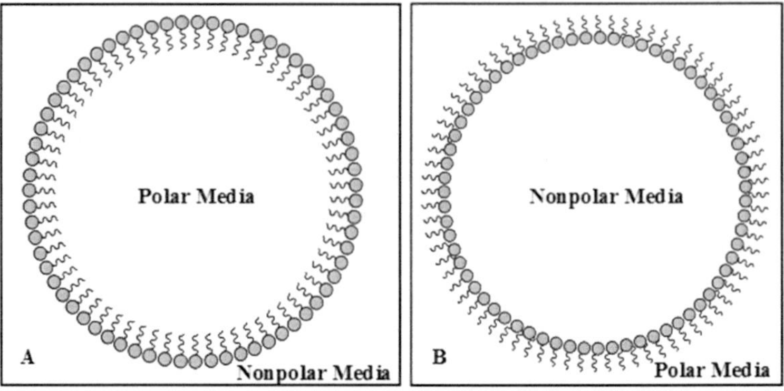

Fig. 1 Structure of micelles. (**a**) Normal micelles. (**b**) Reverse micelles

surrounding aqueous solvent, thereby, impounding the hydrophobic tail regions towards the micelle center (in case of organic in aqueous emulsion) [7]. However, in case of reverse micelles the association is exactly opposite to the aforesaid (Fig. 1). Theoretically, the micelle formation is based upon the principle of free energy, where the system tries to form a stable structure, e.g., micelles by decreasing its free energy [6]. Thus, micelles are formed when the contact of the hydrophobic fragments of the polymer/ surfactant with the aqueous environment is minimized. This restores the hydrogen bond network in aqueous phase and decreases the free energy of the system, thereby forming a stable formation viz., micelles [6, 8].

As mentioned above, the micelles can be prepared using either a surfactant (surfactant micelles) or block copolymers (polymeric micelles). However, in either of the cases the theory of micelle formation remains the same [9–11]. The structural units, i.e., the block polymers or the surfactants have a hydrophilic fragment A and a hydrophobic fragment B [12]. The hydrophobic core for a block copolymer can be poly(b-benzyl-L-aspartate), poly(DL-lactic acid), poly(e-caprolactone), etc., which provide an excellent protection to the water insoluble moiety from coming in the vicinity of the aqueous phase [13–15]. The hydrophilic fragments can be poly vinyl alcohol, poly (aspartic acid), etc., which remain in the hydrophilic boundaries, thereby providing stability to the system [16–18]. The primary focus of the pharmaceutical research is A–B diblock micellar structure. However, the current trend is moving towards multiblock copolymers with A–B–A conformation such as poly(ethylene oxide)-poly(propylene oxide)-poly(ethylene oxide), which are also known to form excellently stable micellar formations [19, 20].

Besides CMC, which determines the concentration of structural units at which micelles are formed (*see* **Notes 1** and **2**), other formulation considerations include viscosity of the micellar core, which influences its physical stability and drug release (*see* **Note 3**) and; micelle size and size distribution, etc., which contribute to a stable micellar formulation (*see* **Notes 4** and **5**) [21].

Micelles, in this present era, find excellent applications as active and passive targeting moieties, site-specific targeting moieties, diagnostic agents, solubility enhancers, etc. [6]. This chapter aims at developing a protocol for the preparation of micellar carriers loaded with a serotonin 5-HT1B/1D receptor agonist, sumatriptan.

2 Materials and Instruments

2.1 Preparation of Buffers

1. Citrophosphate buffer pH 5.8: Prepare 2.1% w/v solution of citric acid in ultrapure water. Prepare 7.15% w/v solution of disodium hydrogen phosphate in ultrapure water. Mix 36.8 ml of the above prepared citric acid solution with 63.2 ml of disodium hydrogen phosphate solution. Adjust pH to 5.8 with 0.01 N HCl, if required.
2. Sodium dihydrogen orthophosphate buffer pH 6.5 (0.3 M): Dissolve 36 g of sodium dihydrogen orthophosphate in 700 ml ultrapure water. Adjust the pH to 6.5 with dilute sodium hydroxide solution and makeup the volume to 1000 ml with ultrapure water.

2.2 Formulation of Sumatriptan Micellar Nanocarrier

1. Cyclo mixer (Remi, Mumbai, India).
2. Sumatriptan stored at 25 °C.
3. Diethylene glycol monoethyl ether (Transcutol P®).
4. Benzyl alcohol.
5. Poly[ethylene oxide]–poly[propylene oxide] block copolymer (Pluronic® F127).
6. Polyethylene glycol-400 (PEG-400).
7. Vitamin E–D-α-tocopheryl polyethylene glycol 1000 succinate (TPGS).
8. Citrophosphate buffer, pH 5.8: Disodium hydrogen phosphate, citric acid.
9. Ultrapure water (Dispensed from Milli Q Plus system (Millipore), Bedford, USA).

2.3 CMC Determination

1. Malvern Autosizer 4800 employing 7132 digital correlator at 25 °C (Malvern, Worcestershire, UK).
2. Ultrapure water (Dispensed from Milli Q Plus system (Millipore), Bedford, USA).

2.4 Dynamic Light Scattering Method for Determining Particle Size (PS) and Polydispersity Index (PDI)

1. Malvern Autosizer 4800 instrument employing 7132 digital correlator, at 25 °C (Malvern, Worcestershire, UK), light source: an argon ion laser (Coherent, Innova, USA) operated at 514.5 nm with a maximum output power of 2 W. The scattering angle for the routine measurements was 90°; however, for multi-angle measurement it was varied between 30° and 150° and the correlation functions were analyzed by the method of cumulants.
2. Ultrapure water (Dispensed from Milli Q Plus system (Millipore), Bedford, USA).

2.5 Small-Angle Neutron Scattering (SANS)

1. SANS diffractometer at the Dhruva reactor with a beryllium oxide filtered beam of mean wavelength (l) 5.2 Å and accessible wave vector transfer ($Q = 4p\sin q/l$, where $2q$ is the scattering angle) range 0.02–0.3 Å (Bhabha Atomic Research Centre, Mumbai, India).
2. Deuterium oxide (D_2O).

2.6 Cryo-transmission Electron Microscopy (Cryo-TEM)

1. FEI T12 G2 microscope (FEI Company, Eindhoven, The Netherlands).
2. Leica Cryo Plunger CMC (Leica, Germany).
3. Gatan CP3 cooling holder (Gatan Inc., USA).
4. Ultrapure water (Dispensed from Milli Q Plus system (Millipore), Bedford, USA).
5. Filter membrane (0.45 μm).

2.7 High Performance Liquid Chromatography (HPLC) Analysis for Drug Content of Sumatriptan Micelles

1. Plus Intelligent HPLC pump PU-2080 (Jasco, Tokyo, Japan), equipped with a ultraviolet-2075 Intelligent UV/VIS detector (Jasco, Tokyo, Japan), a Rheodyne 7725 injector (Rheodyne, Cotati, CA, USA), and a Jasco ChromaPass Chromatography Data System Software (Version 1.8.6.1) (Jasco, Tokyo, Japan).
2. Inertsil ODS-3 RP-18 column (4.6 × 250 mm, 5 μm) (Waters Corp., Milford, USA).
3. Ultrapure water (Dispensed from Milli Q Plus system (Millipore), Bedford, USA).
4. 0.03 M sodium dihydrogen orthophosphate buffer (pH 6.5): Sodium dihydrogen orthophosphate, sodium hydroxide.
5. HPLC grade acetonitrile.

3 Methods

3.1 Formulation of Sumatriptan Micellar Nanocarrier

The micellar nanocarriers loaded with sumatriptan are formulated using a block copolymer with a basic structure as Ethylene Oxide–Propylene Oxide–Ethylene Oxide (EO–PO–EO). The same has been reported to exhibit high potentials of forming self-assembled polymeric micelles in aqueous solutions [22]. Association of PEG-400 and TPGS with the aforementioned EO–PO–EO block copolymer is reported to form stable micellar nanostructures, therefore, the two are employed as micelle-forming aid [23]. Transcutol P® is employed owing to its excellent potential of aiding penetration and absorption of drug. Transcutol P® along with benzyl alcohol serves as a solubilization aid for the drug.

1. To prepare drug loaded micelles, weigh accurately 180 mg of Transcutol P® and cyclomix it with benzyl alcohol (100 mg).
2. To the above, add 20 mg of accurately weighed sumatriptan and cyclomix until clarity is obtained.
3. Weigh 50 mg of Pluronic® F127 and dissolve (*see* **Note 6**) in pH 5.8 citrophosphate buffer (500 mg) along with heating aid (40 °C). Further, weigh PEG-400 (100 mg) and TPGS (50 mg) and dissolve the same in the above Pluronic® F127 solution.
4. Mix **step 3** with **step 2** by means of agitation and further cyclomix it for 10 min to obtain optical clarity.

3.2 CMC Determination

The micellar formulation upon dilution with water at diverse molar fractions, is analyzed for changes in the light scattering intensity. An impulsive escalation in the light scattering intensity due to the abrupt changes of the associated physical properties upon micelle formation is perceived and the same is regarded as the indicator and recorded [21, 24].

1. Dilute the formulation with water in the ratios (formulation: Water) 10:0, 7:3, 5:5, 3:7, and 0:10 (*see* **Note 7**).
2. The above ratios are further analyzed using dynamic light scattering method and the light intensities are recorded for each ratios.
3. The CMC values for the diluted micelles of each ratios are shown in Table 1.

Table 1
CMC determination of zolmitriptan micellar nanocarrier

Concentration (M)	Formulation[a]:water (molar ratio)				
	10:0	7:3	5:5	3:7	0:10
CMC (M)	3.2×10^{-4}	2.8×10^{-4}	7.6×10^{-5}	2.3×10^{-4}	3.0×10^{-4}

[a]Formulation: pluronic F 127: TPGS: TCP: BA: PEG 400: sumatriptan (10:10:36:20:20:4)

Table 2
Multi-angle dynamic light scattering of sumatriptan micellar nanocarrier

Angle (°)	PS (nm)	PDI
50	25.5	0.213
70	24.1	0.125
90	23.1	0.127
110	21.5	0.114
150	20.4	0.112

3.3 Dynamic Light Scattering Method for Determining PS and PDI

The PS and the PDI are measured using dynamic light scattering method. The scattering angle used for the routine measurements is 90°; however, for the multi-angle measurements the same is varied between 50° and 150° and the correlation functions are analyzed by the method of cumulants.

1. For PS and PDI analysis, dilute the micellar formulations with water in the ratios (water:formulation) 0:100, 90:10, and 99:1 (*see* **Note 7**).
2. Record the PS and PDI using quartz cuvette at 25 °C.
3. A typical record of the micelles at various angles is shown in Table 2.

3.4 SANS Analysis

SANS (*see* **Note 8**) exhibits dramatic increments in the forward neutron scattering, occurring at phase transitions and therefore is used to characterize the complex structure of the prepared nano-sized micelles. It is also known to provide valuable information over a wide variety of scientific and technological applications including chemical aggregation, defects in materials, surfactants, colloids, polymers, proteins, biological membranes, and macromolecules. SANS analysis of the data gives insight on size, shape, and morphology [25].

1. Dilute the micellar formulation with D_2O (formulation: D_2O) in the ratios 50:50, 30:70, and 10:90 v/v.
2. Measure the differential scattering by using a quartz sample holder of 0.5 cm thickness at 30 °C.
3. The graph for the intensity at various dilution of micelles with D_2O for SANS is shown in Fig. 2 (*see* **Note 9**) while the SANS profile for the micelles is expressed in Table 3 (*see* **Note 10**).

3.5 Cryo-TEM Analysis

The samples for Cryo-TEM are prepared in the controlled environment vitrification system (CEVS) at 25 °C and 100% relative solvent saturation.

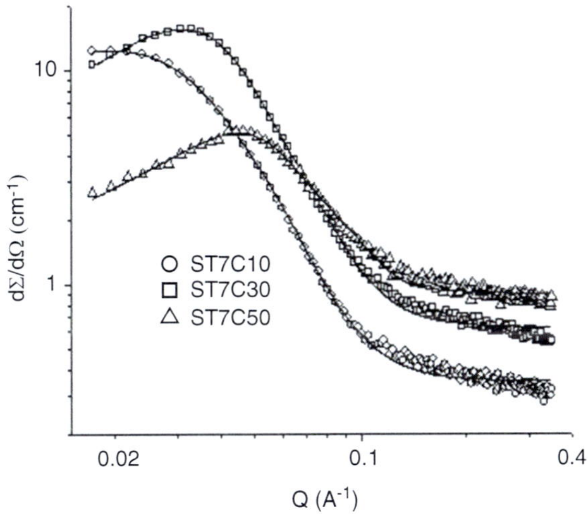

Fig. 2 Intensity graph of SANS (ST 7C 10 = micelle: D_2O ratio of 10:90 v/v, ST 7C 30 = micelle: D_2O ratio of 30:70 v/v, and ST 7C 50 = micelle: D_2O ratio of 50:50 v/v)

Table 3
SANS profile of sumatriptan micellar nanocarriers

Sample	Core radius (Å)	Hard sphere radius (Å)
Micelle: D_2O ratio of 10:90 v/v	33.8	107
Micelle: D_2O ratio of 30:70 v/v	30.8	83
Micelle: D_2O ratio of 50:50 v/v	27.0	62

1. Cryo-TEM was precooled to −176 °C using liquid nitrogen.
2. Dilute the micellar formulation up to ten times using filtered (0.2 μm filter) ultrapure water.
3. Place a drop of the above liquid onto a perforated carbon film, supported by copper grid and held by tweezers.
4. Blot the sample with a filter paper to form a thin liquid film and immediately plunge it into liquid ethane held into Leica Cryo Plunger CMC to freeze the sample at −183 °C.
5. Transfer the vitrified sample into Gatan CP3 cooling holder and maintain the temperature under −174 °C to −176 °C.
6. Transfer Gatan 626 cooling holder under the FEI T12 G2 microscope maintaining temperature between −174 °C and −176 °C.
7. Scan the samples and acquire the images recorded at nominal underfocus of 1–2 μm to enhance the phase-contrast using Gatan UltraScan 1000 high-resolution cooled-CCD camera.
8. Images for the scans of sumatriptan micelles are shown in Fig. 3.

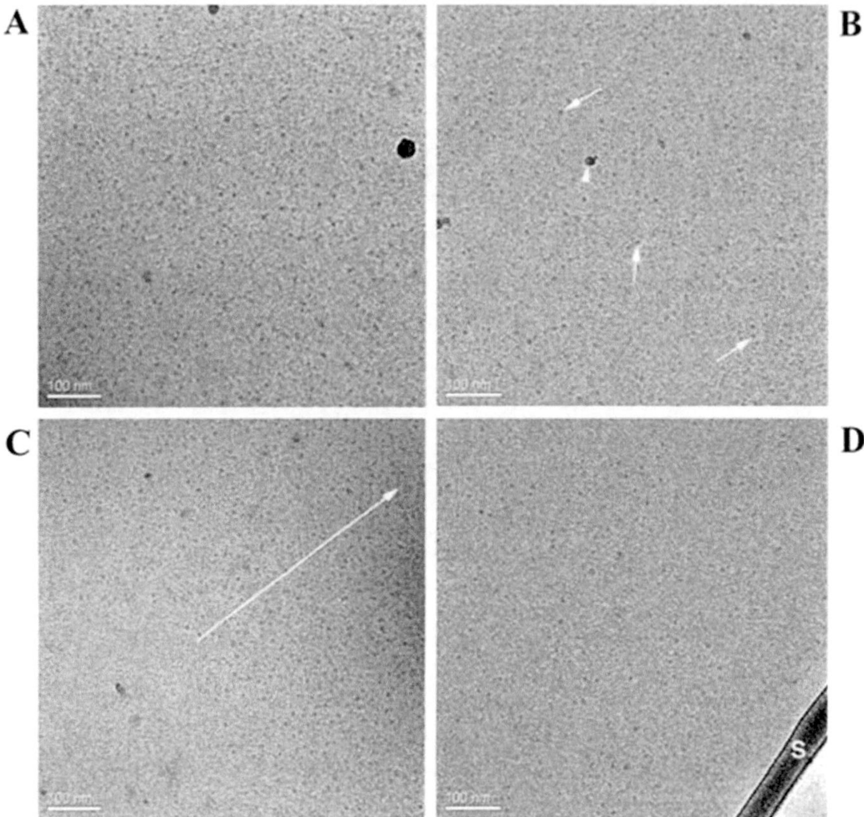

Fig. 3 Cryo-TEM images of (**A**) Sumatriptan micellar nanocarrier. (**B**) Placebo micellar nanocarrier. (**C**) Image showing thickness gradient, and (**D**) Micellar nanocarriers with perforated carbon film

3.6 HPLC Analysis for Drug Content of Sumatriptan Micelles

The drug content of the micelles is analyzed using a HPLC method [26].

1. Prepare 0.3 M sodium dihydrogen orthophosphate buffer (pH 6.5) in ultrapure water.
2. To prepare the mobile phase, mix the above buffer with acetonitrile in the ratio 75:25 v/v, filter through 0.45 μm filter, and sonicate for 15 min to degas the solution.
3. Dissolve 1 ml of micellar formulation in 10 ml of mobile phase, filter through 0.45 μm filter, and inject 20 μl in rheodyne injector.
4. Analyze the sample at $\lambda_{282 \text{ nm}}$ using ultraviolet-2075 Intelligent UV/VIS detector with flowrate of 1.5 ml min^{-1} and maintaining column temperature at 25 °C.
5. The retention time for sumatriptan is expected to be 4.5 min.
6. Determine the concentration using calibration curve prepared initially using six concentrations of sumatriptan in the concentration range of 25–150 μg ml^{-1}.

4 Notes

1. Determination of CMC can be performed using various techniques, which are principally based upon monitoring the abrupt changes of the associated physical properties upon micelle formation [10].
2. The various techniques that can be used to determine CMC include dynamic light scattering, which monitors the particle size and poly dispersity index [27], tensiometry, which monitors the surface tension [28], spectroflourometry, which measures fluorescence [10, 29], conductometry, which measures electrical conductance [29, 30], sound velocity [31], and static light scattering [32].
3. Viscosity of the micellar core can be measured employing fluorescent probes such as bis(1-pyrenyl-methyl)ether [33], 1,2-(1,1′-dipyrenyl)propane or 1,6-diphenyl-1,3,5-hexatriene, which are sensitive to viscosity changes in their local environment [34]. ^{1}H-nuclear magnetic resonance can also be used for determining the viscosity of the core [35].
4. One of the most interesting features of micelles is their size, which ranges between 10 and 100 nm. Such small size, besides allowing extravasation, permits sterilization of the micellar preparation using simple economic technique such as filtration as well as minimizing the risk of embolism of blood capillaries [36].
5. Micellar size and distribution can be monitored employing dynamic light scattering method or other methods such as atomic force microscopy, transmission electron microscopy, and scanning electron microscopy. [6, 24].
6. Before adding other ingredients ensure that Pluronic® F127 is completely dissolved in citrophosphate buffer.
7. Cyclomix well to ensure uniformity.
8. SANS refers to an experimental technique, which employs neutrons scattered at very small scattering angles in order to examine geometry and structures of particles at mesoscopic scale of 1–1000 nm.
9. The graph in Fig. 2 represents a graphical model fit analysis of micelles at varied concentrations and confirms their spherical geometry. The model fit is based upon the transformation of neutron scattering angle and intensity distance data to represent an equation that reveals specific geometry and size.
10. The core radius and the hard sphere radius are determined due to the difference in the densities at the core and the surface of the micelles. The same can be measured owing to the penetrative properties of the neutrons.

Acknowledgments

The authors are thankful to the Board of Research in Nuclear Sciences (Sanction No. 2006/35/11/BRNS), Department of Atomic Energy, Govt. of India, for providing research funding. The authors are also thankful to Dr. Krishanu Ray and Mr. Lalit Borade from Tata Institute of Fundamental Research, Mumbai, Dr. P. Hassan and Dr. Vinod Aswal from Bhabha Atomic Research Centre, and Prof. Ishi Talmol, Haifa, Israel for their help towards successful completion of this project.

References

1. Loftsson T, Brewster ME (2010) Pharmaceutical applications of cyclodextrins: basic science and product development. J Pharm Pharmacol 62(11):1607–1621
2. Hodgson J (2001) ADMET-turning chemicals into drugs. Nat Biotechnol 19:722–726
3. Kalepu S, Nekkanti V (2015) Insoluble drug delivery strategies: review of recent advances and business potential. Acta Pharm Sin B 5 (5):442–453
4. Savjani KT, Gajjar AK, Savjani JK (2012) Drug solubility: importance and enhancement techniques. ISRN Pharm 2012:1–10
5. Riess G (2003) Micellization of block copolymers. Prog Polym Sci 28(7):1107–1170
6. Jones MC, Leroux JC (1999) Polymeric micelles-a new generation of colloidal drug carriers. Eur J Pharm Biopharm 48 (2):101–111
7. Xu W, Ling P, Zhang T (2013) Polymeric micelles, a promising drug delivery system to enhance bioavailability of poorly water soluble drug. J Drug Deliv 2013:1–15
8. Yokoyama M (2005) Polymeric micelles for the targeting of hydrophobic drugs. Polym Drug Deliv Syst 148:533–576
9. Kwon GS, Kataoka K (1995) Block copolymer micelles as long circulating drug vehicles. Adv Drug Deliv Rev 16(1–2):295–309
10. Astafieva I, Zhong X, Eisenberg FA (1993) Critical micellization phenomena in block polyelectrolyte solutions. Macromolecules 26 (26):7339–7352
11. Price C (1983) Micelle formation by block copolymer in organic solvents. Pure Appl Chem 55(10):1563–1572
12. Torchilin VP (2001) Structure and design of polymeric surfactant based drug delivery systems. J Control Release 73(2–3):137–172
13. La SB, Okano T, Kataoka K (1996) Preparation and characterization of the micelle-forming polymeric drug indomethacin-incorporated poly(ethylene oxide)-poly(b-benzyl l-aspartate) block copolymer micelles. J Pharm Sci 85 (1):85–90
14. Connor J, Norley N, Huang L (1986) Biodistribution of immunoliposomes. Biochim Biophys Acta 884:474–481
15. Shin IL, Kim SY, Lee YM et al (1998) Methoxy poly(ethylene glycol)/e-caprolactone amphiphilic block copolymeric micelle containing indomethacin: I. Preparation and characterization. J Control Release 51(1):1–11
16. Yokoyama M, Miyauchi M, Yamada N et al (1990) Characterization and anticancer activity of the micelle-forming polymeric anticancer drug adryamicin-conjugated pol(ethylene glycol)-poly(aspartic acid) block copolymer. Cancer Res 50(6):1693–1700
17. Yokoyama M, Kwon GS, Okano T et al (1992) Preparation of micelle-forming polymer-drug conjugates. Bioconjug Chem 3:295–301
18. Yokoyama M, Okano T, Sakurai Y et al (1996) Introduction of cisplatin into polymeric micelles. J Control Release 39(2–3):351–356
19. Malmsten M, Lindman B (1992) Self-assembly in aqueous block copolymer solutions. Macromolecules 25(20):5440–5445
20. Prasad KN, Luong TT, Florence AT et al (1979) Surface activity and association of ABA polyoxyethylene-polyoxypropylene block copolymers in aqueous solution. J Colloid Interface Sci 69(2):225–232
21. Jain R, Nabar S, Patravale V et al (2010) Formulation and evaluation of novel micellar nanocarrier for nasal delivery of sumatriptan. Nanomedicine 5(4):575–587
22. Oh KT, Bronich TK, Kabanov AV (2004) Micellar formulations for drug delivery based on mixtures of hydrophobic and hydrophilic Pluronic® block copolymers. J Control Release 94(2–3):411–422

23. Ivanova R, Lindman B, Alexandridis P (2002) Effect of pharmaceutically acceptable glycols on the stability of the liquid crystalline gels formed by Poloxamer 407 in water. J Colloid Interface Sci 252(1):226–235
24. Topel O, Cakir BA, Budama L et al (2013) Determination of critical micelle concentration of polybutadiene-block-poly(ethyleneoxide) diblock copolymer by fluorescence spectroscopy and dynamic light scattering. J Mol Liq 177:40–43
25. Schillen K, Brown W, Johnson RM (1994) Micellar sphere-to-rod transition in an aqueous triblock copolymer system. A dynamic light scattering study of translational and rotational diffusion. Macromolecules 27(17):4825–4832
26. United States Pharmacopoeial Convention: United States Pharmacopeia (2003). Asian edition, USP 27. Rockville, MD
27. Gao Y, Li LB, Zhai G (2008) Preparation and characterization of Pluronic/TPGS mixed micelles for solubilization of camptothecin. Colloids Surf B Biointerfaces 64(2):194–199
28. Goon P, Manohar C, Kumar VV (1987) Determination of critical micelle concentration of anionic surfactants: comparison of internal and external fluorescent probes. J Colloid Interf Sci 189:177–180
29. Nakahara Y, Nakatsuji Y, Lida T et al (2005) New fluorescence method for the determination of the critical micelle concentration by photosensitive monoazacryptand derivatives. Langmuir 21(15):6685–6695
30. Dominguez A, Fernandez A, Gonzalez N et al (1997) Determination of critical micelle concentration of some surfactants by three techniques. J Chem Educ 74(10):1227–1231
31. Zielinski R, Ikeda S, Nomura H et al (1987) Adiabatic compressibility of alkyltrimethylammonium bromides in aqueous solutions. J Colloid Interf Sci 119(2):398–408
32. Unal HI, Price C, Budd PM et al (1994) Block copolymer of isoprene and tertiary butyl acrylate: synthesis, characterization and micelle formation. Eur Polym J 30(9):1037–1041
33. Winnik FM, Davidson AR, Hamer GK et al (1992) Amphiphilic poly (N-isopropylacrylamides) prepared by using a lipophilic radical initiator: synthesis and solution properties in water. Macromolecules 25 (7):1876–1880
34. Ringsdorf H, Venzmer J, Winnik FM (1991) Fluorescence studies of hydrophobically modified poly(N-isopropylacrylamides). Macromolecules 24(7):1678–1686
35. Nakamura K, Endo R, Takeda M (1997) Study of molecular motion of block copolymers in solution by high-resolution proton magnetic resonance. J Polym Sci Polym Phys Ed 15 (12):2095–2101
36. Kwon GS, Okano T (1996) Polymeric micelles as new drug carriers. Adv Drug Deliv Rev 21 (2):107–116

Chapter 3

Anionic and Cationic Vitamin E-TPGS Mixed Polymeric Phospholipid Micellar Vehicles

Mingyi Yao and Tamer Elbayoumi

Abstract

Berberine (Brb) is an active isoquinoline alkaloid occurring in various common plant species, with well-known potential for cancer therapy. Earlier reports has shown that Brb not only augments the efficacy of antineoplastic chemotherapy and radiotherapy, but it also exhibits direct anti-mitotic, and pro-apoptotic activities, plus significant anti-angiogenic and anti-metastatic activities in a variety of solid tumors. Notwithstanding its low systemic toxicity, a few pharmaceutical limitations severely hamper the application of Brb in cancer therapy (namely, very slight aqueous solubility and exceedingly low membrane permeability; combined with poor systemic pharmacokinetic, PK, profile).

Lipid-based nanocarriers, amphiphilic mixed micelles (Mic) composed of polymeric phospholipid conjugates and PEG-succinate ester of tocopherol were investigated as promising strategy, to improve Brb delivery into tumors. Following physicochemical characterization of micellar Brb, in vitro release studies in simulated physiological media were performed, combined with PK-simulation and in vitro assays of cytotoxicity and direct apoptosis induction in different human prostate cancer cell lines (PC3 and LNPaC).

Optimized stealth PEG-PE/TPGS-mixed micelles achieved efficient solubilization of Brb to potentially improve its systemic PK profiles (>30-fold). Our mixed micellar platform resulted in significant enhancement of the pro-apoptotic action and overall anticancer efficacy of Brb, against various in vitro (monolayer and spheroid) models of prostate cancers.

Key words Isoquinoline alkaloid, Berberine, Anti-mitotic, Pro-apoptotic, Anti-angiogenic, Pharmacokinetic profile, Mixed micelles, Polymeric phospholipid conjugates, PEG-succinate ester of tocopherol

1 Introduction

Majority of natural compounds with significant anticancer activities exhibit very low aqueous solubility, and in turn typically suffer from poor systemic bioavailability and tissue biodistribution [1]. Furthermore, their therapeutic effectiveness is often compromised by their short plasma half-life and systemic toxicity. Such biologically active molecules may also need one or more lipophilic groups to acquire a sufficient affinity towards the appropriate target receptor [2]. Over the past few decades, great advances in nanomedicine have been

Volkmar Weissig and Tamer Elbayoumi (eds.), *Pharmaceutical Nanotechnology: Basic Protocols*, Methods in Molecular Biology, vol. 2000, https://doi.org/10.1007/978-1-4939-9516-5_3,

realized to circumvent these pharmaceutical limitations and improve the therapeutic benefits of natural anticancer therapeutics [3–5].

One of the common strategies for sparingly soluble anticancer compounds has relied on the utilization of organic co-solvency and certain surfactants in their formulations. Yet, the administration of many co-solvents or surfactants carried the risks of toxicity hypersensitivity, or other undesirable adverse effects. Therefore, polymeric and phospholipid-based micellar nanocarriers have been regarded as ideal solution for such hydrophobic active agents [2, 5]. They are biocompatible and biodegradable drug carriers with very small particle size (5 nm < typically < 100 nm), and offer high loading capacity, extended circulation time, and the ability to spontaneously accumulate in pathological tumorous sites in the body [6]. The micellar formulation of drug not only offers protection from potential inactivation within biological milieu, and minimization of eliciting side effects on non-intended tissues, but it can also enhance the drug's permeability across physiological barriers, thus substantially improving its overall biodistribution [2, 6–8].

Compared to conventional detergent-based micelles, amphiphilic copolymer micelles are often more stable, with their CMC values as low as 10^{-5}-to-10^{-6} M. This class of micelles is formed by block-copolymers, consisting of hydrophilic blocks, most notably poly(ethylene glycol), PEG with molecular weight ranging from 1 to 10 KD, which constitutes the micelle's hydrophilic corona [5, 7, 8]. This surrounds the smaller hydrophobic core block copolymers, which efficiently incorporate the poorly soluble drug cargo. Diacyl phospholipid residues, end-conjugated with hydrophilic PEG polymers, have been successfully used as hydrophobic core-forming groups and are recognized as "phospholipid-core polymeric micelles." Micellar formulations comprised of diacyl lipid-PEG carry the additional advantages of uniform size distribution plus increased particle stability and hydrophobic drug payload, over other PEG-containing amphiphilic block-copolymers due to the existence of two fatty acid acyls, which might contribute considerably to an increase in the hydrophobic interactions between the polymeric chains in the micelle's core [7, 8]. Micelles made of PEG–phosphatidylethanolamine (PEG–PE) conjugates (Fig. 1) were successfully loaded with various poorly soluble drugs (tamoxifen, paclitaxel, camptothecin, porphyrins, polyphenols, etc.) and have demonstrated good in stability and longevity, in vitro and in vivo respectively, along with the ability to spontaneously accumulate in pathological areas of the body (e.g. infarcts and tumors) with compromised vasculature [2, 5, 6, 9–11].

To carry further this advantage, mixed micelles made of PEG-PE and other micelle-forming components, namely vitamin E-TPGS (PEG-succinate ester of α-tocopherol) have been reported

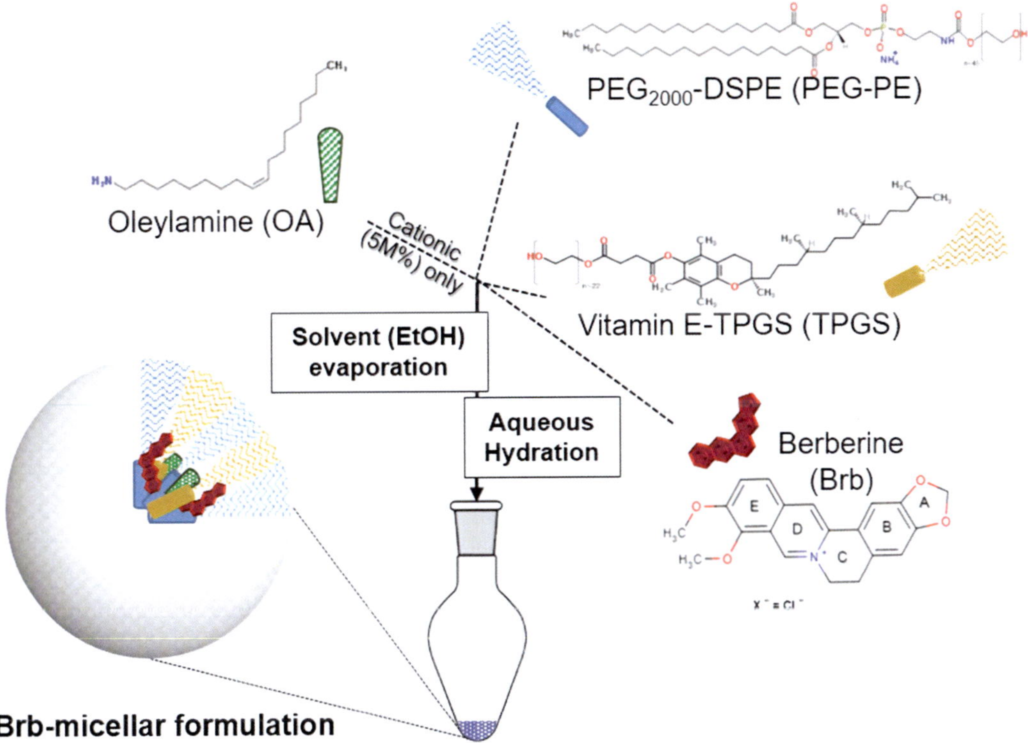

Fig. 1 Schematic diagram of anionic/cationic micelle formulation of berberine HCl: Illustrating the chemical structures of mixMic components, PEG-PE and TPGS, plus OA cationic moiety, and that of Brb active drug

to provide even better solubilization of certain poorly soluble drugs, especially polycyclic compounds, due to the increase in the capacity of the hydrophobic core (Fig. 1) [6, 12]. In several examples, anticancer drugs incorporated in such mixed lipid-core polymeric micelles were associated with micelles firmly enough that after dialysis against aqueous buffer at sink conditions, all confirmed formulations demonstrated at least 80% retentions of loaded active compound, post-24 h incubation [2, 5, 6].

Common to many plant species, the biologically important alkaloid skeleton of Berberine, Brb, has recently attracted much attention owing to its diverse pharmacological effects including anti-inflammatory, antimicrobial, antipyretic, and anti-hyperlipidemic activities (Fig. 1). Thus far, a variety of reports have investigated Brb as a possible medicinal agent in a broad spectrum of therapeutic applications, such as hyperlipidemia, diabetes, metabolic syndrome, obesity, and mycotic infections. Furthermore, over the past decade, accumulated preclinical studies have strongly demonstrated marked antineoplastic activities of Brb, such as inhibition of proliferation, induction of apoptosis, arrest of angiogenesis, and suppression of metastasis, in a variety of solid tumors [13–18]. The significant impact of Brb on cancer progression and

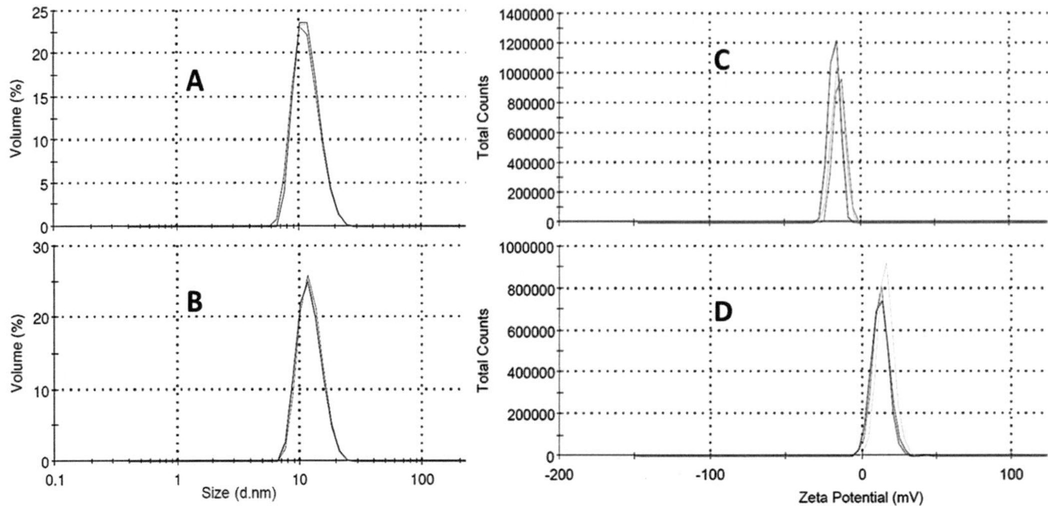

Fig. 2 Representative physicochemical characterization of anionic and cationic berberine-mixed micelles. Mean (**a**, **b**) droplet size and (**c**, **d**) interfacial electrical charge (measured as ζ-potential) of Brb loaded in either anionic 3:1 PEG-PE:TPGS mixMic (upper panels) or cationic 5%OA-mixMic (lower panels)

metastasis have been mainly attributed to direct inhibition of NF-kβ, MMP-1, -2, and -9, and associated with activation of AMPK signaling and reduction of ERK and COX-2 activities [14, 16, 17, 19–22].

Fine-tuned incorporation of tocopheryl succinate moieties into the unsaturated lipid core of our PEG2000-DSPE micelles, in 1:3 M ratio respectively (Fig. 2a, c), played a significant role in improving Brb-vehicle incorporation, contributing to higher stability of drug (up to 300%) at physiological conditions, consequently improving its systemic bioavailability and pharmacokinetic profile. Furthermore, at the cellular level, Brb incorporation in our TPGS-mixed PEG-PE micelles (mixMic) also allowed for enhanced intracellular delivery of the drug, in both time- and dose-dependent kinetics [22], especially since vitamin E moieties are known to assist as penetration enhancers into biological membranes [12, 23, 24]. Furthermore, our mixMic design takes advantage of the pro-apoptotic activity of vitamin E-TPGS—optimized as 25–30% wt of micelles forming material—to supplement the overall anticancer efficacy of co-loaded Brb drug cargo [9, 12, 24].

Using about 5% wt of cationic lipid residue, oleylamine (OA), positively charged mixMic particles (also about 20–25 nm approx. size range) were produced successfully, without negative impact on the loading capacity for Brb (Fig. 2b, d), thus increasing the capability of cancer cell association and subsequent Brb-internalization [12, 24]. Mechanistically, thanks to markedly higher cellular uptake of Brb-containing mixed micelles, our Brb-mixMic nano-formulation dramatically amplified apoptosis and overall cytotoxic effectiveness against monolayer and spheroid cultures of human prostate carcinomas [12]. Altogether, our data strongly proposes TPGS-mixed phospholipid micelles as an effective pharmaceutical system for systemic administration of Brb, and the potential for further development of these Brb-mixMic nano-preparations for anticancer drug applications.

2 Materials

2.1 Preparation and Characterization of Empty and Brb-Loaded Anionic and Cationic PEG-PE/TPGS Mixed Micelles

1. Chloroform (100%, dry).
2. 1,2-Distearoyl-*sn*-glycero-3-phosphoethanolamine-*N*-[methoxy(polyethylene glycol)-2000] (mPEG$_{2000}$-DSPE, or PEG-PE) (Avanti Polar Lipids, Inc., Alabaster, AL). Dissolve 0.143 M PEG-PE in 20 mL chloroform, to make 20 mg/mL PEG-PE stock solution. Store at −20 °C.
3. Speziol® TPGS-Pharma (NF-grade Vitamin E polyethylene glycol succinate, TPGS) (Cognis, Cincinnati, OH).
4. Ethanol (200% proof, denatured).
5. Oleyl amine cationic lipid.
6. Berberine Hydrochloride (Brb HCl). Prepare 2 mg/mL stock solution by dissolving 5.4 mM of Brb per 1 mL of warmed ethanol (200 proof) (*see* **Note 1**).
7. 15–20 mL pear-shaped glass flasks that fit rotary evaporator spout, for organic/co-solvent evaporation.
8. Rotary evaporator with vertical coiled condenser, RE100-Pro (Scilogix, LLC, Rocky Hill, CT) with rotation speed control, connected to a dry-vacuum pump capable of providing at least 100 mtorr of vacuum.
9. Stock HEPES (2×) buffered saline, pH 7.05. Dissolve 280 mM Sodium Chloride (16.4mg, 50 mM HEPES, free acid (11.9 mg), 1.5 mM Na_2HPO_4 (0.21 mg) in 100 mL of MQ water. Titrate to pH 7.1 with 5 M NaOH, adjust final volume to 1 L. Store at 4 °C.
10. 0.22 μm pore size polycarbonate membrane filters.
11. Benchtop lyophilyzer,
12. Weigh balance (up to 0.001 mg in precision for accuracy).
13. Pipette(s) capable of dispensing at 10, 500, and 1 mL.
14. 5 mL glass vials
15. Milli-Q (MQ) water.
16. Inert gas (N_2 or Ar) source, with flow meter regulator.
17. Malvern Zetasizer Nano ZS (Malvern Instruments, Westborough, MA).

18. Disposable folded capillary (electrophoretic) cells for zeta potential measurements (Malvern Instruments, Westborough, MA).
19. Disposable 12 mm square polystyrene cuvettes, for particle size analysis.

2.2 Determination of Critical Micelle Concentration (CMC) of PEG-PE/TPGS Mixed Micelles

1. 15 mL capacity polypropylene centrifuge tubes, with conical bottom and cap.
2. Pyrene, 98%, ACROS Organics (Thermo Fisher Scientific, Hampton, NH).
3. Speed adjustable/digital vortex mixer (Thermo Fisher Scientific, Hampton, NH).
4. Temperature-controlled orbital shaking incubator (Thermo Fisher Scientific, Hampton, NH).
5. 0.22 μm pore size polycarbonate membrane filters.
6. Weigh balance (up to 0.001 mg in precision for accuracy).
7. 96-well microplates: Opaque/black-walled polystyrene plates with clear flat bottom (300 μL well capacity), compatible with fluorimeter (Corning Inc., Corning, NY).
8. Fluorescence plate reader with excitation 380–410 nm and emission 495–525 nm filter pair, Victor X3 Multi-label microplate reader (PerkinElmer, Santa Clara, CA).

3 Methods

3.1 Preparation of Empty Anionic and Cationic PEG-PE/TPGS Mixed Micelles

Prepare all micellar formulations (Mic) using only clean glassware. Thoroughly clean the glassware and spatulas with concentrated nitric acid followed by ethanol. Make sure no residue of whitish phospholipids or drug remains in the glassware. Furthermore, use MQ water during the entire formulation processes to guarantee purified grade final product

1. Turn on the hot plate and adjust to 30 °C. Warm clean 25 mL beaker on the hot plate for 5 min, filled with 1X HEPES buffered saline adjusted to pH 7.4.
2. In a 20 mL pear-shaped glass flask, add polymeric phospholipid surfactants, 42.6 mMol PEG-PE (as 2.13 mL from 20 mg/mL PEG-PE stock solution).
3. Connect pear-shaped glass flask to the rotary evaporator, and slowly evaporate organic solvent under 100 mtorr (26 Hg) vacuum, set at 50–60 rpm rotation, and 40 °C water bath temperature, for approx. 30 min (*see* **Notes 2** and **3**).
4. Release vaccum pressure, and disconnect flask, then, add 19.7 mMol of vitamin E-TPGS (TPGS, as 29.75 mg), followed by

2 mL of denatured ethanol directly to the warm lipid film inside pear-shaped glass flask (For cationic OA-mixMic, *see* **Note 4**).

5. Using an air-heat gun, mix surfactant components and the oils using vortex mixer, while monitoring the oily mixture temperature not to exceed 40 °C. Final micelle matrix composition is TPGS: PEG-PE as 1:3 molar ratio.
6. Reconnect pear-shaped glass flask to the rotary evaporator and slowly evaporate organic solvent under 100 mtorr (26 Hg) vacuum, set at 50–60 rpm rotation and 40 °C water bath temperature, for approximately 45 min (*see* **Note 5**).
7. Using 1 mL pipette gradually add 3 mL of warm 1× HEPES buffered saline, pH 7.4 onto the warm mixture inside the 25 mL pear-shaped glass flask, mixing thoroughly but slowly, using vortex mixer at about 800 rpm for 10 min, or until all lipid films on the glass has been dispersed in buffered solution (*see* **Notes 6** and **7**).
8. Filter mixMic dispersions with a 0.22 μm pore size polycarbonate membrane.
9. Seal aliquots of Mic filtrates in 5 mL glass vials under argon and store at 4 °C until use.
10. Alternatively, freeze Mic filtrates in aliquots in liquid nitrogen and vacuum-freez drying with FreeZone 4.5 Lyophilyzer with $p < 200 \times 10^{-3}$ mbar, condenser temperature ≤ -50 °C).

3.2 Preparation of Brb-Loaded Anionic and Cationic PEG-PE/TPGS Mixed Micelles

Perform all the procedures at 40 °C, unless specified

1. Turn on the hot plate and adjust to 30 °C. Warm the clean 25 mL beaker on the hot plate for 5 min, filled with 1× HEPES buffered saline adjusted to pH 7.4.
2. In a 20 mL pear-shaped glass flask, add polymeric phospholipid surfactants, 42.6 mMol PEG-PE (as 2.13 mL from 20 mg/mL PEG-PE stock solution).
3. Connect pear-shaped glass flask to the rotary evaporator, and slowly evaporate organic solvent under 100 mtorr (26 Hg) vacuum, set at 50–60 rpm rotation, and 40 °C water bath temperature, for approx. 30 min (*see* **Notes 2** and **3**).
4. Release vaccum pressure, and disconnect flask, then, add 19.7 mMol of vitamin E-TPGS (TPGS, as 29.75 mg), followed by 3 mL of 16.2 mM Brb dissolved in ethanol (2 mg/mL), directly to the warm lipid film inside pear-shaped glass flask (For cationic OA-mixMic, *see* **Note 4**).
5. Using an air-heat gun, mix all surfactant components and Brb in the organic solvent using vortex mixer, until clear homogenous mixture is obtained while monitoring temperature not to exceed 45 °C.

6. Cover the pear-shaped glass flask with aluminum foil, then reconnect to the rotary-evaporator, and slowly evaporate solvent under 100 mtorr (26 Hg) vacuum, set at 50–60 rpm rotation, and 40 °C water bath temperature, for approximately 30 min (*see* **Note 5**).
7. Using 1 mL pipette, gradually add 3 mL of warm 1× HEPES buffered saline, pH 7.4 onto the warm mixture inside the 25 mL pear-shaped glass flask, mixing thoroughly but slowly, using vortex mixer at about 800 rpm for 10 min, or until all lipid films on the glass has been dispersed in buffered solution (*see* **Notes 6** and 7).
8. Filter mixMic dispersions with a 0.22 μm pore size polycarbonate membrane.
9. Seal aliquots of Mic filtrates in 5 mL glass vials under argon and store at 4 °C until use.
10. Alternatively, freeze Mic filtrates in aliquots in liquid nitrogen and vacuum-freez drying with FreeZone 4.5 Lyophilyzer with $p < 200 \times 10^{-3}$ mbar, condenser temperature ≤ -50 °C).

3.3 Physical Characterization of Drug-Loaded Anionic and Cationic PEG-PE/TPGS Mixed Micelles

Produced Mic formulations are characterized for particle size and size distribution using the dynamic light scattering (DLS) technique with a Malvern Zetasizer Nano ZS (Malvern Instruments, Westborough, MA) at 273° fixed angle and at 23 °C temperature.

1. Dilute Mic formulation, for particle size analysis, using MQ water at about 100-fold vol./vol., in disposable polystyrene cuvettes. The numbered average particle hydrodynamic diameter and the polydispersity index (DPI) will be determined (Fig. 2a, b).
2. For the zeta potential, dilute Mic samples in MQ water, pH 6.8, at 1000-fold, then employ a 1 mL syringe, horizontally, to insert the almost transparent solution carefully inside the folded capillary electrophoretic cell of the Malvern Zetasizer Nano ZS, while making sure to avoid inserting any air bubbles. The average surface charge will be measured (Fig. 2c, d).

3.4 Determination of Critical Micelle Concentration (CMC) of PEG-PE/TPGS Mixed Micelles

The CMC value of both PEG-PE micelle types, TPGS-mixed or pure, can be estimated via the pyrene method, utilizing the Victor X3 fluorescence microplate reader (PerkinElmer, Santa Clara, CA).

1. Exactly disperse 10 mL vol. of 10^{-3} to 10^{-7} M of micellar solution of PEG_{2000}-DSPE, TPGS and 1:3 molar mix of TPGS/PEG_{2000}-DSPE in HBS to 15 mL centrifuge tubes, each containing 1 mg of pyrene crystals (4.9 mM), prepare each sample in triplicates (*see* **Note 7**).

2. Transfer all tubes into temperature-controlled orbital shaker and incubate at room temperature (RT) for 24 h with continuous shaking at 200 rpm.
3. Remove free pyrene by filtering all Mic dispersions through a 0.22 μm pore size polycarbonate membrane.
4. Carefully transfer 100 μL vol. from each sample tube into corresponding well on the black-walled polystyrene plates with clear flat bottom.
5. Measure the fluorescence of these filtrated Mic samples (λ excitation/emission: 390/505 nm) using Victor X3 Multilabel microplate reader (PerkinElmer, Santa Clara, CA). CMC values corresponding to the concentration of the polymer at which a sharp increase in fluorescence is observed.

4 Notes

1. Use 20 mL glass vial with stopper to make Brb HCl-stock solution. If some Brb particles are still visibly suspended in solvent after 10 min of vortexing (at 2500 rpm ($\sim$56.0 $\times$ g)), put glass vial in bath sonicator for 5–15 min, then vigorously mix the vial contents for an additional 5–15 min. Confirm that Brb has completely dissolved in organic solvent, before proceeding to next steps.
2. While the vacuum is best adjusted based on each solvent, for optimized solvent removal via the rotary evaporator system, apply the 20/40/60 "technical" rule, which correlates to at least 20 °C difference in temperature between the system's main components. Use operating bath temperature of at least 40 °C, to yield a solvent vapor temperature of 20 °C, which is subsequently condensed at about 0 °C (using ice-water to cool the condenser).
3. For efficient and complete evaporation of organic solvent, make sure to keep the connected pear-shaped flask tilted at about 60–45° angle to the plane of the surrounding warm water bath. For improved ethanol evaporation, also increase the flask's rotation speed to about 90 rpm.
4. For cationic mixMic containing 5 M% of oleylamin (OA) formulations, add 3.28 mM of OA (as 0.9 mg) directly on the dried phospholipid film, along with vitamin E TPGS, and mixing all together within the later added ethanol solvent.
5. Complete removal of organic/alcoholic solvent is confirmed when a clear translucent off-white dry film residue remains in the flask, which may get somewhat more opaque as the flask temperature cools down. The dried Mic film must be clear from any suspending yellow Brb drug precipitates. Otherwise,

redissolve in another 3 mL of ethanol and repeat the evaporation process using slightly lower water bath temperature and vacuum settings.

6. Optional: briefly put pear-shaped flask in bath sonicator (only for 2–3 min) to dislodge resistant dried Mic lipid film remains present on the glass walls of flask.
7. To prevent incomplete dispersion of Mic-forming components, it is best to minimize formation of bubbles in the formed mixMic dispersion. Thus, make sure to add warm aqueous buffer (approximately 35 °C) gradually in 1 ML increments and to thoroughly mix dried Mic film, keeping the vortex mixer at low speeds (<1000 rpm (9.0 × g)).

References

1. Wong HL, Bendayan R, Rauth AM, Li Y, Wu XY (2007) Chemotherapy with anticancer drugs encapsulated in solid lipid nanoparticles. Adv Drug Deliv Rev 59:491–504
2. Torchilin VP (2007) Micellar nanocarriers: pharmaceutical perspectives. Pharm Res 24:1–16
3. Yallapu MM, Jaggi M, Chauhan SC (2013) Curcumin nanomedicine: a road to cancer therapeutics. Curr Pharm Des 19:1994–2010
4. Shi J, Kantoff PW, Wooster R, Farokhzad OC (2017) Cancer nanomedicine: progress, challenges and opportunities. Nat Rev Cancer 17:20–37
5. Elbayoumi TA, Pabba S, Roby A, Torchilin VP (2007) Antinucleosome antibody-modified liposomes and lipid-core micelles for tumor-targeted delivery of therapeutic and diagnostic agents. J Liposome Res 17:1–14
6. Mu L, Elbayoumi TA, Torchilin VP (2005) Mixed micelles made of poly(ethylene glycol)-phosphatidylethanolamine conjugate and d-alpha-tocopheryl polyethylene glycol 1000 succinate as pharmaceutical nanocarriers for camptothecin. Int J Pharm 306:142–149
7. Sawant RR, Torchilin VP (2010) Multifunctionality of lipid-core micelles for drug delivery and tumour targeting. Mol Membr Biol 27:232–246
8. Sawant RR, Torchilin VP (2010) Polymeric micelles: polyethylene glycol-phosphatidylethanolamine (PEG-PE)-based micelles as an example. Methods Mol Biol 624:131–149
9. Pham J, Brownlow B, Elbayoumi T (2013) Mitochondria-specific pro-apoptotic activity of genistein lipidic nanocarriers. Mol Pharm 10:3789–3800
10. Musacchio T, Laquintana V, Latrofa A, Trapani G, Torchilin VP (2009) PEG-PE micelles loaded with paclitaxel and surface-modified by a PBR-ligand: synergistic anticancer effect. Mol Pharm 6:468–479
11. Sawant RR, Torchilin VP (2009) Enhanced cytotoxicity of TATp-bearing paclitaxel-loaded micelles in vitro and in vivo. Int J Pharm 374:114–118
12. Shen R, Kim JJ, Yao M, Elbayoumi TA (2016) Development and evaluation of vitamin E d-alpha-tocopheryl polyethylene glycol 1000 succinate-mixed polymeric phospholipid micelles of berberine as an anticancer nano-pharmaceutical. Int J Nanomedicine 11:1687–1700
13. Yi T, Zhuang L, Song G, Zhang B, Li G, Hu T (2015) Akt signaling is associated with the berberine-induced apoptosis of human gastric cancer cells. Nutr Cancer 67:523–531
14. Kim JS, Oh D, Yim MJ, Park JJ, Kang KR, Cho IA, Moon SM, Oh JS, You JS, Kim CS, Kim DK, Lee SY, Lee GJ, Im HJ, Kim SG (2015) Berberine induces FasL-related apoptosis through p38 activation in KB human oral cancer cells. Oncol Rep 33:1775–1782
15. Yan K, Zhang C, Feng J, Hou L, Yan L, Zhou Z, Liu Z, Liu C, Fan Y, Zheng B, Xu Z (2011) Induction of G1 cell cycle arrest and apoptosis by berberine in bladder cancer cells. Eur J Pharmacol 661:1–7
16. Wang N, Feng Y, Zhu M, Tsang CM, Man K, Tong Y, Tsao SW (2010) Berberine induces autophagic cell death and mitochondrial apoptosis in liver cancer cells: the cellular mechanism. J Cell Biochem 111:1426–1436
17. Patil JB, Kim J, Jayaprakasha GK (2010) Berberine induces apoptosis in breast cancer cells

(MCF-7) through mitochondrial-dependent pathway. Eur J Pharmacol 645:70–78

18. Choi MS, Oh JH, Kim SM, Jung HY, Yoo HS, Lee YM, Moon DC, Han SB, Hong JT (2009) Berberine inhibits p53-dependent cell growth through induction of apoptosis of prostate cancer cells. Int J Oncol 34:1221–1230
19. Barzegar E, Fouladdel S, Movahhed TK, Atashpour S, Ghahremani MH, Ostad SN, Azizi E (2015) Effects of berberine on proliferation, cell cycle distribution and apoptosis of human breast cancer T47D and MCF7 cell lines. Iran J Basic Med Sci 18:334–342
20. Refaat A, Abdelhamed S, Yagita H, Inoue H, Yokoyama S, Hayakawa Y, Saiki I (2013) Berberine enhances tumor necrosis factor-related apoptosis-inducing ligand-mediated apoptosis in breast cancer. Oncol Lett 6:840–844
21. Yip NK, Ho WS (2013) Berberine induces apoptosis via the mitochondrial pathway in liver cancer cells. Oncol Rep 30:1107–1112
22. Ho YT, Lu CC, Yang JS, Chiang JH, Li TC, Ip SW, Hsia TC, Liao CL, Lin JG, Wood WG, Chung JG (2009) Berberine induced apoptosis via promoting the expression of caspase-8, -9 and -3, apoptosis-inducing factor and endonuclease G in SCC-4 human tongue squamous carcinoma cancer cells. Anticancer Res 29:4063–4070
23. Pham J, Nayel A, Hoang C, Elbayoumi T (2015) Enhanced effectiveness of tocotrienol-based nano-emulsified system for topical delivery against skin carcinomas. Drug Deliv 1–11
24. Pham J, Grundmann O, Elbayoumi T (2015) Mitochondriotropic nanoemulsified genistein-loaded vehicles for cancer therapy. Methods Mol Biol 1265:85–101

Chapter 4

Phosphopholipid Micelles for Peptide Drug Delivery

Karina Esparza, Dulari Jayawardena, and Hayat Onyuksel

Abstract

Sterically stabilized micelle (SSM) is a self-assembled nanoparticle ideal for the delivery of therapeutic peptides. The PEGylated phospholipid forming the particle, DSPE-PEG_{2000}, is a safe, biocompatible, and biodegradable ingredient already approved for human use in the marketed product Doxil®. SSM can overcome formulation difficulties such as instability associated with peptide drugs, enabling their development for clinical application. The key advantage of this lipid-based nanocarrier is its simple preparation even at large scales, which allows easy transition to the clinics and the pharmaceutical market. In this chapter, we describe methods for preparation and characterization of peptides self-associated with SSM (peptide–SSM). We also discuss approaches to evaluate the biological activity of the peptide nanomedicines in vitro and in vivo.

Key words Sterically stabilized micelles, DSPE-PEG_{2000}, Peptide drug delivery, Micellar nanocarrier, Vasoactive intestinal peptide, Glucagon-like peptide 1, Pancreatic polypeptide, Neuropeptide Y, Pituitary adenylate cyclase activating polypeptide, Glucose-dependent insulinotropic peptide

1 Introduction

Therapeutic peptides have gained interest over the past few years due to their high biological specificity, efficacy, and tolerability upon administration [1]. However, clinical use of peptide drugs remains challenging due to intrinsic limitations such as poor physicochemical stability and short biological half-life [2]. In this regard, our laboratory has developed a self-assembled phospholipid nanocarrier termed sterically stabilized micelle (SSM) as a method of overcoming these barriers.

SSM is a self-assembled nanostructure of about 15 nm in diameter composed of PEGylated phospholipids. The main phospholipid used in the system is the sodium salt of 1,2-distearoyl-*sn*-glycero-3-phosphoethanolamine-*N*-[amino(polyethylene glycol)-2000] (DSPE-PEG_{2000}), a biocompatible pharmaceutical ingredient present in an already FDA-approved marketed product, Doxil®. In aqueous solution, above their critical micellar concentration (0.5–0.1 μM), PEGylated phospholipids organize themselves into

Volkmar Weissig and Tamer Elbayoumi (eds.), *Pharmaceutical Nanotechnology: Basic Protocols*, Methods in Molecular Biology, vol. 2000, https://doi.org/10.1007/978-1-4939-9516-5_4, © Springer Science+Business Media, LLC, part of Springer Nature 2019

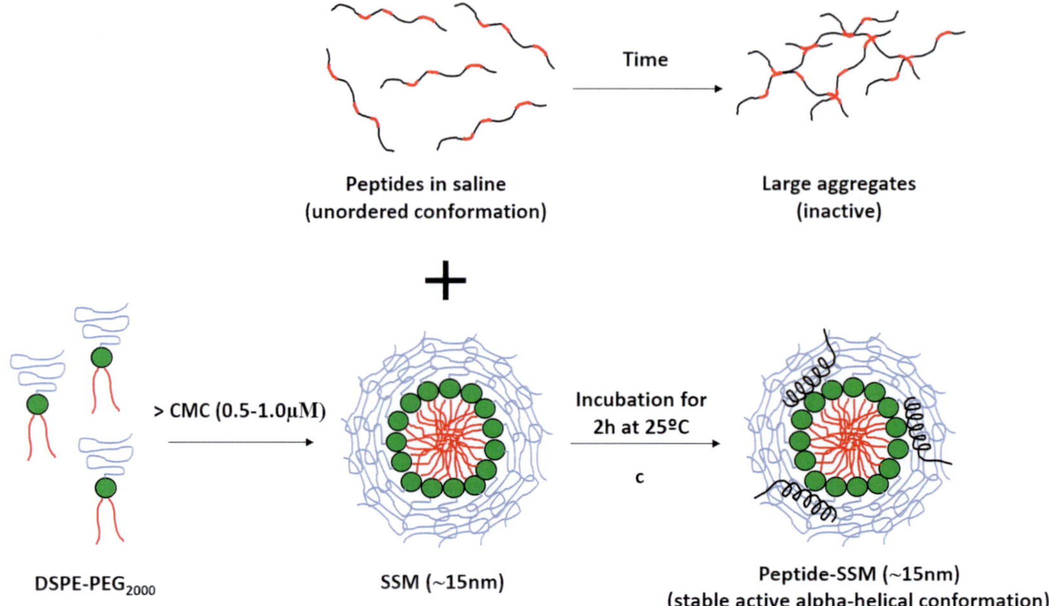

Fig. 1 Schematic representation of peptide association with sterically stabilized micelles (SSM). (**a**) DSPE-PEG2000 molecule is composed of two nonpolar acyl chains (red), a polar phosphate head (green), and a semi-polar, but mostly hydrophilic, 2000 Da PEG moiety (blue). This amphiphilic monomer self-assembles with others above its critical micellar concentration (CMC) forming SSM with three distinct regions: hydrophobic core (red), phosphate ionic interface (green), and PEG corona (blue). (**b**) Amphiphilic peptides dissolved in saline solutions are found in an unordered conformation and tend to aggregate over time due to hydrophobic interactions (red regions). (**c**) Amphiphilic peptides in their monomeric form self-associate with SSM after incubation driven by molecular interactions. Note that this is a graphical representation only and does not reflect actual scale

a nanoparticle with a hydrophobic core, an ionic interface, and a semipolar PEG corona (Fig. 1) [3]. In isotonic buffers, we found that each micelle consists of approximately 90 monomers [4], and they can encapsulate hydrophobic and amphiphilic drug molecules in various regions of SSM to address delivery issues [5, 6].

Phospholipid micelles improve the safety and efficacy of drugs through several mechanisms [5, 6]. First, the PEG corona present in the outer layer of the nanoparticle interacts with surrounding aqueous media to form a structured layer of water. This "water coat" prevents nanoparticles from being recognized by immune cells, thus increasing their circulation half-life in vivo. The steric barrier also impedes access of enzymes to the encapsulated drug molecules, reducing their potential enzymatic degradation. In addition, SSM promotes passive targeting of drugs to diseased tissues due to its optimal nanosize (~15 nm). Drugs associated with nanoparticles refrain from extravasating to healthy tissues but reaches tissues with leaky vasculatures such as tumors and inflamed tissue [6]. This phenomenon of passive accumulation not only increases drug availability and efficacy at the target site but also prevents off-target effects and toxicity [7, 8]. SSM also prevents drug aggregation in aqueous media by accommodating individual molecules in their interior. This association improves the drug solubility and stabilizes peptide drugs in their active molecular form. Finally, SSM nanoparticles can be fully dispersed in the aqueous media without aggregation due to the hydrated PEG layer on the micelle surface. All these properties render SSM an ideal platform for safe and effective delivery of peptide-based drugs.

In the absence of micelles, amphiphilic peptides at high concentrations undergo aggregation in aqueous media (Fig. 1). This behavior is caused by the gradual clustering of peptide hydrophobic domains to prevent contact with the aqueous environment and attain a lower energy state. The aggregation leads to loss of peptide activity and increased immunogenicity in vivo, limiting its clinical application [9]. However, when monomeric peptides are incubated with SSM, they self-associate with the semipolar PEG corona of micelles by molecular interactions (Fig. 1) [10]. This association is driven by the peptide's internal energy and does not require any chemical conjugation. This association not only protects the peptide from aggregation and degradation but also induces peptide transition from the unstable random coil to a stable alpha-helical conformation, which is the preferred conformation of peptides in the secretin/glucagon superfamily for receptor interaction [11]. Peptide drugs circulate long hours in the blood when associated with SSM [7], most probably with their charged receptor-binding domain exposed on the surface of micelles. Once the nanomedicine reaches the target site and it is close enough to its specific receptor, the peptide initiates strong interaction with the receptor and leaves the delivery system to acquire the lowest energy state (Fig. 2).

Over the past two decades, our laboratory has tested several peptides for their suitability to be delivered in SSM. We found that the following optimal features are necessary to achieve a successful peptide–SSM association: (a) approximate length range between 17 and 42 amino acids, (b) distinct hydrophobic and hydrophilic regions within the peptide molecule, and (c) open and flexible random coil structure in water which transitions into an active alpha-helix upon association with micelles. In addition, the presence of intrinsic fluorescent amino acids on the peptide sequence (tryptophan, tyrosine, or phenylalanine) allows easier nanomedicine characterization. For example, we have successfully incorporated the following peptides in SSM: LP17 (TREM-1 blocking peptide) (17 amino acids) [12], secretin (27 amino acids) [13, 14], vasoactive intestinal peptide (VIP) (28 amino acids) [7, 15, 16], glucagon-like peptide-1 (GLP-1) (30 amino acids) [16–18], neuropeptide Y (NPY) (36 amino acids) [19], pancreatic polypeptide (PP) (36 amino acids) [20, 21], pituitary

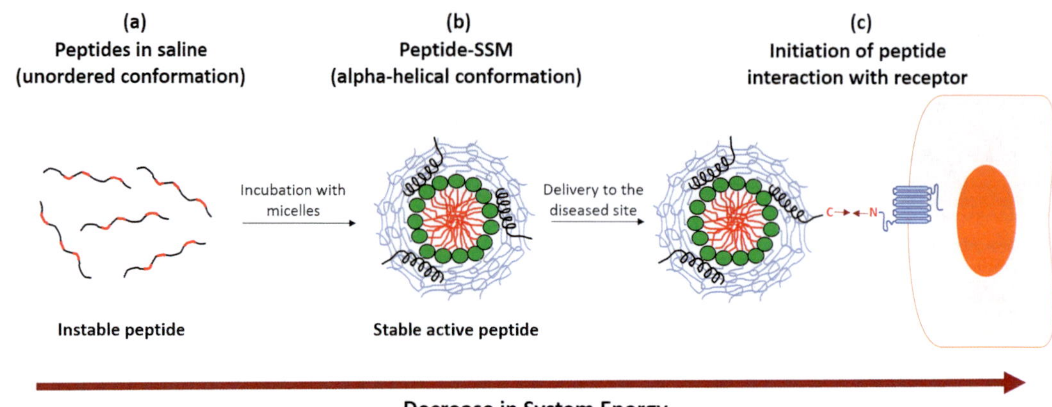

Fig. 2 System energy of amphiphilic peptide drugs at different states. (**a**) Peptides in saline have the highest energy state and tendency to aggregation due to the exposure of hydrophobic domains to the aqueous environment. (**b**) Association of peptides with semipolar PEG corona of SSM reduces their internal energy and improves their stability. Peptides also transition to an active alpha-helical conformation which is usually required for receptor interaction. (**c**) Peptides are strongly attracted to their specific receptor when in close proximity to achieve the lowest energy state. Taking VIP as a model, the initial attraction between the C-terminal portion of the peptide with the N-terminal portion of the membrane receptor triggers translocation of peptide from the micelle to the receptor to show its activity. Since peptides are not covalently attached to SSM, they can easily leave the nanocarrier. Note that this is a graphical representation only and does not reflect actual scale

adenylate cyclase activating polypeptide (38 amino acids) ($PACAP_{1-38}$) [22], and glucose-dependent insulinotropic peptide (GIP) (42 amino acids) [16, 23].

In this chapter, we describe the methods of preparation, lyophilization, and characterization of water-soluble amphiphilic peptides in SSM. In addition, we discuss common problems encountered during micelle preparation and ways to avoid them. We believe that the simple and reproducible preparation technique of peptide–SSM nanomedicines along with the safety and long-term stability of the freeze-dried formulation offers great promise for peptide drugs to be used in clinical practice in the near future.

2 Materials

All materials and equipment cited in this chapter are based on our research group's work. However, similar instruments can be substituted with necessary modifications if needed.

2.1 Peptide–SSM Preparation

1. 1,2-distearoyl-*sn*-glycero-3-phosphoethanolamine-*N*-[amino (polyethylene glycol)-2000] sodium salt ($DSPE\text{-}PEG_{2000}$)
2. Purified peptide in freeze-dried form.
3. Filtered isotonic aqueous media prepared with deionized water. In our laboratory, we have successfully used sterile 1×

phosphate-buffered saline and normal saline (0.9% NaCl injection USP).

4. Methanol for washing.
5. Deionized water for washing.
6. Glass vials with screw caps.
7. 0.22 μm syringe filter units
8. Pressurized argon gas.
9. Incubator set at 25 °C.
10. Sonicator.
11. Mini vortex.
12. Microbalance.

2.2 Peptide–SSM Lyophilization

1. Freeze-dryer.
2. Freezer −80 °C.
3. Liquid nitrogen.
4. Lint-free wipes (Kimwipes).
5. Rubber bands.

2.3 Peptide–SSM Characterization

2.3.1 Particle Size by Dynamic Light Scattering

1. Dynamic light scattering instrument.
2. Drop-in vials.
3. Lint-free wipes (Kimwipes).
4. Parafilm.

2.3.2 Saturation Molar Ratio by Fluorescence Spectroscopy

1. Spectrofluorimeter.
2. 1 cm path length quartz cuvette
3. Glass vials with screw caps.
4. SigmaPlot (Systat Software Inc., San Jose, CA).

2.3.3 Peptide Secondary Structure by Circular Dichroism Spectropolarimetry

1. Spectropolarimeter.
2. 1 cm path length quartz cuvette
3. Glass vials with screw caps.

2.3.4 Bioactivity

Materials will depend on the peptide-specific activity assays in vitro and in vivo.

3 Methods

3.1 Preparation of Peptide–SSM Nanomedicine

1. Rinse vial caps with deionized water and dry with compressed air. Rinse glass vials thoroughly with methanol, dry with compressed air, and recap them immediately to avoid entry of dust (*see* **Note 1**).

2. Prepare stock solutions of DSPE-PEG_{2000} by transferring weighed amounts of phospholipid into a glass vial (*see* **Note 2**). Dissolve the material in the required volume of isotonic solution (*see* **Note 3**). Use gentle swirling or vortexing followed by brief sonication if necessary (*see* **Note 4**).
3. Flush with argon gas to replace air in the void space of vials and incubate the sealed vials for 1 h at 25 °C in the dark protected from light (*see* **Note 5**).
4. Prepare a stock solution of peptide immediately before mixing it with phospholipid dispersion. Transfer the weighted amount of peptide into a glass vial (*see* **Note 6**) and dissolve it in the same type of aqueous medium as used for micelle preparation. Use gentle swirling and brief sonication if necessary (*see* **Note 4**).
5. Mix predetermined amounts of DSPE-PEG_{2000} dispersion and peptide solution into a third vial to obtain the desired molar ratio of peptide: DSPE-PEG_{2000}. Flush sample with argon gas and incubate it at 25 °C for 2 h protected from light.
6. If needed, sterilize the micellar formulation by filtration through a 0.22 μm syringe filter (*see* **Note 7**).
7. Freeze-dry the formulation for longer shelf stability (*see* Subheading 3.2).

3.2 Long-Term Storage of Peptide–SSM Nanomedicine

Pharmaceutical formulations must demonstrate reasonable storage stability to become a viable marketed product. In the case of peptide–SSM, peptides remain stable in micellar liquid dispersion for about 7 days at 25 °C [7]. To increase the shelf stability of peptide nanomedicines, freeze-drying can be employed to remove water and dissolved oxygen. This will decrease the rate of degradation of nanomedicine by hydrolysis and oxidation during the storage period. Peptide–SSM formulations can be lyophilized without any lyo- or cryoprotectant [16]. The optimal phospholipid concentration to form lyophilized cakes with minimum shrinkage and acceptable particle size ranges from 5 to 15 mM (*see* **Note 8**) [16]. Furthermore, to avoid oxidation, freeze-dried product should be stored protected from light under inert gas.

General steps to freeze-dry peptide–SSM formulation using a Labconco Freezone 4.5 freeze-drier are described below, but other laboratory-scale freeze-drying equipment can also be used with appropriate modifications.

1. Prepare peptide–SSM formulation as described in Subheading 3.1.
2. Transfer 1 mL of formulation into a 2 mL vial (*see* **Note 9**).
3. Freeze samples at −80 °C overnight.

4. Replace vial caps with a non-dust material such as double-folded Kimwipes tied with a rubber band.
5. Dip samples in liquid nitrogen for few minutes to ensure that samples are completely frozen.
6. Freeze-dry samples overnight using a freeze-dryer according to the manufacturer's protocol.
7. Next day, remove samples from freeze-dryer, gently flush with inert gas, and recap vials. Samples can be stored in the freezer until use.
8. To reconstitute the samples, add 1 mL of sterile water and gently swirl in a circular motion until complete dissolution. Incubate for 2 h at 25 °C protected from light before use to ensure sample equilibration.

3.3 Peptide–SSM Nanoparticle Characterization

The characterization of peptides–SSM is based mainly on the particle size, maximum peptide molecules per micelle (saturation molar ratio), and alpha-helicity of peptide in SSM. In addition, depending on the specific activity of the nanomedicine, bioactivity can also be determined using in vitro assays and/or animal models. In this section, glucagon-like peptide 1 associated with SSM (GLP1-SSM) is used as a model peptide nanomedicine to illustrate the application of each characterization technique. However, data for other peptides successfully used in SSM in our laboratory can be found in the literature [12–15, 17, 19–22].

3.3.1 Particle Size by Dynamic Light Scattering

Dynamic light scattering (DLS) is the preferred nondestructive technique to measure the hydrodynamic radius of nanoparticles dispersed in aqueous media. This method is applied to confirm the presence of homogeneous micelle population in the formulation and to detect possible peptide precipitates as separate populations. For example, our group verified that the peptide GLP-1 formed large aggregates in saline over time (Fig. 3). However, when GLP-1 was incubated with DSPE-PEG_{2000} above its

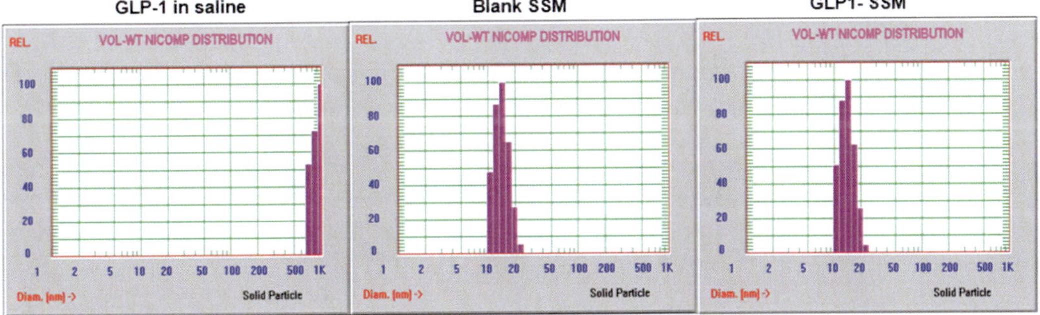

Fig. 3 Representative particle size distribution of glucagon-like peptide-1 (GLP-1) in saline (20 μM), blank SSM (5 mM DSPE-PEG_{2000}), and GLP-1-SSM (Reproduced from ref. 17 with permission from Springer)

saturation molar ratio (*see* Subheading 3.3.2), due to GLP-1 association with micelles in its monomeric form, aggregates were no longer detectable (Fig. 3) [17]. This same behavior was observed for other peptides such as NPY [19] and PP [20].

In this characterization technique, the hydrodynamic radius of dispersed particles is measured based on their diffusivity caused by the Brownian motion. A laser beam is directed towards the sample, and the scattered light from particles is detected perpendicularly. The pattern of scattered light is analyzed by the correlator which calculates the diffusion coefficient of particles using the correlation function of time versus intensity of scattered light. Small molecules travel faster and therefore exhibit higher diffusivity while bigger particles move slower and present lower diffusivity. The Stokes-Einstein equation is then applied to convert particle diffusion coefficient into hydrodynamic radius (Eq. 1) [24].

$$D = \frac{\kappa T}{3\pi d\eta} \tag{1}$$

where D is the particle translational diffusion coefficient, k is the Boltzmann constant (1.38×10^{-16} erg K^{-1}), T is the absolute temperature, d is the hydrodynamic diameter, and η is the media viscosity.

In our laboratory, we use NICOMP 380 ZLS equipped with a 5 mW helium-neon laser at 632.8 nm. We employ default parameters for aqueous media at 23 °C where viscosity (η) is 0.933 cP and refraction index is 1.333.

The specific procedure to measure the particle size will depend on the equipment used, but general steps for NICOMP 380 ZLS particle size analyzer are describe below. Examples for other equipment that could be used include Malvern Zetasizer Systems (Malvern, United Kingdom) and Brookhaven (Holtsville, NY)

1. Turn on the equipment and let it warm up for about 10 min.
2. Verify if the drop-in vial fits the equipment vial holder.
3. Rinse drop-in vials with methanol and dry with filtered pressurized air. Keep vials facing down on a clean stand to prevent dust entry.
4. Transfer at least 400 μL of sample into the drop-in vial and seal it with parafilm.
5. Clean vial surface with a lint-free paper tissue.
6. Insert the vial sample into the cell holder of the equipment.
7. Initialize the instrument and adjust the filter position to obtain photo pulse rate around 300 kHz.
8. Measure particle size using the NICOMP intensity and volume-weighted distribution for 5 cycles of 3 min each.
9. Confirm the presence of peptide–SSM with diameter of approximately 15 nm (*see* **Notes 10** and **11**).

3.3.2 Saturation Molar Ratio by Fluorescence Spectroscopy

The saturation molar ratio is the maximum amount of peptide that can self-associate with each phospholipid micelle assuming that, in equilibrium, the amount of peptide molecules associated with each micelle is constant. This ratio can be determined by measuring the fluorescence of peptides containing intrinsic fluorophores (tryptophan, tyrosine, or phenylalanine) in the presence of increasing concentrations of phospholipid. Peptides tend to aggregate in aqueous medium causing the fluorescence signal to reduce due to energy transfer, a phenomenon known as quenching [25]. However, as PEGylated phospholipids are added to the medium above its critical micellar concentration, more peptides interact with micelles in a monomeric form instead of aggregating, and this results in an increase of the emission signal. The saturation molar ratio is detected when the peak fluorescence becomes constant, indicating that all peptide molecules are interacting with micelles [13, 16, 17, 19, 20]. For example, in a previous study, the peptide GLP-1 exhibited a saturation molar ratio of 13:1 phospholipid: peptide (Fig. 4), indicating that approximately 6 GLP-1 molecules interact with each micelle [17].

The specific procedure to measure the fluorescence will depend on the equipment used, but general steps for determination of DSPE-PEG$_{2000}$: peptide saturation molar ratio are described below. In our laboratory, we use the spectrofluorimeter SLM

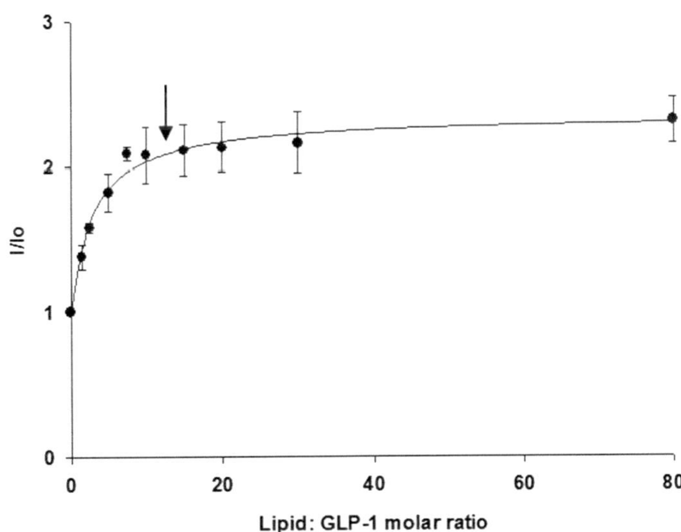

Fig. 4 Representative saturation molar ratio of DSPE-PEG$_{2000}$: peptide determined by fluorescence spectroscopy. The fluorescence enhancement ratio (I/I0) of 5 μM of glucagon-like peptide-1 (GLP-1) with increasing amounts of DSPE-PEG$_{2000}$ reaches a plateau when all GLP-1 molecules are associated with SSM (saturation molar ratio). Arrow indicates the experimental molar ratio in which saturation of peptides in SSM occurs (Reproduced from ref. 17 with permission from Springer)

Aminco 8000 (Rochester, NY), but other spectrofluorimeters with similar sensitivity and detection range can also be used.

1. Prepare stock solutions of peptide and DSPE-PEG$_{2000}$ in isotonic media (*see* Subheading 3.1).
2. Keeping the final peptide concentration constant (any amount between 4 and 20 μM), add varying amounts of the DSPE-PEG$_{2000}$ stock to obtain ratios of phospholipid: peptide from 1:1 to 150:1.
3. Prepare the same concentration of free peptide (0:1 phospholipid: peptide).
4. Prepare blank samples (without peptide) at each ratio with DSPE-PEG$_{2000}$.
5. Incubate samples at 25 °C for 2 h protected from light.
6. Transfer samples to a 1 cm path length quartz cuvette taking care to not introduce air bubbles.
7. Wipe the optical surface with soft lint-free tissue (*see* **Note 12**).
8. Measure the fluorescence intensity in triplicate (*see* **Note 13**).
9. Use maximal emission intensity for calculations (*see* **Note 14**). Subtract blank signals for each sample to obtain the corrected fluorescence values.
10. Divide the fluorescence signal of DSPE-PEG$_{2000}$: peptide samples (I) by the signal of free peptide (I_0).
11. Plot I/I_0 versus DSPE-PEG$_{2000}$: peptide molar ratio using SigmaPlot software.
12. Determine best-fit curve using SigmaPlot (*see* **Note 15**).
13. The saturation molar ratio is defined as the lowest molar ratio of phospholipid: peptide in which the fluorescence is not significatively different from the plateau (with an excess of phospholipid).
14. Given that each micelle is composed of approximately 90 monomers [4], the maximum number of peptides that can be associated with each micelle is calculated by dividing 90 by X, where X is the saturation molar ratio determined by fluorescence.

3.3.3 Peptide Secondary Structure by Circular Dichroism Spectropolarimetry

Circular dichroism spectropolarimetry is used to detect changes in the peptide secondary conformation upon association with PEGylated micelles. In this technique, samples are subjected to circularly polarized light in the visible and ultraviolet regions. Differences in absorbance of the left-hand and right-hand circularly polarized light at specific wavelengths indicate changes in secondary conformation of the chiral peptide. Peptides tested in our laboratory were found in a predominantly unordered conformation in aqueous

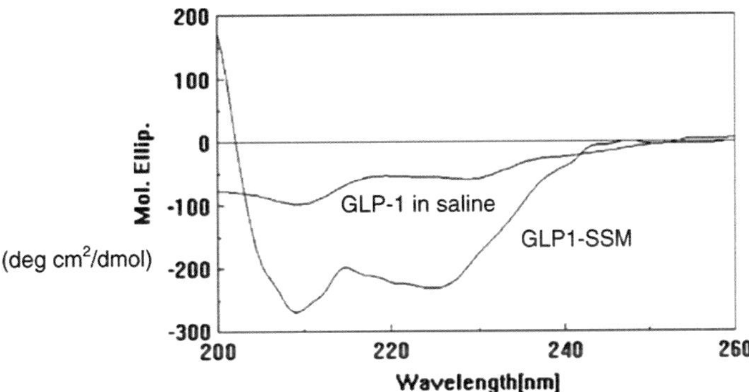

Fig. 5 Representative circular dichroism spectra of peptide in the presence or absence of DSPE-PEG$_{2000}$ micelles. Glucagon-like peptide-1 (GLP-1) (20 μM) in saline exhibits a predominantly random coil structure, but incubation with 5 mM of DSPE-PEG$_{2000}$ promotes the formation of alpha helical secondary structures as indicated by the characteristic negative bands at 208 and 222 nm (Reproduced from ref. 17 with permission from Springer)

media, but after interacting with SSM, they transitioned to an alpha-helix conformation which was the active conformation for the peptides we used [13, 17, 19, 20]. For example, the peptide conformation transition can be clearly observed in our experiment with the peptide GLP-1, in which the addition of micelles caused the circular dichroism spectra to exhibit negative bands at 208 and 222 nm, which are characteristic for alpha-helical conformation (Fig. 5) [17].

The specific procedure to characterize peptide–SSM interaction by circular dichroism will depend on the equipment used, but general steps adopted in our laboratory using the spectropolarimeter Jasco 710 are described below.

1. Prepare peptide–SSM samples as described in Subheading 3.1 (*see* **Note 16**).
2. Prepare DSPE-PEG$_{2000}$ alone as a blank sample.
3. Prepare peptide solution at the same concentration as peptide–SSM in the isotonic media for comparison.
4. Transfer samples to a quartz cuvette taking care to not introduce air bubbles. Wipe the optical surface with soft lint-free tissue.
5. Obtain spectra over three runs at room temperature using following parameters: wavelengths between 190 and 260 nm, 1 nm bandwidth, 50 nm/min speed, 2 s response, 50 mdeg sensitivity, and 0.1 cm pathlength.
6. Use the manufacturer's Savitzky Golay algorithm to smooth the spectra.

7. Subtract respective background scans (blank samples).
8. Use the equipment's software to calculate the percentage of alpha-helix in free peptide and peptide–SSM samples.

3.3.4 Bioactivity

A peptide-specific assay must be employed to determine if peptide, after self-association with SSM, retains its biological activity. Since several therapeutic peptides act on G-protein coupled receptors present on the cell membrane, second messengers such as cAMP or final products of the signal transduction pathway can be used to assess the peptide activity [17, 19, 20]. Bioactivity can be performed in vitro in cell culture and also in vivo based on the activity under investigation of the peptide. For example, to determine if the peptide GLP-1 retained its activity after self-association with SSM, our group quantified the production of cAMP and insulin caused by the stimulation of GLP-1 receptors in rat insulinoma cells (RINm5F) [17]. The cAMP levels were quantified by enzyme-linked immunosorbent assay (ELISA) and insulin quantified by radioimmunoassay. We observed that both free peptide and peptide–SSM exhibited similar in-vitro bioactivity on GLP-1 receptor stimulation. Furthermore, GLP-1-SSM exhibited in vivo anti-inflammatory activity in acute lung injury mouse model [17] and dextran sodium sulfate (DSS) induced colitis [18]. There are other examples for peptide nanomedicine biological activity on animal models from our laboratory. In an animal model of rheumatoid arthritis, VIP-SSM reduced the incidence and severity of collagen-induced joint inflammation at a much lower dose than free peptide [7]. In another study, using an animal model of pancreatogenic diabetes, we showed better improvement of glucose tolerance, insulin resistance, and hepatic glucogen content after treatment with PP-SSM compared to free peptide [21].

4 Notes

1. Dust and other particulate matters can interfere with dynamic light scattering experiments in SSM formulations. Therefore, measures needs to be taken to prevent entry of foreign materials, including: (1) rinsing glass vials with methanol followed by drying with filtered pressurized air and recapping vials immediately, (2) avoiding prolonged exposure of uncapped vials and bottles to air, (3) using only recently opened or freshly prepared aqueous media, (4) filtering aqueous media through 0.22 μm membrane regularly, and (5) using clean, autoclaved pipette tips for solution transfers.
2. Prevent degradation by oxidation of solid DSPE-PEG_{2000} by storing material at −20 °C with an inert gas (argon) and protected from light.

3. Micelle structure varies according to the ionic strength of the media [3]. In pure water, repulsion between phosphate groups of phospholipids cause micelles to have a lower aggregation number, reduced micelle diameter, and higher critical micelle concentration compared to formulation prepared in isotonic buffer. Conversely, abundant free counterions in ionic solutions neutralize phosphate head groups, resulting in a micellar system composed of approximately 90 monomers with optimal drug loading capabilities. Therefore, our group has successfully prepared SSM formulations in isotonic buffers such as 0.9% normal saline or phosphate-buffered saline at DSPE-PEG$_{2000}$ concentrations between 1 and 15 mM.
4. Care must be taken to avoid excessive foam formation during the preparation of peptide–SSM. Samples should be dissolved by gentle swirling in a circular motion or, if necessary, use vortexing and sonication for 2–5 min each.
5. Wrap vials in aluminum foil to protect samples from light during experiments.
6. Glass vials are preferred over plastic Eppendorf tubes due to the possible interaction of peptides with plastic material.
7. Physicochemical characteristics such as size, fluorescence, concentration, and alpha-helicity must be assessed before and after filtration to ensure that peptide does not interact significantly with the filter membrane. Also, it is recommended to filter a higher volume than needed for the experiment since part of the volume is retained in the filtration apparatus.
8. Properties such as size, fluorescence, and biological activity must be assessed for each peptide–SSM before and after freeze-drying to guarantee that product stays stable during lyophilization and resuspension processes.
9. Select a vial size compatible with the volume of sample for freeze-drying. Keep sample volume about half of the vial volume for proper formation of lyophilized cake.
10. If the amount of peptides in the formulation exceeds the maximum association capacity of SSM, large peptide aggregates will be observed (>100 nm).
11. Optimal concentration of DSPE-PEG$_{2000}$ is 1–15 mM for dynamic light scattering measurements at NICOMP 380 ZLS. Samples that are too diluted (<1 mM) will have poor particle detection and greater background signal, while excessively concentrated samples (>15 mM) will exhibit higher viscosity due to intermicellar interaction.
12. Wash quartz cuvette between samples with warm water and neutral detergent, rinse with 0.1 N hydrochloric acid, and rinse thoroughly with deionized water.

13. Excitation and emission wavelengths will depend on the presence of each fluorescent amino acid in the primary structure of peptides (tryptophan, tyrosine, and phenylalanine). If peptide contains a mixture of amino acids, tryptophan (Exλ 280 nm) will dominate the total fluorescence, followed by tyrosine (Exλ 274 nm) and phenylalanine (Exλ 257 nm). After determining the fluorescent amino acids present in the peptide, perform emission spectral scan (approximately between 290 and 350 nm) to determine the maximal emission intensity for the peptide.
14. When peptides interact with DSPE-PEG_{2000} micelles, a shift in peak fluorescence might occur due to the change of peptide microenvironment polarity. Therefore, use the peak emission intensity for calculations.
15. In our lab, data points are fitted in a nonlinear regression using the single rectangular *I*,3 parameter with the hyperbola equation ($y = y_0 + ax/(b + x)$) on SigmaPlot software.
16. Suggested peptide concentration is 4–20 μM with DSPE-PEG_{2000} above the saturation molar ratio to ensure complete peptide association.

References

1. Craik DJ, Fairlie DP, Liras S, Price D (2013) The future of peptide-based drugs. Chem Biol Drug Des 81(1):136–147
2. Fosgerau K, Hoffmann T (2015) Peptide therapeutics: current status and future directions. Drug Discov Today 20(1):122–128
3. Vukovic L, Khatib FA, Drake SP, Madriaga A, Brandenburg KS, Král P, Onyuksel H (2011) Structure and dynamics of highly PEG-ylated sterically stabilized micelles in aqueous media. J Am Chem Soc 133(34):13481–13488
4. Ashok B, Arleth L, Hjelm RP, Rubinstein I, Önyüksel H (2004) In vitro characterization of PEGylated phospholipid micelles for improved drug solubilization: effects of PEG chain length and PC incorporation. J Pharm Sci 93(10):2476–2487
5. Banerjee A, Onyuksel H (2012) Peptide delivery using phospholipid micelles. Wiley Interdiscip Rev Nanomed Nanobiotechnol 4 (5):562–574
6. Lim SB, Banerjee A, Önyüksel H (2012) Improvement of drug safety by the use of lipid-based nanocarriers. J Control Release 163(1):34–45
7. Sethi V, Rubinstein I, Kuzmis A, Kastrissios H, Artwohl J, Onyuksel H (2013) Novel, biocompatible, and disease modifying VIP nanomedicine for rheumatoid arthritis. Mol Pharm 10(2):728–738
8. Khaja FA, Koo O, Onyuksel H (2012) Nanomedicines for inflammatory diseases. Methods Enzymol 508:355–375
9. Bak A, Leung D, Barrett SE, Forster S, Minnihan EC, Leithead AW, Cunningham J, Toussaint N, Crocker LS (2015) Physicochemical and formulation developability assessment for therapeutic peptide delivery—a primer. AAPS J 17(1):144–155
10. Vuković L, Madriaga A, Kuzmis A, Banerjee A, Tang A, Tao K, Shah N, Král P, Onyuksel H (2013) Solubilization of therapeutic agents in micellar nanomedicines. Langmuir 29 (51):15747–15754
11. Krishnadas A, Onyuksel H, Rubinstein I (2003) Interactions of VIP, secretin and PACAP1–38 with phospholipids: a biological paradox revisited. Curr Pharm Des 9 (12):1005–1012
12. Yuan Z, Syed M, Panchal D, Joo M, Bedi C, Lim S, Onyuksel H, Rubinstein I, Colonna M, Sadikot RT (2016) TREM-1-accentuated lung injury via miR-155 is inhibited by LP17 nanomedicine. Am J Physiol Lung Cell Mol Physiol 310(5):L426–L438

13. Gandhi S, Rubinstein I, Tsueshita T, Onyuksel H (2002) Secretin self-assembles and interacts spontaneously with phospholipids in vitro. Peptides 23(1):201–204
14. Gandhi S, Tsueshita T, Önyüksel H, Chandiwala R, Rubinstein I (2002) Interactions of human secretin with sterically stabilized phospholipid micelles amplify peptide-induced vasodilation in vivo. Peptides 23 (8):1433–1439
15. Önyüksel H, Ikezaki H, Patel M, Gao X-P, Rubinstein I (1999) A novel formulation of VIP in sterically stabilized micelles amplifies vasodilation in vivo. Pharm Res 16 (1):155–160
16. Lim SB, Rubinstein I, Önyüksel H (2008) Freeze drying of peptide drugs self-associated with long-circulating, biocompatible and biodegradable sterically stabilized phospholipid nanomicelles. Int J Pharm 356(1):345–350
17. Lim SB, Rubinstein I, Sadikot RT, Artwohl JE, Önyüksel H (2011) A novel peptide nanomedicine against acute lung injury: GLP-1 in phospholipid micelles. Pharm Res 28 (3):662–672
18. Anbazhagan AN, Thaqi M, Priyamvada S, Jayawardena D, Kumar A, Gujral T, Chatterjee I, Mugarza E, Saksena S, Onyuksel H (2017) GLP-1 nanomedicine alleviates gut inflammation. Nanomedicine 13(2):659–665
19. Kuzmis A, Lim SB, Desai E, Jeon E, Lee B-S, Rubinstein I, Önyüksel H (2011) Micellar nanomedicine of human neuropeptide Y. Nanomedicine 7(4):464–471
20. Banerjee A, Onyuksel H (2012) Human pancreatic polypeptide in a phospholipid-based micellar formulation. Pharm Res 29 (6):1698–1711
21. Banerjee A, Onyuksel H (2013) A novel peptide nanomedicine for treatment of pancreatogenic diabetes. Nanomedicine 9(6):722–728
22. Tsueshita T, Gandhi S, Önyüksel H, Rubinstein I (2002) Phospholipids modulate the biophysical properties and vasoactivity of PACAP-(1—38). J Appl Physiol 93(4):1377–1383
23. Rubinstein I, Lim SB, Jeon E, Onyuksel H (2007) Human GLP-1 {alpha} and GIP {alpha}: novel, long-acting nanomedicines for type II diabetes mellitus. FASEB J 21(5):A434
24. Kaszuba M, McKnight D, Connah MT, McNeil-Watson FK, Nobbmann U (2008) Measuring sub nanometre sizes using dynamic light scattering. J Nanopart Res 10 (5):823–829
25. Chung LA, Lear JD, DeGrado WF (1992) Fluorescence studies of the secondary structure and orientation of a model ion channel peptide in phospholipid vesicles. Biochemistry 31 (28):6608–6616

Chapter 5

Design, Preparation, and Characterization of Peptide-Based Nanocarrier for Gene Delivery

Mohsen Alipour and Saman Hosseinkhani

Abstract

The delivery of nucleic acid to eukaryotic cells is challenging due to presence of various intra- and extracellular barriers and need to delivery carriers. However, current gene delivery carriers, including PLL, PEI, and liposome, suffer from nanocarrier associate toxicity, low efficiency and polydispersity, and non-biodegradability. Here we describe our strategy for developing safe, efficacious, and monodisperse peptide-based carrier for gene delivery. We explain the use of genetic engineering technology for integrating various functional motifs in a single peptide, with less than 100 nm size, which facilitated gene delivery into mammalian cell.

Key words Gene delivery, Peptide, Nanocarrier, Genetic engineering, Hemolysis, Transfection, Gel retardation

1 Introduction

Delivery vehicles have attracted a great attention to improve the clinical efficacy of many pharmaceutical compounds. On the other hand, most of the gene therapy approaches suffer from lack of a suitable delivery vehicle, which has been known as the main shortcoming of human gene therapy during last decades [1, 2]. This significant challenge promoted several research groups around the world to design, engineer, and utilize various materials for gene delivery. However, none of these materials could satisfy the safety and performance requirements for use in clinic [3].

The complexation of anatomo-physiological barriers along with various cellular barriers like cell membrane and endosome remains the major hurdle for design of gene delivery vehicles [4]. Moreover, the mentioned barriers made a big wall against the clinical translation of powerful gene modulation technology, like CRISPR/Cas9- and SiRNA-based gene silencing [5].

A good clinical outcome for protein-based therapy like monoclonal antibody, nab technology (Abraxane), and peptide-

Volkmar Weissig and Tamer Elbayoumi (eds.), *Pharmaceutical Nanotechnology: Basic Protocols*, Methods in Molecular Biology, vol. 2000, https://doi.org/10.1007/978-1-4939-9516-5_5,

functionalized material (like RGD) as well as decoding the mechanism of infectious viruses and identifying the diverse functions of their capsids navigate the research to exploiting the extraordinary potential of protein- and peptide-based material for designing intelligent nanovehicles [6, 7].

Understanding the extracellular and intracellular barriers is an important key factor for developing a gene delivery vehicle. DNA as a giant macromolecule with a negatively charged backbone is unable to diffuse into cell membrane, and moreover endosome/lysosome as defense mechanism of eukaryotic cell can entrap and enzymatically degrade the nucleic acid-based material [8]. Moreover, entry to nucleus membrane also is a rate-limiting step in gene delivery process. According to our previous finding peptide-based nanocarriers can possess unique feature for gene therapy [9].

The peptide-based material as a unique ancient material can be used to fabricate multifunctional materials using chains of amino acids.

These programmable materials can be used for design of many bio-based materials, which can be produced in a reproducible manner. The peptide-based materials by programing on their primary amino acid sequence can possess various forces in their different parts [6].

Finally, here we have provided an approach for design, production, and characterization of peptide-based nanomaterials, a collection of main steps, which have been developed by our group. Indeed, we introduce how recombinant peptide technology helped us to provide gene delivery materials.

2 Material

2.1 Material for Peptide Expression and Purification

1. LB medium: Prepare LB medium by adding tryptone (10 g), yeast extract (5 g), and NaCl (10 g) to 1 l of deionized water. Sterilize by autoclaving at 121 °C for 15 min and cool to 45 °C and then add appropriate antibiotic.
2. Lysis buffer: Prepare lysis buffer, by adding 20 mM Tris–HCl, 500 mM NaCl, 8 M urea, and 5 mM imidazole to 100 ml deionized water and adjust pH to 12.
3. Washing buffer: Prepare washing buffer by adding 20 mM Tris, 500 mM NaCl, and 20 mM imidazole to 100 ml deionized water and adjust pH to 8.
4. Elution buffer: Prepare elution buffer by adding 20 mM Tris, 150 mM NaCl, and 250 mM imidazole.
5. Dialysis bag, Ni-Sepharose column.

2.2 Materials for SDS Page

1. Stock acrylamide solution: Prepare this solution by adding 30 g acrylamide and 0.8 g *bis*-acrylamide to 100 ml distilled water. Filter the solution and store at 4 °C.
2. Separating gel buffer: Prepare 100 ml of 1.5 M Tris–HCl, and adjust its pH to 8.8.
3. Stacking gel buffer: Prepare 100 ml of 0.5 M Tris–HCl, and adjust its pH to 6.8.
4. APS 10% solution: Add 100 mg ammonium persulfate in 1 ml water to prepare this solution.
5. TEMED 10%: Dilute 100 μl of *N,N,N,N*-tetramethylethylenediamine (TEMED) to 1 ml distilled water.
6. Sample buffer (5×): Mix 3.1 ml of 1 M Tris–HCl (pH 6.8), 0.5 ml of a 1% solution of bromophenol blue, 5 ml of glycerol, and 1.4 ml of water. Samples should be diluted with sample buffer to give a solution that is 1× sample buffer.
7. Electrophoresis buffer: Dissolve 3.0 g of Tris base and 14.4 g of glycine in water and adjust the volume to 1 l.
8. Staining solution: Dissolve 0.25 g Coomassie brilliant blue R250, in 125 ml of methanol, then add 25 ml of glacial acetic acid, and reach the volume to 100 ml by adding deionized water.
9. Destaining solution: Add 100 ml methanol and 100 ml glacial acetic acid to 800 ml water.

2.3 Materials for Bradford's Assay

1. Bradford's reagent: Dissolve 100 mg of Coomassie Blue G250 in 50 ml of 95% ethanol, then mix with 100 ml of 85% phosphoric acid, and make up to 1 l with distilled water.

   However, durıng this time dye may precipitate from solution and so the stored reagent should be filtered before use.
2. Protein standard: Bovine serum albumin at concentrations of 100 μg/ml to 1 mg/ml in distilled water is used as a stock solution.

2.4 Material for Serum Stability

1. Serum: Prepare serum by centrifugation of whole-blood cell from rats, for 15 min at 4 °C.
2. SDS 1%: Prepare SDS 1% by adding 1 g SDS to 1 ml deionized water.
3. Agarose 1%: Dissolve 100 mg of agarose in 100 ml TAE buffer. Heat to 100 °C for 5 min and then cool to 50 °C.

2.5 Material for Hemolysis Assay

1. The red blood cells (RBC): Collect whole blood from healthy volunteers, with a written consent, into heparinized vacutainers under aseptic conditions. Centrifuge the whole blood at 5000 rpm (4200 × *g*) for 5 min. Discard the supernatants and wash the red blood cell (RBC) pellet with PBS three times.

2. Triton ×100 1%: Add 10 μl Triton ×100 to 990 μl of distilled water.
3. Acetate buffer: Dissolve 5.44 g sodium acetate in 98.8 ml distilled water, add 1.2 ml glacial acetic acid, and then adjust the pH to 5.0 with 10 N NaOH.

2.6 Material for Transfection and Luciferase Assay

1. Culture medium: According to cell line requirement use RPMI or DMEM medium and supplement with FBS 10% and antibiotic (Penstrep) 1%.
2. Transfection medium: Prepare transfection medium similar to culture medium, without addition of FBS.
3. Cell lysis buffer: Prepare cell lysis buffer by mixing 37 mg of EDTA, 231 mg of KH_2PO_4, 3.18 g of K_2HPO_4, 10 ml of glycerol, 1 ml of triton ×100 1%, and 48 μl of 7 mm 2-mercaptoetanol and adjust the volume up to 100 ml by distilled water.
4. Luciferase substrate: Prepare luciferase substrate by mixing 2 mM luciferin, 4 mM ATP and 100 mM $MgSO_4$ and then adjust its pH to 7.8.

3 Methods

Preparation of Chimeric Peptide:

3.1 Cloning

There are two approaches for preparation of peptide. The first is solid-phase synthesis, which is useful for small peptide amino acids (AA < 30). The second approach is recombinant method, which is suitable for production of long peptide (AA > 30), like peptide-based carriers. The details of both methods have been explained elsewhere.

The production of recombinant nanocarrier is performed according to the following protocol:

1. Design coding sequence of peptide-based nanocarrier and examine the properties of designed sequence using bioinformatics software. Evaluate the stability of the corresponding mRNA secondary structure by M-fold web server, which is available at http://mfold.rna.albany.edu/?q=mfold/RNA-Folding-Form. Check the codon usage of sequence using the following link: http://www.bioinformatics.org/sms2/codon_usage.html.

   The sequence and physicochemical properties are available at ExPASy's server. For example, compute their amino acid composition by the following link: http://www.expasy.ch/tools/protparam.html.

2. Synthesize the coding sequence by solid-phase synthesis and ordering to a commercial company.
3. Synthesize the optimized coding sequence by ordering to one of the gene synthesizer companies and select a suitable bacterial expression vector like pET28a or pET21b with hexahistidine tag for easy purification using Ni-NTA column chromatography.
4. Modify the sequence by polymerase chain reaction (PCR) if necessary, and clone the desired sequence (20 ng) in a suitable (100 ng) expression vector like pET28a by addition of 5 units ofT4 DNA ligase enzyme in the presence of tango buffer.
5. Transform the ligated plasmid (5 μl) in BL21 chemically competent *E. coli* as a cloning host. 5 μl of plasmid should be mixed with 100 μl of bacteria, incubate on ice for 30 min, place the mixture in water bath with 42 °C for 45 s, incubate on ice for 5 min, add 1 ml LB, incubate at 37 °C for 1 h, centrifuge for 5 min at 4000 rpm (2600 × *g*), and transfer to a LB agar plate.
6. Pick a single clone and inoculate into 10 ml LB medium containing kanamycin for 16 h at 37 °C incubator.
7. Add 1 ml of pre-culture into 250 ml medium containing kanamycin for 4 h at 37 °C incubator.
8. Add 100 mM IPTG (1 ml) into medium for protein expression for 4 h at 37 °C.
9. Prepare bacterial pellet by centrifugation at 4000 rpm (2600 × *g*) for 15 min at 4 °C.

3.2 Purification by the Following Step

One of the main factors in the production of recombinant peptides is choosing a suitable purification strategy to obtain the highest amount of peptide. The hexahistidine tag in N- or C-terminal of peptide made a strong affinity in recombinant peptide for Ni-NTA column chromatography (commercial) and facilitated the purity of peptide with suitable reagent according to its solubility.

1. Lyse bacterial pellet in lysis buffer; pellet of 50 ml bacteria should be suspended in 1 ml of lysis buffer.
2. Centrifuge the bacterial cell lysate at 13,000 rpm (28000 × *g*) for 30 min and incubate the supernatant on Ni-NTA column chromatography containing Ni Sepharose for 5 min.
3. Remove the impurity by addition of 15 ml of wash buffer.
4. Eluted the desired peptide by addition of 1 ml elution buffer ten times with 5-min interval.
5. Check the purity of peptide by 17.5% sodium dodecyl sulfate/polyacrylamide gel electrophoresis (SDS PAGE gel) and Coomassie blue staining. Mix 20 μl of elution with 5 ml of sample buffer 5×.

3.3 Method for SDS-PAGE

1. Set up the gel cassette.
2. Prepare the separating gel, by mixing 7.5 ml stock acrylamide solution, 7.5 ml separating gel buffer, 14.85 ml water, and 150 ml 10% ammonium persulfate and add 15 μl of TEMED to initiate the polymerization reaction.
3. Prepare stacking gel solution by mixing 1.5 ml stock acrylamide solution, 3.0 ml stacking gel buffer, 7.4 ml water, and 100 μl 10% ammonium persulfate.
4. Add separating gel solution and add the stacking gel solution to the gel cassette until the solution reaches the cutaway edge of the gel plate.
5. Place the well-forming comb into this solution and leave to set. This will take about 30 min.
6. Mark the positions of the bottoms of the wells on the glass plates with a marker.
7. Remove the comb from the stacking gel, remove any spacer from the bottom of the gel cassette, and assemble the cassette in the electrophoresis tank.
8. Fill the top reservoir with electrophoresis buffer and load sample onto the gel.
9. Continue electrophoresis for 3 h until bromophenol blue reaches the bottom of the gel.
10. Stop electrophoresis and remove the gel from the cassette.
11. Immerse the gel in stain solution, with shaking, for a minimum of 2 h,
12. Replace the stain solution with destain solution; stronger bands will be immediately apparent and weaker bands will appear as the gel destains.

3.4 Buffer Exchange

1. Following purification, the eluted peptide should be transferred to a suitable buffer, like PBS or HEPES. For this purpose, dialysis bag with proper cutoff according to peptide size should be used.
2. Put the eluted peptide on ice.
3. Add glycerol up to 10 % of total volume.
4. Transfer the mixture to dialysis bag.
5. Dialyze for 48 h; the volume of dialysis should be 100 times more than the volume of eluted peptide.

3.5 Determination of Peptide Concentration

The concentration of peptide can be measured by different methods, including UV absorbance at 219 nm or 280 nm, BCA, and Bradford. The Bradford's method due to producing of a colorimetric signal can be used for fast screening.

1. Pipet triplicate samples containing a range of dilutions (1, 1/10, 1/100, 1/1000) of peptide.
2. For the calibration curve, pipet duplicate volumes of 10, 20, 40, 60, 80, and 100 μl of standard protein, and adjust the volume to 100 μl with PBS. Pipet 100 μl of PBS into an additional tube to provide blank.
3. Add 1 ml of the Bradford's reagent to each tube.
4. Measure the absorbance of each sample for 5 min after addition of the peptide. The *A*595 value of a sample containing 10 μg of BSA is about 0.45.

3.6 Preparation of Nanoparticles

Nanoparticles were produced by complexation of nanocarriers and plasmids at different N/P ratios. The N/P ratio refers to the molar ratio of the amine groups of the carriers to the phosphate groups of plasmid. Theoretical N/P ratios were calculated according to Eq. 1:

$$N/P = \frac{\frac{\textit{Weight of peptide } (\mu g)}{(\textit{Molecular weight of peptide})/(\textit{Number of positive charge})}}{\frac{\textit{Weight of } \mathrm{DNA}\ (\mu g)}{\textit{Mean of molecular weight of dNMPs}}}. \tag{1}$$

where the charges of Lys and Arg residues were considered as +1, while those of His residue as 0.

1. Mix 1 μg of plasmid with 1.5 μg of the peptide carrier, which contain 23 Lys residues to prepare nanoparticle at N/P ratio of 1.
2. Mix the complexes immediately and incubate at 25 °C for 45 min.

3.7 Gel Retardation Assay

The gel retardation assay can be used to determine the DNA binding capacity of the peptide-based nanocarrier; the basis of this method is that electrophoretic movement of DNA is completely retarded when the net charge of this macromolecule becomes zero. Theoretically at N/P ratio of 1, the system shows an equal number of negative and positive charges which leads to a zero net charge and retardation of electrophoretic movement.

1. Prepare nanoparticle at different N/P ratios and incubate for 25 min at 25 °C.
2. Load the nanoparticle on agarose gel (1%) and electrophorese for 1 h at 80 V.
3. Place the agarose gel in diluted ethidium bromide solution and visualize with UV illumination. Retarded nanoparticle should remain in wells; meanwhile the naked DNA shows a movement on agarose gel, and the SDS can uncomplex the nanoparticle into components (Fig. 1).

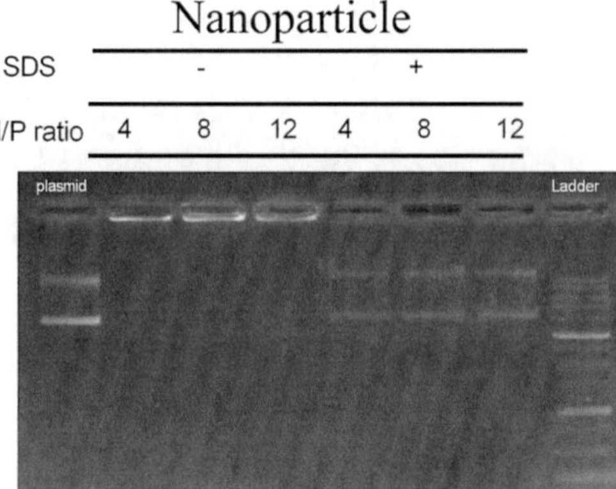

Fig. 1 The release of pGL3 plasmid from peptide-based nanoparticles. Agarose gel retardation assay of nanocarrier/pDNA complex in the function of N/P ratio. The nanoparticles contained 300 ng pGL3 plasmid and similar amount of this plasmid was used as control pDNA DNA ladder. SDS treatment released pDNA from nanoparticles. Ladder: 1 kb

3.8 Analysis of Charge and Zeta Potential

Peptide-based nanocarriers should have a positive charge to electrostatically bind to DNA and neutralize its electrostatic charge. Moreover, condensing the structure of DNA macromolecule is a critical factor for cellular uptake of peptide-plasmid complex.

Characterize the zeta potential and size of nanoparticles as below:

1. Prepare three independent batches of nanoparticle containing 1 μg DNA at different N/P ratios by complexation of peptide-DNA for 30 min at 25 °C.
2. Add cold PBS up to 1 ml to each NPs to prevent unwanted particle aggregation.
3. Perform DLS analysis to examine the size and zeta potential of nanoparticle.

3.9 Serum Stability Assay

In order to understand the stability of DNA–peptide complex in the blood circulation the serum stability assay was performed.

1. Prepare nanoparticle at different N/P ratios and incubate for 25 min at 25 °C.
2. Add human serum 10% to nanoparticles and naked DNA and incubate at 37 °C for 1 h.
3. Add SDS 10% to nanoparticle-serum mixture and incubate at 37 °C for 1 h.
4. Load the resultant on agarose gel 1% and stain with ethidium bromide.

3.10 Endosome Escaping Potency Assay

Endosome entrapment: Red blood cells are usually used as an ex vivo model of endosome; endosome entrapment is one of the main barriers in gene delivery, which is followed by lysosomal degradation. A peptide-based nanocarrier should be able to overcome these barriers and protect their cargo from degradation. For this purpose RBC membrane lytic active can be used as a sign of endosome-escaping potency, which is performed as follows:

1. Centrifuge 1 ml fresh blood at 3000 rpm (1500 × g) for 15 min at 4 °C.
2. Isolate the supernatant for serum stability assay.
3. Wash the RBC cells with PBS three times.
4. Dilute the RBC in 1 ml PBS or other low-pH buffer like acetate buffer.
5. Incubate the peptide-based nanocarrier on RBC and adjust the volume up to 1 ml by appropriate buffer.
6. Shake the sample for 1 h at 37 °C and centrifuge at 13,000 rpm (28000 × g) for 15 min.
7. Triton ×100 can be used as a positive control.
8. Measure the absorbance at 540 nm using spectrophotometer.
9. Calculate the hemolysis percent according to the following formula:
   Hemolysis (%) = (Abs of sample − Abs negative control)/(Absorbance positive control − Absorbance negative control) × 100

3.11 Transfection and Analysis of Gene Delivery and Expression

1. Seed the HEK 293 T (2 × 104) seed at 24 well in DMEM medium plus FBS for 48 h.
2. Prepare nanoparticles at different N/P ratios containing 1 μg of luciferase-encoded plasmid.
3. Replace the medium with 100 μl of serum-free DMEM medium supplemented with nanoparticles at different N/P ratios for 4 h.
4. After 4 h, add 400 μl DMEM medium supplemented with FBS (1%) and PenStrep antibiotic (1%).
5. After 2 days, remove the medium completely, add 25 μl CCLR buffer to each well, and shake for 15 min at 4 °C.
6. Add 10 μl of cell lysate to 10 μl luciferase substrate and measure the luciferase activity using a luminometer.

4 Notes

1. Some studies have shown that functionalization of peptide-based nanocarrier with targeting motifs increases the transfection rate in cells, which overexpress the receptor for mentioned targeting motifs [10].

2. As a model 16 mer histone H1 sequence as lysine-rich sequence can be used for equipping nanocarrier with DNA-binding element. Gp41 sequence of HIV virus can be used as an endosome-disrupting element and SV40 NLS can be used as nuclear localization signal [9].
3. Any motif, which can bind to pDNA and condense its structure, can be used instead of histone H1 but its effect on transfection efficiency should be evaluated along with an appropriate control. In case of endosome disruption and nuclear localization various alternative motifs can also be used but similar to DNA-binding motif their effect on intracellular delivery efficiency of nanocarrier should be revealed.
4. Prepare a large scale (1 l) of bacterial culture to ensure a high concentration of peptide-based nanocarrier after purification.
5. The bacterial growth should be visible with naked eyes, but some form of nanocarriers due to bacterial toxicity may kill their host and decrease the turbidity of bacteria.
6. These nanocarriers use hexahistidine tag to facilitate the purification of peptide; other purification strategies can be used, which may be needed for procedure optimization [11].
7. Dialyzed peptides should be dispensed into 1.5 ml aliquot tube and stored at −20 °C. Thaw the peptide stocks on ice before use.
8. Soluble form of peptide may kill the bacteria, so induction of expression at 37 °C was performed to produce inclusion body and prevent bacterial death.
9. The increase of N/P ratio increases the charge of peptide-based particles and decreases their size [6, 11].
10. We have found that the sequence of endosome-escaping peptide can ascertain the delivery efficiency of carriers.
11. Dialyzed peptides are extremely sensitive to aggregation, particularly at room temperature. However, formation of DNA peptide noncomplex can prevent the formation of aggregated particle.
12. Gel retardation assay can be performed using single-strand or double-strand DNA; the main factor that should be considered is the DNA weight.
13. Many methods use starvation for transfection. We have found that serum deprivation at transfection leads to similar result for peptide-based transfection.
14. We have found that at N/P ratio higher than 16 the intensity of retarded plasmid becomes weak presumably through less ethidium bromide accessibility [9].

15. Note that this procedure can also be used to evaluate the effect of various function elements on the final delivery efficiency of nanocarrier.
16. If bacterial expression did not produce the expected amount of peptide, the first solution is plating the plasmid harboring bacterial stock on LB agar and picking single colony for nanocarrier production.
17. The bacterial pellet can be stored at −20 °C for up to a year and thawed on ice and simultaneously add lysis buffer before use.
18. In some cases the high amount of nanocarrier due to its compacted structure decreases the transfection efficiency.

Acknowledgments

The authors would like to thank Elaheh Emamgholizadeh, Fatemeh Rabbani, Behenam Hajipour, Farzad Yousefi, Roya Cheraghi, and Asia Majidi, for help in developing the various methods for characterization of peptide-based nanocarriers.

References

1. Fonseca SB, Pereira MP, Kelley SO (2009) Recent advances in the use of cell-penetrating peptides for medical and biological applications. Adv Drug Deliv Rev 61(11):953–964
2. Davidson BL, McCray PB (2011) Current prospects for RNA interference-based therapies. Nat Rev Genet 12(5):329–340
3. Kim DH, Rossi JJ (2007) Strategies for silencing human disease using RNA interference. Nat Rev Genet 8(3):173–184
4. Bertrand N, Leroux JC (2012) The journey of a drug-carrier in the body: an anatomo-physiological perspective. J Control Release 161(2):152–163
5. Cheraghi R, Alipour M, Nazari M, Hosseinkhani S (2017) Optimization of conditions for gene delivery system based on PEI. Nanomed J 4(1):8–16
6. Sadeghian F, Hosseinkhani S, Alizadeh A, Hatefi A (2012) Design, engineering and preparation of a multi-domain fusion vector for gene delivery. Int J Pharm 427(2):393–399
7. Temming K, Schiffelers RM, Molema G, Kok RJ (2005) RGD-based strategies for selective delivery of therapeutics and imaging agents to the tumour vasculature. Drug Resist Updat 8 (6):381–402
8. Xiang S et al (2012) Uptake mechanisms of non-viral gene delivery. J Control Release 158:371–378
9. Majidi A, Nikkhah M, Sadeghian F, Hosseinkhani S (2016) Development of novel recombinant biomimetic chimeric MPG-based peptide as nanocarriers for gene delivery: imitation of a real cargo. Eur J Pharm Biopharm 107:191–204
10. Cheraghi R, Nazari M, Alipour M, Majidi A, Hosseinkhani S (2016) Development of a targeted anti-HER2 scFv chimeric peptide for gene delivery into HER2-positive breast cancer cells. Int J Pharm 515(1–2):632–643
11. Alipour M, Hosseinkhani S, Sheikhnejad R, Cheraghi R (2017) Nano-biomimetic carriers are implicated in mechanistic evaluation of intracellular gene delivery. Sci Rep 7:41507

Chapter 6

Gelatin Nanoparticles

Rania M. Hathout and Abdelkader A. Metwally

Abstract

Currently, gelatin nanoparticles are gaining more grounds in drug and gene delivery throughout all the available several routes of administration. Yet, the homogenous and less disperse preparation of this type of nanoparticles is still a challenging task due to the variation of the gelatin quality according to its source and due to its variable molecular weight. Accordingly, several methods were proposed from which the double-desolvation method has been proven to yield optimum results regarding particle size and homogeneity. Thereby, we describe in this chapter a detailed procedure of this method. We also introduce our protocols of the cationization of this kind of nanoparticles as it is extensively needed in case of loading genetic materials or proteins. Additionally, FITC labeling of gelatin nanoparticles that is usually utilized for purposes of imaging or bio-distribution studies is also introduced step by step.

Key words Gelatin, Nanoparticles, Desolvation, Cationization, FITC, Labeling, Drug delivery, Gene

1 Introduction

Biomaterials and naturally derived chemicals and compounds are now under extensive research to explore their several merits in drug and gene delivery, tissue engineering, and membrane adsorption [1]. In this context, one of the emerging classes of biomaterials that were proven to highly accommodate newly synthesized therapeutic molecules and genetic materials at the nano level is the protein carriers. Protein carriers possess many additional advantages such as biodegradability, biocompatibility, non-antigenicity, low cost, and availability. Furthermore, the surface of protein nanocarriers can be modified with site-specific ligands, cationized with amine derivatives, or coated with polyethylene glycols to achieve targeted and sustained-release drug delivery [2]. Compared to other colloidal carriers, protein nanocarriers are better stable in biological fluids to provide the desired controlled and sustained release of entrapped drugs or genetic materials [3]. Among the commonly used proteins, gelatin resides as a cornerstone due to its availability from abundant resources and its safe plasma and tissue profile.

Volkmar Weissig and Tamer Elbayoumi (eds.), *Pharmaceutical Nanotechnology: Basic Protocols*, Methods in Molecular Biology, vol. 2000, https://doi.org/10.1007/978-1-4939-9516-5_6, © Springer Science+Business Media, LLC, part of Springer Nature 2019

Gelatin is a denatured protein that is usually obtained from a natural source such as collagen by acid or alkaline hydrolysis. Gelatin possesses unique physicochemical characteristics. It is a polyampholyte having both cationic and anionic groups together with hydrophobic ones in the approximate ratio of 1:1:1. The gelatin molecule is ~13% positively charged due to the presence of basic amino acids such as lysine and arginine, and ~12% negatively charged due to acidic counterparts such as glutamic and aspartic acids. Moreover, approximately 11% of the gelatin chain is hydrophobic in nature due to the presence of leucine, isoleucine, methionine, and valine amino acids that are known for their lipophilicity [4]. Other amino acids, glycine, proline, and hydroxyproline, constitute the rest of the chain. This unique structure and nature pose its success as a carrier system that can be suitable for a wide variety of drug molecules.

However, compared to other polymeric biomaterials such as chitosan, pluronics, PLGA, and albumin, studies adopted on gelatin are noticeably less [5–9]. This observation was previously ascribed to the relatively tedious common methods of preparation.

The commonly reported methods of gelatin nanoparticle preparation were the single and double desolvation [5, 10, 11], coacervation-phase separation [12], emulsification-solvent evaporation [13], reverse-phase microemulsion [14], nanoprecipitation [15, 16], self-assembly of gelatin molecules, and normal ionic-gelation method which was also successful for the preparation of cationized gelatin nanoparticles [17].

Although all of the aforementioned methods have several advantages, there are some limitations. In case of the emulsification techniques, large amounts of surfactants are required to produce the small-sized gelatin nanoparticles, which needs a complicated post-process and may impose an obstacle for delivery through several routes of administration such as the ocular route [18, 19]. The coacervation method is a process of phase separation followed by cross-linking step. This method usually leads to non-homogeneous cross-linking with unsatisfactory loading efficiency [20]. Moreover, gelatin nanoparticles prepared by many of these methods were found to be large in size and have a high polydispersity index (PDI) due to heterogeneity in molecular weight of the gelatin polymer [21]. Recently, a nanoprecipitation technique was also introduced. It is considered to be rapid, easy, and straightforward compared to other methods and leads to the formation of stable, small-sized, and homogenous particles [16]. Yet, it still requires two miscible solvents in which the polymer is soluble in one of them (the solvent, for example water), but not in the other (the nonsolvent, for example ethanol). The polymer in the solvent phase is then added to the nonsolvent containing a stabilizer which is usually a poloxamer. Macromolecules were successfully formulated in gelatin nanoparticles prepared by this method

Table 1
Drawbacks of some methods used in the preparation of gelatin nanoparticles

The method	Drawback(s)	Reference (s)
Emulsification techniques	– Large amounts of surfactants are required to produce the small-sized gelatin nanoparticles – Complicated post-process – Not suitable for some routes of drug administration such as the ocular route	[18, 19]
Coacervation	– Nonhomogeneous cross-linking – Unsatisfactory loading efficiency	[20]
Single desolvation	– Large-sized nanoparticles – High polydispersity index (PDI) of the produced nanoparticles	[21]
Nanoprecipitation	– Requires two miscible solvents in which the polymer (gelatin) is soluble in one but not the other – Less yield	[15]

[15]. However, compared to the other methods of preparation, the nanoparticle yield was always less.

As a conclusion, the two-step "double-desolvation" method was developed that enabled the production of the nanoparticles with a reduced tendency for aggregation [5]. This method is now the first choice for the preparation of gelatin nanoparticles because it solves the drawbacks of the former single-desolvation method where after the first desolvation step the low-molecular-weight gelatin fractions (not expected to produce nanoparticles) present in the supernatant are removed by decanting and subsequently the high-molecular-weight fractions present in the sediment are redissolved [5]. Table 1 shows the common methods used in fabricating gelatin nanoparticles and their drawbacks.

To this end, we introduce in this chapter a detailed methodology of the double-desolvation technique used for the preparation of gelatin nanoparticles where usually only a brief description of the method is included in manuscripts. A protocol for cationization of these valuable nanoparticles, aiming to enhance their accommodating ability to negatively charged molecules such as the genetic materials and antigens, is also introduced. Moreover, a labeling protocol for the nanoparticles using FITC is explained.

2 Materials

1. Gelatin A from porcine skin (300 g Bloom).
2. Gelatin B from beef nails (175 g Bloom).
3. Glutaraldehyde solution in water, 25% (w/v) (used for cross-linking the gelatin).

4. Glycine (used to block the reaction between the aldehyde groups of glutaraldehyde and gelatin).
5. Acetone (gelatin desolvating agent).
6. Dimethyl sulfoxide (DMSO) (used for dissolving FITC).
7. 5 mM 4-(2-Hydroxyethyl) piperazine-1-ethanesulfonic acid (HEPES, pH 7.4) buffer (used for zeta potential measurements of the prepared nanoparticles).
8. Cholamine, (2-aminoethyl)-trimethylammonium chloride hydrochloride (used for the cationization of the gelatin nanoparticles).
9. EDC, 1-ethyl-3-(3-dimethyl-aminopropyl) carbodiimide hydrochloride (used in the cationization reaction between cholamine and gelatin).
10. FITC, fluorescein isothiocyanate (used for labeling of the prepared nanoparticles).
11. 5 M HCl, 5 M NaOH, and ultrapure water (e.g., Milli-Q water, MQ).
12. Falcon tubes, 15 ml, and Nanoseps® (pore size 10 nm).
13. Parafilm®.

3 Methods

General Notice: All good laboratory practice and safety protocols should be followed strictly. Material safety datasheets should always be revised prior to any lab work.

3.1 Preparation of Gelatin Nanoparticles Using the Double-Desolvation Method

3.1.1 Preparation of the Gelatin Solution

1. Dissolve 1.25 g gelatin A or B in 25 ml deionized water at 50 °C.
2. Seal the solution carefully using parafilm.
3. Stir for 30 min at 250 rpm.

3.1.2 First Desolvation Step

1. To the previously prepared solution add 25 ml cold acetone and leave for 60 min at room temperature.
2. Seal carefully with parafilm (*see* **Note 1**).
3. After 60 min, discard the supernatant (*see* **Note 2**).
4. Redissolve the precipitated or sedimented gelatin in 25 ml deionized water at 50 °C.
5. Reseal again the glassware using parafilm.
6. Stir for 30 min at 250 rpm.

3.1.3 The Second Desolvation Step

1. Prepare 5 M HCl or NaOH solution.
2. Use the previously prepared solutions to adjust the pH of the gelatin solution to 2.5 or 12 (*see* **Note 3**).
3. Add cold acetone slowly (2 ml/min) to the above solution (*see* **Note 4**):
   (a) 80 ml at 50 °C for gelatin B
   (b) 75 ml at 40 °C for gelatin A (*see* **Note 5**)
4. Once the addition of acetone starts, stir the gelatin solution vigorously at 16,500 × *g*.

3.1.4 Glutaraldehyde Cross-Linking

1. Remove the gelatin dispersion from the heated stirrer to a non-heated one (at room temperature).
2. Start adding dropwise the cross-linker (25% w/v glutaraldehyde solution in water) (*see* **Note 6**).
3. Seal the dispersion firmly using parafilm.
4. Stir at 600 rpm overnight (*see* **Note 7**).

3.1.5 Glycine Washing

1. Prepare glycine solution in water (100 mM, i.e., 0.1 M).
2. Add an equal volume of this glycine solution to the gelatin nanoparticle dispersion (*see* **Note 8**).
3. Seal carefully with parafilm.
4. Stir for 1 h (*see* **Note 9**).
5. For long-term storage, the nanoparticle dispersion can be stored at 4 °C.

3.1.6 The Washing Step

1. Transfer the nanoparticle suspension to 50 ml Falcon tubes.
2. Centrifuge at 16,500 × *g* for 1 h.
3. Resuspend the nanoparticles using 15 ml of deionized water by the help of vortexing (*see* **Note 10**).
4. At least three times of washing should be carried out.

3.1.7 Freeze-Drying

1. Transfer the nanoparticles to 15 ml Falcon tubes (fill to only 5 ml in each tube).
2. Freeze the above tubes to −80 °C by means of dipping in liquid nitrogen.
3. Immediately transfer the frozen tubes to the freeze-drying chamber, and leave it overnight (at least ~16 to 24 h) (*see* **Note 11**).

3.1.8 Dynamic Light Scattering (DLS) and Zeta Potential Measurements

1. DLS size measurements are performed in deionized water.
2. Zeta potential measurements are usually performed after medium enrichment with a buffer solution (very low strength such as HEPES buffer, 5 mM at pH 7).

3.2 Cationization of Gelatin Nanoparticles

1. Adjust the aqueous dispersion of the nanoparticles to pH 4.5.
2. Add 50 mg of cholamine per each 500 mg nanoparticle under constant stirring.
3. After 5 min of incubation, add 50 mg of EDC to the nanoparticle suspension to activate the free carboxyl groups on the particles to react with cholamine.
4. Leave the nanoparticles to react with cholamine for 1 h.
5. The nanoparticles are purified by threefold centrifugation and re-dispersion, analogous to the purification of plain nanoparticles.

3.3 Protocol for Labeling Gelatin Nanoparticles with FITC

1. Prepare 5 mg/ml nanoparticle suspension in MQ water.
2. Adjust the suspension to pH 9.5 using 100 μl carbonate buffer.
3. Prepare 5 mg/ml stock solution of FITC in DMSO (FITC does not dissolve in water and is unstable in aqueous solutions).
4. Take 19.4 μl from this stock solution and add it to the nanoparticle suspension.
5. Stir the suspension for 3–4 h at room temperature.
6. Purify the nanoparticles by centrifugation and then filter using Nanoseps® (10 nm pore size).

4 Notes

1. Make sure that the acetone is in contact with the whole-gelatin solution and not only in contact with its surface by performing gentle swirling of the glassware containing the solution.
2. This step is performed to get rid of the low-molecular-weight gelatin that does not form gelatin nanoparticles. The high-molecular-weight counterpart is precipitated.
3. Adjusting of the pH with 5 molar solution of HCL or NaOH is adopted in order to use only four or five drops to reach the desired pH to avoid adding too much extra water to the prepared system.
4. This step can be performed using a burette to ensure the dropwise addition of gelatin.
5. The end point of addition of acetone is the formation of a thick white-colored colloidal dispersion. Adding small excess of acetone could also increase the yield.
6. In case of using a molar ratio of the amino groups of gelatin to glutaraldehyde, [NH_2:GA] equals 1:1, then 185 μl of glutaraldehyde is needed for gelatin B. In case of gelatin A, only 165 μl is needed.

7. Usually the overnight stirring time is taken as 16 h.
8. Molecular weight of glycine = 75.07 g/mole.

   For gelatin B: The total volume of the dispersion is ~105 ml. Therefore, the required amount of glycine in 105 ml = $0.105 \times 0.1 \times 75.07 = 788$ mg.

   For gelatin A: The total volume of the dispersion is ~100 ml. Therefore, the required amount of glycine in 100 ml = $0.1 \times 0.1 \times 75.07 = 751$ mg.
9. This step is performed in order to block the unreacted aldehyde groups of glutaraldehyde and stop the cross-linking reaction.
10. Sometimes, probe sonication is needed in this step.
11. No cryoprotectants are needed for this step.

Acknowledgments

The authors would like to thank Dr. Reza Nejadnik, Leiden/Amsterdam Centre for Drug Research, for introducing us to several important points in the preparation of gelatin nanoparticles.

References

1. Abozeid SM, Hathout RM, Abou-Aisha K (2016) Silencing of the metastasis-linked gene, AEG-1, using siRNA-loaded cholamine surface-modified gelatin nanoparticles in the breast carcinoma cell line MCF-7. Colloids Surf B Biointerfaces 145:607–616
2. Hathout RM, Omran MK (2016) Gelatin-based particulate systems in ocular drug delivery. Pharm Dev Technol 21(3):379–386
3. Sahoo N, Sahoo RK, Biswas N, Guha A, Kuotsu K (2015) Recent advancement of gelatin nanoparticles in drug and vaccine delivery. Int J Biol Macromol 81:317–331
4. Elzoghby AO (2013) Gelatin-based nanoparticles as drug and gene delivery systems: reviewing three decades of research. J Control Release 172(3):1075–1091
5. Coester CJ, Langer K, van BH, Kreuter J (2000) Gelatin nanoparticles by two step desolvation—a new preparation method, surface modifications and cell uptake. J Microencapsul 17(2):187–193
6. Farid MM, Hathout RM, Fawzy M, Abou-Aisha K (2014) Silencing of the scavenger receptor (Class B - Type 1) gene using siRNA-loaded chitosan nanoparticles in a HepG2 cell model. Colloids Surf B Biointerfaces 123:930–937
7. Mehanny M, Hathout RM, Geneidi AS, Mansour S (2016) Bisdemethoxycurcumin loaded polymeric mixed micelles as potential anti-cancer remedy: preparation, optimization and cytotoxic evaluation in a HepG-2 cell model. J Mol Liq 214:162–170
8. El-Marakby EM, Hathout RM, Taha I, Mansour S, Mortada ND (2017) A novel serum-stable liver targeted cytotoxic system using valerate-conjugated chitosan nanoparticles surface decorated with glycyrrhizin. Int J Pharm 525(1):123–138
9. Mehanny M, Hathout RM, Geneidi AS, Mansour S (2017) Studying the effect of physically-adsorbed coating polymers on the cytotoxic activity of optimized bisdemethoxycurcumin loaded-PLGA nanoparticles. J Biomed Mater Res A 105(5):1433–1445
10. Ofokansi K, Winter G, Fricker G, Coester C (2010) Matrix-loaded biodegradable gelatin nanoparticles as new approach to improve drug loading and delivery. Eur J Pharm Biopharm 76(1):1–9
11. Lu Z, Yeh TK, Tsai M, Au JL, Wientjes MG (2004) Paclitaxel-loaded gelatin nanoparticles for intravesical bladder cancer therapy. Clin Cancer Res 10(22):7677–7684
12. Mohanty B, Aswal VK, Kohlbrecher J, Bohidar HB (2005) Synthesis of gelatin nanoparticles

via simple coacervation. J Surf Sci Technol 21 (3–4):149–160
13. Li JK, Wang N, Wu XS (1998) Gelatin nanoencapsulation of protein/peptide drugs using an emulsifier-free emulsion method. J Microencapsul 15(2):163–172
14. Cascone MG, Lazzeri L, Carmignani C, Zhu Z (2002) Gelatin nanoparticles produced by a simple W/O emulsion as delivery system for methotrexate. J Mater Sci Mater Med 13 (5):523–526
15. Khan SA, Schneider M (2013) Improvement of nanoprecipitation technique for preparation of gelatin nanoparticles and potential macromolecular drug loading. Macromol Biosci 13 (4):455–463
16. Lee EJ, Khan SA, Lim KH (2011) Gelatin nanoparticle preparation by nanoprecipitation. J Biomater Sci Polym Ed 22(4-6):753–771
17. Zorzi GK, Contreras-Ruiz L, Parraga JE, Lopez-Garcia A, Bello RR, Diebold Y, Seijo B, Sanchez A (2011) Expression of MUC5AC in ocular surface epithelial cells using cationized gelatin nanoparticles. Mol Pharm 8(5):1783–1788
18. Ethirajan A, Schoeller K, Musyanovych A, Ziener U, Landfester K (2008) Synthesis and optimization of gelatin nanoparticles using the miniemulsion process. Biomacromolecules 9 (9):2383–2389
19. Azimi B, Nourpanah P, Rabiee M, Arbab S (2014) Producing gelatin nanoparticles as delivery system for bovine serum albumin. Iran Biomed J 18(1):34–40
20. Akhter KF, Zhu J, Zhang J (2012) Nanoencapsulation of protein drug for controlled release. J Phys Chem Biophys S11:001–005
21. Metwally AA, El-Ahmady SH, Hathout RM (2016) Selecting optimum protein nano-carriers for natural polyphenols using chemoinformatics tools. Phytomedicine 23(14):1764–1770

Chapter 7

Green Synthesis of Chitosan-Silver/Gold Hybrid Nanoparticles for Biomedical Applications

Ibrahim M. El-Sherbiny and Mohammed Sedki

Abstract

Silver and gold nanoparticles (NPs) attract great attention nowadays because of their unique characteristics that entitle them for various biomedical applications. However, there is still a need for successful green synthesis methods for these two metal NPs, especially in a hybrid form, as well-established protocols. On the other hand, chitosan (Cs) is a polysaccharide of great promise for green synthesis of metal NPs, especially in the presence of some plant/fruit extracts. Together, Cs and the appropriate natural products in the extracts play the roles of both capping and reducing agents toward the green synthesis and stabilization of the silver/gold hybrid NPs. In this chapter, we introduce a well-established protocol for the green synthesis of Cs-Ag/Au hybrid NPs which could incorporate a therapeutic agent. In this approach, Cs plays the role of a carrier for the therapeutic agent, in addition to its capping/reducing contributions.

Key words Chitosan, Silver, Gold, Hybrid NPs, Green synthesis, Natural extract

1 Introduction

Metal nanoparticles (NPs), particularly silver (Ag) and gold (Au), and their hybrid nanocomposites have attracted a great attention in recent years due to their unique physical and chemical properties that entitle them for various biomedical applications [1]. For instance, Ag NPs are well known for their significant antimicrobial activity to overcome the multidrug resistance of contemporary antibiotics. Besides, Ag and Au NPs have demonstrated wonderful responses in theranostics, which is mainly associated with their surface plasmons [2]. In addition to their superior biomedical applications, Ag and Au NPs are also highly applicable in other areas such as catalysis, electrochemical applications, DNA sequencing, and surface-enhanced Raman spectroscopy (SERS) [3, 4]. Different methodologies have been reported for the fabrication of Ag/Au NPs, such as chemical and photoreduction [5], radiation chemical reduction [6], and thermal decomposition in organic solvents [7]. However, most of these approaches are not green

Volkmar Weissig and Tamer Elbayoumi (eds.), *Pharmaceutical Nanotechnology: Basic Protocols*, Methods in Molecular Biology, vol. 2000, https://doi.org/10.1007/978-1-4939-9516-5_7, © Springer Science+Business Media, LLC, part of Springer Nature 2019

which negatively affects the environmental ecological systems. Hence, there is a pressing need for well-established protocols toward the green synthesis of these NPs.

Many studies have explored the green/biosynthesis of metal NPs, some of which used fungi, bacteria, and actinomycetes, while others applied whole plants [8–10]. However, these methods have many limitations such as the difficulty of scaling up and conserving cell cultures, when using microorganisms [11]. Plant extracts have been heavily considered for the synthesis of metal NPs as this approach is green, practical, and scalable [11]. For instance, many studies have described the use of seeds, leaves, latex, roots, and fruits [12–15]. In this regard, Sedki et al. [16] have successfully phyto-synthesized Ag NPs and Ag-reduced graphene oxide (Ag-rGO) nanocomposite using *Potamogeton pectinatus L.* extract and AbdelHamid et al. [17] have synthesized the Ag-Au alloy using the same plant extract.

Chitosan (Cs) is a polysaccharide consisting of linearly connected *D*-glucosamine units, and it is obtained via deacetylation of *N*-acetyl-*D*-glucosamine (chitin) which is the central component of cell wall in crustacean shells [18, 19]. Cs, as a natural polymer, has many superior properties such as biodegradability, biocompatibility, low cost, and hydrogel-formation ability [20]. Besides, Cs-based nano- and microparticles are greatly used in drug delivery systems due to the aforementioned properties as well as its sustained drug release ability. Moreover, Cs is useful in the green synthesis of metal NPs due to its capping (as a large molecule) and reducing properties, which is attributed to the presence of electron-rich amine groups [21, 22].

El-Sherbiny et al. have introduced a new hybrid nanostructure using Cs/grape leaf aqueous extract (Cs/GLE) or Cs/GLE NPs. In both cases, Cs and GLE work as reducing and stabilizing (capping) agents [23]. This method enables the green synthesis of hybrid metal NPs, particularly Ag and Au NPs, either as metal NPs coated with Cs/GLE molecules or as metal NPs directly attached to the surface of GLE-loaded Cs NPs (GLE-Cs NPs/Ag NPs). The green-synthesized nanostructures are of a great importance, and were found to be promising anticancer agents as applied on HpG2 cells. In the next sections, we focus on the methodologies to develop such nanostructures and the technical points behind their synthesis. We also try to generalize the technique for the synthesis of therapeutic agent-loaded polymer-metal NPs.

2 Materials

1. Chitosan powder (Mw = 600,000 × g/mol) with a degree of deacetylation (DD) of 75%.
2. Sodium triphosphate pentabasic (STPP) of practical grade 90–95%.

3. Silver nitrate, $AgNO_3$ (AR grade, 99.5% purity).
4. Grape leaves (GL) (Metro Co., Egypt).
5. Acetic acid and other solvents were of analytical grade.
6. Deionized water with resistivity >2 × 108 Ω·cm was applied for all sample preparations.

3 Methods

3.1 Preparation of Cs/GLE Aqueous Mixture and Cs/GLE NPs

The preparation protocol of Cs/GLE and Cs/GLE NPs can be summarized in four steps: (1) preparation of GLE solution, (2) preparation of Cs solution, (3) mixing GLE and Cs solution to form Cs/GLE mixture at different ratios, and (4) synthesis of Cs/GLE NPs.

3.1.1 Preparation of GLE Solution

1. Grape leaves (GL) were washed with deionized water several times.
2. 50 g of the washed GL was heated for 1 h in 250 ml of deionized water at 60 °C (*see* **Note 1**).
3. The grape leaf extract (GLE) was then obtained by centrifugation of the leaf solution at 6000 rpm (685 RCF) for 10 min followed by filtration using Whatman filter papers (*see* **Note 2**).
4. The GLE (filtrate) was stored at 4 °C (*see* **Note 3**) for future use.

3.1.2 Preparation of Cs Solution

1. Cs solutions (2% w/v) were formed by dissolving the calculated amount of Cs into 1% acetic acid aqueous solution via stirring for 24 h until complete dissolution.
2. The produced viscous solution was left overnight to remove any air bubbles.

3.1.3 Preparation of Cs/GLE Mixtures at Different Ratios (see Note 4)

1. The Cs/GLE mixtures were achieved by mixing different volumes: volume ratios (3:1, 1:1, and 1:3) of Cs and GLE solutions, respectively.

3.1.4 Synthesis of GLE-Cs NPs

1. The GLE-Cs NPs were synthesized via the ionotropic gelation technique.
2. 5 ml of STPP solution (25% w/w with respect to the Cs content) was added dropwise to 45 ml of the different Cs/GLE mixture ratios (3:1, 1:1, and 1:3) with pulsed sonication (5 s on and 5 s off) at a power of 60 W for 10 min, (*see* **Note 5**).

3.2 Synthesis of Ag NPs Coated with Cs/GLE, and the GLE-Cs/Ag NPs

The prepared Ag NPs in both cases were evaluated using UV-Vis spectrophotometry and FTIR analysis.

1. *The Ag NPs coated with Cs/GLE* were synthesized in the presence of Cs/GLE aqueous solutions as follows:
2. 20 ml of the different v/v ratios of the Cs/GLE solutions were added with stirring to 40 ml of $AgNO_3$ (1 mM).
3. After stirring for 5 min (*see* **Note 6**), the mixture was exposed to UV irradiation for 50 min (*see* **Note 7**).
4. The solution color will change into yellow with time due to the plasmonic effect of Ag NPs.
5. *The GLE-Cs/Ag NPs* were synthesized using the same procedure of Ag NPs -coated with Cs/GLE, with the same quantities; all the change is in replacing Cs/GLE mixture by GLE-Cs NPs suspension.
6. The prepared Ag NPs in both cases were evaluated using UV-Vis spectrophotometry and FTIR analysis (*see* **Notes 8–11**).

3.3 Applications of the Developed GLE-Cs/Ag NPs

1. The prepared hybrid Ag NPs were tested for their anticancer activity against HpG2 cells, and induced cellular apoptosis by downregulating *BCL2* gene and upregulating *P53*.
2. The proposed formula showed a strong anticancer activity.

4 Notes

1. The applied temperature upon preparing the grape leaf extract (GLE) should not highly exceed 60 °C to avoid any thermal degradation of the extract active components.
2. The active components of GLE are mainly phenolic compounds with capping and reducing ability that helps in the synthesis of metal NPs.
3. The GLE must be stored at low temperatures (around 4 °C) to avoid microbial growth in the extract, as it is a rich medium for microbes.
4. Different ratios of Cs/GLE were explored to reach the optimum ratio which can achieve the highest stability for the developed hybrid metal NPs. This is attributed to the difference in reducing and capping properties between Cs and GLE. Then, for the development of any other newly introduced polymer/extract mixtures, several trials should be conducted to reach the optimum ratio, but with simply maintaining the Cs concentration at 2% (w/v).

5. In the synthesis of GLE-Cs NPs, the procedure can be easily modified to obtain microparticles via increasing the Cs concentration and controlling the addition rate of STTP. That depends particularly on the anticipated application of the developed hybrid particles.
6. In the synthesis of Ag NPs in the presence of Cs/GLE solution, the added Ag ions should be stirred well but gently for no less than 5 min to produce well-dispersed ions in the viscous Cs solution, and to avoid the aggregation of the resulting NPs.
7. Upon synthesis of Ag NPs, the maximum/optimum exposure time to UV irradiation is determined by monitoring the UV-Vis results of the reaction solution at different time intervals. The reaction mixture is exposed to UV light until no significant increase in the absorbance peak of the prepared plasmonic NPs is attained, which means that there are no more ions to be reduced into NPs.
8. The UV-Vis measurements are really helpful in monitoring the synthesis of plasmonic metal NPs, for instance (1) the red shift means increasing in size, (2) higher absorption intensity reflects more NP formation, (3) shorter intensity after continuous increase would mainly reflect aggregation and precipitation of the resulting NPs, etc.
9. In synthesis of GLE-Cs NPs, the sonication process should be done in an ice bath, especially in the case of using probe sonicator, to avoid the thermal degradation of extract active components, and also to keep the formed Cs NPs with no deformations.
10. The main advantage behind this protocol is that it is applicable for Ag, Au, or other metals with similar reduction potentials, and also it works as a general formula for encapsulating different drugs or biologically active plant extracts.
11. In the case of synthesis of Au NPs using the above protocol, the reaction time will be a bit increased as the reduction rate of gold ions is normally slower than that of silver ions.

References

1. Guzmán MG, Dille J, Godet S (2009) Synthesis of silver nanoparticles by chemical reduction method and their antibacterial activity. Int J Chem Biomol Eng 2:3
2. Baker C et al (2005) Synthesis and antibacterial properties of silver nanoparticles. J Nanosci Nanotechnol 5(2):244–249
3. Cao YW, Rongchao J, Chad AM (2001) DNA-modified core–shell Ag/Au nanoparticles. J Am Chem Soc 123(32):7961–7962
4. Matejka P et al (1992) The role of triton X-100 as an adsorbate and a molecular spacer on the surface of silver colloid: a surface-enhanced Raman scattering study. J Phys Chem 96 (3):1361–1366
5. Pileni MP (2000) Fabrication and physical properties of self-organized silver nanocrystals. Pure Appl Chem 72(1–2):53–65
6. Henglein A (2001) Reduction of Ag (CN) 2-on silver and platinum colloidal nanoparticles. Langmuir 17(8):2329–2333
7. Esumi K, Keiichi M, Kanjiro T (1995) Preparation of rodlike gold particles by UV

irradiation using cationic micelles as a template. Langmuir 11(9):3285–3287

8. Klaus T et al (1999) Silver-based crystalline nanoparticles, microbially fabricated. Proc Natl Acad Sci 96(24):13611–13614
9. Ahmad A et al (2003) Intracellular synthesis of gold nanoparticles by a novel alkalotolerant actinomycete, rhodococcus species. Nanotechnology 14(7):824
10. Beattie IR, Richard GH (2011) Silver and gold nanoparticles in plants: sites for the reduction to metal. Metallomics 3(6):628–632
11. Song JY, Beom SK (2008) Biological synthesis of bimetallic Au/Ag nanoparticles using persimmon (Diospyros kaki) leaf extract. Korean J Chem Eng 25(4):808–811
12. Kumar V, Subhash CY, Sudesh KY (2010) *Syzygium cumini* leaf and seed extract mediated biosynthesis of silver nanoparticles and their characterization. J Chem Technol Biotechnol 85(10):1301–1309
13. Dubey SP et al (2010) Bioprospective of *Sorbus aucuparia* leaf extract in development of silver and gold nanocolloids. Colloids Surf B: Biointerfaces 80(1):26–33
14. Das RK et al (2011) Synthesis of gold nanoparticles using aqueous extract of *Calotropis procera* latex. Mater Lett 65(4):610–613
15. Ahmad N et al (2010) Rapid synthesis of silver nanoparticles using dried medicinal plant of basil. Colloids Surf B: Biointerfaces 81 (1):81–86
16. Sedki M et al (2015) Phytosynthesis of silver--reduced graphene oxide (Ag–RGO) nanocomposite with an enhanced antibacterial effect using *Potamogeton pectinatus* extract. RSC Adv 5(22):17358–17365
17. AbdelHamid AA et al (2013) Phytosynthesis of Au, Ag, and Au–Ag bimetallic nanoparticles using aqueous extract of sago pondweed (*Potamogeton pectinatus* L.). ACS Sustain Chem Eng 1(12):1520–1529
18. Muzzarelli RAA (2009) Chitins and chitosans for the repair of wounded skin, nerve, cartilage and bone. Carbohydr Polym 76(2):167–182
19. Thakur VK, Manju KT (2014) Recent advances in graft copolymerization and applications of chitosan: a review. ACS Sustain Chem Eng 2 (12):2637–2652
20. An J et al (2013) Electrochemical study and application on rutin at chitosan/graphene films modified glassy carbon electrode. J Pharmaceut Anal 3(2):102–108
21. Bhumkar DR et al (2007) Chitosan reduced gold nanoparticles as novel carriers for transmucosal delivery of insulin. Pharm Res 24 (8):1415–1426
22. Huang H, Xiurong Y (2004) Synthesis of chitosan-stabilized gold nanoparticles in the absence/presence of tripolyphosphate. Biomacromolecules 5(6):2340–2346
23. El-Sherbiny IM et al (2016) Newly developed chitosan-silver hybrid nanoparticles: biosafety and apoptosis induction in HepG2 cells. J Nanopart Res 18(7):1–13

Chapter 8

Methods of Fabrication of Chitosan-Based Nano-in-Microparticles (NMPs)

Ibrahim M. El-Sherbiny and Amr Hefnawy

Abstract

Chitosan nano-in-microparticles (NMPs) are promising carrier systems that have gained recently more interest aiming to combine advantages of both the nano- and microsystems. They have been employed for various purposes including sustained pulmonary delivery of drugs and pulmonary delivery of peptides, proteins, or genes or as injectable scaffolds for simultaneous delivery of stem cells and supporting growth factors. Among these delivery systems, chitosan was a common ingredient due to its biocompatibility, biodegradability, and ability to sustain the release of drugs and improving their bioavailability. Here we introduce a method for the development of chitosan self-assembly nanoparticles and the incorporation of these nanoparticles into chitosan microparticles via spray drying.

Key words Chitosan, Nano-in-micro, NMPs, Pulmonary, Spray drying

1 Introduction

Nano-in-microparticles (NMPs) have recently gained increasing interest due to their ability to combine the advantages of both the nano- and microsystems. Pulmonary delivery of various agents is among the most common reported applications of NMPs. For example, Du, El-Sherbiny, and Smyth loaded ciprofloxacin antibiotic into hydrogel NMPs for local delivery to the lungs. Self-assembling chitosan derivative was used to form the nanoparticles which were then loaded into swellable alginate microparticles. The use of nanoparticles alone for pulmonary delivery proved to be inefficient as they would be easily cleared by exhalation. The microparticles would rather provide suitable aerodynamic diameter for delivery into the deep lungs (0.5–5 μm); however, they would be cleared by macrophage uptake. This was in turn avoided by the use of swellable hydrogels which increase the size of the microparticles in the moist environment to a size larger than the macrophage can uptake [1, 2]. The same system was also proved to efficiently deliver peptide molecules using bovine serum albumin as a model drug

Volkmar Weissig and Tamer Elbayoumi (eds.), *Pharmaceutical Nanotechnology: Basic Protocols*, Methods in Molecular Biology, vol. 2000, https://doi.org/10.1007/978-1-4939-9516-5_8,

[3]. NMPs were also used for pulmonary delivery of insulin to the systemic circulation by loading into chitosan nanoparticles which were then incorporated into microparticles by spray drying with mannitol solution. This system provided higher hypoglycemic effect compared to the negative control indicating the successful delivery of active insulin to the circulation [4]. Grenha and her coworkers also developed NMP delivery system for the delivery of insulin where it was loaded into chitosan/tripolyphosphate nanoparticles that were in turn spray dried into microparticles. The presence of phospholipids supports controlled release of the loaded insulin and adequate aerodynamic diameter for delivery into the deep lungs [5]. Chitosan-based NMPs were also used for delivery of the anticancer agent, capecitabine, which was formulated using a method combining emulsification and electrospraying. Although the formula successfully sustained the release of the loaded drug, the maximum encapsulation efficiency reached was less than 20% with cumulative release of less than 30%. This shows that the formula is promising for sustained pulmonary delivery but needs further optimization or it might be more efficient with different active agents [6].

There is a wide variety of NMPs that can be developed depending on the materials or polymers used for preparation of the nanoparticles and microparticles. This chapter focuses on chitosan-based NMPs. The chapter introduces a method for preparation of self-assembly chitosan-derived nanoparticles, and then it shows the method for incorporating these nanoparticles into chitosan microparticles using spray drying method.

2 Materials

2.1 Self-Assembling Chitosan Nanoparticles

1. Chitosan (Cs) of molecular weight 400–500 kDa with deacetylation percentage around 75%.
2. Monomethoxy-poly(ethylene glycol) (mPEG) of average Mw 5 kDa.
3. Phthalic anhydride.
4. 1-Hydroxybenzotriazole (HOBt).
5. Succinic anhydride.
6. 4-Dimethylaminopyridine (DMAP).
7. 1-Ethyl-3-(3-dimethylaminopropyl) carbodiimide hydrochloride (EDC·HCl).
8. Triethylamine.

2.2 Cs or PEG-g-Cs Microparticles via Spray Drying

1. Chitosan (Cs) of molecular weight 354 kDa.
2. Monomethoxy-poly(ethylene glycol) (mPEG) of average Mw 5 kDa.
3. 1-Hydroxybenzotriazole (HOBt).
4. Succinic anhydride.
5. 4-Dimethylaminopyridine (DMAP).
6. 1-Ethyl-3-(3-dimethylaminopropyl) carbodiimide hydrochloride (EDC·HCl).
7. Triethylamine.
8. Büchi Mini spray dryer B-290 (Büchi, Switzerland).

3 Methods

3.1 Preparation of Self-Assembly Chitosan Nanoparticles [2, 3]

The process can be divided into four main steps: (a) protection of chitosan amino groups using phthalic anhydride to form *N*-phthaloyl chitosan (*N*-PhCs), (b) conversion of mPEG into mPEG-COOH, (c) conjugation of mPEG-COOH and *N*-PhCs, and (d) deprotection of the amino group to yield the polymer conjugate PEG-g-Cs.

3.1.1 Protection of Chitosan Amino Groups

1. Prepare *N*-PhCs by reacting of 10 g chitosan with 5 mole equivalent (44.8 g) of phthalic anhydride (relative to pyranose ring of chitosan). Dissolve the reactants in 200 ml of DMF, and allow the reaction mixture to stir at 130 °C for 8 h (*see* **Note 1**).
2. Let reaction to cool to room temperature before pouring it on ice water while stirring (*see* **Note 2**).
3. Separate the product using filtration or centrifuge at 6000 rpm (685 RCF) for 10–15 min.
4. Wash the product extensively three times with ethanol and then once with methanol to remove excess phthalic acid and phthalic anhydride.
5. Dry the final product at 40 °C overnight or until obtaining pale brown product (*see* **Note 3**).
6. Characterization: Successful synthesis of *N*-PhCs can be confirmed by the appearance of absorbance peaks at 1395 and 732 cm^{-1} in the Fourier transform infrared (FTIR) spectrum of the product. These peaks are attributed to the aromatic (C=C) and aromatic (C–H) of the phthaloyl groups, respectively. The grafting percentage can be estimated using elemental analysis.

3.1.2 Conversion of mPEG into mPEG-COOH

1. Synthesis of mPEG-COOH is done by reacting 100 g of mPEG with equimolar amount of succinic anhydride (2.4 g) in dimethylformamide (60 ml) at 60 °C overnight in the presence of catalytic amount of pyridine (*see* **Note 4**).
2. The reaction is left to cool to room temperature followed by addition of 400 ml of diethyl ether to precipitate the product (*see* **Note 5**).
3. The precipitate is redissolved in CCl_4, filtered, and then reprecipitated using diethyl ether.
4. The product is then separated by vacuum filtration and dried by air-drying.
5. Characterization: FTIR absorption for mPEG-COOH should confirm successful synthesis by showing a sharp peak at around 3500 cm^{-1} with shoulder at around 2900 cm^{-1} characteristic of carboxylic OH (*see* **Note 6**).

3.1.3 Conjugation of mPEG-COOH and N-PhCs

1. Mix 37.9 g of mPEG-COOH with *N*-PhCs (5 g) in 75 ml of DMF.
2. Add 3.4 g of HOBt to catalyze the reaction with stirring at room temperature.
3. After the solution gets clearer, add 4.25 g of EDC·HCl and continue stirring at room temperature overnight (*see* **Note 7**).
4. The product is purified by dialysis against distilled water and then washing with ethanol.

3.1.4 Deprotection of the Amino Group to Yield PEG-g-Cs

1. PEG-g-*N*-PhCs (4.0 g) is dissolved in 15 ml of DMF and heated to 110 °C with stirring under nitrogen.
2. 20 ml of hydrazine monohydrate is added, and the reaction is continued for 2 h.
3. Purify the resulting PEG-g-Cs copolymer via dialysis against a mixture of deionized ethanol and water (1:1) and then dried under vacuum at 40 °C.

3.2 Preparation of Self-Assembly Chitosan-Based Nanoparticles

1. The PEG-g-Cs graft copolymer was dispersed at the concentration of 1% (w/v) in distilled water followed by probe sonication for 2 min using 60 W power.
2. Sonication process was done in an ice bath while using pulse 5 s on/5 s off to avoid excessive heating of the copolymer solution.
3. The process is repeated several times till the optimum size is achieved [3] (*see* **Note 8**).

3.3 Preparation of Microparticles of Cs Using Spray Drying

1. Prepare suspension of the nanoparticles prepared from the previous step at the concentration of 0.33% (w/v).

2. Add the prepared suspension dropwise to 0.5% (w/v) Cs solution dissolved in 0.06% acetic acid solution with homogenization at 10,000 rpm.
3. Spray dry, with the aid of a mini spray dryer, the prepared mixture using 0.7 mm two-fluid pressurized atomizer at a feed rate of 6 ml/min [7] (*see* **Notes 9–10**).

4 Notes

1. Phthalic anhydride reacts slowly with hot water forming phthalic acid; however, in the presence of excess phthalic acid in the reaction this should not represent a problem. A study reported that this may even be beneficial to increase the selectivity of the reaction toward *N*-phthylation rather *O*-phthylation. The study reported that the reaction was completely selective in the presence of 5% water in the reaction medium [8].
2. Solubility of phthalic acid and phthalic anhydride in water is low (6.965 g/l and 6.2 g/l at 25 °C, respectively) [9] which means that precipitation of the reaction using ice water will virtually precipitate excess unreacted reactants.
3. Centrifuge or vacuum filtration may yield large bulks of material. In this case drying should be done on 2 days, and the product should be dried for 24 h followed by grinding of the product, washing with methanol, and then drying for another 24 h.
4. Succinic anhydride decomposes in water into succinic acid which is relatively more soluble. Consequently, it is recommended to use dried solvents and reagents for this reaction. Drying can be achieved either using drying agents as anhydrous Na_2SO_4 or using other methods as distillation.
5. Succinic anhydride solubility in ether is relatively low (around 0.64 g/100 ml) [10]. Accordingly, large amount of the solvent should be used in the precipitation step to ensure that unreacted succinic anhydride is dissolved and that the precipitated product is purified.
6. It has been observed that the use of attenuated total reflection (ATR) FTIR for analysis of mPEG-COOH may not show the peak of the COOH group despite its presence. In this case, it would be more accurate to use the older FTIR instruments that employ compressed KBr pellets for analysis of the samples. According to the optical theory, the sampling depth using ATR-FTIR at the frequency range of 3000–4000 cm^{-1} is only one-tenth of that at the frequency range of 500–1500 cm^{-1} resulting in weaker peaks. This effect is not

observed when using FTIR spectrometers with the aid of compressed KBr cells [11].

7. EDCI·HCl is a water-soluble carbodiimide that acts as a catalyst for the formation of peptide bonds. Its optimum activity is achieved in aqueous environment at pH 4–5. However, it can also be used in DMSO and DMF. In this case other supporting agents should be used as HOBt which improves the activity of EDC as it inhibits many side reactions such as the formation of *N*-acylurea or racemization [12].
8. Increasing sonication time has been reported in several studies to reduce the particle size but also reduce the entrapment efficiency of loaded substances. This observation was also noted in research work done by our group that has not been published yet [13–15].
9. The size of microparticles obtained from the spray drying process is dependent on several factors including the working frequency of the device, type of atomizer or nozzle used, feed rate, viscosity, and surface tension of the polymer solution [16–18].
10. In this chapter we described the preparation of chitosan-derived self-assembly nanoparticles and their subsequent loading into chitosan microparticles. This was guided by the focus of the chapter on methods for preparation of chitosan nano- and microparticles. However, it is not recommended to load nanoparticles into microparticles of the same nature particularly in the absence of cross-linking of the inner particles. In this case, nanoparticles would probably lose their integrity into the matrix of the microparticles. Alternatively, the described protocol should be regarded as two separate processes; that is, the nanoparticles may be loaded into various types of microparticles and the microparticles might be loaded with other different types of nanoparticles.

References

1. Houtmeyers E et al (1999) Regulation of mucociliary clearance in health and disease. Eur Respir J 13(5):1177–1188
2. Du J, El-Sherbiny IM, Smyth HD (2014) Swellable ciprofloxacin-loaded nano-in-micro hydrogel particles for local lung drug delivery. AAPS PharmSciTech 15(6):1535–1544
3. El-Sherbiny IM, Smyth HDC (2010) Biodegradable nano-micro carrier systems for sustained pulmonary drug delivery: (I) selfassembled nanoparticles encapsulated in respirable/swellable semi-IPN microspheres. Int J Pharm 395(1–2):132–141
4. Al-Qadi S et al (2012) Microencapsulated chitosan nanoparticles for pulmonary protein delivery: in vivo evaluation of insulin-loaded formulations. J Control Release 157 (3):383–390
5. Grenha A et al (2008) Microspheres containing lipid/chitosan nanoparticles complexes for pulmonary delivery of therapeutic proteins. Eur J Pharm Biopharm 69(1):83–93
6. Liu Y et al (2013) Preparation of embolic NEMs loading capecitabine. J Mater Sci Mater Med 24(1):155–160
7. El-Sherbiny IM, Smyth HDC (2011) Controlled release pulmonary administration of

curcumin using swellable biocompatible microparticles. Mol Pharm 9(2):269–280

8. Kurita K et al (2002) Chemoselective protection of the amino groups of chitosan by controlled phthaloylation: facile preparation of a precursor useful for chemical modifications. Biomacromolecules 3(1):1–4
9. Yalkowsky SH, He Y, Jain P (2016) Handbook of aqueous solubility data. CRC Press, Boca Raton
10. Furia TE (1973) CRC handbook of food additives, vol 1. CRC Press, Cleveland, OH
11. Yamamoto K, Ishida H (1994) Optical theory applied to infrared spectroscopy. Vib Spectrosc 8(1):1–36
12. Pottorf RS, Szeto P (2001) Encyclopedia of reagents for organic synthesis. John Wiley & Sons, Hoboken, NJ
13. Song X et al (2008) PLGA nanoparticles simultaneously loaded with vincristine sulfate and verapamil hydrochloride: systematic study of particle size and drug entrapment efficiency. Int J Pharm 350(1–2):320–329
14. Mainardes RM, Evangelista RC (2005) PLGA nanoparticles containing praziquantel: effect of formulation variables on size distribution. Int J Pharm 290(1–2):137–144
15. Bilati U, Allémann E, Doelker E (2003) Sonication parameters for the preparation of biodegradable nanocapsulesof controlled size by the double emulsion method. Pharm Dev Technol 8(1):1–9
16. Cal K, Sollohub K (2010) Spray drying technique. I: hardware and process parameters. J Pharm Sci 99(2):575–586
17. Gharsallaoui A et al (2007) Applications of spray-drying in microencapsulation of food ingredients: an overview. Food Res Int 40 (9):1107–1121
18. Estevinho BN et al (2013) Microencapsulation with chitosan by spray drying for industry applications–a review. Trends Food Sci Technol 31(2):138–155

Chapter 9

Fabrication of Mucoadhesive-Dendrimers as Solid Dosage Forms

Nidhi Raval, Rahul Maheshwari, Kiran Kalia, and Rakesh Kumar Tekade

Abstract

Mucoadhesion has a potential role in the delivery of pharmaceutical medicaments via various routes of administration, viz. oral, nasal, vaginal, and buccal. Mucoadhesion provides controlled drug delivery, sustained drug delivery, and local or site-specific drug delivery. This chapter focuses on the mechanism of bio-adhesion to glycoprotein layer of mucosal membrane. Some of the gastric mucoadhesive solid dosage forms of nanocarrier, viz. nanoparticle, microsphere, and nanofibers, undergo evaluation of mucoadhesive parameters. That includes mucoadhesive strength, tensile strength, swelling index, stability studies, in vivo study, etc. The oral route is the most desirable way among intravenous, subcutaneous, intramuscular, intranasal, intravaginal, etc. for drug delivery and because of patient compliance. One of the novel approaches is where nanocarrier is loaded in the solid dosage form for effective drug action and enhanced local delivery of a drug. Mainly this chapter explains about dendrimer-based oral solid dosage form (tablet) employing mucoadhesive polymers with an aim to improve retention time of drug at desired sites. Dendrimer-loaded mucoadhesive tablets promise controlled drug delivery with a gastro-retentive property, higher drug incorporation, ease of formulation development, and accessible absorption, owing to adjacent interaction with a biological membrane and prolonged retention to mucosa providing higher bioavailability of drugs.

Key words Mucoadhesive dosage form, Biological membrane, Controlled drug delivery, Dendrimer, Tablets, Gastro-retentive property

1 Introduction

Transformation of the medical research from the progression of a new chemical moiety to the establishment of a new drug carrier system of available drugs is a very cost effective and time-saving idea [1]. The development of innovative drug delivery tools may also maximize the effectiveness and clinical outcomes of existing biomolecules [2, 3]. Moreover, when we talk about innovative drug delivery systems, most of the efforts have been made to accurately target the drug molecules or genes to the particular sites in the body [4, 5]. The area of drug targeting has been explored extensively in the last few years, not limited to the localization of

Volkmar Weissig and Tamer Elbayoumi (eds.), *Pharmaceutical Nanotechnology: Basic Protocols*, Methods in Molecular Biology, vol. 2000, https://doi.org/10.1007/978-1-4939-9516-5_9,

medicines but also to improve the control over release pattern of drugs at the desired sites [6–8]. The selection of delivery route is also an important step, and as a fact most of the commercial products available exist as oral formulations and mostly as tablet dosage form [9, 10]. Tablet dosage form is the most popular route because it is easy to ingest and easy to swallow by adults, and because of pain avoidance and patient compliance [11, 12]. Tablets comprise biologically active substances, often drugs and a suitable blend of excipients, mostly in powder form, compressed into an appropriate shape [13]. The excipients are meant to achieve efficient tableting and include a broad range of classes such as diluents, binders or granulating agents, glidants and lubricants disintegrants, sweeteners or flavors, and pigments. A polymeric capping is also sometimes used to impart different characteristics as per the requirement of formulators such as to sustain or modify the drug release, to give the responsive properties (pH-responsive), to alter the shelf life, or to enhance the tablet's appearance [14–16]. The orally administered dosage form comprises oral fast-dispersing tablets, film-coated tablets, and sugar-coated tablets, buccal and sublingual tablets, dispersible tablets, controlled and sustained drug-releasing tablets, vaginal tablets, plus enteric coated tablets [17–23].

From the last few years, the mucoadhesive drug delivery system had gained considerable attention. This book chapter was aimed to provide the overview of the modified mucoadhesive dosage form in that mucoadhesive polymer was conjugated with the dendrimer, their factors affecting mucoadhesion, evaluation of conjugated polymeric dosage form, and a drug discharge from the formulation. Here dendrimer is used as a carrier for active ingredients and improves the loading of the drug. Mucoadhesive polymers exist in the form of chemically synthesized or bio-originated polymers/ biopolymers which are attached to the mucosal film capping the mucosal epithelial membrane and particles comprising a critical portion of epithelial mucus [24].

For decades, nano-medicines have gained a lot more attention [25–27]. A mucoadhesive nanocarrier via oral route is a novel strategy of the drug delivery. Mucoadhesion usually proceeds via the following mechanisms namely (1) intimate contact between a membrane mucoadhesive and a membrane (wetting or swelling phenomenon) and (2) penetration of the mucoadhesive into the tissue or into the surface of the mucous membrane (interpenetration) which is dictated in Fig. 1a [28]. Primary modes of adhesion with biological tissue include wetting or adsorption or electronic adhesion (Fig. 1b, c) or attachment of polymeric molecule. In addition to it, the theories that are involved in mucoadhesion are presented in brief in Fig. 2. This nanocarrier-based system had many advantages such as they tailored the cellular interaction via

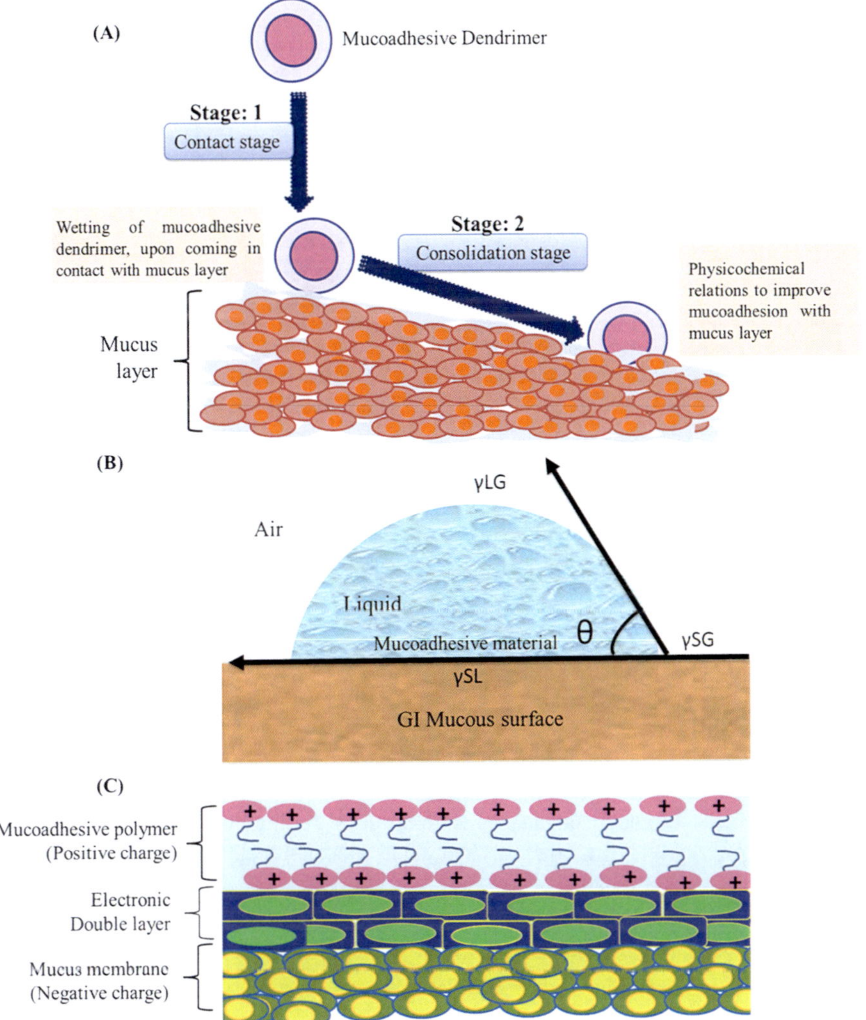

Fig. 1 (**a**) Pictorial representation of stages involved in mucoadhesion mechanism of mucoadhesive dendrimer with a mucoadhesive layer (stage 1: contact angle: designate contact between dosage form and mucosal membrane; stage 2: consolidation stage: make stronger and lengthen mucoadhesion with the help of physicochemical interaction). (**b**) An electric double layer between a mucoadhesive polymer and mucus membrane from an electrical theory of mucoadhesion. (**c**) Contact angle between mucoadhesive system and mucosal surface at the interface

tailoring the formulation, dual-drug delivery for synergistic action, etc. [29–31].

Chitosan-containing nanocarriers were also explored by a researcher for gastric drug delivery in H. pylori treatment. These provide complete safety of the drug when it reaches to the acidic environment via adhering to the mucous membrane, reside for a longer time, and also allow the drug to enter the infection site via diffusion process [32]. Bhalekar et al. had developed thiolated

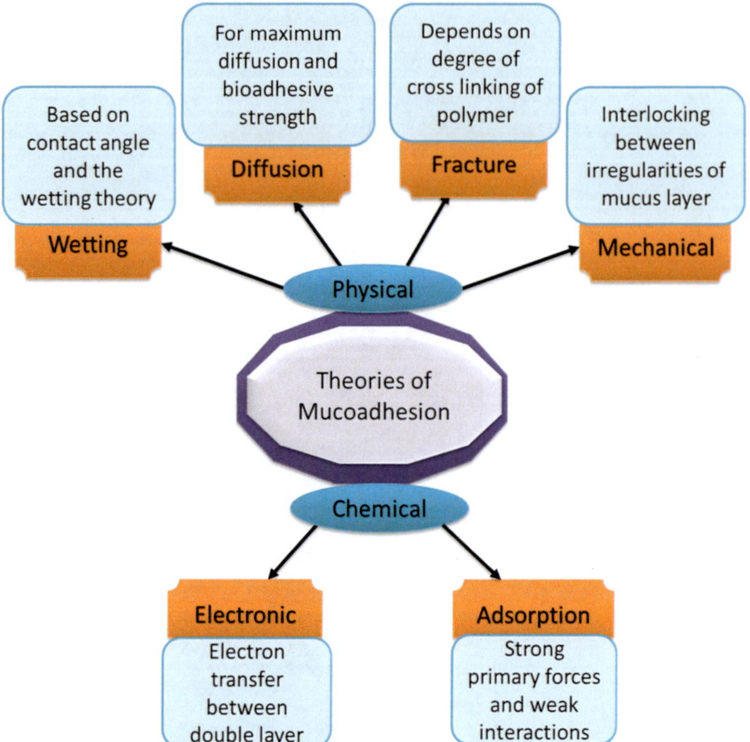

Fig. 2 Theories of mucoadhesion. Physical methods and chemical approaches. Physical contains (1) wetting method, (2) diffusion theory, (3) fracture theory, (4) and mechanical theory and chemical methods involve (1) electronic theory and (2) adsorption theory

xyloglucan-containing mucoadhesive gastro-retentive tablets. It provides great mucoadhesion, controlled drug release, as well as retention of the pharmaceutical carrier up to 7 h with high area under the curve (AUC) in rabbits as shown in Fig. 3 [33].

Modified dendrimer, e.g., PAMAM dendrimer, can be used as a potential carrier for the mucoadhesive drug delivery. Anionic surfaced dendrimer was quickly interacting with the cationic mucin layer [34]. Advantages of dendrimers include that they have nanoscopic molecule dimension run from 1 to 100 nm, which makes them less prone to reticuloendothelial system (RES) take-up. Because of accurate control of amide amalgamation, they have brought down polydispersity file.

Those dendritic polymeric carriers have also proved as solubility enhancers of drugs based on their generation, size, pH, charge, and temperature sensitivity. Some chemical interactions like ionic interaction, H-bonding, and hydrophobic interface interaction are promising phenomena for satisfactory solubility through dendrimer [35]. Dendrimer has static micellar-like properties. Researchers had successfully compared dendrimers with cyclodextrin regarding

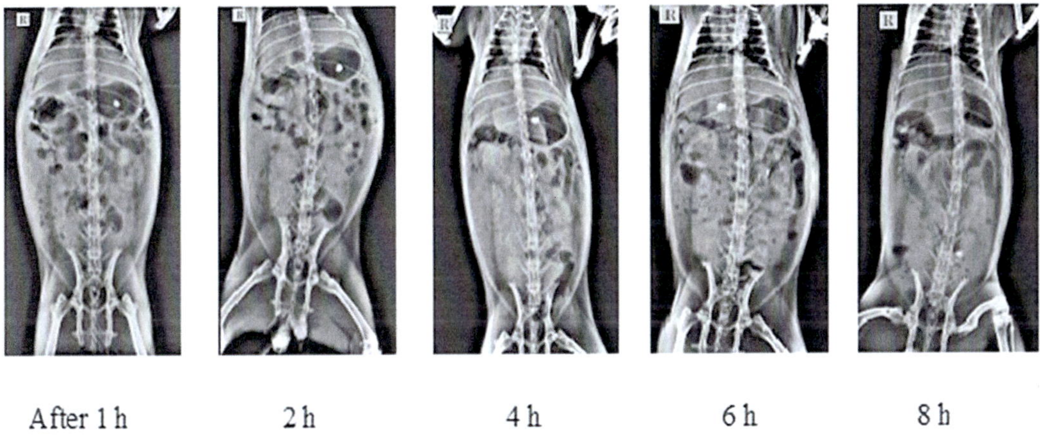

Fig. 3 X-ray imaging of the thiolated xyloglucan gastro-retentive system. Adapted from [33] without any changes

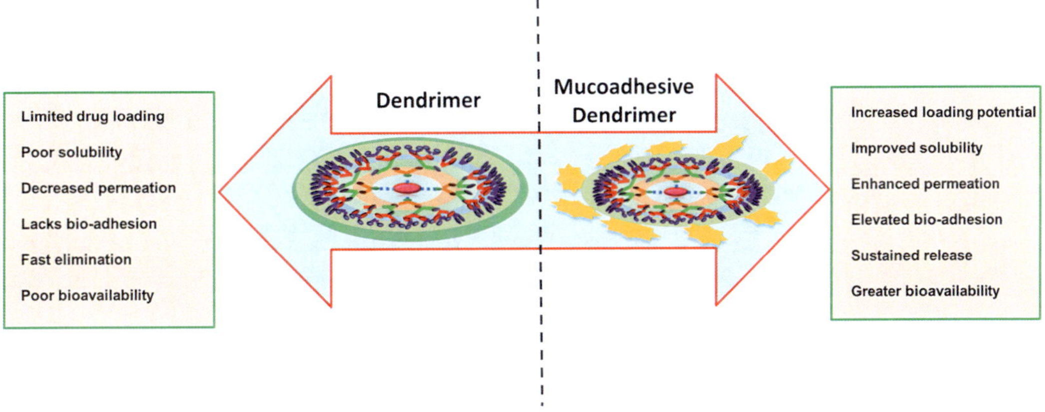

Fig. 4 Advantages of mucoadhesive polymer-encrusted dendrimeric system over dendrimer-loaded carrier system

solubility enhancement of hydrophobic substances [36]. Despite numerous benefits dendrimers contain inherent toxicity. In addition dendrimers have highly positive charge which easily interacts with negatively charged cell membrane in vivo. The marks in nanoholes are due to disruption of the phospholipid bilayer, membrane weakening, and erosion. It results in cytotoxicity, hemolytic toxicity, and hematological toxicity. Surface modification is one of the approaches to reduce the toxicity of cationic surface group via neutralization of charges [37]. It is achieved by PEGylation [38], acetylation [39], carbohydrate [40], and peptide conjugation [41, 42]. Furthermore, advantages of polymer-encompassed dendrimeric system are as shown in Fig. 4.

In this context, our following section describes in detail the procedure of preparation of fifth-generation polypropylene imine dendrimer (5G), chitosan-encompassed dendrimers, albendazole-comprised tablets of mucoadhesive chitosan dendrimers [43], and various characterization processes of prepared tablet formulation.

2 Materials

1. Ethylenediamine (EDA) solution (1.0 mol): Take 6.0212 g of EDA in 100 ml volumetric flask. Add distilled water to make up the volume of solution.
2. EDA solution (5.0 mol) in acetonitrile (ACN) solution (2.5 molar times per NH_2 of EDA: Take 30.106 g EDA and transfer to the volumetric flask) (100 ml). Add ACN to make up the volume (100 ml).
3. Treated Raney nickel (catalyst): Add 160 g sodium hydroxide (NaOH) in 100 ml deionized water in Erlenmeyer flask. Stir the solution for 30 min using magnetic stirrer at 800 rpm. Cool the solution using ice bath. Add Raney nickel-aluminum alloy powder (125 g) in small parts with 25–30-min interval to the cooled solution. Stir the solution for 60 min at 300 rpm.
4. Chitosan solution: Dissolve 10 mM chitosan in 0.5%v/v of glacial acetic acid.
5. *Tert*-butoxycarbonyl (*t*-BOC) (12 mM): 0.261 g *t*-BOC weighed utilized in the above chitosan solution.
6. Carbonyl-di-imidazole (CDI) (10 mM) solution: Add 1.621 g of CDI in 100 ml of deionized water.
7. Formic acid.
8. Albendazole.
9. Microcrystalline cellulose.
10. Starch.
11. Guar gum.
12. Talc.
13. Magnesium stearate.
14. Electronic balance machine (0.1 mg sensitivity).
15. Roche friabilator.
16. Simulated gastric fluid (SGF), pH 1.2: Add 2 g of NaCl and 3.2 g of pepsin derived from stomach mucosa in 7 ml concentrated HCl in a 1000 ml volumetric flask. Dropwise add deionized/distilled water up to 1000 ml. Adjust the pH using pH meter to 1.199.

17. Simulated intestinal fluid (SIF), pH 6.8: Dissolve dipotassium phosphate (KH_2PO_4) (68.05 g) and NaOH (8.96 g) in deionized water to make the volume10 l.
18. Goat intestinal mucosa.
19. 0.9% w/v Saline solution: Dissolve 9 g NaCl in 700 ml deionized water. Make the volume up to 1000 ml in a volumetric flask.
20. Starch solution (10% w/v): Take 1 g of starch in 250 ml beaker. Add 100 ml of deionized water stirrer for 5 min.

3 Methods

3.1 Half-Generation Dendrimer Synthesis

1. Add ACN (5.0 mol; 2.5 molar times per NH_2 of EDA) to a solution of 1.0 mol EDA in water (*see* **Notes 1–3**).
2. Observe rise in the temperature (*see* **Note 4**).
3. Reflux the reaction mixture at 80 °C for 1 h to complete the addition reaction.
4. Remove excess ACN using vacuum distillation (*see* **Note 5**).
5. Collect the half-generation dendrimers obtained as white crystalline solid powder (*see* **Note 6**).

3.2 Fifth-Generation Dendrimer Synthesis

1. Take 5 g of Raney nickel (catalyst) which is pretreated with NaOH (*see* **Note 7**) in a flask.
2. Add 5 ml of water and above-prepared half-generation dendrimer (EDA-dendrimer-$(CN)_4$) in 20 ml of methanol in a flask containing Raney nickel.
3. Hydrogenate the mixture for 1 h at 70 ± 4 °C, 40 atm hydrogen pressure.
4. Cool the reaction mixture and evaporate solvent from reaction mixture via vacuum evaporator (*see* **Notes 8** and **9**).
5. Illustrative representation of mentioned dendrimer is given in the following section (*see* **Note 10**).

3.3 Synthesis of Chitosan-Secured Polypropylene Imine (PPI) Dendrimer

1. Take 10 mM chitosan and dissolve in 0.5% glacial acetic aqueous acid solution with the help of magnetic stirrer (800 rpm for 30 min).
2. Add 12 mM *t*-BOC and stir (600 rpm) for 48 h (*see* **Note 11**).
3. Confirm the product obtained using copper sulfate test (*see* **Note 12**).
4. Add 10 mM CDI to the above reaction mixture of chitosan (*see* **Note 13**).
5. Stir for 1 h at 30 °C on a magnetic stirrer at 200 rpm.

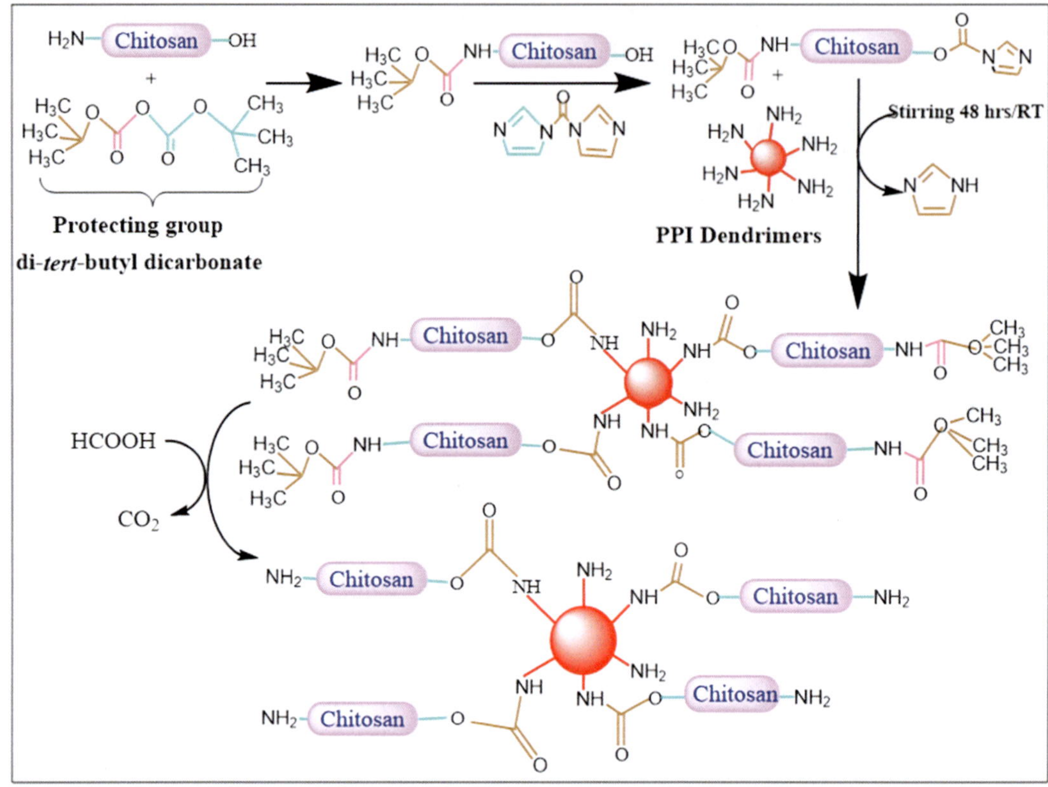

Fig. 5 Synthesis of the chitosan-encrusted PPI dendritic architect as modified mucoadhesive polymer. Adapted from [43]

6. Take 1 mM of PPI dendrimer and solubilize in 10 ml DMSO solvent.
7. Add 25 mM *t*-BOC-chitosan solution to previously prepared PPI dendrimer DMSO solution.
8. Kept in the dark while stirring for 48 h at room temperature.
9. The obtained product is confirmed by copper sulfate test.
10. Dissolve the confirmed product slowly in 2 ml formic acid with continuous stirring condition (*see* **Note 14**).
11. Purify the final product concentrate (*see* **Notes 15** and **16**).
12. Detailed stepwise representation of chitosan-secured polymeric PPI dendrimeric system is given in Fig. 5.

3.4 Fabrication of Mucoadhesive-Dendrimer-Comprised Solid Dosage Forms

1. Mix 50 mg of albendazole and 50 mg mucoadhesive chitosan-dendrimer (Subheading 3.3) and 219.5 mg microcrystalline cellulose (as diluent).
2. Add 5 ml of 10% starch solution.
3. Add 35 mg of Guar gum (10–20% w/w) and mix.

4. Pass through 18 mesh sieve followed by 14 mesh sieve.
5. Add 10% w/v starch solution until soft wet mass is formed.
6. Pass the wet mass through 20 mesh sieve.
7. Collect the granules and dry at 50 °C temperature for 1 h.
8. Pass the dried granules through 18 mesh sieve and lubricate with talc (7 mg) and magnesium stearate (3.5 mg) (*see* **Note 17**).
9. Take lubricated granule for tablet preparation at punch force 4500–5500 kg on a multistation tableting machine (*see* **Note 18**).
10. Punch the tablet to obtain the final weight (350 mg).
11. A diagramatic view of the method is depicted in Fig 8 (*see* **Note 19**).

3.5 Drug Entrapment in Modified Polymeric Dendrimers

1. Take 1:10 ratio of albendazole:mucoadhesive dendrimer.
2. Dissolve in phosphate buffer solution (PBS; pH 7.4) and stir on magnetic stirrer to remove additional solvent from the formulation (*see* **Note 20**).
3. Dialyze the formulations against PBS pH 7.4 to remove unentrapped or free drug (*see* **Notes 21** and **22**).

3.6 Friability, Weight Variation

1. Weigh 20 tablets on a random basis to initiate weight variation test (*see* **Note 23**).
2. Note the individual weight of all tablets and calculate average weight (*see* **Note 24**).
3. Calculate the percentage of deviation of its weight from the average weight determined for each tablet.
4. Match the deviation of each tablet from average tablet and report (whether pass or fail) (*see* **Note 25**).
5. Weigh and calculate the total weight and average weight of the 20 tablets to initiate friability test.
6. Place all the tablets in friabilator.
7. Rotate the drum for 100 rotations and remove tablets.
8. Calculate the loss of weight from the initial weight and report (*see* **Note 26**).

3.7 In Vitro Release Kinetics

1. Set U.S. Pharmacopoeia (USP) dissolution apparatus for 50 rpm and at 37 ± 0.5°C (*see* **Note 27**).
2. Place the tablet formulation at the bottom of the beaker.
3. Add 900 ml dissolution media (simulated gastric fluid (SGF), pH 1.2) initially for 2 h and then change with simulated intestinal fluid (SIF), pH 6.8.

4. Take an aliquot of 5 ml after predetermined time intervals (30 min, 1 h, 2 h, 4 h, 6 h, 8 h, 24 h) from the media and add the same amount of fresh media every time (*see* **Note 28**).
5. Analyze the aliquots for drug concentration using a suitable spectroscopic technique or chromatographic techniques.

3.8 Determination of Mucoadhesive Strength of Tablets

1. Take fresh goat intestinal mucosa (1 cm diameter) and fit the upper side of the cleaned glass vial.
2. Fit tablet (sample) at lower glass vial.
3. Fix the height of vial (*see* **Note 29**).
4. Apply force in incremental order starting from 0.5 kg weight.
5. Note the specific weight which detaches the tablet from a mucous membrane.
6. Calculate the force of adhesion (*see* **Note 30**).

3.9 Swelling Index

1. Weigh six tablets and place in a beaker containing 100 ml of 0.1 N hydrochloric acid (HCl) (*see* **Note 31**).
2. Remove the tablets from beaker at intervals of 1, 2, 4, 6, 8, and 12 h and weigh.
3. Calculate the swelling index (*see* **Note 32**).

3.10 Stability Studies

1. Place six tablets in the amber-color vial (4 set).
2. Place another six tablets in a transparent glass vial (4 set).
3. Kept the individual vials at different temperatures 5 ± 0.5 °C and 25 ± 2 °C for 3 months and 45 ± 2 °C and $75 \pm 5\%$ relative humidity (RH) for 15 days in a stability chamber.
4. Check samples for stability, crystallinity, color change consistency, etc.

3.11 Hemolytic Toxicity Study

1. Collect the human blood and centrifuge at $4200 \times g$ for 10 min.
2. Separate red blood cells (RBC) from blood (*see* **Note 33**).
3. Wash RBC with 0.9% w/v saline solution and again centrifuge at the same condition.
4. Collect 5 ml pellet of RBC and dilute with 100 ml 0.9% saline solution to get 5% RBC suspension.
5. Take 1 ml of RBC suspension (5% v/v) and add 5 ml deionized water for 100% hemolysis, 5 ml of 0.9% saline solution (control), and 0.5 ml of diluted previously prepared dendrimer solution with 4.5 ml 0.9% saline solution and 1 ml RBC suspension (sample to analyze), individually.
6. Incubate the suspension for 2 h at 37 ± 2 °C.
7. Centrifuge all test samples and collect supernatant.

8. Determine the absorbance at 540 nm of supernatant through UV-visible spectrophotometry and calculate the percentage of hemolysis (*see* **Note 34**).

4 Notes

1. Dendrimer synthesis is done through a divergent method that involves Michael addition reaction followed by amidation reaction to form esters by reacting with ethylene diamine.
2. Another method recognizes a convergent synthesis which starts with the peripheral dendritic units attached to additional building blocks to shape the branching architects, with dendrons developing along these lines from the surface toward the core focal point. General preparation method of dendrimers is depicted here in Fig. 6.
3. In this approach, EDA was used as dendrimer core.
4. Due to the exothermic reaction temperature might rise.
5. In vacuum distillation pressure is maintained at 16 mbar and bath temperature 40 °C to remove excess ACN as water.
6. After removing excess ACN crystalline powder remained as half generation of dendrimer.
7. Full generation is prepared from half-generation dendrimer through catalytic hydrogenation in a lab with the help of catalytic hydrogenator.
8. After evaporation of solvent first-generation dendrimer [EDA-dendrimer-$(NH_2)_4$] was prepared.

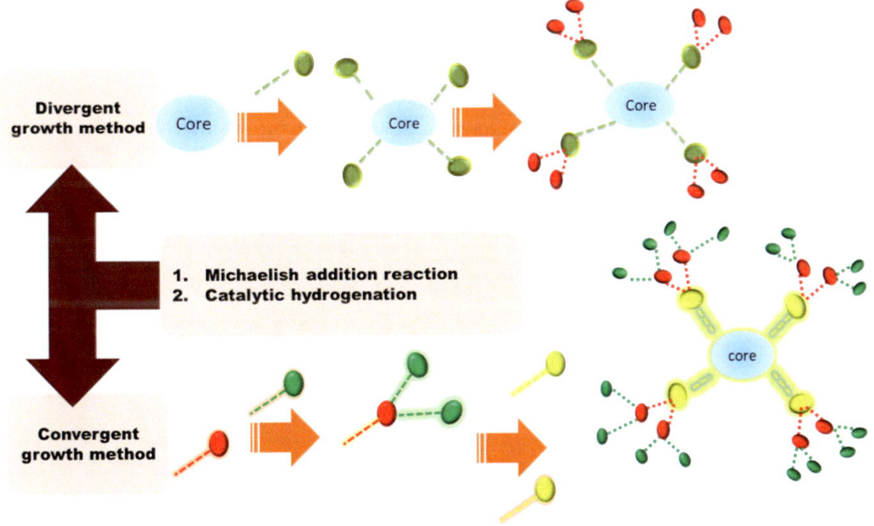

Fig. 6 Synthesis method for dendrimers; divergent growth method and convergent extension method

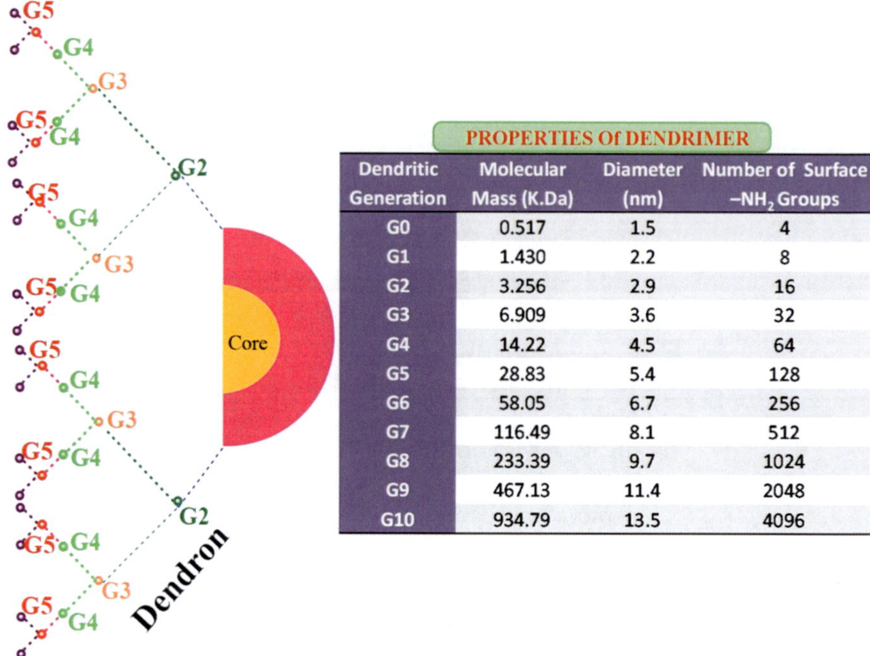

PROPERTIES Of DENDRIMER

Dendritic Generation	Molecular Mass (K.Da)	Diameter (nm)	Number of Surface $-NH_2$ Groups
G0	0.517	1.5	4
G1	1.430	2.2	8
G2	3.256	2.9	16
G3	6.909	3.6	32
G4	14.22	4.5	64
G5	28.83	5.4	128
G6	58.05	6.7	256
G7	116.49	8.1	512
G8	233.39	9.7	1024
G9	467.13	11.4	2048
G10	934.79	13.5	4096

Fig. 7 Diagrammatic representation of dendrimeric architect with generation along with several molecular mass and size

9. Prepared product is characterized by infrared (FTIR) and 1H NMR spectroscopy as well as transmission electron microscopy for confirmation of final product.
10. Representation of prepared dendrimeric structure is as shown in Fig. 7.
11. The amino group of chitosan should be protected with protection group such as *t*-BOC to prevent undesired reaction of the amino group. And only hydroxyl group is available for the response.
12. Copper sulfate test is for the confirmation of free amino group.
13. The second step includes preparation of intermediate substance, which includes converting of the free hydroxyl group of chitosan into the imidazole carbamate intermediates.
14. Continuous stirring is maintained until CO_2 bubble appears after 30 s.
15. Purification of the reaction mixture is done to remove impurities of chitosan by cellulose dialysis membrane (MWCO 12 kDa) against ultrapure water.
16. Powder product is obtained through lyophilization of the final purified product.
17. Take talc and magnesium stearate in 2:1 w/v ratio.

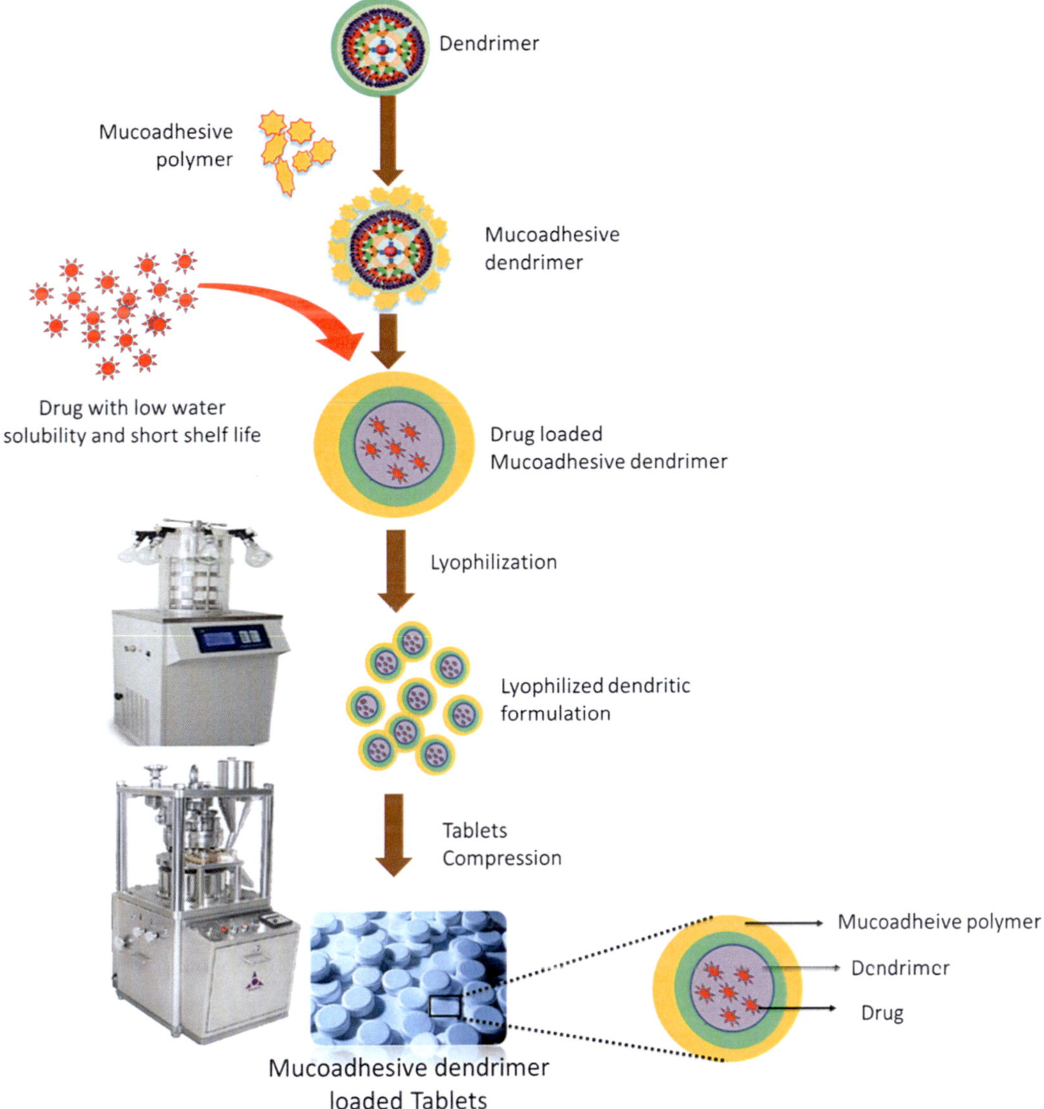

Fig. 8 A schematic procedure of drug-loaded mucoadhesive dendrimer compression in a tablet. Adapted from [43]

18. 6 mm, 246 round, flat, and plain punches are used for tablet preparation.
19. Detailed process of tablet preparation from mucoadhesive chitosan–PPI dendrimer is shown in Fig. 8.
20. Drug entrapped is detected by equilibrium dialysis method. That unentrapped drug concentration is analyzed through spectroscopic technique at a particular wavelength:

$$\%\text{Entappment} = X - \frac{X_1 + X_2}{X} \times 100 \quad (1)$$

Here, X = drug taken (mg), X_1 = drug remaining in a dialysis membrane (mg), and X_2 = final concentration of drug when dialysis bag was washed (mg).

21. Left the assembly for 2 h and then take the absorbance for the estimation of free drug.
22. The specific diameter of a tablet is necessary for oral administration. It is measured by Vernier caliper for both tablet thickness and width of a tablet.
23. Weight variation is performed using electronic balance.
24. Reference table for weight variation calculations:

The average weight of tablets	Deviation (%)	No. of tablets
Less than 80 mg	±10.0 ±20.0	Min. 18 Max. 2
80 to 250 mg	±7.5 ±15.0	Min. 18 Max. 2
More than 250 mg	±5.0 ±10.0	Min. 18 Max. 2

25. Percentage of friability of the tablets of a batch can be found by using the following formula: Percentage friability = $W_1 - W_2 / W_1 \times 100$, where W_1 = weight of tablets before testing, and W_2 = weight of tablets after testing.
26. In vitro release of the drug is performed through dissolution USP apparatus.
27. Conservation of the sink condition during the release profile is necessary to maintain the concentration of drug throughout the release experiments.
28. Set the height of the vial such that tablet is adhered to the mucosa of the upper vial.
29. The force of adhesion was subsequently calculated using the following formula:

$$\text{Force of adhesion } (N) = \frac{9.81 \times \text{mucoadhesion strength}}{1000} \quad (2)$$

$$\text{Bond strength } \left(N/m^2\right) = \frac{\text{Adhesion force } (N)}{\text{Surface area of tablet } (m^2)} \quad (3)$$

30. Swelling index shows the capacity of a polymer under different environment systems which allows the entanglement of the biological molecule inside the mucosal layer.
31. Swelling index is measured in terms of percentage weight gain of the dosage form:

$$\text{Swelling index (S.I.)} = (W_t - W_o)/W_o \quad (4)$$

where S.I. = swelling index, W_t = weight of tablet at time t, and W_o = tablet weight before placing in the beaker.

32. The hemolytic toxicity of dendrimers for their biomedical applications is measured using red blood cell (RBC) suspensions.
33. The following equation calculates the percent hemolysis:

$$\text{Hemolysis (\%)} = \text{Ab}_\text{T} - \text{Ab}_\text{S0}/\text{Ab}_\text{S100} - \text{Ab}_\text{S0} \times 100 \quad (5)$$

where Ab_T is the absorbance of the sample, Ab_S100 is the absorbance for 100%, and Ab_S0 is the absorbance for 0% hemolysis.

34. The percentage of hemolysis should not be more than 5%.

Acknowledgments

The author RKT would like to acknowledge Science and Engineering Research Board (Statutory Body Established Through an Act of Parliament: SERB Act 2008), Department of Science and Technology, Government of India, for the award of early carrier research grant (File Number: ECR/2016/001964) and DST-NPDF to Dr. Maheshwari (PDF/2016/003329) in Dr. Tekades's lab. Authors would also like to thank NIPER-Ahmedabad for providing research support for research on cancer and arthritis.

Reference

1. Tiwari G, Tiwari R, Sriwastawa B et al (2012) Drug delivery systems: an updated review. Int J Pharm Investig 2(1):2
2. Maheshwari R, Tekade M, Sharma PA et al (2015) Nanocarriers assisted siRNA gene therapy for the management of cardiovascular disorders. Curr Pharm Des 21(30):4427–4440
3. Sharma PA, Maheshwari R, Tekade M et al (2015) Nanomaterial based approaches for the diagnosis and therapy of cardiovascular diseases. Curr Pharm Des 21(30):4465–4478
4. Tekade RK, Maheshwari R, Soni N et al (2017) Nanotechnology for the development of nanomedicine. In: Nanotechnology-based approaches for targeting and delivery of drugs and genes, p. 1
5. Tekade RK, Maheshwari R, Soni N et al (2017) Carbon nanotubes in targeting and delivery of drugs. In: Nanotechnology-based approaches for targeting and delivery of drugs and genes, p. 389
6. Lalu L, Tambe V, Pradhan D et al (2017) Novel nanosystems for the treatment of ocular inflammation: current paradigms and future research directions. J Control Release 268:19–39
7. Maheshwari RG, Tekade RK, Sharma PA et al (2012) Ethosomes and ultradeformable liposomes for transdermal delivery of clotrimazole: a comparative assessment. Saudi Pharm J 20 (2):161–170
8. Maheshwari RG, Thakur S, Singhal S et al (2015) Chitosan encrusted nonionic surfactant based vesicular formulation for topical administration of ofloxacin. Sci Adv Mater 7 (6):1163–1176
9. Boddupalli BM, Mohammed ZN, Nath RA et al (2010) Mucoadhesive drug delivery system: an overview. J Adv Pharm Technol Res 1 (4):381
10. Wang X, Li S, Shi Y et al (2014) The development of site-specific drug delivery nanocarriers

based on receptor mediation. J Control Release 193:139–153

11. Zaman M, Qureshi J, Ejaz H et al (2016) Oral controlled release drug delivery system and characterization of oral tablets; a review. Pak J Pharm Res 2(1):67–76
12. Bandari S, Mittapalli RK, Gannu R (2014) Orodispersible tablets: an overview. Asian J Pharma 2(1)
13. Rahul M, Piyoosh S, Tekade M et al (2017) Microsponge embedded tablet for sustained delivery of nifedipine. Pharm Nanotechnol
14. Fujihara K (2017) Pharmaceutical composition, Google Patents
15. Ngwuluka NC, Choonara YE, Kumar P et al (2015) A novel pH-responsive interpolyelectrolyte hydrogel complex for the oral delivery of levodopa. Part II: characterization and formulation of an IPEC-based tablet matrix. J Biomed Mater Res A 103(3):1085–1094
16. Dudhat SM, Kettler CN, Dave RH (2017) To study capping or lamination tendency of tablets through evaluation of powder rheological properties and tablet mechanical properties of directly compressible blends. AAPS PharmSciTech 18(4):1177–1189
17. M. P. Ratnaparkhi, G. Mohanta and L. Upadhyay (2015) Review on: fast dissolving tablet. J Pharm Res 5–12
18. Asif F, Sultana T, Sohel MD et al (2014) In vitro dissolution pattern of metronidazole film coated tablet in presence of fruit juice. Am J Pharm Sci 2(2):32–36
19. Sakata Y, Higuchi M (2014) Sugar-coated preparation and production method for the same, Google Patents
20. Meng-Lund E, Jacobsen J, Müllertz A et al (2016) Buccal absorption of diazepam is improved when administered in bioadhesive tablets—an in vivo study in conscious Göttingen mini-pigs. Int J Pharm 515(1):125–131
21. Abbas R, Palumbo D, Walters F et al (2016) Single-dose pharmacokinetic properties and relative bioavailability of a novel methylphenidate extended-release chewable tablet compared with immediate-release methylphenidate chewable tablet. Clin Ther 38(5):1151–1157
22. Mohammed KAB, Ibrahim HK, Ghorab MM (2016) Effervescent tablet formulation for enhanced patient compliance and the therapeutic effect of risperidone. Drug Deliv 23 (1):297–306
23. McConville C, Major I, Devlin B et al (2016) Development of a multi-layered vaginal tablet containing dapivirine, levonorgestrel and acyclovir for use as a multipurpose prevention technology. Eur J Pharm Biopharm 104:171–179
24. Takeuchi H, Thongborisute J, Matsui Y et al (2005) Novel mucoadhesion tests for polymers and polymer-coated particles to design optimal mucoadhesive drug delivery systems. Adv Drug Deliv Rev 57(11):1583–1594
25. Maheshwari R, Tekade M, Gondaliya P et al (2017) Recent advances in exosome-based nanovehicles as RNA interference therapeutic carriers. Nanomedicine (Lond) 12 (21):2653–2675
26. Soni N, Soni N, Pandey H et al (2016) Augmented delivery of gemcitabine in lung cancer cells exploring mannose anchored solid lipid nanoparticles. J Colloid Interface Sci 481:107–116
27. Tekade RK, Maheshwari R, Tekade M et al (2017) Chapter 8 - solid lipid nanoparticles for targeting and delivery of drugs and genes A2. In: Vijay M, Kesharwani P, Amin MCIM, Iyer A (eds) Nanotechnology-based approaches for targeting and delivery of drugs and genes. Academic Press, pp 256–286
28. Smart JD (2005) The basics and underlying mechanisms of mucoadhesion. Adv Drug Deliv Rev 57(11):1556–1568
29. Parayath NN, Nehoff H, Taurin S et al (2016) Prospects of nanocarriers for oral delivery of bioactives using targeting strategies. Curr Pharm Biotechnol 17(8):683–699
30. Sosnik A, das Neves J, Sarmento B (2014) Mucoadhesive polymers in the design of nano-drug delivery systems for administration by non-parenteral routes: a review. Prog Polym Sci 39(12):2030–2075
31. Raval N, Khunt D, Misra M (2018) Microemulsion based delivery of triamcinolone acetonide to posterior segment of eye using chitosan and butter oil as permeation enhancer: an in vitro and in vivo investigation. J Microencapsul (just-accepted):1–37
32. Gonçalves IC, Henriques PC, Seabra CL et al (2014) The potential utility of chitosan micro/nanoparticles in the treatment of gastric infection. Expert Rev Anti-Infect Ther 12 (8):981–992
33. Bhalekar MR, Bargaje RV, Upadhaya PG et al (2016) Formulation of mucoadhesive gastric retentive drug delivery using thiolated xyloglucan. Carbohydr Polym 136:537–542
34. Yandrapu SK, Kanujia P, Chalasani KB et al (2013) Development and optimization of thiolated dendrimer as a viable mucoadhesive excipient for the controlled drug delivery: an acyclovir model formulation. Nanomedicine 9 (4):514–522

35. Dwivedi N, Shah J, Mishra V et al (2016) Dendrimer-mediated approaches for the treatment of brain tumor. J Biomater Sci Polym Ed 27(7):557–580
36. Gupta U, Agashe HB, Asthana A et al (2006) Dendrimers: novel polymeric nanoarchitectures for solubility enhancement. Biomacromolecules 7(3):649–658
37. Tambe V, Thakkar S, Raval N et al (2017) Surface engineered Dendrimers in siRNA delivery and gene silencing. Curr Pharm Des 23(20):2952–2975
38. Liao H, Liu H, Li Y et al (2014) Antitumor efficacy of doxorubicin encapsulated within PEGylated poly (amidoamine) dendrimers. J Appl Polym Sci 131(11)
39. Yan C, Gu J, Lv Y et al (2017) Improved intestinal absorption of water-soluble drugs by acetylation of G2 PAMAM dendrimer nanocomplexes in rat. Drug Deliv Transl Res 3(7):408–415
40. Roy R, Shiao TC (2015) Glyconanosynthons as powerful scaffolds and building blocks for the rapid construction of multifaceted, dense and chiral dendrimers. Chem Soc Rev 44 (12):3924–3941
41. Duncan R, Izzo L (2005) Dendrimer biocompatibility and toxicity. Adv Drug Deliv Rev 57 (15):2215–2237
42. Jain K, Kesharwani P, Gupta U et al (2010) Dendrimer toxicity: let's meet the challenge. Int J Pharm 394(1):122–142
43. Mansuri S, Kesharwani P, Tekade RK et al (2016) Lyophilized mucoadhesive-dendrimer enclosed matrix tablet for extended oral delivery of albendazole. Eur J Pharm Biopharm 102:202–213

Chapter 10

In Situ Vaccination of Tumors Using Plant Viral Nanoparticles

Abner A. Murray, Mee Rie Sheen, Frank A. Veliz, Steven N. Fiering, and Nicole F. Steinmetz

Abstract

Viral nanoparticles are self-assembling units that are being developed and applied for a variety of applications. While most clinical uses involve animal viruses, a plant-derived virus, *cowpea mosaic virus* (CPMV) has been shown to have antitumor properties in mice when applied as in situ vaccine. Here we describe the production and characterization of CPMV and its use as in situ vaccines in the context of cancer. Subsequent analyses to obtain efficacy or mechanistic data are also detailed.

Key words *Cowpea mosaic virus* (CPMV), In situ vaccine, Immunotherapy, Melanoma, B16F10, Flow cytometry, Luminex multiplex, Cytokine

1 Introduction

CPMV is a 30 nm-sized icosahedral virus with $T = 3$ symmetry, which has been extensively used as a biomaterial for various applications in biotechnology and medicine. In the context of cancer therapeutics, we have recently demonstrated potent efficacy of CPMV as an in situ vaccination platform. When introduced into the tumor microenvironment, the CPMV-based in situ vaccine functions as an immune activator to prime an antitumor immune response; the immune response is systemic and generates immune memory thus protecting from outgrowth or recurrence of the disease. We have demonstrated efficacy in mouse models of melanoma, breast cancer, ovarian cancer, and colon cancer [1].

Here we describe the methods for obtaining CPMV and its use as in situ vaccine. It should be noted that in our previous work, eCPMV, an RNA-free version of CPMV was used, while here we describe the application of native CPMV for in situ vaccination of melanoma. CPMV particles are obtained through infection of

Volkmar Weissig and Tamer Elbayoumi (eds.), *Pharmaceutical Nanotechnology: Basic Protocols*, Methods in Molecular Biology, vol. 2000, https://doi.org/10.1007/978-1-4939-9516-5_10, © Springer Science+Business Media, LLC, part of Springer Nature 2019

Vigna unguiculata plants followed by extraction and purification from the infected leaf tissue.

We describe the application of CPMV as in situ vaccine in a mouse model of melanoma. While tumor burden is the primary read-out to determine efficacy, we also provide protocols allowing the characterization of the immune cell profiles and chemo/cytokines. Multicolor flow cytometric analysis is used to determine cellular populations and changes amongst those populations in the tumor microenvironment. In this situation, this can be achieved by creating single cell tumor suspensions, which are probed with fluorescently labeled antibodies targeted towards specific cell differentiating surface markers. In addition, the in situ vaccination-mediated immunological changes are facilitated through communication via cytokines and chemokines. The interaction can be quantified by isolating the protein mediators from all cellular components and quantifying their levels in a high-throughput approach using a Luminex assay.

In this chapter, we describe these steps in detail including propagation, purification, and characterization of CPMV, B16F10 tumor cell culture and dermal tumor establishment, in situ vaccination, and immunological analyses using the B16F10 model. It should be noted that these methods could be applied to other plant viruses and tumor models.

2 Materials

2.1 CPMV Propagation, Purification, and Characterization

1. *Vigna unguiculata* seeds (California Blackeye No. 5).
2. Pro Mix BX potting Soil.
3. Plant incubators, e.g. Geneva Scientific E-41L2 or Conviron A100.
4. Carborundum.
5. Avanti J-E centrifuge with JLA 10.500 rotor and JLA 16.25 rotor.
6. Optima L-90K ultracentrifuge with 50.2 Ti rotor and SW 32 Ti rotor.
7. Tabletop centrifuge.
8. Sucrose.
9. Chloroform.
10. 1-Butanol.
11. Sodium chloride (NaCl).
12. PEG [8000 MW].
13. Potassium phosphate dibasic.

14. Potassium phosphate monobasic.
15. Miracloth or cheesecloth.
16. Blender.
17. Spectrophotometer.
18. AKTA Explorer 100 chromatograph with Superose6 column.
19. NuPAGE SDS sample buffer (4×).
20. NuPAGE 4–12% Bis-Tris gel.
21. NuPAGE MOPS SDS running buffer (1×).
22. Novex SeeBlue Plus2 pre-stained protein standard.
23. Safestain.
24. 2% (w/v) uranyl acetate in water.
25. Carbon-coated TEM grids.
26. TEM, e.g. FEI Tecnai F30 300 kV transmission electron microscope.

2.2 B10F10 Tumor Cell Culture and Dermal Tumor Establishment

1. B16F10 mouse melanoma cells.
2. Complete RPMI 1640 medium: RPMI 1640 medium supplemented with 10% (v/v) fetal bovine serum, 2 mM L-glutamine, 1 mM sodium pyruvate, 0.1 mM MEM nonessential amino acid, and 1% (v/v) Penicillin/Streptomycin, and 0.05 mM 2-Mercaptoethanol.
3. Phosphate-buffered saline.
4. Trypsin-EDTA (0.05%), phenol red.
5. Dual-chamber PolyPro bath.
6. Biological safety cabinet.
7. CO_2 incubator.
8. C57BL/6 mice (Jackson Labs, Bar Harbor, ME).
9. BD Lo-Dose™ U-100 insulin syringes.
10. Isoflurane.

2.3 In Situ Vaccination

1. BD Lo-Dose™ U-100 insulin syringes.

2.4 Tumor Homogenation for Cytokine Analysis

1. T-PER™ tissue protein extraction reagent.
2. Complete™ protease inhibitor cocktail.
3. HBSS, no calcium, no magnesium, no phenol red.
4. PBS.
5. Handheld homogenizer.
6. Laboratory balance.

7. Cryotube.
8. Clear 6-well plates—untreated; 6-well plate.
9. Clear polystyrene 96-well plates—untreated; well: V-shaped.

2.5 Flow Cytometry

1. 40 μm cell strainer
2. RPMI.
3. Ethanol.
4. Trypan blue.
5. Na_2-EDTA.
6. Potassium bicarbonate.
7. Ammonium chloride.
8. Purified anti-mouse CD16/32 Antibody to block Fc.
9. Pacific Blue™ anti-mouse CD45 antibody.
10. Pacific Blue™ Rat IgG2b, κ isotype ctrl antibody.
11. PE anti-mouse/human CD44 antibody.
12. PE Rat IgG2b, κ isotype ctrl antibody.
13. APC/Cy7 anti-mouse CD3ε antibody.
14. APC/Cy7 Armenian hamster IgG isotype ctrl antibody.
15. PE/Cy7 anti-mouse CD62L antibody.
16. PE/Cy7 Rat IgG2a, κ isotype ctrl antibody.
17. APC anti-mouse CD8a antibody.
18. APC Rat IgG2a, κ isotype ctrl antibody.
19. FITC anti-mouse CD4 antibody.
20. FITC Rat IgG2b, κ isotype ctrl antibody.
21. FITC anti-mouse/human CD11b antibody.
22. PE anti-mouse CD80 antibody.
23. PE Armenian hamster IgG isotype ctrl antibody.
24. PE/Cy7 anti-mouse CD86 antibody.
25. APC anti-mouse I-A/I-E antibody.
26. APC/Cy7 anti-mouse Ly-6G antibody.
27. APC/Cy7 Rat IgG2a, κ isotype ctrl antibody.
28. UltraComp eBeads.
29. Dead cell marker: fluorophore-conjugated viability dyes, propidium iodide, or 7-aminoactinomycin D.
30. LSR II flow cytometer or similar 8+ color flow cytometer.

3 Methods

3.1 CPMV Propagation, Purification, and Characterization

3.1.1 Plant Growth

1. Fill a plant tray with 3¾″ square pots, filling each pot with Pro-Mix BX Biofungicide + Mychorrhizae soil.
2. Place 3–4 California Blackeye No. 5 cowpea seeds into each pot, approximately 1½–2″ apart.
3. Water each plant pot enough to keep the soil moist (*see* **Note 1**).
4. Place plant trays into an incubator or plant room, providing ~15 h of sunlight with 25 °C with 50% humidity. For the night cycle, maintain the same temperature and humidity, with lights off.
5. Water every 2–3 days for approximately 10 days.

3.1.2 Plant Infection with CPMV

1. To work with or propagate CPMV or any other plant virus, USDA-approved protocols and facilities need to be established.
2. Prepare a 0.1 mg/mL CPMV in 0.1 M potassium phosphate (KP) buffer pH 7.0.
3. After approximately 10 days, when the seedlings have grown and the trifoliates leaves are starting to grow dust the primary leaves lightly with carborundum (*see* **Note 2**).
4. Pipette onto each leaf ~50 μL of 0.1 mg/mL CPMV in 0.1 M KP buffer pH 7.0 and gently spread the droplet over the leaf. Gently rubbing the leaves in combination with the carborundum dust will create lesions in the leaf tissue allowing CPMV to enter and start its replication process. Repeat for all primary leaves in the plant tray.
5. Continue watering the infected cowpea plants until the infection is established; at least an additional 10 days. The infection will be detectable based on the typical mosaic symptoms in the primary and trifoliate leaves.

3.1.3 Harvest and Storage of Infected Cowpea Leaves

1. Once the typical mosaic patterns are detectable on the leaves, collect the leaves and place in a Ziploc bag. Discard the pots with stems into biohazard waste.
2. Weigh the bag with leaves and note the date and weight. Once a bag reaches 100 g of leaves, begin collecting leaves in a new Ziploc bag.
3. Infected cowpea leaves should be stored at −80 °C (*see* **Note 3**).

3.1.4 Purification

1. By hand, pulverize the leaves by squeezing the bag (*see* **Note 4**).
2. Homogenize the pulverized leaves in a blender with approximately 3× volumes of ice-cold 0.1 M KP buffer pH 7.0. Filter homogenate through 2–3 layers of miracloth or cheesecloth into an autoclaved, sterile beaker. To improve filtration and

recovery, carefully squeeze the miracloth or cheesecloth to force the filtered homogenate through. Discard miracloth or cheesecloth into biohazard waste (*see* **Note 5**).

3. Centrifuge the filtered plant homogenate using an Avanti J-E Centrifuge and JLA 10.500 rotor at 18,000 × *g* for 20 min at 4 °C. Collect the supernatant into autoclaved, sterile beaker.
4. Add a stir bar to the plant sap and place the beaker into an ice bath sitting on a stir plate in a fume hood. Into the beaker, add 0.7 volumes of 1:1 (v/v) chloroform:1-butanol. Stir the mixture for 20–30 min, avoiding the formation of bubbles from turbulent mixing (*see* **Note 6**).
5. Centrifuge the mixture using an Avanti J-E Centrifuge and a JLA 10.500 rotor at 6600 × *g* for 10 min at 4 °C. Remove the centrifuge bottles carefully as to not mix the separated aqueous and organic phases. Collect the upper aqueous phase using a 20 mL syringe and transfer the aqueous phase into an autoclaved, sterile beaker (*see* **Note** 7). Discard organic waste in liquid chemical waste.
6. Add NaCl to the aqueous phase to give a final molarity of 0.2 M NaCl. In addition, add 8% (w/v) PEG 8000 to the solution. Mix the solution in the beaker with NaCl and PEG 8000 in an ice bath using a stir plate for a minimum of 30 min and store the beaker at 4 °C for at least 2 h (*see* **Note 8**).
7. Place solution in an autoclaved, sterile 250 mL centrifuge bottle and centrifuge the solution in a JLA 16.25 rotor at 30,000 × *g* for 15 min at 4 °C. Discard supernatant and resuspend pellet(s) with 10 mM KP buffer pH 7.0 by pipetting up down repeatedly (*see* **Note 9**).
8. Centrifuge resuspended pellet in a JLA 16.25 rotor at 13,500 × *g* for 15 min at 4 °C and collect supernatant.
9. Purify the sample over a 10–40% sucrose gradient using ultra clear tubes for a SW32 rotor. Run SW32 rotor at 133,000 × *g* for 3 h at 4 °C. In a dark room, shine a light through the tube to visualize the CPMV bands. Remove the light scattering bands using a pipette and place in ultracentrifuge tube (Part number 337901, Beckman—polycarbonate tubes with cap assembly).
10. Fill the remaining volume of the tube with 0.1 M KP, if needed. Centrifuge in a Type 50.2 Ti ultracentrifuge rotor at 210,000 × *g* for 3 h at 4 °C. Discard supernatant and resuspend pellet with 1 mL 0.1 M KP. Recover resuspended pellet into a sterile 1.5 or 2 mL microcentrifuge tube.
11. Clearing spin: Centrifuge resuspended pellet in a tabletop centrifuge at 10,000 × *g* for 10 min. Recover supernatant in a new, sterile microcentrifuge tube.

3.1.5 Characterization by UV-Vis Absorbance

1. Determine concentration of purified CPMV using UV-Vis absorbance (e.g. using a Nanodrop instrument). Measure the absorbance (A) of the CPMV solution at 260 nm (RNA) and 280 nm (protein). The ratio of A_{260}/A_{280} is a good indication of purity and should be as close to 1.8 as possible (*see* **Notes 10** and **11**).
2. Using the Beer-Lambert law ($A = \varepsilon cl$), the concentration can be determined. A is absorbance at 260 nm, ε is the extinction coefficient of CPMV (8.1 $mL{\cdot}cm^{-1}{\cdot}mg^{-1}$), c is concentration (mg/mL), and l is path length (cm).

3.1.6 Characterization by Size Exclusion Chromatography (SEC)

1. Prepare 200 μL of 0.5 mg/mL CPMV in 0.1 M KP pH 7.0.
2. Inject CPMV solution into an AKTA explorer FPLC system using a Superose 6 10/300 GL size exclusion column (GE Lifesciences).
3. Run the AKTA explorer FPLC system at 0.5 mL/min, setting the absorbance readings at 260 nm and 280 nm. Intact CPMV elutes in the 10–15 mL fraction on a Superose6 column (*see* **Note 12**).

3.1.7 Characterization by Sodium Dodecyl Sulfate Polyacrylamide Gel Electrophoresis (SDS PAGE)

1. Prepare samples by mixing: 2 μL of 3 mg/mL CPMV, 7 μL of 0.1 M KP pH 7.0, and 3 μL of NuPage LDS sample buffer (4×). Denature the proteins by boiling for 5–7 min at 100 °C.
2. Prepare a NuPAGE 4–12% Bis-Tris gel in a gel tank. For this particular gel, fill the inner chamber completely with 1× MOPS and fill a small volume of the 1× MOPS on the outer chamber.
3. To serve as a molecular weight reference, load 10 μL of the Novex SeeBlue Plus2 pre-stained protein standard into a well.
4. Next to the well containing the protein standard, load the CPMV samples along with any known control sample in the remaining wells.
5. Run the gel electrophoresis for ~45 min at 200 V.
6. Once gel run is complete, remove the gel from the plastic casing and stain the gel using Safetstain according to the manufacturer's instructions (*see* **Note 13**).

3.1.8 Characterization by Transmission Electron Microscopy

1. Dilute CPMV sample to a concentration of 0.1 mg/mL in DI water (*see* **Note 14**).
2. On a piece of parafilm, place a 20 μL droplet of each of the following: CPMV dilution, water, water, 2% (w/v) uranyl acetate, and water.
3. Place TEM grid on CPMV dilution droplet for 2 min. Dry the TEM grid by carefully wicking away the liquid on filter paper.

4. Very briefly, wash the TEM grid by placing the TEM grid on the two water droplets and dry by wicking with filter paper.
5. Place TEM grid on the 2% (w/v) uranyl acetate droplet for 2 min.
6. Rinse the TEM grid by briefly placing on the last water droplet. Store TEM grid in grid holder until analysis.
7. Image the TEM grids using a FEI Tecnai F30 300 kV transmission electron microscope.

3.2 B16F10 Tumor Cell Culture and Dermal Tumor Establishment

1. Acquire approval from the institutional animal care and use committee (IACUC) prior to initiating any studies involving animals. The procedures described here were approved by IACUC at Case Western Reserve University.
2. Thaw cryopreserved B16F10 cells from liquid nitrogen tank, wash with PBS or complete culture medium to remove DMSO, and transfer cells into an appropriate culture vessel in relation to the number of cells thawed (for one million cells use a T175 tissue culture flask containing 20 mL of complete RPMI medium).
3. Place a culture flask with cells into 37 °C gassed (5% CO_2) cell incubator and culture cells, changing the media every 2–3 days.
4. Maintain cells until confluent (usually takes 4–5 days when cells seeded at 1/10 ratio).
5. Once cells form a confluent monolayer, split, and expand as follows: every 4–5 days, remove old media and rinse cells with 10 mL of PBS. Discard PBS, then add 3 mL of 0.25% Trypsin/2.21 mM EDTA, and place the flask in the incubator at 37 °C for 2–5 min to allow for cell dissociation. Observe cells under the inverted microscope to confirm that cells are released. Then immediately add fresh complete RPMI medium containing FBS to stop the trypsin and collect the dissociated cells from the flask (*see* **Note 15**). Gently mix cell suspension by pipetting the solution up and down, place cells in 50 mL tube, spin the cells down by centrifugation at 500 × *g* for 5 min. Wash cells once with 10 mL of PBS or complete culture medium. Resuspend cells in complete culture medium, and then distribute the cells at the desired dilution into new tissue culture flask using complete RPMI for the dilution (*see* **Note 16**).
6. For tumor inoculation, harvest tumor cells when cells reach no more than 85% confluence to ensure good viability.
7. Prepare cells in plain RPMI without serum at a concentration of 1.25×10^5 cells in 30 μL per mouse, and aliquot cells in 1.5 mL tubes for easy loading of syringe and to avoid overmixing of cells while tumor challenging. Keep cells on ice during tumor cell inoculation and invert or pipet to mix prior to drawing into syringe.

8. After mice were anesthetized with isoflurane. Inject 30 μL of the cell suspension intradermally using 0.5 mL insulin syringes (*see* **Notes 17** and **18**). Intradermal tumor growth is autochthonous for melanoma and facilitates tumor observation. The injections must be done very slowly with excellent control to make sure the inoculation is within the dermis and tumor cells do not leak out. Practice is advised.
9. Monitor mice for development of tumors (*see* **Note 19**).

3.3 In Situ Vaccination

1. Tumor volume is measured using caliper and calculated using the following formula, $V = (\text{length} \times \text{width}^2)/2$, where V is tumor volume, W is tumor width (shorter dimension) and L is tumor length.
2. After 7–10 days, when tumors a volume of 40–80 mm^3 prepare for in situ vaccination. CPMV in PBS is injected intratumorally at a concentration of 100 μg CPMV in 20 μL. Control groups are treated with 20 μL of PBS. Again, patience and good control is required to make sure the treatment is fully incorporated into the tumor and does not leak out.
3. Monitor tumor growth at regular intervals. Euthanize mice once tumors reach 1000 mm^3 or according to IACUC-approved protocols.

3.4 Euthanasia

1. Euthanize mice according to IACUC-approved protocol.
2. CO_2 inhalation is described here. Briefly, place mice in a CO_2 box for 5 min with CO_2 infusion.
3. Turn off CO_2 and leave mice for another 5 min in the chamber.
4. Remove mice from chamber and conduct cervical dislocation to assure of euthanasia.

3.5 Tumor Homogenation for Cytokine Analysis

1. For follow-up immunological investigation, remove tumors and homogenize using the following protocol (*see* **Note 20**).
2. Extract tumor by cutting with surgical scissors along the base/ margin of the tumor making sure to remove connective tissue and overlying skin (*see* **Note 21**). Weigh and note tumor mass.
3. Dissolve one tablet of complete protease inhibitor in 8 mL of T-PER buffer at room temperature before use.
4. Add a collected tumor sample to tissue grinder or a well on a six well plate on ice.
5. Add 1 mL of the tissue extraction reagent (made in step three) to the well per 100 mg of tissue sample.
6. Homogenize the tissues on ice using a homogenizer (*see* **Note 22**).
7. Collect the tissue lysate into a 1.5 mL microcentrifuge tube. Maintain tubes on ice.

8. Add 0.5 mL of HBSS buffer to rinse tissue homogenizer and collect HBSS buffer to the microcentrifuge tube in the previous step.
9. Centrifuge the sample at 9000 × *g* for 10 min at 2–8 °C to pellet the tissue debris.
10. Collect the supernatant, taking care to avoid the fat layer floating on the top (*see* **Note 23**).
11. Aliquot the cleared lysate into clean microcentrifuge tubes noting the aliquot volume (*see* **Note 24**).
12. Between the samples, run tissue homogenizer sequentially in 70% ethanol (v/v) and 1× PBS to prevent contamination between the samples (*see* **Note 25**).
13. Measure protein concentration of the cleared lysate using BCA or Bradford assay.
14. To store samples, freeze aliquots at −80°C (*see* **Note 26**).

3.6 Luminex

1. Transfer 30 μL of the supernatant to 96-well plate (*see* **Note 27**). Include blank tissue extraction reagent as a negative control.
2. Take plate to core providing Luminex services. A typical starting place is the Mouse Cytokine/Chemokine 32plex panel. Panels can also be customized for individual needs.
3. Luminex results will be in pg/mL and should be analyzed taking the total protein concentration of the tumor as determined by the BCA assay.

3.7 Preparation of Single Cell Suspensions for Flow Cytometric Analysis

1. Euthanize and remove tumor as described above.
2. Excise established intradermal tumor mass, remove connective tissue and fat around tumor mass, and cut the tumor into smaller pieces using surgical scissors.
3. Place pieces of tumor mass into a 40 μm cell strainer and add 1 mL of HBSS or tissue culture media such as complete RPMI medium to wet the tumor sample.
4. Using the plunger end of a syringe (*see* **Note 28**), mash the tumor mass through the cell strainer into the 50 mL falcon tube.
5. Rinse strainer with 5 mL of HBSS or complete culture medium (*see* **Note 29**) and then discard the strainer.
6. Spin cells down by centrifugation at 200 × *g* for 5 min at 4 °C.
7. Discard supernatant and resuspend pellet of tumor cells in 5 mL of ACK lysis buffer allowing for red blood cell (RBC) lysis.
8. Incubate for 5 min on ice with occasional shaking.
9. Stop the reaction by diluting the ACK lysis buffer with 10–20 mL of 1× PBS and spin cells down at 200 × *g* for 5 min.

10. Repeat RBC lysis procedure until all RBCs are completely lysed and the supernatant is clear (*see* **Note 30**).
11. Resuspend pellet in 1× PBS and volume for use in the next step of experimental procedure.
12. Count cells using an automatic cell counter or hemocytometer and aliquot 1 × 10^6 cells per well for flow cytometry analysis (*see* **Note 31**).
13. To limit counting time, dilute10 μL cell suspension in 90 μL of trypan blue and incubate for 5 min. Load 10 μL of the sample into the hemocytometer for counting.

$$\text{Total cells/mL} = \left(\frac{\sum \text{Cells in quadrants}}{\text{Number of quadrants counted}}\right) \times 10{,}000 \times \text{Dilution factor}$$

3.8 Flow Cytometry

1. Aliquot 1 × 10^6 cells for each condition (*see* **Notes 32** and **33**).
2. Wash cells with 500 μL of ice-cold 1× PBS and centrifuge at 400–600 × *g* for 5 min at 4 °C.
3. Discard supernatant and disperse the cell pellet by tapping and resuspending in 50 μL of FACS staining buffer (*see* **Note 34**) or 1× PBS.
4. Prepare controls including the unstained control, single color controls, isotype controls, compensation controls (using beads), and fluorescence minus one (FMO) controls.
5. For Fc receptor blocking, pre-incubate the cells with purified anti-CD16/CD32 antibody (≤ 1.0 μg per 10^6 cells) for 15–30 min on ice and in the dark prior to immunostaining.
6. Wash cells with 500 μL of ice-cold 1× PBS and centrifuge at 400–600 × *g* for 5 min at 4 °C.
7. Discard supernatant and disperse the cell pellet by tapping and resuspending in 50 μL of chilled PBS.
8. Prepare the stain solutions with recommended titer of each antibody (*see* **Note 35**). We have included a list of a useful panel we have used in the past, but it is just one option as an example.
9. Add antibodies to cells and incubate with slow rock at 4 °C and in the dark for 30 min (*see* **Note 36**).
10. Wash cells with twice 500 μL of ice-cold 1× PBS and centrifuge at 400–600 × *g* for 5 min at 4 °C.
11. Discard supernatant and disperse the cell pellet by tapping and resuspending in 300 μL of FACS buffer or ice-cold 1× PBS (*see* **Note 37**).

12. If needed, fix cells after staining by adding cell fixation buffer (*see* **Note 38**) and incubate for 10–15 min at room temperature.
13. Analyze stained cells using a multicolor flow cytometer. Number of colors needed depend on number of receptor targets. Here we used the LSR II Flow Cytometer.
14. Analyze cell populations with preferred flow cytometry analysis software. Here we used FlowJo®. Cell population filtration and exclusion strategies are dependent on receptors being probed.

4 Notes

1. Avoid waterlogging as this can stunt the growth of the plants.
2. After mechanical inoculation keep the plants in the dark for several hours, then wash the leaves with tap water to remove carborundum and avoid "burning" of the leaves, prior to placing the plants into a plant incubator.
3. Leaves stored in −80 °C can be stored indefinitely.
4. Pulverizing the leaves by hand is most efficient when the leaves are frozen.
5. Filtering through cheesecloth may indicate a false positive of LPS if using the *Limulus* Amebocyte Lysate (LAL) assay. This is due to cheesecloth being made of cotton, a cellulose-based material.
6. Due to the dangerous fumes given off by chloroform, this step should be performed in a fume hood.
7. Extract as much aqueous phase as possible near the interface between aqueous and organic phases as this is where the majority of CPMV particles may rest following centrifugation. If organic phase is extracted, simply wait for the two phases to separate while in the syringe, and discard the organic phase once it has settled.
8. Alternatively, this step can be done inside of a cold room.
9. Additional resuspension can take place overnight at 4 °C on a nutator if needed.
10. A_{260}/A_{280} ratio that is lower than 1.8 may indicate that there is less RNA packaged within the CPMV particles. A_{260}/A_{280} ratio that is higher than 1.8 may indicate the presence of contaminating proteins. A deviating A_{260}/A_{280} ratio may also indicate presence of solvent or plant material contaminants.
11. Protein concentration of the purified CPMV sample can also be determined by Bradford Assay or Lowry Assay.

12. A single absorbance peak for the 260 nm and 280 nm settings should form at an elution volume around 10–15 mL. Additional peaks at a lower elution volume may indicate aggregates and peaks at larger elution volumes, such as around 20 mL, may indicate broken particles.
13. Due to the denaturing step, the CPMV sample should show two distinct bands. The "L" capsid protein should show a band at approximately 42 kDa, and the "S" capsid protein should show a band at approximately 24 kDa.
14. Diluting CPMV in a buffer, such as phosphate buffer, will create precipitates on the TEM grid.
15. 6 mLs of fresh DMEM are usually added to yield at ~10 mLs of cells. This allows for easier splitting to the desired ratio.
16. A 1:10 split is recommended for cell maintenance. Cells should be monitored daily and are typically split every 4–5 days when grown in a T175 flask.
17. Flank of mice should be shaved at least 1 day prior to allow for injections of tumor cells.
18. To assure uniform tumor growth, gently mix the cell suspension prior to loading of the syringe, and by gently tapping the syringe in between injections.
19. Tumors are usually visible and palpable about 8 days post inoculation.
20. Homogenation of tumors should be done soon after harvesting to prevent cytokine degradation. Freezing of tissues prior to homogenation is not recommended.
21. Tumor is encapsulated in fascia, which should not be punctured.
22. 5 min of homogenation on a medium setting is sufficient.
23. If there is fat contamination, centrifuge again at 9000 × g for 10 min at 2–8 °C to pellet the remaining tissue debris.
24. Total protein concentration of the lysate can be determined using a bicinchoninic acid assay (BCA assay), Bradford assay, or by measuring protein at A280. All cytokine levels should be normalized to total protein levels.
25. Depending on the number of samples, these solutions will need to be replaced with fresh solutions.
26. Avoid multiple freeze thaw cycles.
27. Consult with core to determine actual volumes required. Depending on concentration ranges, samples might need to be diluted.
28. Diameter of plunger should be less than diameter of the strainer. Five cc syringes work well for 40 μm cell strainer.

29. Make sure to collect all cells; cells may be stuck on the sidewalls of the strainer.
30. For dermal tumor, 1–2 times is usually enough.
31. Higher cells/well might be needed if a rare population is being studied.
32. Cells can be aliquoted in a variety of tubes or plates as long as an appropriate centrifuge is available. This is also dependent on the capabilities of the flow cytometer as to which tubes it can handle. For multiple samples, untreated v-bottom or u-bottom 96-well plates are preferred.
33. Make sure not to have any clumps. If cell clumps are visible, filter the cells through a 40 μm cell strainer.
34. Buffer made with 1× PBS, 1% (v/v) BSA (fraction IV, protease free) or 5–10% (v/v) FBS and 0.01% (w/v) sodium azide.
35. Dilutions should be made in PBS or FACS staining buffer.
36. Propidium Iodide (PI) (0.1–10 μg/mL) or 7-Aminoactinomycin D) (1 mg/mL) can be added here for live/dead staining and exclusion.
37. Make sure not to have any clumps. If cell clumps are visible, filter the cells through a 40 μm cell strainer.
38. Chilled PBS supplemented with 1–2% (v/v) paraformaldehyde, 1% BSA (w/v) (fraction IV, protease free) or 3% (v/v) FBS and 0.01% (w/v) sodium azide. Aim for final paraformaldehyde concentration between 0.5% and 1% (v/v).

Acknowledgements

This work was funded in part by a CWRU CAHH award and NIH U01-CA218292 (to N.F.S.). A.A.M. was supported in part by NIH grants T32 GM007250 and TL1 TR000441.

Reference

1. Lizotte PH, Wen AM, Sheen MR, Fields J, Rojanasopondist P, Steinmetz NF, Fiering S (2016) In situ vaccination with cowpea mosaic virus nanoparticles suppresses metastatic cancer. Nat Nanotechnol 11:295–303

Chapter 11

Bioconjugation in Drug Delivery: Practical Perspectives and Future Perceptions

Perihan Elzahhar, Ahmed S. F. Belal, Fatema Elamrawy, Nada A. Helal, and Mohamed Ismail Nounou

Abstract

For the past three decades, pharmaceutical research has been mainly converging to novel carrier systems and nanoparticulate colloidal technologies for drug delivery, such as nanoparticles, nanospheres, vesicular systems, liposomes, or nanocapsules to impart novel functions and targeting abilities. Such technologies opened the gate towards more sophisticated and effective multi-acting platform(s) which can offer site-targeting, imaging, and treatment using a single multifunctional system. Unfortunately, such technologies faced major intrinsic hurdles including high cost, low stability profile, short shelf-life, and poor reproducibility across and within production batches leading to harsh bench-to-bedside transformation.

Currently, pharmaceutical industry along with academic research is investing heavily in bioconjugate structures as an appealing and advantageous alternative to nanoparticulate delivery systems with all its flexible benefits when it comes to custom design and tailor grafting along with avoiding most of its shortcomings. Bioconjugation is a ubiquitous technique that finds a multitude of applications in different branches of life sciences, including drug and gene delivery applications, biological assays, imaging, and biosensing.

Bioconjugation is simple, easy, and generally a one-step drug (active pharmaceutical ingredient) conjugation, using various smart biocompatible, bioreducible, or biodegradable linkers, to targeting agents, PEG layer, or another drug. In this chapter, the different types of bioconjugates, the techniques used throughout the course of their synthesis and characterization, as well as the well-established synthetic approaches used for their formulation are presented. In addition, some exemplary representatives are outlined with greater emphasis on the practical tips and tricks of the most prominent techniques such as click chemistry, carbodiimide coupling, and avidin–biotin system.

Key words Bioconjugation, Antibody–drug conjugates (ADCs), Antibody-radionuclide conjugates (ARCs), Polymeric, Peptide-based, Peptidomemetic, Nanoparticles-based conjugates, Dialysis, Crystallization, Rotary evaporation, Freeze-drying (lyophilization), Thin-layer chromatography (TLC), Liquid column chromatography (LC), High-performance liquid size exclusion chromatography (HPLC-SEC), Nuclear magnetic resonance (NMR), Matrix-assisted laser desorption ionization-time of flight mass spectrometry (MALDI-TOF MS), Infrared (IR) spectroscopy, Click chemistry, Nucleophilic substitutions, Carbodiimide crosslinkers, EDC/Sulfo-NHS crosslinker, DCC carbodiimide crosslinker, (Strept)Avidin–biotin system

Volkmar Weissig and Tamer Elbayoumi (eds.), *Pharmaceutical Nanotechnology: Basic Protocols*, Methods in Molecular Biology, vol. 2000, https://doi.org/10.1007/978-1-4939-9516-5_11,

1 Introduction and Basics

Bioconjugation can be defined as the methodology of connecting two molecules through a stable covalent bond. At least one of the two molecules can be a biomolecule or a biomolecule recognition element (Fig. 1). The nature of the other molecule depends on the purpose of bioconjugation. With the ultimate goal of biosensing, the second molecule usually is a signal transducer. On the other hand, the second molecule can simply be a drug with the final goal of drug targeting and/or pharmacokinetic or dynamic modulation. The chemical nature of the whole conjugate or a part of it can be synthetic or semisynthetic or genetically encoded. The main aim of bioconjugation is to form a stable biologically cleavable covalent link between two molecules, at least one of which is a biomolecule [1]. Bioconjugation is a form of functionalization, which aims to increase stability, protect drug from proteolysis, or enhance the targeting properties of the delivery system [1, 2]. In spite of the historic fact that bioconjugates are older than nanoparticles, research is currently being diverted back to it [1]. This could be attributed to its ease of synthesis, high-scale-up yield, ease of bench-to-bedside transformation, ease of formulation, and final formulation stability [1]. As any delivery system, bioconjugates are usually tailor-designed to provide the function of interest. The active drug entity can be linked to a diagnostic agent; targeting moiety; pharmacokinetics-modifying agent, such as Polyethylene Glycol (PEG/PEGylation), bio-responsive or stimuli-sensitive

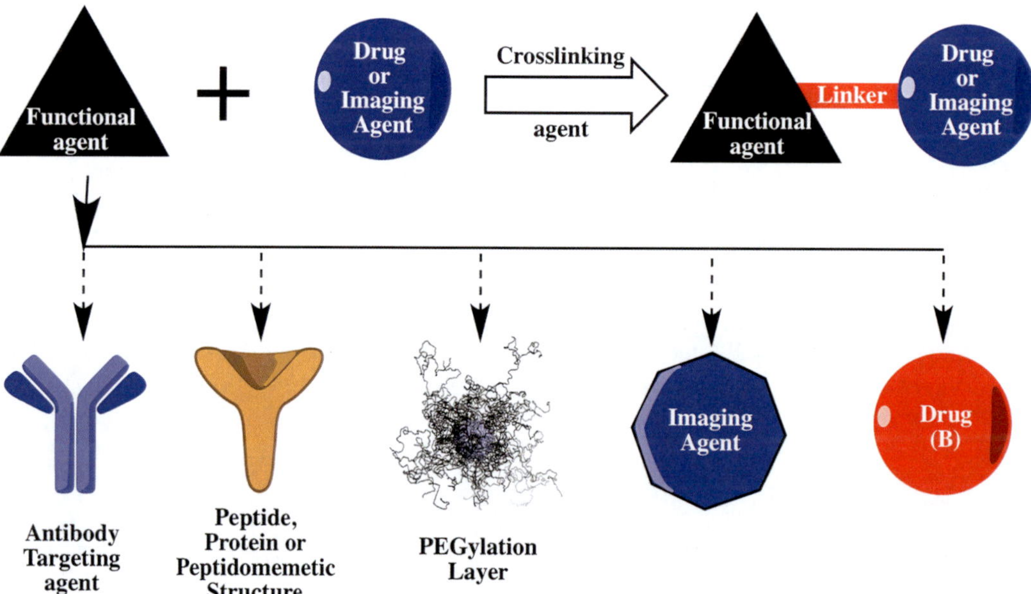

Fig. 1 Basic diagrammatic representation of a bioconjugates' design and structure

agent, an aptamer, or antibody. Furthermore, the choice of the proper linker can impart new functions and smart characteristics to the bioconjugate system (Fig. 1).

Bioconjugation reactions are generally categorized by the general reactivity or functional group that is involved in the associated conjugation process, such as amine reactions, thiol reactions, carboxylate reactions, hydroxyl reactions, aldehyde and ketone reactions, active hydrogen reactions, photochemical reactions, or cycloaddition reactions [1]. The design of a useful bioconjugate will depend mainly on its use, purpose, and the desired properties needed [3, 4]. Thus, one could choose a suitable molecule and suitable crosslinker to form personalized and custom-designed bioconjugates [3, 5, 6]. An important key to forming a successful bioconjugate is choosing the suitable crosslinker between the molecules [1].

In this chapter, we will focus more on the synthetic bioconjugation strategies. It is not our intent to exclusively address all the available bioconjugation synthetic techniques; we would rather focus on the most famous and practically feasible (common) techniques.

2 Types of Bioconjugates

Tracking the historical evolution of synthetic bioconjugation techniques, three main realms of applications are encountered; namely, antibody–drug conjugates (ADCs); polymeric, peptide-based, peptidomimetic, and nanoparticles-based conjugates, along with lesser extent, antibody-radionuclide conjugates (ARCs). Among the largest and most important application areas for bioconjugate techniques is the field of human therapeutics and diagnostics. The strategy in creating therapeutic bioconjugates for cancer is to design a final complex that has high specificity for the intended cells combined with high efficacy in killing the tumor being targeted, with no off-target effects or toxicity [1].

The conjugation of synthetic polymers with biological molecules has been a focus in pharmaceutical and medical research and in the development of biomaterials for many years. Polymer bioconjugates are being used in new applications such as biosensors, electronic nanodevices, biometrics, and artificial enzymes [7]. Bioconjugates have been implied for a variety of biological applications, including drug and gene delivery applications, biological assays, imaging, and biosensing [8].

Recombinant DNA and monoclonal antibody technology have created a biotech revolution that is providing a growing number of peptide, protein, and antibody-based drugs. Most of these proteins are limited in their clinical applications because of unexpectedly low therapeutic effects. The reason for this limitation is that these proteins are immediately decomposed by various proteases in vivo and are rapidly excreted from the blood circulation, leading to a short plasma half-life. Furthermore, they are limited by poor stability and, for proteins, immunogenicity. Consequently, frequent administration at an excessively high dose is required to reveal their therapeutic effects in vivo. As a result, homeostasis is destroyed and unexpected side effects occur. Therefore, there has been a continuing search for improved alternatives. Bioconjugation with water-soluble polymers improves the plasma clearance and body distribution, resulting in increased therapeutic effects and decreased side effects [9].

The limited clinical efficacy of anticancer drugs, whether used alone or in combination, could be attributed to the insufficient therapeutic window of these compounds [10–12]. Targeted therapies offer the potential to generate agents that will be selectively cytotoxic to tumor cells, coupled with lower toxicity to the host, resulting in a larger therapeutic index. The major areas of focus include monoclonal antibodies (mAbs) and their inclusion in antibody–drug conjugates (ADCs).

2.1 Antibody–Drug Conjugates (ADCs)

The first monoclonal antibody for the treatment of cancer, rituximab, was approved by the United States Food and Drug Administration (U.S. FDA) in 1997 for use in patients with relapsed or refractory, CD-20 positive, B-cell, low-grade or follicular non--Hodgkin's lymphoma. Rituximab is a chimeric antibody that binds to the CD20 antigen expressed on the surface of a majority of B-cell lymphomas [13].

Keep in mind, unconjugated mAbs possess modest antitumor efficacy as single agents; hence, combination therapy with the mAb and a chemotherapeutic drug is intuitively an effective strategy to achieve higher therapeutic efficacy. Taking the advantage of antibody specificity to biological targets of interest, the first generation of ADCs was developed to deliver anticancer chemotherapeutic drugs such as doxorubicin.

The first generation of ADCs was BR96-Dox, in which a chemotherapeutic agent, doxorubicin, is linked to the chimeric BR96 antibody through an acid-labile hydrazone bond (Fig. 2). The conjugate was advanced to a Phase II human clinical trial in metastatic breast cancer [13, 14]. In this randomized trial, patients were treated either with conjugate or free doxorubicin. The toxicity profile of the conjugate was markedly different from that of the unconjugated doxorubicin, suggesting that antibody-mediated delivery can indeed alter the biodistribution of the drug. Unexpectedly, despite the strong preclinical data wherein the conjugated doxorubicin was shown to be superior to free doxorubicin, the conjugate failed to demonstrate clinically meaningful therapeutic activity. Although BR96 was a chimeric antibody, the conjugate elicited an immune response in about 50% of the evaluable patients.

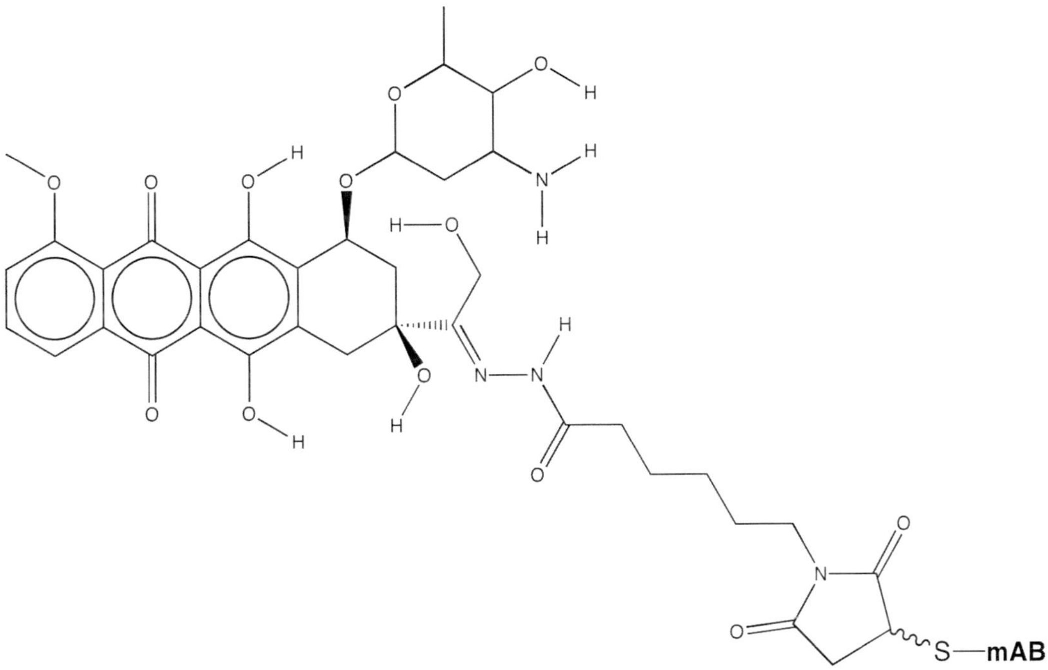

Fig. 2 Structure of BR96-Doxorubicin (first-generation antibody–drug conjugates)

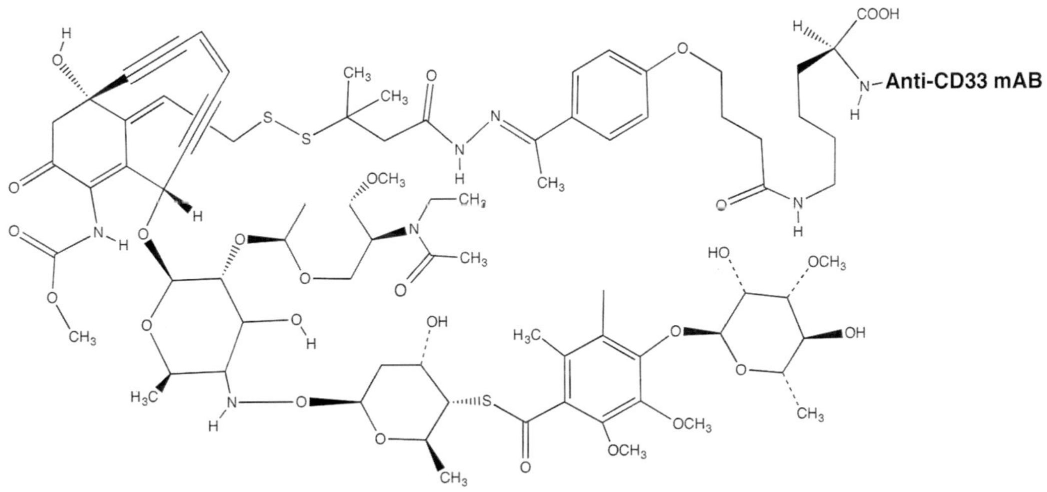

Fig. 3 Gemtuzumab ozogamicin (Mylotarg®) for CD33-positive acute myelogenous leukemia treatment (FDA-approved in 2000; withdrawn in 2010) [14, 15]

This issue has been addressed with the advancement of antibody engineering technology for the generation of humanized and fully human antibodies [13, 14].

Second-generation ADCs, depicted by Gemtuzumab ozogamacin (Mylotarg®, Fig. 3), are anti-CD33 mAbs conjugated to calicheamicin as the payload via an acid-labile hydrazone linker. It

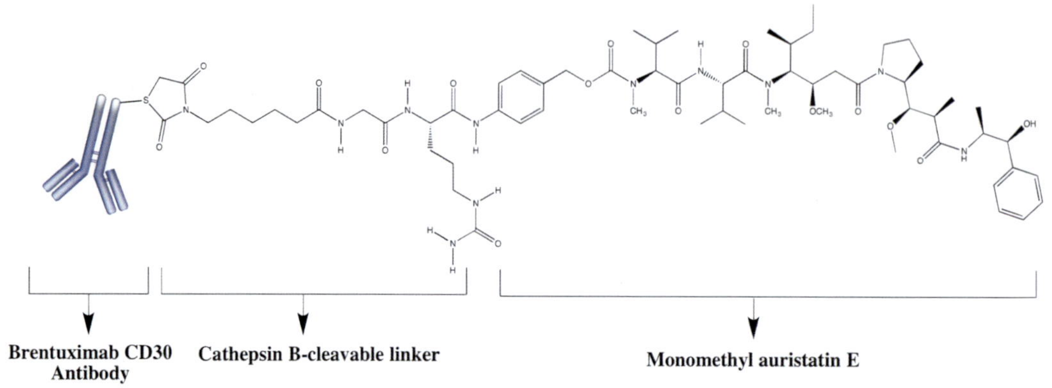

Brentuximab CD30 Antibody
Cathepsin B-cleavable linker
Monomethyl auristatin E

Fig. 4 Brentuximab vedotin (Adcetris®) for CD30-positive relapsed or refractory Hodgkin's lymphoma treatment (FDA-approved in 2011)

is considered a second-generation ADC; but, it was the first ADC drug to reach the market. It was given accelerated approval for treatment of acute myeloid leukemia (AML) during the first relapse of patients >60 years of age. However, it was voluntarily withdrawn from the market in 2010 due to relative therapeutic benefit concerns associated with hepatic veno-occlusive disease (VOD) and lack of sufficient activity [14, 15].

Another interesting example of a further generation ADC is brentuximab vedotin (Adcetris®, Fig. 4). Adcetris® consists of monomethyl auristatin E (MMAE), linked to the chimeric CD30 antibody cAC10 at cysteine residues by a valine–citrulline dipeptide linker. The valine–citrulline dipeptide linker contains a p-aminobenzylcarbamate (PABC) self-immolative spacer. The valine–citrulline dipeptide is designed to be cleaved in lysosomes (by cathepsin B), leading to self-immolation of the PABC moiety and release of MMAE. The unconjugated anti-CD30 antibody cAC10 was previously tested in the clinic, but did not show sufficient activity as a single agent to progress beyond phase I and II clinical trials (8% overall response rate). In contrast, the compelling clinical activity of its MMAE conjugate, brentuximab vedotin, led to accelerated approval by the FDA in 2011. It is indicated for use in Hodgkin's lymphoma and anaplastic large cell lymphoma (ALCL) [13].

Ado-trastuzumab emtansine (T-DM1, Kadcyla®, Fig. 5) is another example of ADCs. T-DM1 is composed of trastuzumab, a humanized monoclonal antibody targeting the oncogene HER2, linked to lysine residues with the maytansinoid DM1 by the noncleavable SMCC thioether linker. It received approval by the US Food and Drug Administration (FDA) in 2013 for treatment of HER2 metastatic breast cancer in patients who had previously received trastuzumab and a taxane [13].

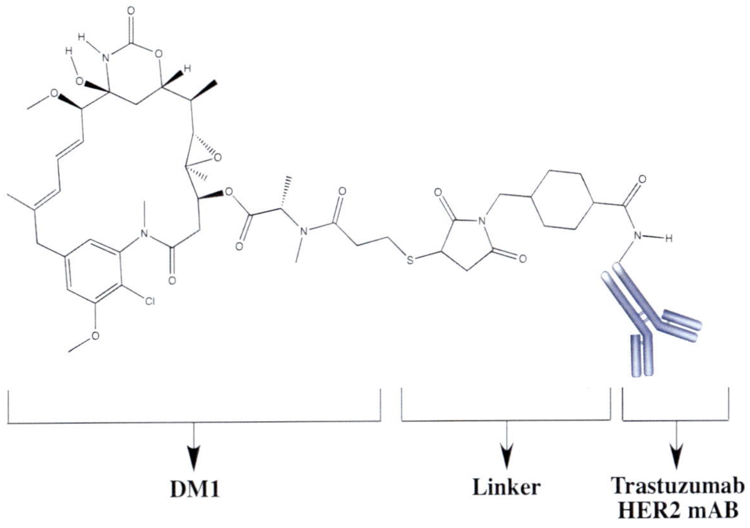

DM1 Linker Trastuzumab HER2 mAB

Fig. 5 Trastuzumab emtansine (Kadcyla®) for Her2-positive breast cancer treatment (FDA-approved in 2013)

A number of ADCs are in various stages of clinical evaluation. A majority of them employ microtubule-disrupting compounds (maytansinoids or auristatins) as the payload. The two approved ADCs, ado-trastuzumab emtansine and brentuximab vedotin, are undergoing additional clinical trials to broaden the treatment indications [13].

2.2 Antibody-Radionuclide Conjugates (ARCs)

There are only two FDA-approved ARCs, ibritumomab tiuxetan (Zevalin®) and tositumomab-I131 (Bexxar®). Both of these ARCs were approved in early 2000; and both are anti-CD20 mAbs conjugated to β-emitting radionuclides, ^{90}Y and ^{131}I, respectively. Although both Zevalin® and Bexxar® utilize a murine-derived antibody, there are no other successful Zevalin® and Bexxar® antibodies (e.g., humanized or human mAb backbone) or other ARCs with FDA approval, despite the evidence for greater clinical efficacy compared to the unconjugated mAb. One of the possible explanations is the challenge associated with handling and scalability of ARCs and the potential effects of radioactivity accumulation in normal cells [14].

2.3 Polymeric, Peptide-Based, Peptidomemetic, and Nanoparticles-Based Conjugates

The notion of linking a biologically active molecule through a chemically stable bond to a polymer has passed through different progressive phases starting from the fifties of the last century and continuing up till the moment. In the next few paragraphs, we are going to shed light on the design and chemistry of drug polymer conjugates, as well as their potential applications and current developmental status. Bioconjugation of biocompatible polymers to active pharmacological entity has been conceptualized by the

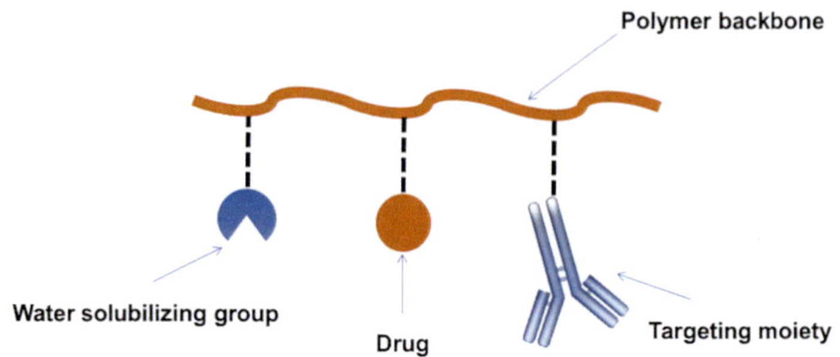

Fig. 6 Ringsdorf's model for drug-delivery systems based on synthetic polymers

introduction of Helmut Ringsdorf's model of drug delivery systems based on synthetic polymers (Fig. 6). This model comprises three components: an aqueous solubility enhancer, a drug or active pharmacological entity and a targeting moiety. It highlights the main purposes sought from including a drug in a polymer conjugate, which are improving water solubility, providing spatial and temporal control of delivery into the biological target, and altering drug pharmacodynamics and biodistribution [16–18].

Anticancer drugs are among the most famous examples of classes of drugs that attracted much attention in the polymer drug conjugate arena. This can be attributed to their poor water solubility, high toxicity, and lack of selectivity. Besides, solid tumors suffer from erratic lymphatic drainage and enhanced vascular permeability that lead to beneficial retention of drug–polymer conjugate, which enhances the therapeutic index significantly [19].

However, designing successful drug–polymer conjugates and its reflections on clinical practices suffers from a lot of challenges and caveats. Including a drug in a polymer conjugate makes it a New Chemical Entity (NCE), which necessitates thorough safety and efficacy studies to be conducted. Moreover, sometimes the stability of the conjugate becomes unpredictable; too early or too late drug release leads to drug toxicity or under-medication, respectively. In brief, various factors are to be included and attention to very fine details should be implemented to get a fruitful and custom-tailored design [20–22].

There is a lot of versatility in chemistries and architectures of the polymers used in drug conjugations. Poly (ethylene) Glycol (PEG) and *N*-(2-Hydroxypropyl)methacrylamide (HPMA) are among the most widely investigated biocompatible linear polymers. PEG is one of the simplest polymers and is available with a variety of functional groups that facilitate its conjugation (Fig. 7) [23].

PEG protects against both enzyme degradation and reticuloendothelial uptake, which enhances drugs pharmacodynamics and biodistribution [24, 25]. Numerous PEGylated proteins have been approved by the FDA for clinical practice, such as PEG-interferon α-2a (Pegasys) [26], PEG-interferon α-2b

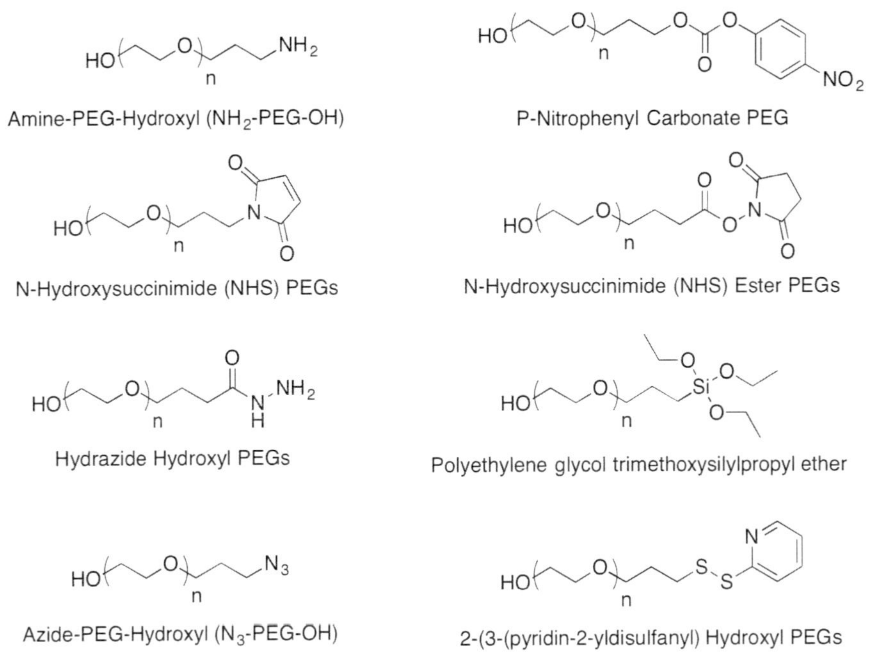

Amine-PEG-Hydroxyl (NH_2-PEG-OH)

P-Nitrophenyl Carbonate PEG

N-Hydroxysuccinimide (NHS) PEGs

N-Hydroxysuccinimide (NHS) Ester PEGs

Hydrazide Hydroxyl PEGs

Polyethylene glycol trimethoxysilylpropyl ether

Azide-PEG-Hydroxyl (N_3-PEG-OH)

2-(3-(pyridin-2-yldisulfanyl) Hydroxyl PEGs

Fig. 7 Examples of commercially available, functionalized PEGs

(PEG-Intron) [27], PEG-granulocyte colony-stimulating factor (Neulasta) [28], and PEG-growth hormone receptor antagonist (Somavert) [29]. However, their use is limited by their nonbiodegradability and slow renal filtration that might lead to their accumulation and, hence, toxicity. Furthermore, low drug-loading is another key limiting disadvantage in PEGylated proteins transition from bench to bedside [30, 31].

N-(2-Hydroxypropyl)methacrylamide (HPMA) copolymer is one of the most investigated and advanced polymers used in therapeutics. HPMC is a hydrophilic and biocompatible polymer. It is used extensively in the formulation of polymeric drug carriers [32–34]. Presence of α-carbon substitution and amide linkage ensures hydrolytic stability. Besides, the monomer exists in a crystalline form, as compared to the liquid hydroxyethyl methacrylate-type esters. The drug-loading is found to be high in HPMA, as compared to PEG, which is attributed to the presence of side reactive group (e.g., amine, carboxyl, or hydroxyl) for coupling. Reactive side groups also enabled the combination of drug, targeting moiety, and imaging agent simultaneously with relative synthetic ease. A variety of drugs included in HPMA conjugate have been reported in literature like taxanes [20], camptothecin [35], platinates [36], dexamethasone [37], gemcitabine [38], and geldanamycin [39].

Other soluble polymer conjugates designed as drug carriers include polyvinylpyrrolidone (PVP), alginate, chitosan, hyaluronic acid, poly(vinyl alcohol) (PVA), and polyionic complexes (PICs) [23].

One of the key evolutionary steps in drug–polymeric conjugate history is the development of dendrimers. Dendrimers are synthetic polymeric macromolecules with a branching tree-like structure. Their first appearance in the drug delivery arena was afforded by the work of Tomalia et al. [40, 41], who first described the synthesis of poly (amido amine) (PAMAM) dendrimers in 1985. They showed that stepwise addition of a branching, also known as generation, displayed a linear growth in size and an exponential growth in surface area with each successive "generation." Dendritic polymers have been successfully utilized as multifunctional nanocarriers, drug bearers [42], imaging agents [43], and/or targeting moieties [44]. Nonetheless, many concerns were raised over their biocompatibility and toxicity since they showed substantial affinities for metal ions, lipids, bile salts, proteins, and nucleic acids, resulting in the disruption of biological processes and toxicity. All of these factors imply that the biggest challenge in the area of dendrimer drug conjugate would be the design of biocompatible, safe, and cost-effective dendrimers with necessary surface modifications to enhance biocompatibility [45].

Another important milestone in the development and evolution of drug polymeric conjugates was the use of biodegradable polymers. Biodegradable polymers allowed the proper biodistribution of the drug conjugate via slow degradation rate and, at the same time, prevented the excessive accumulation of drugs; hence their long-term adverse effects [31]. A number of biologically degradable bonds have been described. Biodegradation generally occurs via hydrolysis, enzymatic cleavage, or reductive degradation. Biodegradable polymers have been described [46], which include poly(α-amino acids) such as poly(L-lysine) [47], poly(L-glutamic acid) [48], and poly ((*N*-hydroxyalkyl)glutamine) [49], as well as carbohydrate polymers such as dextrins [50], hydroxyethylstarch (HES) [51], polysialic acid [52], and the polyacetal Fleximer [53].

Another contemporary approach in drug polymeric conjugates was the use of stimuli-sensitive polymers or smart polymers, which respond via conformational and/or electrostatic changes to environmental stimuli such as pH, ionic strength, temperature or externally applied heat, magnetic, or electric fields, or ultrasound [54]. The low pH of diseased areas (6.5 vs. 7.4 of human blood) like tumors and infarcts could enable targeting to those affected areas [55]. pH-sensitive moieties like carboxylic, sulfonic, and ammonium salts can undergo protonation or deprotonation in response to changes in pH and can be conveniently introduced into a polymeric backbone [56]. Also, pH-sensitive chemical bonds can be used for conjugating a drug to a polymer and can result in site-specific drug delivery. For example, hydrazones exhibit hydrolysis under mildly acidic conditions (pH 5–6), such as that present in lysosomes, while maintaining stability at pH values found in blood [57]. On the other hand, water-soluble, temperature-sensitive polymers, such as those based on poly(*N*-isopropylacrylamide) (poly(NIPAAM)), undergo a lower critical solution temperature (LCST) phase transition, wherein polymer chains collapse and aggregate at temperatures above their LCST as a result of the reversible dehydration of hydrocarbon side chains. Temperature-controlled drug release can be induced via elevated temperature that is associated with diseased tissues or by external application of hyperthermia [58].

3 Various Techniques Used During Synthesis and Characterization of Bioconjugates

3.1 Dialysis

In 1861, chemist Thomas Graham used the process of dialysis to separate colloidal particles from dissolved ions or molecules [59]. Dialysis is based on diffusion during which the mobility of solute particles between two liquid spaces is restricted, mostly according to their size. The term "selective diffusion" describes the diffusion of molecules across a semipermeable membrane to separate molecules based on size. In this process, colloidal particles cannot pass through a parchment or cellophane membrane while the ions of the electrolyte can pass through it. Membranes used have a molecular weight cut off (MWCO). The diffusion of molecules near the MWCO will be slower compared to molecules significantly smaller than the MWCO [59].

Dialysis can be easily affected by several factors such as the concentration and the hydrophobicity of molecules. These factors can influence the ability of the dialysate to diffuse through a dialysis membrane. The temperature, volume, agitation rate, and frequency of exchange of the external buffer are also important factors. However, dialysis is a clean technique, shows low consumption of energy, and has low installation and operating cost. Besides, the integrations of dialysis with other methods can increase the processing capability and efficiency, as acids and other metal salts can be successfully recovered [60].

Dialysis can be implemented in four modes: passive or conventional dialysis; active dialysis in the Donnan mode; active dialysis in the electrodialysis mode; and microdialysis (MD), which is also a passive mode. The first and last of which are the most widely used [61]. Sandwich and tubular (hollow-fiber) membrane separation modules are used in dynamic dialysis. The sandwich type comprises two blocks made of Perspex™, Teflon™, aluminum, or some other material having identical, internally engraved conduits (usually semicircular, triangular, or rectangular grooves 0.1–0.5 mm deep and 0.5–2 mm wide) that make up the inner chamber, the geometry of which varies from model to model. The membrane is placed between the two blocks, which must be joined tightly in order to avoid leakage. Each engraved microconduit has two holes on its ends that connect it with the manifold tubing. The best relative position of donor and acceptor chambers is with the acceptor chamber below the donor chamber in order to favor mass transfer [61, 62]. The tubular module comprises two concentric tubes, the inner one being a porous tube of an appropriate polymer through which the donor stream (the sample) is circulated internally, while the acceptor stream is circulated externally or vice versa [61, 63].

Dialysis is an invaluable tool for purification of synthesized bioconjugates, removing all byproducts and starting components of the bioconjugates, retaining only the bioconjugates inside of the dialysis bag or membrane. The choice of the proper molecular weight cutoff (MWCO) is critical in segregating the reactions' by-products and starting materials. The MWCO should be larger than the size of the starting material and smaller than the size of the final bioconjugates.

3.2 Crystallization

Crystallization is one of the oldest unit operations known to mankind. Namely, the crystallization of salts can be found through the ages. Due to purity issues, industrial techniques have developed over time, resulting in the modern continuous and vacuum-based crystallization apparatus [64]. Crystal formation can be achieved by various methods such as cooling, evaporation, addition of a second solvent to reduce the solubility of the solute (technique known as antisolvent or drown-out), solvent layering, sublimation, changing the cation or anion, as well as other methods [65].

The development of an industrial crystallization process is driven by a variety of considerations such as properties of the moiety to be crystallized, the technique with which the supersaturation can be generated, product properties such as particle size, and finally the crystallization to be performed in a batch or in a continuous mode [66, 67].

Crystallization is a valuable unit operation that can help in separating the pure form of the synthesized bioconjugates from the reaction mixture. The process requires a suitable solvent. A suitable solvent is one which readily dissolves the bioconjugate when the solvent is hot, but not when it is cold. The best solvents exhibit a large difference in solubility over a reasonable range of temperatures. For example, water can be a crystallization solvent between 0 °C and 100 °C. Hydrocarbon solvents such as hexanes or petroleum ether have a different temperature range since they can be cooled below 0 °C but boil below 100 °C. The choice of the proper solvent for crystallization depends on the physiochemical parameters and molecular weight of the synthesized bioconjugates along with the polarity of the solvent [64].

3.3 Rotary Evaporation

Taking advantage of the low boiling points of solvents by creating an environment where the solvent will rapidly boil, leaving the desired compounds, is the basic concept of the rotary evaporator. Typically, the heat is applied through a water bath, accompanied by rotation that ensures the equal distribution of heat. This keeps the solvent from freezing during the evaporation process. The solvent is removed under vacuum, is trapped by a condenser, and is collected for easy reuse or disposal. Dry-ice condensers are used to prevent volatilized solvent from escaping [68]. Rotary evaporation is a valuable technique in concentrating the purified, synthesized bioconjugates. It is usually performed post-dialysis to ensure that only the synthesized bioconjugates are concentrated after the removal of the starting material via dialysis.

3.4 Freeze-Drying (Lyophilization)

Freeze-drying is the removal of ice or other frozen solvents from a material through the process of sublimation and the removal of bound water molecules through the process of desorption. Controlled freeze-drying keeps the product temperature low enough during the process to avoid changes in the dried product appearance and characteristics. It is an excellent method for preserving a wide variety of heat-sensitive materials [69].

Freeze-drying can be used as a late-stage purification procedure because it can effectively remove solvents. Furthermore, it is capable of concentrating substances with low molecular weights that are too small to be removed by a filtration membrane. However, freeze-drying equipment is relatively expensive and has a long process time. Therefore, freeze-drying is often reserved for materials that are heat-sensitive. It is important to note that only water-based solutions of the bioconjugates can be lyophilized. Any traces of organic solvents can damage the lyophilizer. Consequently, lyophilization takes places after rotary evaporation to remove any traces of organic solvents from the synthesized bioconjugates [69].

4 Various Techniques Used for Bioconjugates' Characterization

4.1 Thin-Layer Chromatography (TLC)

Izmailov and Shraiber achieved separations on thin-layers in 1938 [70]. Thin-layers (TLC) is the easiest of chromatographic techniques to perform and requires simple apparatus [71]. It readily provides qualitative information; and, with careful attention to detail, it is often possible to obtain quantitative data [71]. Complex prebiotic reaction mixtures is often performed via thin-layer chromatography (TLC) because of its simplicity and speed [72]. It is also considered as a flexible tool to monitor the progress of the bioconjugates' synthesis and ensure the completion of the reaction.

Compounds with different properties can be separated from one another by exploiting the diverse interactions of the solutes with the sorbent and the mobile phase [73]. This is due to the speed at which the solute moves through the stationary phase which depends on the net attraction force between the mobile phase and the solute. Consequently, the mobile phase dissolves the solute and moves it up the TLC plate. Furthermore, the

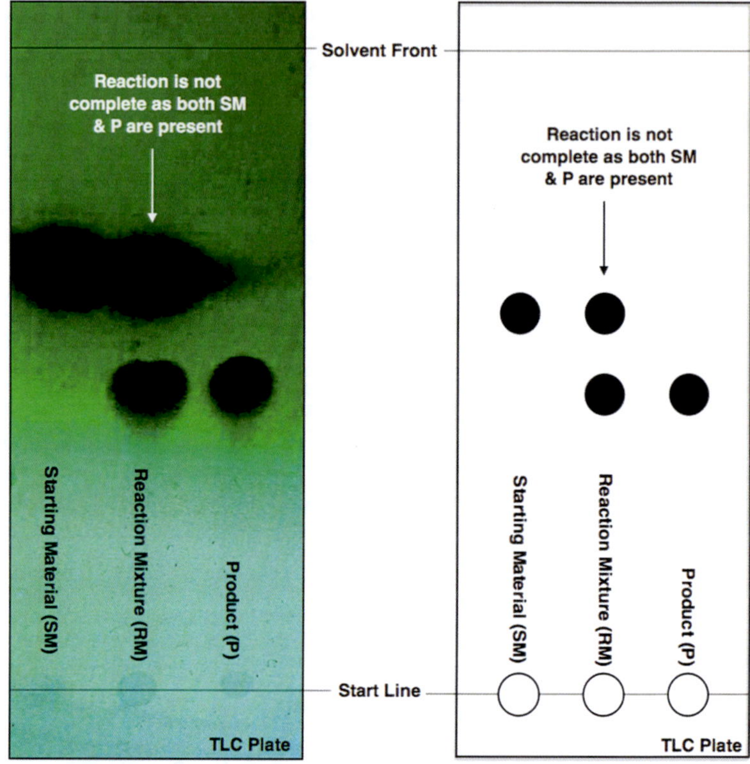

Fig. 8 Representative illustration of thin-layer chromatography (TLC) plate during the progression of the reaction

differential resistance of the sorbent to solute and solvent movements is a rate limiting factor in the separation process and the solute movement speed on the stationary phase as it pulls the solute out of solution and back into the sorbent.

Less equipment needed and high sensitivity acquired are the most significant advantages of using TLC. The little time for separation can be of valuable merit, although it limits the length of separation. The lower detection limit of most analytical samples in TLC is approximately one decimal lower than that of paper chromatography and very small quantities of sample is sufficient for analysis. TLC operates as an open system, so factors such as humidity and temperature can be consequences to the results of the chromatogram.

TLC is used in bioconjugation mainly to check the progress of reactions and to monitor the purity of the final product (Fig. 8).

A sorbent is needed to perform TLC. A thin sorbent layer, usually about 0.10–0.25 mm thick, is applied to a firm backing of glass, aluminum, or plastic sheet to act as a support [74]. The most common TLC sorbents are silica-based. Silica gel (SiO_2) is a white, porous material made by precipitation from silicate solutions by addition of acid [73]. However, there are other successful sorbents including cellulose, aluminum oxide, polyamide, and chemically bonded silica gels [74].

Selectivity of separation is greatly influenced by the choice of solvent or solvent mixture. As a general rule, nonpolar solvents will effect migration of low polarity substances, whilst more polar samples will require more polar solvents on a normal-phase sorbent layer [75, 76]. All solvents should be of chromatographic grade purity with particular attention paid to low impurities, moisture content, and nonvolatile matter. Solvent mixtures should be thoroughly shaken together in order to attain complete homogeneity.

Separated compounds can be detected and visualized by a combination of the above techniques. A nondestructive technique, such as UV irradiation, may be used first, followed by a universal reagent, ammonia vapor, fluorescein, dichlorofluorescein, and iodine. Finally, a functional group-specific method can be used to enhance selectivity and sensitivity. Visualization of developed TLC plates most commonly uses dyes or fluorescence to identify spots corresponding to the separated components [72]. More advanced techniques like mass spectrometry (MS) and nuclear magnetic resonance (NMR) can be used to increase the sensitivity of quantitative and qualitative analysis of the separated chemicals.

4.2 Liquid Column Chromatography (LC)

Liquid chromatography (LC) was the first type of chromatography to be discovered. Liquid–solid chromatography (LSC) was the first subtype of LC to be investigated. It was originally used in the late 1890s by the Russian botanist, Tswett [77] to separate and isolate various plant pigments [78]. In the late 1930s and early 1940s Martin and Synge [79] introduced a form of *liquid–liquid* chromatography by supporting the stationary phase [78].

The basic liquid chromatography consists of six basic units: the mobile phase supply system, the pump and programmer, the sample valve, the column, the detector, and finally a means of presenting and processing the results [78]. Liquid chromatography may be coupled by other techniques to increase sensitivity and selectivity of the analysis method. High-performance liquid chromatography is basically a highly improved form of column chromatography [78]. Instead of a solvent being allowed to drip through a column under gravity, it is forced through under high pressures of up to 400 atmospheres making the technique notably fast [78].

4.3 High-Performance Liquid Size Exclusion Chromatography (HPLC-SEC)

Size Exclusion Chromatography (SEC) is a type of LC and subsequently, solid stationary and liquid mobile phases are used. However, the separation mechanism in SEC relies solely on the size of the polymer molecules in solution, rather than any chemical interactions between particles and the stationary phase.

Size exclusion chromatography (SEC) is also known as gel filtration, gel permeation, or molecular sieve chromatography. The added value of HPLC-SEC is that it allows the determination and quantification of the level of aggregates and fragments of the compound [75].

Grant Henry Lathe and Colin R. Ruthven discovered the principle of SEC in 1955, where they used starch gels as the matrix [80]. Later, Jerker Porath and Per Flodin introduced dextran gels [81]. Currently, more SEC solid phases are in use such as polyacrylamide sieves [82] and granulated agar [83].

SEC is the simplest and mildest of all the chromatography techniques and separates molecules on the basis of differences in size (Molecular Weight, MWT) [84]. However, this method may produce erroneous results if appropriate attention is not paid to the conditions of measurement and to the data processing [85]. Another restraint of SEC is its speed. Exclusion processes are fast, but the overall rate of SEC analyses is limited. Present "high-speed" SEC instruments, equipped with specially designed columns, reduce experiment duration down to few minutes. This is still far from the possibility to monitor polyreactions in real time.

High-performance liquid chromatography (HPLC), which uses size-exclusion as well as more precise column-based strategies, can aid in more effective separation. HPLC-SEC can be used in the characterization and purification of bioconjugates.

Moreover, quantum dots bioconjugates could be obtained by SEC purification with twice the yield under the optimized conditions [86]. The use of Low Temperature Evaporative Light Scattering Detectors (LT-ELSD) coupled with SEC could aid in the proper determination of the generated bioconjugates' molecular weight with sensitivity and selectivity comparable to mass spectroscopy [87]. For the determination of the MWT, gel filtration calibration kits are required. The High Molecular Weight (HMW) Kit contains calibrated polymers of different MWTs and provides simple, reliable calibration of gel filtration columns.

4.4 Nuclear Magnetic Resonance (NMR)

Nuclear Magnetic Resonance (NMR) spectroscopy is a unique tool to study molecular interactions in solution, and it became an essential technique to characterize events of molecular recognition [88]. NMR is extremely useful for analyzing samples nondestructively. There are two different approaches for NMR, namely, by looking at the protein (target) spectrum and following the changes in chemical shift by ligand titration, or recording the spectra of a sample of ligand with small amounts of protein. NMR needs a trained specialist as a single incorrectly set parameter can mean the difference between getting an accurate, realistic spectrum and getting a meaningless result [88].

Four types of Information can be obtained from NMR: number of signals, position of signals (chemical shift), relative intensity of signals (integration) and splitting of signals (spin–spin coupling) [89].

4.4.1 ^{1}H (Proton) Nuclear Magnetic Resonance

The 1D ^{1}H (Proton) NMR experiment is the most common NMR experiment. The proton (^{1}Hydrogen nucleus) is the most sensitive nucleus (apart from tritium) and usually yields sharp signals [90, 91]. Proton NMR spectroscopy, as an analytical tool for quantitative analysis, was first reported in 1963 by Jungnickel and Forbes [92].

A routine NMR spectrum yields integrals with an accuracy of ±10%. Accuracies of ±1% can be achieved by increasing the relaxation delay to five times the longitudinal relaxation time (T_1) of the signals of interest. Furthermore, NMR provides information on the number of neighboring hydrogen atoms existing for a particular hydrogen or group of equivalent hydrogens. This phenomenon is called "splitting." In general, an NMR resonance will be splitting into $N + 1$ peaks, where N represents the number of hydrogen atoms on the adjacent atom or atoms. Consequently, the NMR signal can be singlet, doublet, triplet, or multiplet. When multiplets overlap, the total integral of the spectral region may be used [93].

Proton nuclear magnetic resonance (^{1}H NMR) is a spectroscopic technique usually used for structural determination of molecules. The proton NMR chemical shift is affected by closeness to electronegative atoms (O, N, halogen) and unsaturated groups (C=C, C=O, aromatic groups).

^{1}H-NMR can be an invaluable tool in bioconjugates' characterization. It can aid in structural elucidation and determination of the conjugation ratio along with its yield [94]. The ratio of the monomer unit composition in the synthesized PMBN (poly [2-methacryloyloxyethyl phosphorylcholine (MPC)-*co*-*n*-butyl methacrylate (BMA)-*co*-*p*-nitrophenyloxycarbonyl poly(ethylene glycol) methacrylate (MEONP)]) product was determined by ^{1}H-NMR [95]. PMBN is used as a bioconjugated phospholipid polymer biointerface with nanometer-scaled structure for highly sensitive immunoassays [95].

4.4.2 ^{13}Carbon Nuclear Magnetic Resonance (NMR)

The ^{13}C NMR is generated in the same fundamental way as proton NMR spectrum. Only 1.1% of naturally occurring carbon is ^{13}C, and this is actually an advantage because it minimizes the coupling [96]. The 1D ^{13}Carbon NMR experiment is much less sensitive than Proton (^{1}H) but has a much larger chemical shift range. Its low natural abundance (1.108%) and proton decoupling means that spin–spin couplings are seldom observed. This greatly simplifies the spectrum and makes it less crowded. ^{13}C is a low sensitivity nucleus that yields sharp signals and has a wide chemical shift range [97].

^{13}C chemical shift is affected by electronegative effect and steric effect. The ^{13}C–^{13}C spin–spin splitting rarely exists between adjacent carbons because ^{13}C is naturally less abundant (1.1%) [98, 99]. ^{13}C–^{1}H Spin coupling provides useful information about the number of protons attached to a carbon atom. Distortionless Enhancement by Polarization Transfer (DEPT) is a multipulse, multichannel ^{13}C NMR technique [100]. DEPT is an effective means of determining ^{13}C multiplicity that, when combined with other NMR spectra and other experimental techniques (MS, FT-IR, etc.), can be an invaluable tool for the analysis of unknown compounds.

Numerous structure and chemistry drawing software such as PerkinElmer Informatics ChemDraw® 16, ChemOffice® 16, and ChemNMR® can be used to accurately estimate ^{13}C and ^{1}H (proton) chemical shifts. The chemical shifts are displayed on the molecule and the spectrum is linked to the structure so that clicking on a peak in the spectrum highlights the related fragment on the molecule (Fig. 9). Such tools provide researchers with what they should expect with respect to their conjugation products' NMR spectra. Such predicted NMR spectra can be compared with actual NMR spectra to confirm the successful synthesis of the synthesized conjugates.

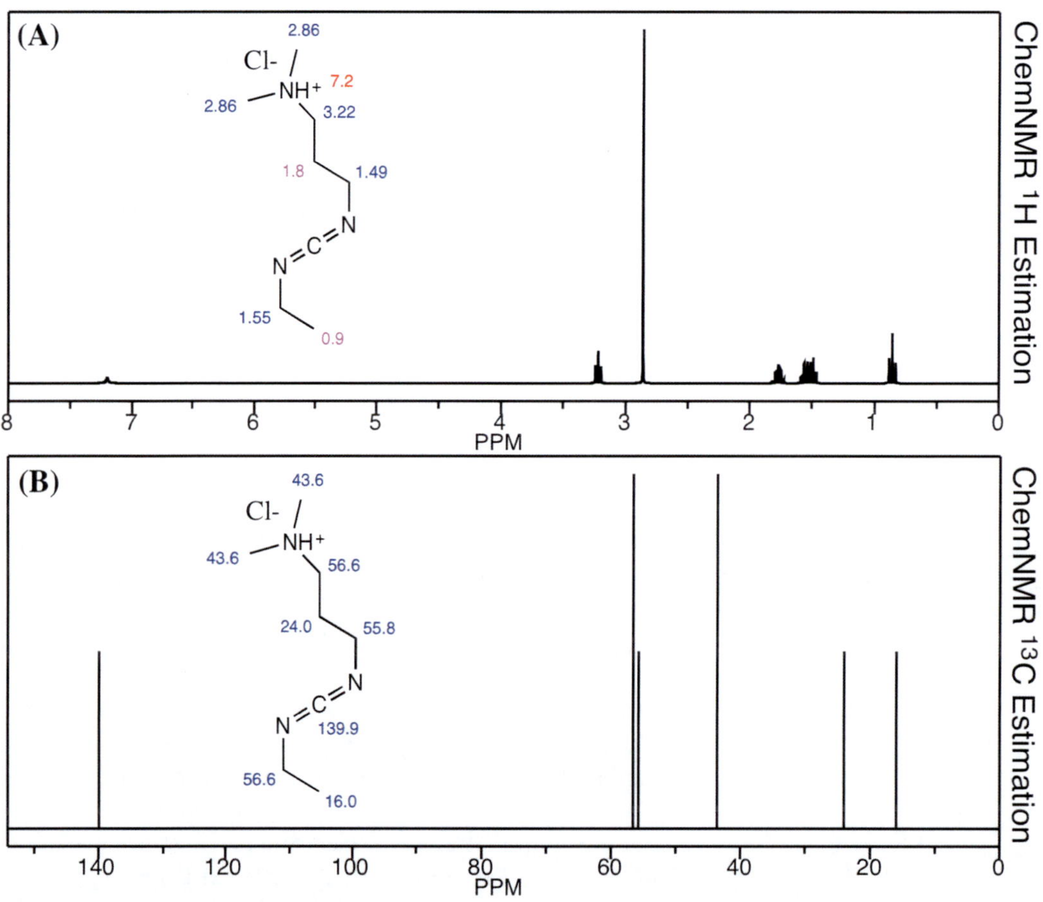

Fig. 9 ^{1}H (proton) (**a**) and ^{13}C (**b**) predicted chemical shifts and NMR spectra of 1-ethyl-3-(3-dimethylaminopropyl)carbodiimide hydrochloride (EDC) as generated by PerkinElmer Informatics ChemNMR® Version 16

4.5 Matrix-Assisted Laser Desorption Ionization-Time of Flight Mass Spectrometry (MALDI-TOF MS)

Mass spectrometry is an analytical technique in which chemical compounds are ionized into charged molecules and ratio of their mass to charge (m/z) is measured [101]. MALDI-TOF MS is referred to as a "soft" ionization technique because it causes minimal or no fragmentation and allows the molecular ions of analytes to be identified, even in complex mixtures of biopolymers [102, 103]. Although this approach has been shown to present high detection sensitivity, drawbacks and limitations frequently arise from the significant background in the mass spectrometric analysis [103, 104].

MALDI TOF-MS has become a versatile and important soft ionization technique in mass spectrometry for the determination of molecular masses of various fragile and nonvolatile samples, including biopolymers [105]. Coupling MALDI with TOF-MS instrumentation allows a "virtually unlimited" mass range to be monitored. Machado et al used MALDI-TOF mass spectrometry (MS) analysis to determine the degree of heterogeneity of the oligosaccharides components in mAbs [106]. Oligosaccharides were released (fragmented) from mAbs upon treatment with PNGase F [106]. The upper mass limit for MALDI is about 350,000 Da [101].

4.6 Infrared (IR) Spectroscopy

Infrared spectrometers have been commercially available since the 1940s [107]. Additionally, IR spectroscopy was applied in the analytical field by Karl Norris in the 1960s [108]. Fourier transform infrared spectroscopy (**FT-IR**) is an easy, fast, and cost-effective analytical tool used in generating an infrared spectrum of absorption or emission of a solid, liquid, or gas. Nowadays, this approach is one of the most well-known spectroscopic techniques that is used in pharmaceutical drug delivery system characterization.

Like all forms of spectroscopy, infrared (IR) depends on the absorption of specific electromagnetic waves in the identification of substances. Infrared spectrum ranges from 2500 to 16,000 nm with respect to wavelength and from 1.9×10^{13} to 1.2×10^{14} Hz with respect to frequency. IR energy is not large enough to excite electrons, but it is sufficient to induce vibrational excitation of covalently bonded atoms and groups [107]. The smallest unit to be analyzed by IR is an asymmetric diatomic molecule. Infrared spectroscopy (IR) is based on the interaction of infrared radiation with the material at a molecular level. IR spectroscopy measures the absorbed or transmitted IR radiation via molecules at different wavelengths [109].

IR spectra are constructed via plotting the absorbance or transmission of each functional group against a range of wavelengths or frequencies. Each functional group has its own spectra which act as a distinct fingerprint. Fortunately, the absorption peak of IR is sharper than the absorption peak of the ultraviolet and visible regions. Thus, IR spectroscopy is a highly sensitive tool for the identification of different functional groups of several inorganic and organic materials [110].

IR spectrometers can be used to analyze a wide variety of products, including powder, liquid, and all kind of solid products saving the product integrity, which is advantageous [111]. However, IR, unlike NMR, analyzes just the presence of bonds, not a distinctive molecular structure. IR is very useful in detecting the purity of subjects under test.

IR analyses have been used extensively in the field of drug delivery and bioconjugation. It is considered a fast, easy, and cost-effective tool in molecular structure confirmation. Tanakaa et al. [112] used Fourier transform infrared spectroscopy (FT-IR) to characterize diblock biodegradable polymeric nanocarriers methoxy polyethylene glycol-polycaprolactone (MPEG-PCL) conjugated with a cytoplasm responsive cell-penetrating peptide (CPP, CH2R4H2C). Taranejoo et al. [113] used attenuated total reflectance (ATR) Fourier transform infrared (FTIR) to detect the grafted glycol chitosan (GCS) into branched low molecular weight polyethyleneimine (PEI) **(GCS-ss-PEI)** nanoparticles using a Perkin Elmer ART FT-IR spectrometer at a range (4000–500 cm^{-1}). In another application, Knapinska et al. [114] characterized a conjugate system of nanodiamonds (NDs) and collagen-derived peptide via FTIR spectroscopy. Furthermore, Imani et al. [115] designed a conjugate of R8 peptide on the surface of nanographene oxide sheets (NGOS) (R8-functionalized NGOS). A successful R8-functionalized NGOS conjugate formation was confirmed via appearance of characteristic band at 1742 cm^{-1} for CONH stretching vibration using FT-IR analysis [115].

5 Mainstream Approaches in Bioconjugates' Synthesis

Many common organic chemical principles can be utilized for the formation of bioconjugates. The next section aims to provide a general overview of the most commonly employed techniques in bioconjugate chemistry namely, click chemistry as bioorthogonal reaction, nucleophilic substitution reaction, carbodiimides conjugation reaction using carbodimides as zero-length crosslinkers and (strept)avidin–biotin system as noncovalent conjugation.

5.1 Click Chemistry as a Bioconjugation Technique

5.1.1 Background

Click chemistry is a broad term used to describe powerful and rapid chemical reactions with excellent bioorthogonality. The most prominent tool in the click chemistry toolbox would be copper (I)-catalyzed azide–alkyne dipolar cycloaddition (CuAAC). Although the classical 1,3-dipolar Huisgen cycloaddition of alkynes and azides has been recognized since the sixties of the last century, its applications till 2001 were limited owing to its high activation energy and the mixture of regioisomers (1,4- and 1,5-disubstituted

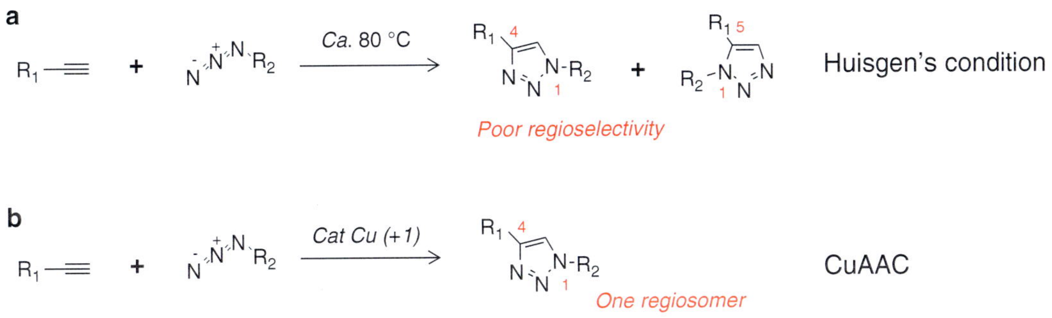

Fig. 10 (**a**) Huisgen cycloaddition reaction under thermal conditions without a catalyst. (**b**) Copper-catalyzed alkyne azide cycloaddition reaction

triazoles) that were often obtained (Fig. 10a). The discovery of the ability of Cu(I) to catalyze Huisgen cycloaddition (CuAAC) was reported independently by the Sharpless-Fokin and the Meldal groups in 2002 [116, 117], and represented a major breakthrough that revolutionized the applications of 1,2,3- triazoles in different branches of life sciences like material science, bioconjugation, DNA labeling, bioanalysis, and even live cell imaging [118].

CuAAC reactions are broad in scope; proceed in high yields under mild conditions, rapidly at low temperatures or even under ambient conditions; are compatible with water and water-miscible solvents such as tetrahydrofuran (THF) and dimethylformamide (DMF); are regiospecific (giving only 1,4-disubstituted derivative) (Fig. 10b); have minimal and/or inoffensive byproducts and are suitable for microscale solution-phase parallel synthesis without the need for protecting group manipulations [118].

Cu(I) salts were generally used as catalysts for this reaction. The source for Cu(I) is typically generated in-situ using Cu(II) in the presence of a reducing agent. The Cu(II) salt, $CuSO_4$, is particularly convenient, as it is readily available and easily converted to Cu (I) with a reducing agent such as sodium ascorbate or tris(2-carboxyethyl)phosphine (TCEP). In solution, Cu(II) is reduced to Cu (I) by ascorbate with concomitant oxidation of ascorbate to dehydroascorbate [119].

Cu(I) species were found to accelerate the rate of azide–alkyne cycloaddition (AAC) by at least 7–8 orders of magnitude compared to purely thermal cycloaddition reaction without metal catalysis. Based on DFT calculations, both Meldal and Sharpless groups proposed a stepwise mechanism involving a cyclic intermediate azide–Cu(I)–alkyne complex, which then goes on to form the 5-membered triazole ring [120] as shown in Fig. 11.

On the light of new evidences, Fokin's group published an intriguing study that revealed that Cu-acetylide complex were only reactive to azides in the presence of exogenous copper catalyst. Moreover, time-of-flight mass spectrometry (TOF-MS)

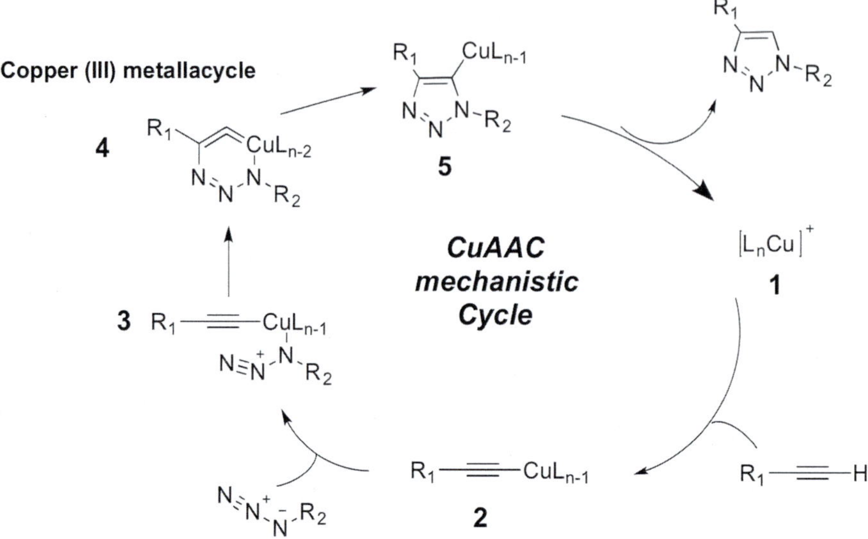

Fig. 11 Sharpless-type catalysis mechanism based on DFT calculations

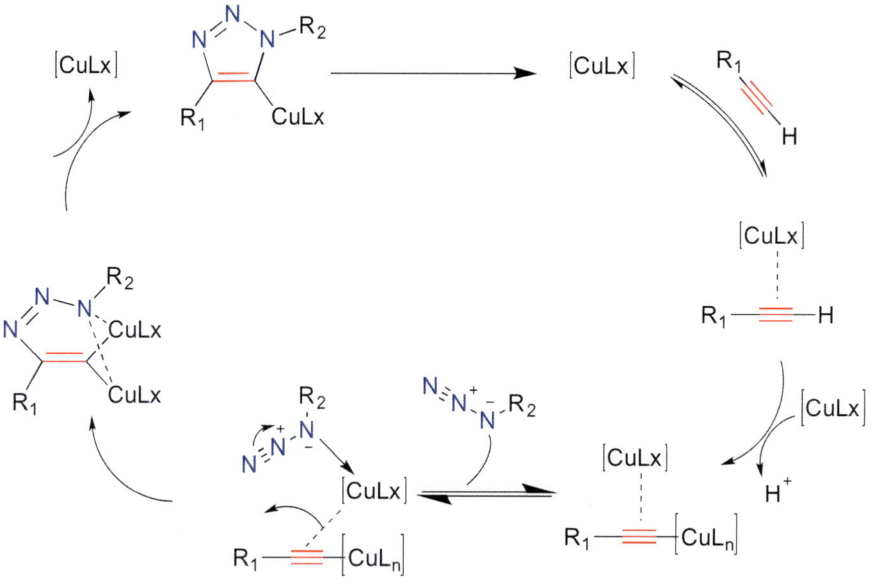

Fig. 12 Fokin's proposed mechanism based on two copper centers

experiments using isotopic enriched copper unequivocally showed that two copper centers were involved in the cycloaddition process to give 1,4-substituted-1,2,3-triazole [121], as depicted in Fig. 12.

Almost complete chemoselectivity with broad functional group tolerance can be offered by this reaction. One of the primary reasons for the increasing popularity of the click chemistry reaction is the bioorthogonal nature of the two reacting groups (alkyne and azide) to a diverse range of functional groups and reaction conditions allowing the assembly of reactants decorated with diverse unprotected functionalities. Alkynes and azides are completely unreactive toward biological molecules and virtually free of side reactions that otherwise would cause reagent instability in aqueous environments. This means that a molecule modified to contain an azide functionality would be able to react specifically with another molecule containing an alkyne group, even in the presence of biological fluids, cells, or cell lysates. In addition, without the presence of Cu(I), the azido-molecule and the alkyne-molecule would not react to an appreciable extent at room temperature even when placed together in solution. Only upon the addition of Cu(I) in sufficient concentration would the cycloaddition reaction take place and a triazole linkage be formed [119].

The 1,2,3-triazole moiety has several advantages. First, its chemical robustness is favorable for adoption in drug discovery and bioconjugation programs. Second, the 1,2,3-triazole moiety has a moderate dipole moment (5 Debye) and the two H-bond acceptors at N_2 and N_3 can interact with biomolecular targets via enhanced H-bonding interactions (compared to the amide, as shown in Fig. 13), dipole interactions, or π–π stacking interactions. Third, the 1,2,3-triazole offers high chemical stability in biological environments, including acidic and basic media and oxidative and reductive conditions. The 1,2,3-triazole structural motif generated

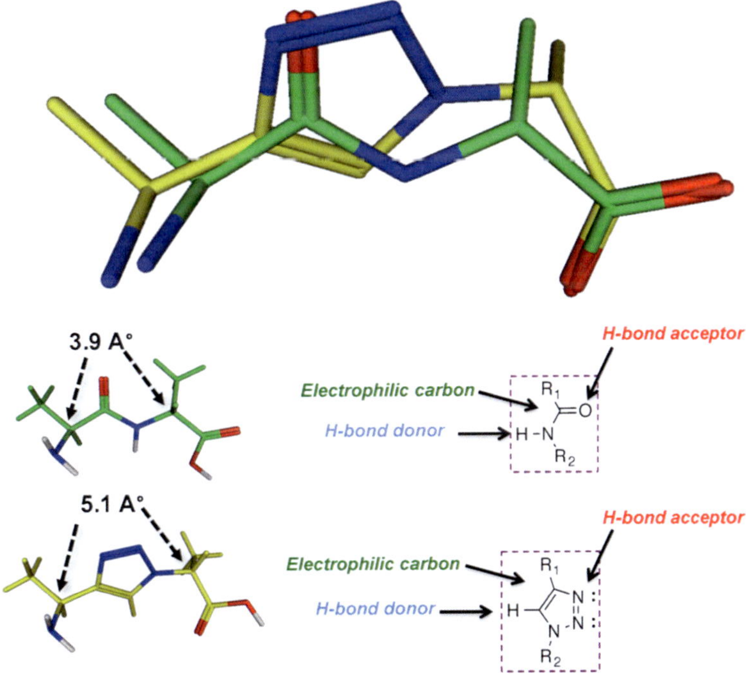

Fig. 13 1,4-Disubstituted 1,2,3-triazole as a good bioequivalent surrogate for the amide bond

in the CuAAC reaction is much more than just a passive linker unit and is considered to be a safe bioequivalent surrogate (nonclassical bioisostere) for amide, which is a widely employed functional group in drug design [118, 122]. All of these features prompted Nobel Laureate K. Barry Sharpless to describe 1,2,3-triazoles as aggressive pharmacophores [123].

Equally important, 1,4-triazoles are metabolically inert, whereas 1,5-triazoles can be metabolized to N-oxides, which are reactive intermediates. Further, 1,4-disubstituted triazoles have a negligible inhibitory effect on the cytochrome P450 family of metabolizing enzymes, compared with the 1,5-isomers; and this is favorable for avoiding drug–drug interactions [118].

For the above reasons, CuAAC has cemented its position at the heart of click chemistry. It paved the way for tremendous applications in organic synthesis and bioconjugation such as small molecule synthesis, protein conjugation, activity-based protein profiling, nucleic acid conjugation, surface modification, and in vivo targeting of molecules on cells [119].

In the field of bioconjugation, the attributes of CuAAC have opened the door to a wide range of unprecedented applications, which enhanced our understanding and manipulation of biomolecular frameworks. These applications include, but are not exclusive to, biomolecule labeling and imaging, activity-based protein profiling and affinity labeling. Azide or alkyne tags (Fig. 14) are often easily introduced into small molecules via chemical synthesis; and hence, bio-monomer analogs containing either azido or alkyne

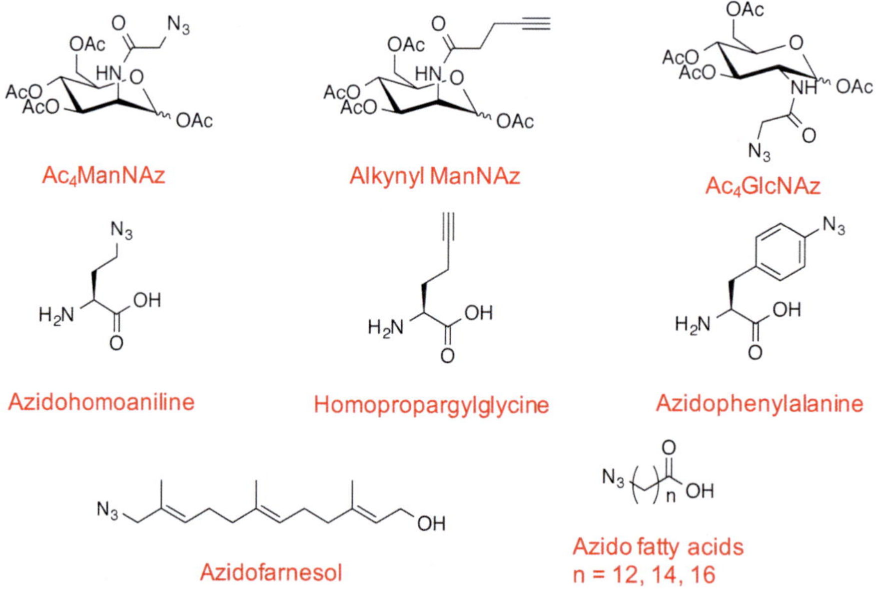

Fig. 14 Monomers of azido and alkyne derivatives of sugars, amino acids, and lipids that can be incorporated into expressed proteins or carbohydrates or lipids and subsequently probed using click chemistry

functionalities can be prepared and then can be incorporated into biopolymers using a cell's native enzymatic machinery. It has been demonstrated that both azido and alkynyl amino acid derivatives could be used as methionine surrogates and get integrated into proteins with nearly the same efficiency as normal methionine. In addition, azide-sugar derivatives have been prepared that are capable of being incorporated into glycans and glycoconjugates using normal enzymatic biosynthetic pathways in cells. Thus, proteins, carbohydrates, and even lipids can be specifically tagged to contain noncanonical amino acids or sugars or lipids for subsequent bioconjugation using reagents containing the opposite click chemistry reactant for detection, crosslinking, or capture [119].

Click chemistry reactant pairs used for surface immobilization have the advantage of being stable to aqueous conditions and long-term storage. Unlike many of the other coupling chemistries used with surfaces (e.g., NHS esters, EDC conjugation), which suffer from hydrolysis and degradation over time, the alkyne or azide components can be used to activate a surface and stored indefinitely until needed. A ligand modified with the opposite reactant can then be spotted on the array surface in the presence of Cu(I) to initiate covalent attachment through triazole ring formation [119].

Nonetheless, the biggest challenge in bioconjugation applications involving CuAAC would be the removal of copper catalyst. Copper is incompatible in vivo and known to cause serious side effects such as hepatitis and neurologic and renal diseases. Hence, removal of copper from any polymeric platforms used in drug delivery is detrimental. Along the same track, live cell imaging protocols cannot tolerate copper catalyst, since it is toxic to both mammalian and bacterial cells. Professor Carolyn R. Bertozzi, of California Berkley, envisioned an intriguing approach to come up with catalyst-free Huisgen cycloaddition, which is ring strain. Inclusion of the alkyne in 8-membered cycloalkyl ring would lead to a massive bond angle deformation of acetylene to 163°, which accounts for nearly 18 Kcal/mol ring strain. Such destabilization factors provide enormous rate acceleration compared to the open chain form. The reaction with an azide lessens the ring strain of the alkyne within the cyclooctyne structure and, thus, drives the reaction without the addition of cytotoxic copper, which constitutes a new click technique that is known as strain-promoted azide–alkyne cycloaddition (SPAAC) [124, 125], as shown in Fig. 15.

5.1.2 Practical Perspective: Click Reaction Between Metronidazole Azido Analogue and S-Propargyl Glutathione

The most commonly used strategy in performing CuAAC is in situ reduction of Cu(II) salts such as $CuSO_4{\cdot}5H_2O$ or copper acetate to form Cu(I) salts that undertake the catalysis. Sodium ascorbate or ascorbic acid is the typical reducing agent used, usually in 3–10 equivalent excess. Usually, copper catalyst is used in 5–10% mole. This strategy proves to be very economic, doesn't need deoxygenated atmosphere, and can be performed in water. Other reducing

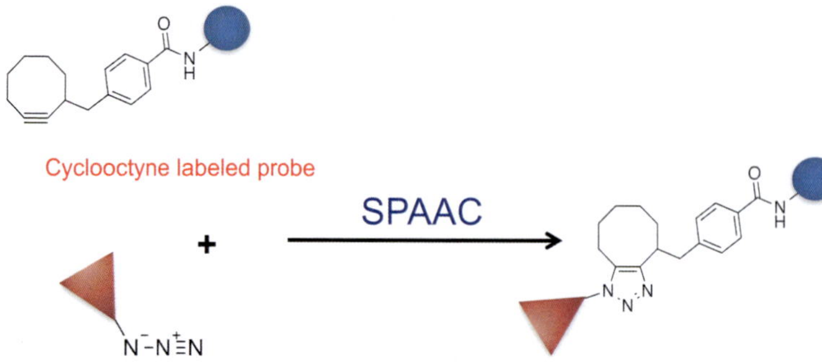

Cyclooctyne labeled probe

SPAAC

Fig. 15 Diagrammatic representation of strain-promoted azide–alkyne cycloaddition (SPAAC)

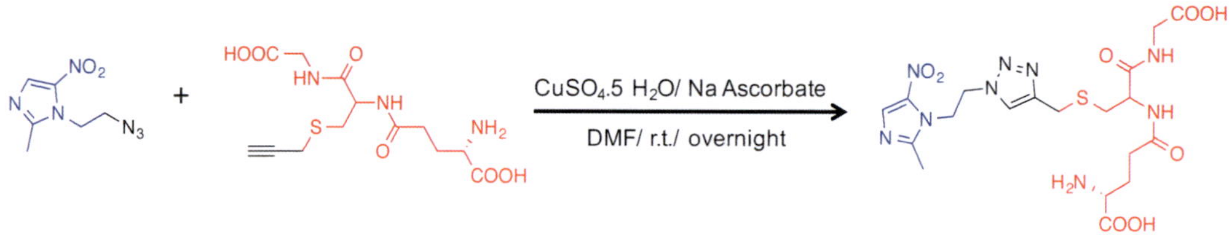

$CuSO_4.5\ H_2O$/ Na Ascorbate

DMF/ r.t./ overnight

Scheme 1 Click reaction between metronidazole azido analogue and *S*-propargyl glutathione

agents can be used such as Hydrazine hydrate. Using Cu(I) salts, such as CuBr and CuI, doesn't require a reducing agent; however, it needs to be carried out in deoxygenated organic solvents and often gives lower yields compared to the aforementioned strategy. Oxidizing copper metal with amines to give active catalyst can be successfully used, yet suffers from some disadvantages: being too expensive, requiring longer time, and being incompatible with acid-sensitive functional groups. Microwave irradiation (MW), in most cases, remarkably decreased reaction times, facilitated work-ups, and enhanced yields. Most water miscible solvents can be used, whether they are protic or aprotic. However, acetonitrile is not recommended since it coordinates with Cu(I) salts. Also, halogenated solvents should be avoided since halides were known to retard CuAAC with iodide being the worst on both yields and rates. For high molecular weight polymers, DMF and DMSO are considered good solvent choice. The following synthesis for coupling metronidazole azido analogue with *S*-propargyl glutathione (Scheme 1) is adopted from Jarrad et al. [126] and Negi et al. [127].

To synthesize metronidazole azido analogue and *S*-propargyl glutathione conjugate, an equimolar mixture of metronidazole azido analogue and *S*-propargyl glutathione are used in DMF, in the presence of a catalytic amount of $CuSO_4{\cdot}5H_2O$/sodium ascorbate. The reaction mixture is stirred at room temperature overnight. The product is separated by adding ice–water mixture, then

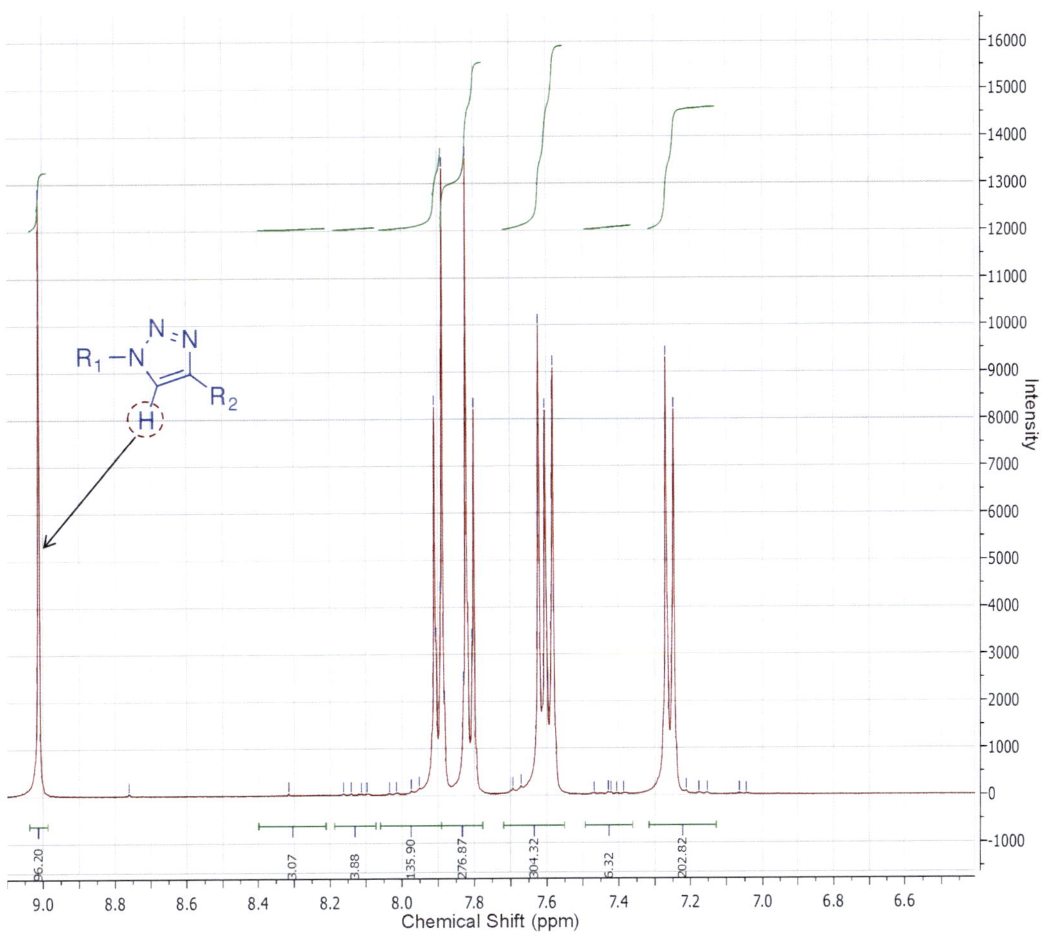

Chart 1 ^{1}H-NMR triazole C_5–H aromatic singlet

extracted with ethyl acetate. The obtained product is purified either by flash chromatography or by crystallization from CH_3OH.

Confirmation of the structure of the expected final triazole product can be done using ^{1}H-NMR, ^{13}C-NMR and IR. The ^{1}H-NMR shows triazole C_5–H aromatic singlet around at 8.7–9 ppm (Chart 1), along with the disappearance of propargylic terminal CH peak Furthermore, ^{13}C-NMR should display the two new triazole C_4 and C_5 peaks at 130–150 ppm along with the disappearance of previously mentioned propargylic CH and quaternary carbon peaks (Chart 2). Moreover, and to a lesser extent, the IR spectrum should be associated with the disappearance of ethynyl CH and C≡C stretching bands.

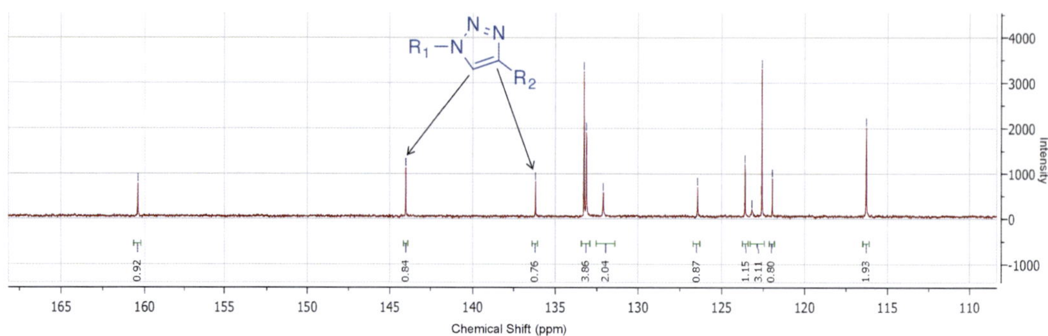

Chart 2 ^{13}C-NMR triazole C_4 and C_5 peaks

Scheme 2 Biotin-PEG-propargyl modification reagent

5.2 Nucleophilic Substitution Reactions as a Bioconjugation Technique

5.2.1 Introducing a Propargyl Group *via Nucleophilic Substitution Reaction*

Propargyl group can be introduced into a biomolecule of interest via a variety of commercially available reagents; for e.g., propargyl bromide, propargyl alcohol, and propargyl amine. Usually introducing a propargyl group occurs via nucleophilic substitution reaction with nucleophilic functional groups like thiol, phenolic OH, and secondary amine; preferably in the presence of base like anhydrous potassium carbonate or triethylamine. Weak nucleophile like alcoholic OH may need a stronger base for activation like sodium hydride (NaH). Special precaution should be considered on reacting propargyl bromide with primary amines to prevent polyalkylation. Adding propargyl bromide dropwise into an excess of amine in presence of ice-bath and carrying out the reaction at room temperature for 1–3 nights instead of heating under reflux might help in affording the monoalkyl product.

Adding an alkyne group to a modification reagent or a cross-linker can be as simple as coupling an activated carbonyl group with propargylamine, which forms the propargylamide linkage and creates a terminal acetylene group for conjugation. Link et al. [128] synthesized a biotin–PEG–alkyne modification reagent using this strategy, which then could be used to modify proteins containing azido-amino acids, as shown in Scheme 2.

The following synthesis involves the reaction of glutathione with propargyl bromide (Scheme 3) and is adopted from Lo Conte et al [129].

Propargyl bromide

Br

NH$_4$OH, CH$_3$OH
r.t., 3 h

Glutathione

Scheme 3 Formation of *S*-propargyl derivative of glutathione

First, personnel performing the chemical synthesis should wear appropriate laboratory clothing during the whole time of chemical synthesis. All experiments should be conducted in appropriate fuming cupboard. All chemical wastes should be appropriately disposed of in accordance with local regulations. Propargyl bromide is flammable and toxic. Personal protective wear like gloves, eye shields, face shields, and full-face respirators are recommended all the time.

From a practical perspective, to synthesize *s*-propargyl derivative of glutathione, propargyl bromide and NH_4OH are added to a stirred and ice-cooled solution of Glutathione in methanol. The reaction mixture is stirred for 1 h at 0 °C and then at room temperature for an additional 2 h. After solvent removal under reduced pressure, water is added to get rid of any inorganic salts. Finally, the residue formed is filtered under vacuum, air-dried, and used directly in the subsequent reaction without any further purification.

Confirmation of the structure of the expected final propargylic product can be done easily using ^{1}H-NMR, ^{13}C-NMR, and IR. The ^{1}H-NMR should show both characteristic triplet and doublet in the range of δ 3.0–5.5 ppm, which corresponds to terminal CH and CH_2, respectively (Chart 3). Both protons usually undergo a long-range coupling of 2–3 Hz. The ^{13}C-NMR shouldshow two acetylenic characteristic peaks at δ 70–80 ppm corresponding to CH and quaternary carbon, respectively (Chart 4). The IR spectra should contain characteristic sharp acetylenic-CH stretching band in the range of 3200–3250 cm^{-1} and C≡C stretching band in the range of 2200–2400 cm^{-1} (Chart 5).

5.2.2 Introducing Azido Group

Azido group can be easily introduced via nucleophilic substitution reaction of alkyl halides with sodium azide or the less hazardous trimethylsilyl azide. Aromatic primary amino can be converted into azido group via diazotization followed by reaction with sodium azide or trimethylsilyl azide. Another way of nucleophilic substitution reaction that can be applied to aliphatic alcohols is to convert the poor leaving alcoholic OH into a better leaving tosylate, followed by reaction with appropriate azide. The following protocol involves the preparation of azido derivative of metronidazole

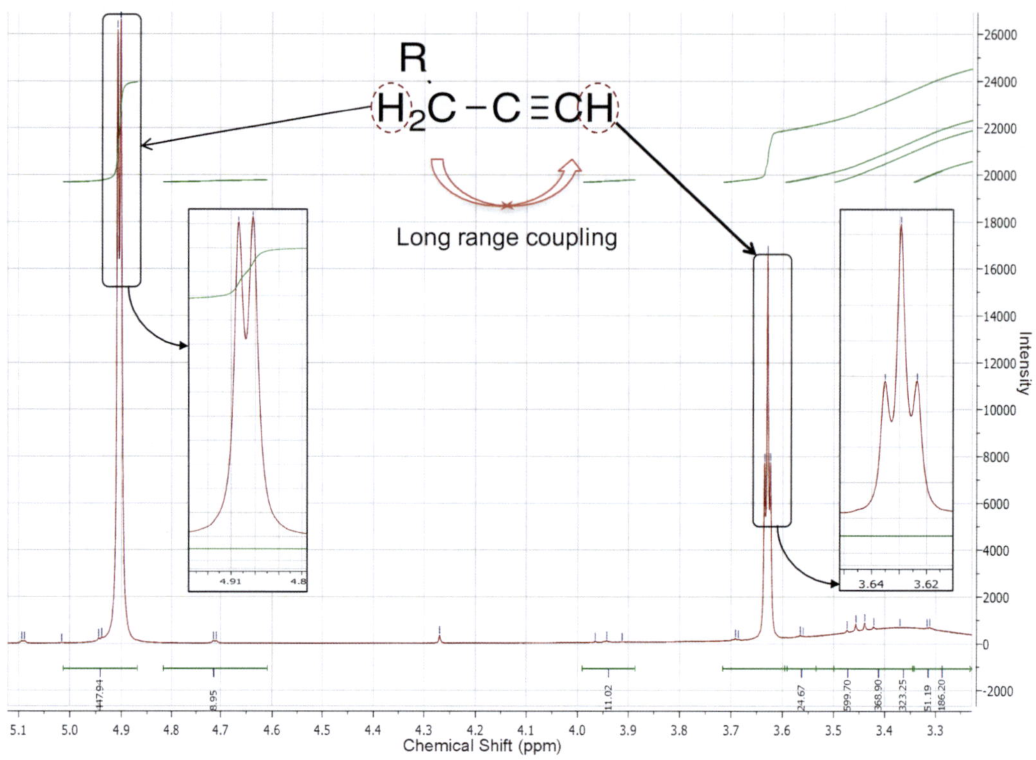

Chart 3 ^{1}H-NMR Characteristic propargylic triplet and doublet. The inner panels show an expansion to the two signals

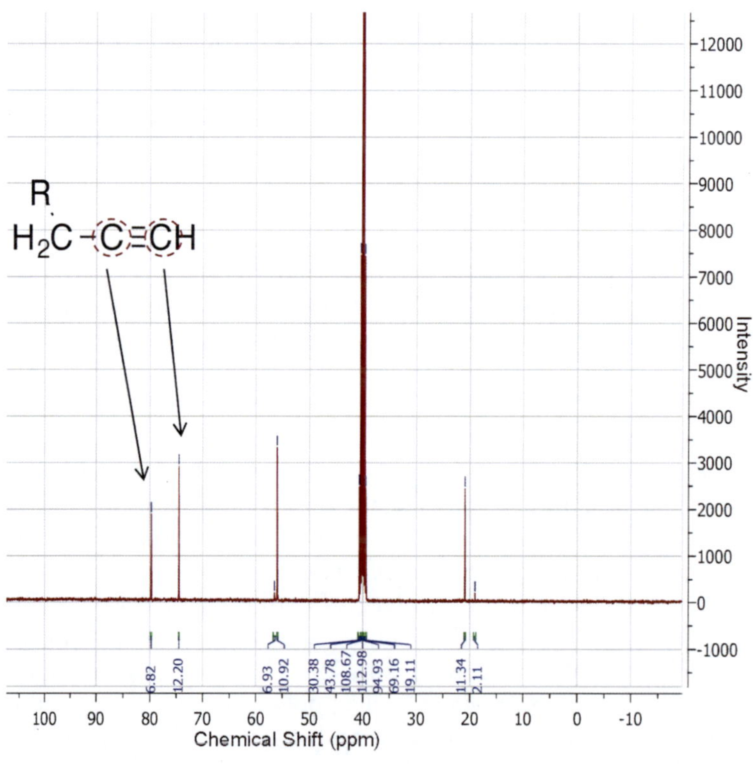

Chart 4 ^{13}C-NMR of the two acetylenic characteristic peaks at δ 70–80 ppm

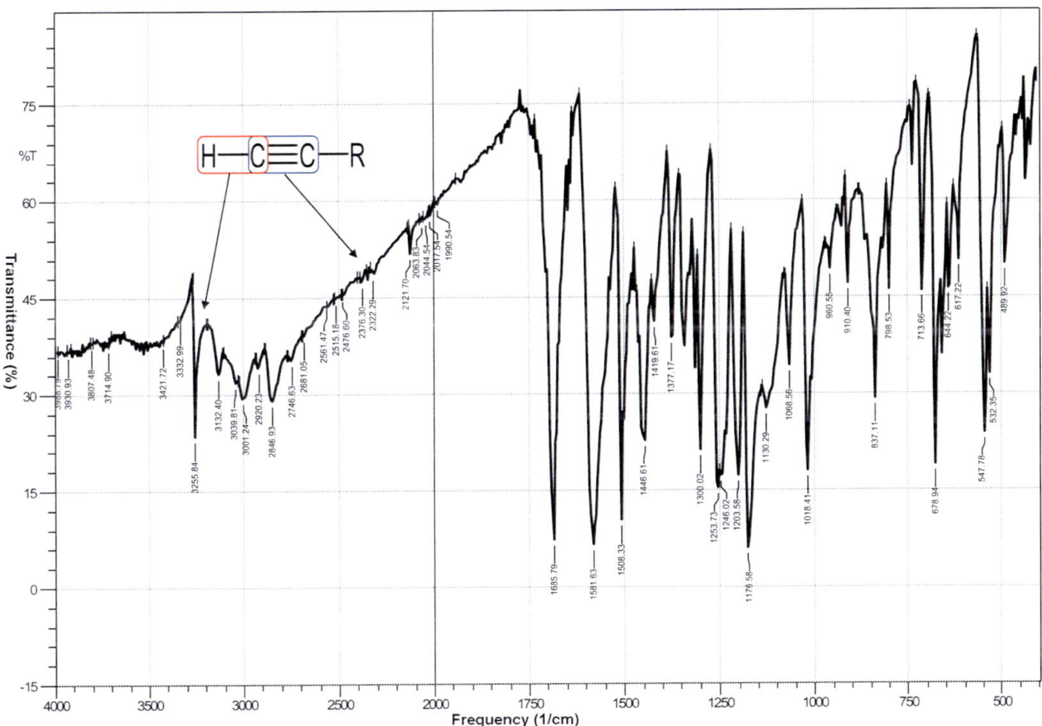

Chart 5 IR characteristic acetylenic–CH and C≡C bands in the ranges of 3200–3250 and 2200–2400 cm^{-1}

Metronidazole + P-TsCl — TEA, DMAP (catalytic), DCM → — NaN_3 / DMF, 60 °C →

Scheme 4 Tosylation of metronidazole followed by reaction with sodium azide

(Scheme 4) and is adopted from both Clayton et al [130] and Zhang et al [131].

From a practical perspective, to synthesis of azido derivative of metronidazole, p-TsCl is added to a magnetically and ice-cooled stirred solution of metronidazole, Triethylamine and DMAP (catalytic) in Dichloromethane portion-wise. The reaction mixture is stirred at room temperature overnight. Afterwards, solvent is removed under reduced pressure using rotary evaporator. Water is added to get rid of any inorganic salts. Subsequently, the formed residue is filtered under vacuum and air-dried. The product (metronidazole tosylate) can be recrystallized from methanol. Consequently, metronidazole tosylate is dissolved in DMF; and then; sodium azide is added to the reaction mixture. Afterwards, the reaction mixture is stirred at 60 °C for 3 h. Finally, ice–water mixture is added to the reaction mixture; and then the residue is filtered and

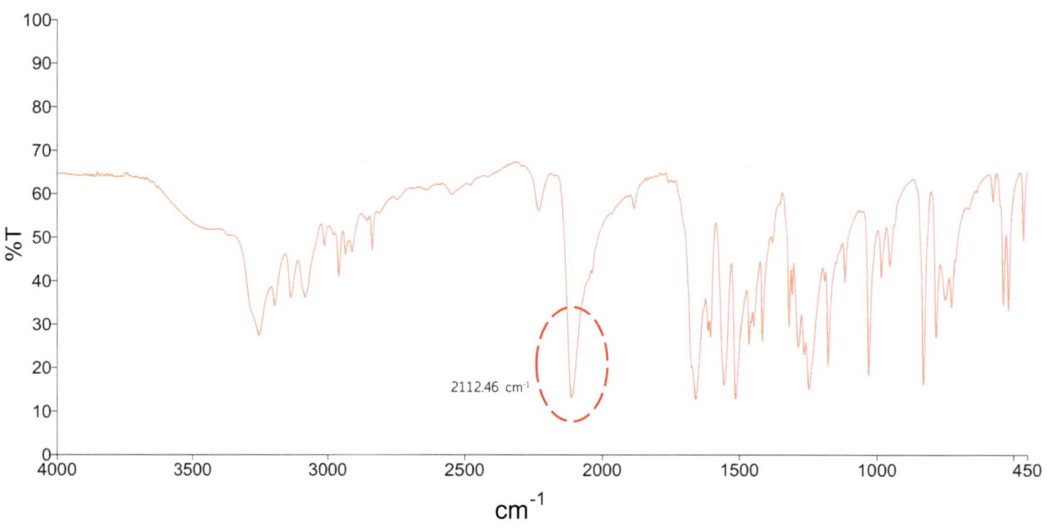

Chart 6 IR characteristic azido sharp band at 2100–2400 cm^{-1}

washed thoroughly with water, air-dried, and used directly in any subsequent reactions without any further purification.

In this synthesis, Infra-Red (IR) can be sufficient for primarily confirming the presence of azido group via a characteristic sharp band in the range of 2100–2400 cm^{-1} as shown in Chart 6.

5.3 Carbodiimide Crosslinkers in Bioconjugation

Carbodiimide compounds provide the most popular and versatile method for labeling or crosslinking to carboxylic acids. The most readily available and commonly used carbodiimides are the water-soluble EDC (1-ethyl-3-(3-dimethylaminopropyl)carbodiimide) for aqueous crosslinking and the water-insoluble DCC (dicyclohexyl carbodiimide) for nonaqueous organic synthesis methods. Carbodiimide conjugation works by activating carboxyl groups for direct reaction with primary amines via amide bond formation. Because no portion of their chemical structure becomes part of the final bond between conjugated molecules, carbodiimides are considered zero-length carboxyl-to-amine crosslinkers [132].

5.3.1 EDC Carbodiimide Crosslinker

EDC (or EDAC; 1-ethyl-3-(3-dimethylaminopropyl)carbodiimide hydrochloride) is the most popular carbodiimide used for conjugating biological substances containing carboxylates and amines. EDC reacts with carboxylic acid groups to form an active *O*-acylisourea intermediate that is easily displaced by nucleophilic attack from primary amino groups in the reaction mixture. The primary amine forms an amide bond with the original carboxyl group, and an EDC by-product is released as a soluble urea derivative. The *O*-acylisourea intermediate is unstable in aqueous solutions; failure to react with an amine results in hydrolysis of the intermediate, regeneration of the carboxyls, and the release of an *N*-unsubstituted urea [133] (Fig. 16).

EDC is water-soluble, which allows for its direct addition to a reaction without prior organic solvent dissolution. Both the reagent itself and the isourea formed as the byproduct of the crosslinking reaction are water-soluble and may be removed easily by dialysis or gel filtration (Fig. 17) [134].

EDC crosslinking is most efficient in acidic (pH 4.5) conditions and must be performed in buffers devoid of extraneous carboxyls and amines. MES buffer (4-morpholinoethanesulfonic acid) is a suitable carbodiimide reaction buffer. Phosphate buffers and neutral pH (up to 7.2) conditions are compatible with the reaction chemistry, albeit with lower efficiency; increasing the amount of EDC in a reaction solution can compensate for the reduced efficiency [132].

An EDC-mediated reaction to form an amide bond in aqueous solution involves a number of potential side reactions that can

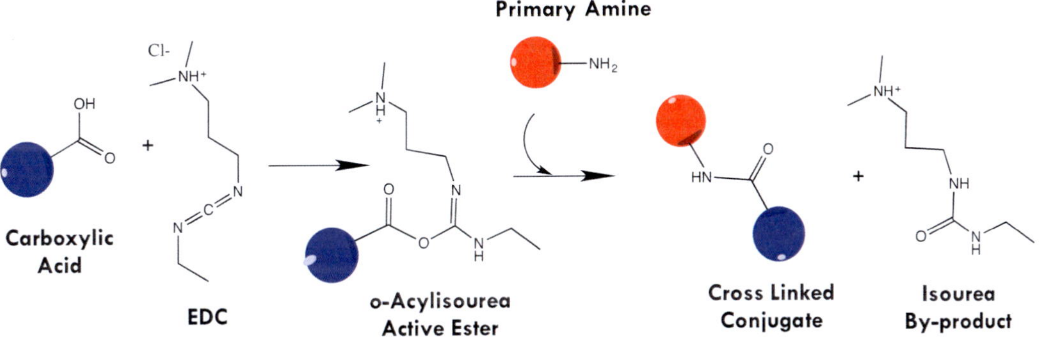

Fig. 16 EDC (carbodiimide) crosslinking reaction scheme

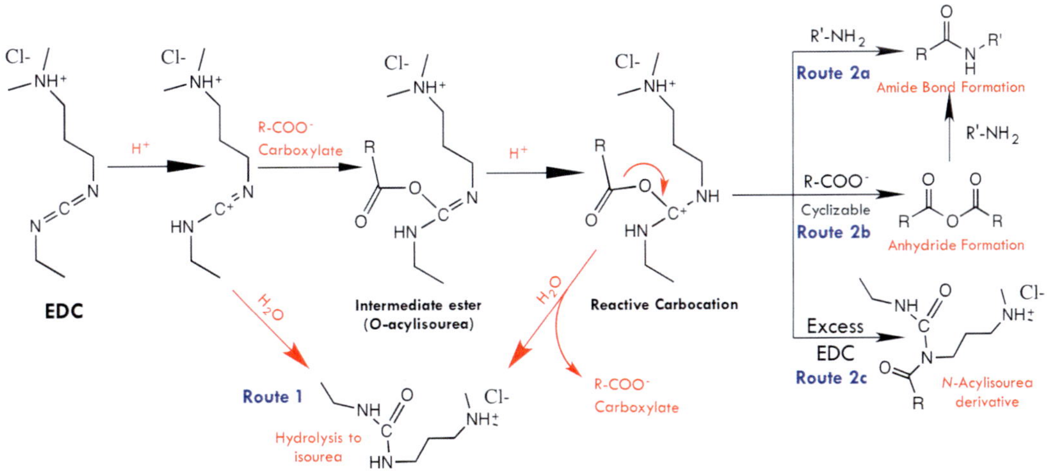

Fig. 17 EDC potential side reactions and their possible routes

occur in addition to the desired conjugation product as shown in Fig. 17. The reaction is initiated by protonation of one of the nitrogens on the imide group of the EDC, which results in the formation of an intermediate carbocation on the central carbon atom. At this point, the modified carbodiimide can itself hydrolyze to form an inactive isourea that no longer can participate in the reaction process (**Route 1**). Alternatively, it can also react with an available ionized carboxylate group to create the desired *O*-acylisourea reactive ester intermediate (**Route 2**). This ester again can accept another proton to form a second carbocation on the central carbon atom, and it is this form of the reactive ester that can react with an amine to create an amide bond (**Route 2a**). If a neighboring carboxylate group is in close proximity to the *O*-acylisourea ester, it may react with it, forming an anhydride intermediate which is also reactive with amine groups. The desired amide bond formation can still occur with at least one of the two carboxylates making up the anhydride (**Route 2b**). In addition, if EDC is in large excess over the amount of carboxylates present, then the intermediate ester may exist for a longer period; and potentially, it can rearrange by reacting with the neighboring secondary amines in the carbodiimide and, thus, form an inactive *N*-acylisourea derivative (**Route 2c**) [135–137].

Despite of the potential side reactions, it is amazing that EDC-mediated amide bond formation can be done with reproducibility, especially when scaling up reactions in production processes. The propensity for EDC to undergo side reactions may be a reason that high variability has been reported using the carbodiimide for particular conjugation reactions [132].

5.3.2 EDC/Sulfo-NHS Crosslinker

N-hydroxysuccinimide (NHS) or its water-soluble analog (sulfo-NHS) is often included in EDC coupling protocols to improve efficiency or create dry-stable (amine-reactive) intermediates. EDC couples NHS to carboxyls, forming an NHS ester that is considerably more stable than the *O*-acylisourea intermediate while allowing for efficient conjugation to primary amines at physiologic pH (Fig. 18).

The advantage of adding sulfo-NHS to EDC reactions is to increase the solubility and stability of the active intermediate, which ultimately reacts with the attacking amine. EDC reacts with a carboxylate group to form an active ester (*O*-acylisourea)-leaving group [132]. Unfortunately, this reactive complex is slow to react with amines and can hydrolyze in aqueous solutions. If the target amine does not find the active carboxylate before it hydrolyzes, the desired coupling cannot occur [138].

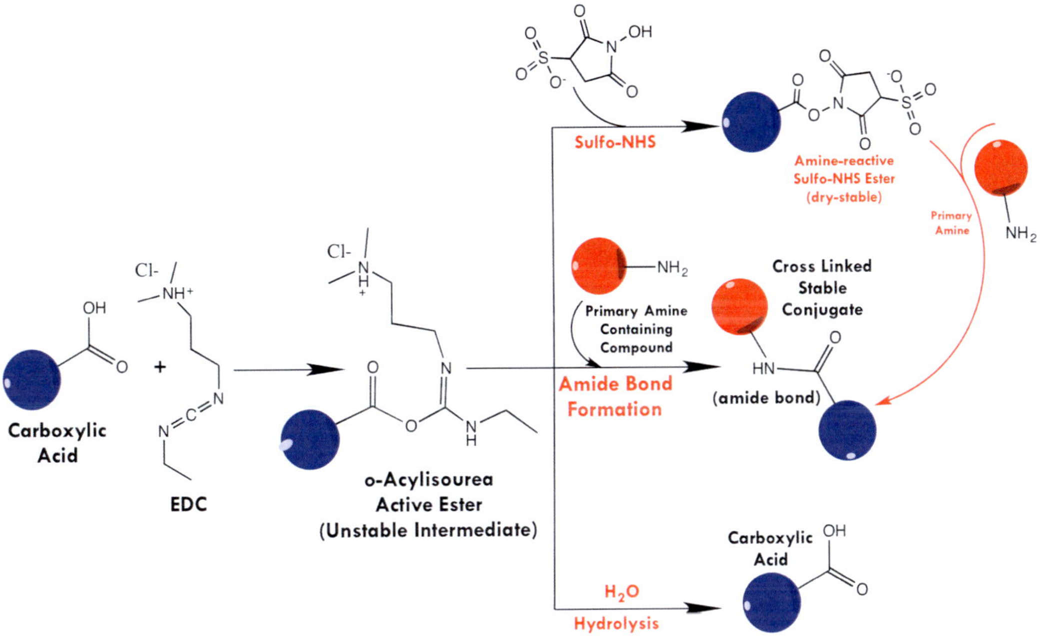

Fig. 18 Sulfo-NHS plus EDC (carbodiimide) crosslinking reaction scheme

5.3.3 DCC Carbodiimide Crosslinker

DCC (dicyclohexyl carbodiimide) is one of the most frequently used coupling agents, especially in peptide synthesis. It is water-insoluble, but it has been used in 80% DMF for the immobilization of small molecules onto carboxylate-containing chromatography supports for use in affinity separations [139, 140]. In addition to forming amide linkages, DCC has been used to prepare active esters of carboxylate-containing compounds using NHS or sulfo-NHS [141].

Unlike the EDC/sulfo-NHS reaction, active ester synthesis with DCC typically is carried out in organic solvent and, therefore, does not have the hydrolysis problems of water-soluble EDC--formed esters. Thus, DCC is most often used to synthesize active ester-containing crosslinking and modifying reagents, not to perform biomolecular conjugations. DCC is a waxy solid that is often difficult to remove from a bottle. Its vapors are extremely hazardous to inhalation and to the eyes. It should always be handled in a fume hood. The isourea byproduct of a DCC initiated reaction, dicyclohexyl urea (DCU) (Fig. 19), is also water-insoluble and must be removed by organic solvent washing. On the other hand, organic solvent washing is less troublesome for peptide synthesis on insoluble matrices because washing of the support material can be performed without disturbing the conjugate coupled to the support. For solution-phase chemistry, however, reaction products must be removed by solvent washings, precipitations, or recrystallizations [132].

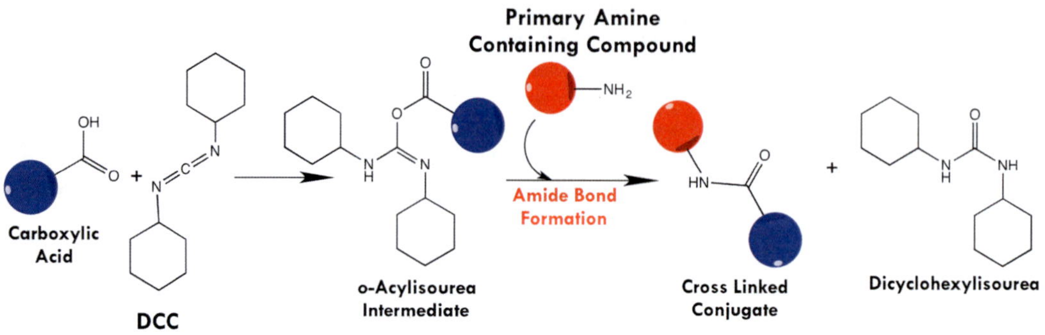

Fig. 19 DCC (carbodiimide) crosslinking reaction scheme

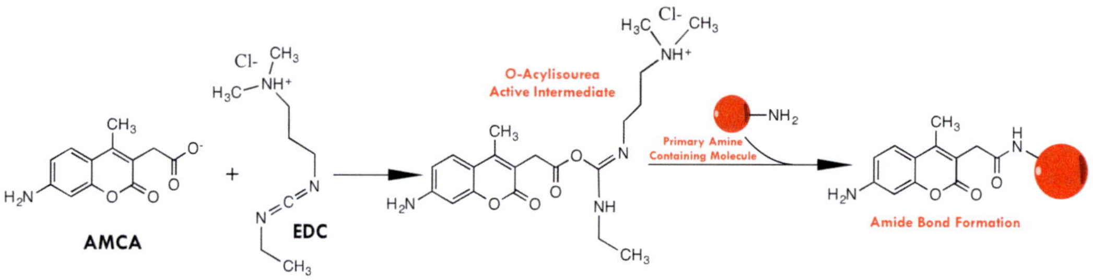

Fig. 20 AMCA linked to amine-containing molecules through its carboxylate group using EDC

5.3.4 Applications

AMCA, or 7-amino-4-methylcoumarin-3-acetic acid, is a fluorescent probe that exhibits a spectacular blue fluorescence. AMCA may be coupled to amine-containing molecules through the use of the carbodiimide reaction using EDC. EDC will activate the carboxylate on AMCA to a highly reactive *O*-acylisourea intermediate. Attack by a nucleophilic primary amine group on the carbonyl of this ester results in the formation of an amide bond (Fig. 20). Derivatization of AMCA off its carboxylate group causes no major effects on its fluorescent properties. Thus, proteins and other macromolecules may be labeled with this intensely blue probe and easily detected by fluorescence microscopy and other techniques [142, 143].

EDC/NHS reaction has also been used for the synthesis of BAP (biotinylated aminopyridine or 2-amino-(6-amidobiotinyl) pyridine), which is a derivative of D-biotin made by reacting the NHS ester of this vitamin with 2,6-diaminopyridine (DAP) in large molar excess. The resultant compound has fluorescent properties due to the presence of the aminopyridine ring, and its remaining free amine group may be used to modify reducing saccharides and glycans by reductive amination (Fig. 21). BAP can be used to label oligosaccharides under mild conditions and without any carbohydrate structural degradation. Carbohydrate structural

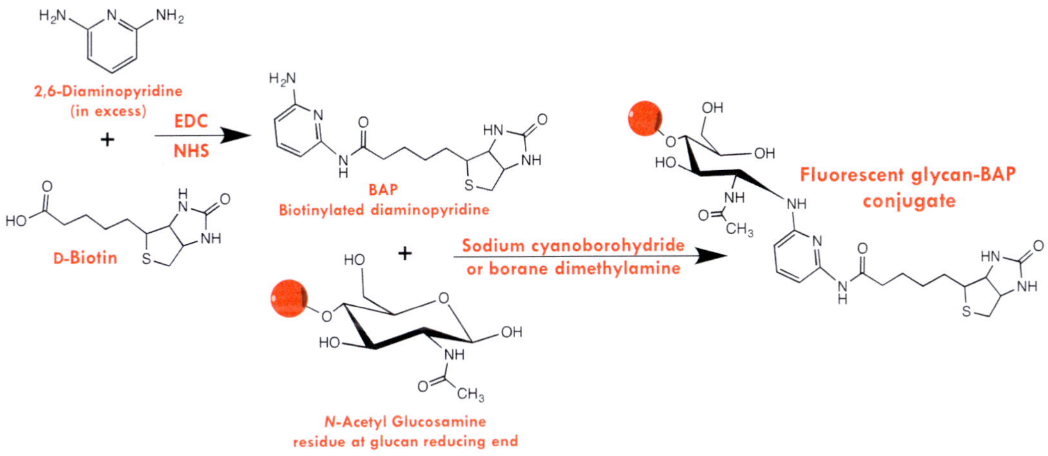

Fig. 21 Synthesis of BAP (by the reaction of an excess of diaminopyridine with biotin in the presence of EDC/NHS) and its use to label the reducing end of released glycans by reductive amination in the presence of a reducing agent

degradation usually results from periodate oxidation of carbohydrates. After modification, the glycans or carbohydrates can be analyzed by chromatography, electrophoresis, or mass spectrometry [144, 145].

5.3.5 Practical Perspective: Synthesis of Cystamine/Drug Conjugate Through Amide Bond Formation

Cystamine is a linear aliphatic diamine composed of a disulfide bridge, generated from the oxidation of two cysteamine residues. Cystamine and cysteamine are both organic compounds which are constitutively related to coenzyme A metabolism in all tissues. Because they coexist in the cell in a redox equilibrium, their metabolism and biological functions are closely related. Cystamine and cysteamine have been shown to protect the liver against acetaminophen poisoning, via their enhancement of the antioxidant glutathione (GSH) system. More recently, they were recognized for their antiviral activity against influenza A and hepatitis A and, more remarkably, against the human immunodeficiency virus (HIV)-1. Apart from being rapidly converted to cysteamine, cystamine can also be metabolized to cysteine, hypotaurine, and taurine, all of which are endogenous cellular components [146].

Cysteamine has been reported as a chemo-sensitization and radioprotective agent, and its antitumor effects have been investigated in various tumor cell lines and chemical-induced carcinogenesis [147].

Moreover, the NSAID (nonsteroidal anti-inflammatory drug) indomethacin, is known to possess anticancer activity against CRC (colorectal cancer) and other malignancies in humans It has been shown that indomethacin selectively activates the dsRNA (double-stranded RNA)-dependent protein kinase PKR in a cyclooxygenase-independent manner, causing rapid

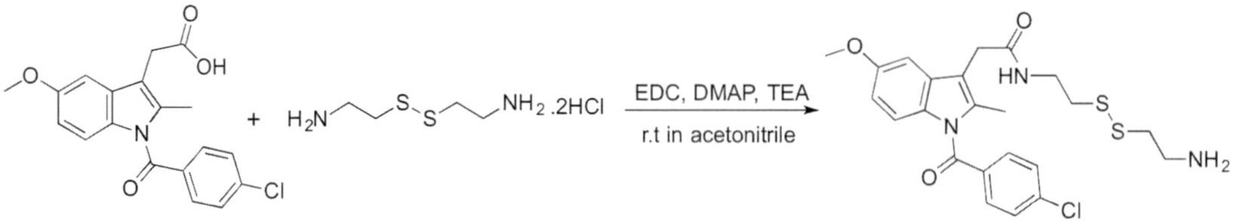

Scheme 5 Synthesis of cystamine/indomethacin amide conjugate

phosphorylation of eIF2α (the α-subunit of eukaryotic translation initiation factor 2) and inhibiting protein synthesis in colorectal carcinoma and other types of cancer cells [148]. Thus, it was thought of interest to merge cystamine and indomethacin into a conjugate aiming at synergistic anticancer activity (Scheme 5).

In this regard, a mixture of indomethacin, EDC and DMAP (catalytic) in Acetonitrile is magnetically stirred at room temperature for 48 h. The reaction mixture is added portion-wise to cooled suspension of cystamine dihydrochloride in Triethyl amine and stirred at room temperature overnight. Solvent is removed under reduced pressure and water is added to get rid of any salts and excess cystamine. Finally, the residue formed is filtered under vacuum and air-dried.

The structure of the final product can be confirmed using ^{1}H-NMR and IR. Comparing ^{1}H-NMR charts for indomethacin and indomethacin/cystamine conjugate, the conjugate chart should show both characteristic singlet and broad multiplet (due to partial double bond character) in the ranges of δ 8.75 and 10.62–10.75 ppm, that correspond to NH_2 and amide NH, respectively (Chart 7a). In addition, the aliphatic region showed the overlapping triplets due to four cystamine CH_2 groups in the ranges of δ 2.27–3.69 ppm (Chart 7b). This is associated with the characteristic peaks for indomethacin appearing at their expected chemical shifts (Chart 7c). As for IR spectra, the characteristic amide C=O and NH stretching bands were shown at 1639 and 2931 cm^{-1}, respectively (Chart 8a). The amide formation could also be confirmed by the disappearance of the carboxylic acid C=O stretching band which could be noticed in indomethacin chart alone (Chart 8b).

5.3.6 Practical Perspective: Synthesis of Folate-Poly(Ethylene Glycol)-Carboxylic Acid Conjugate (FA-PEG$_{3000Da}$-COOH)

PEG modification (PEGylation) is considered one of the most valuable drug delivery carriers and drugs modification. It aids in prolonging the active pharmaceutical ingredient circulation in the bloodstream and prolonging its half-life. Furthermore, PEGylation also provides a flexible platform for further drug/carrier system modifications such as grafting a ligand for specific cell targeting or connecting an imaging agent for dual-acting drug delivery/imaging systems. Targeting ligands, such as antibodies [149], growth

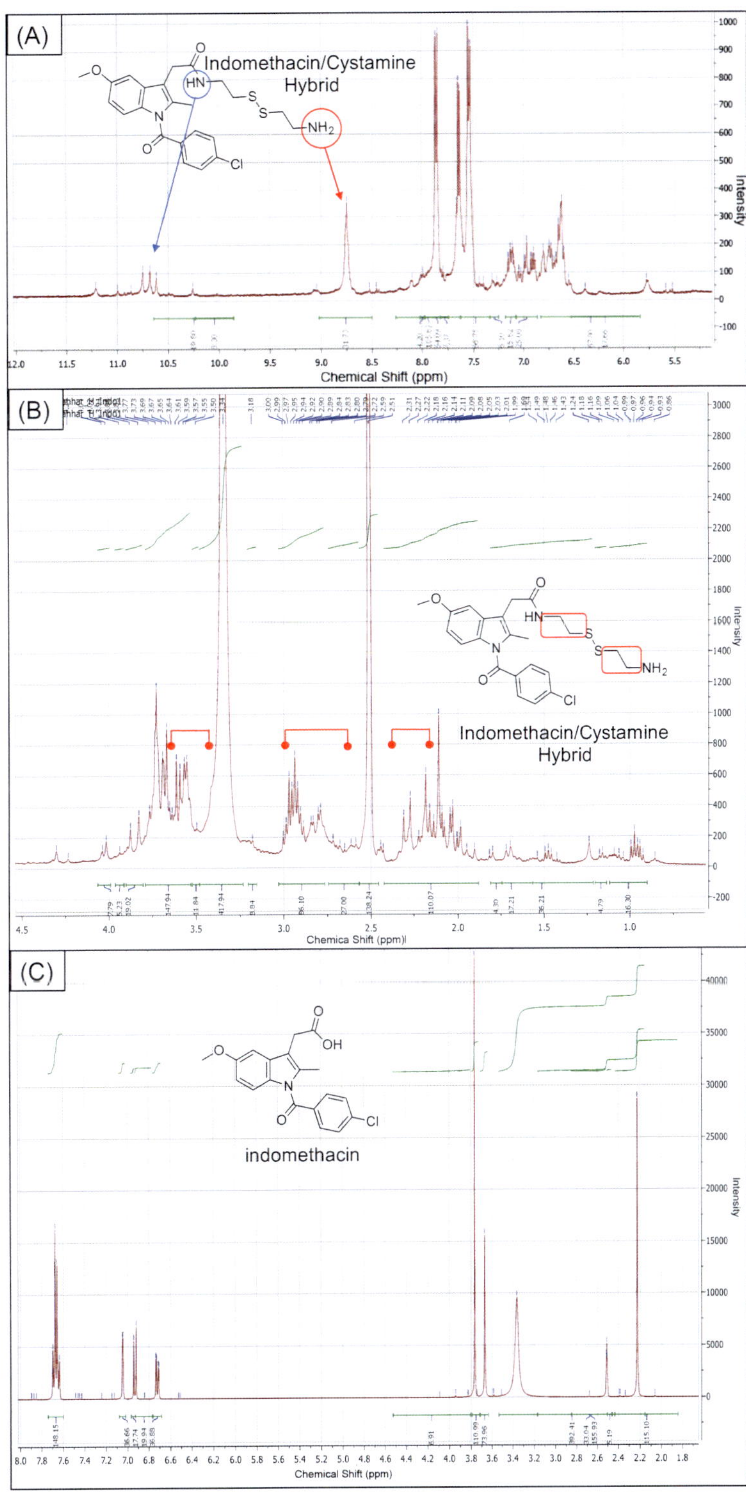

Chart 7 (**a**) NH and NH_2 ^1H-NMR signals of indomethacin/cystamine conjugate, (**b**) Cystamine CH_2 ^1H-NMR signals of indomethacin/cystamine conjugate, (**c**) Characteristic ^1H-NMR signals for indomethacin alone (for comparison)

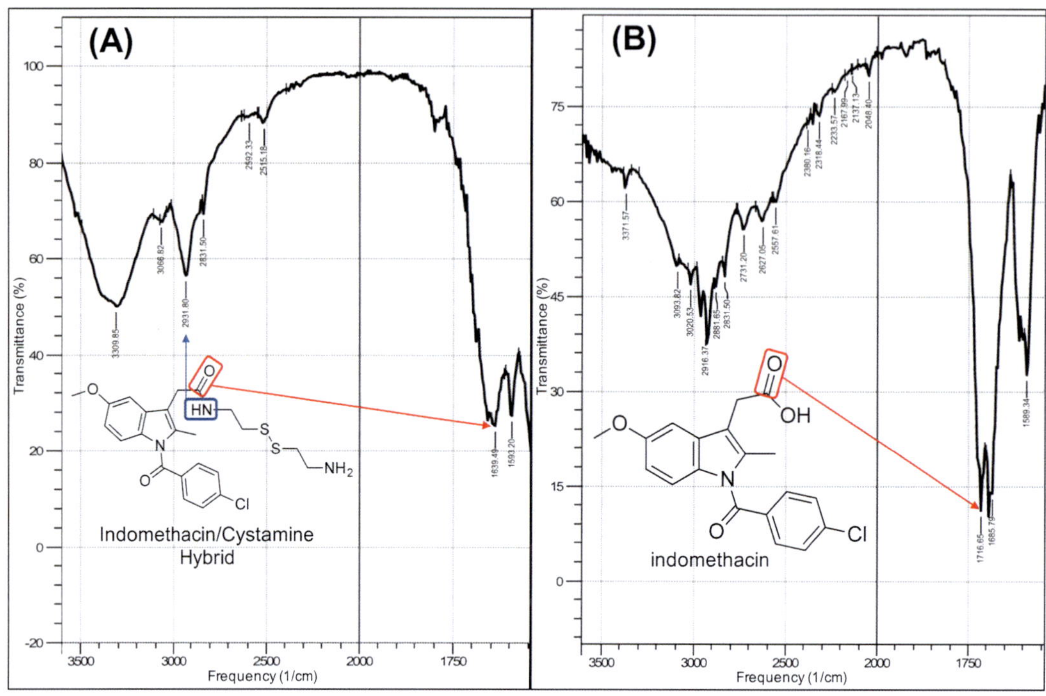

Chart 8 (**a**) Amide IR bands of indomethacin/cystamine conjugate, (**b**) Carboxylic C=O IR band of indomethacin (for comparison)

factors [150], peptides [151], and folic acid [152, 153] have been used extensively in creating smart-targeted bioconjugates. Folic acid is a widespread ligand for anti-cancer drugs and drug delivery systems because folate receptors (FRs) are frequently overexpressed in numerous cancer cells [154]. On the contrary, folate receptors (FRs) are rarely expressed on the normal cell surface [155].

In this protocol, we would synthesize folate-poly(ethylene glycol)-carboxylic acid conjugate (FA-PEG_{3000Da}-COOH), which could be used as a flexible backbone to be conjugated to various drugs or imaging agents for cancer therapy (Scheme 6). In this scheme, DCC/NHS crosslinkers (couplers) are used. The following protocol for coupling folic acid with NH_2-PEG_{3000Da}-COOH (Scheme 6) is adopted from Gabizon et al. [156] and Park et al. [157, 158]. Such FA-PEG_{3000Da}-COOH conjugate would provide the dual smart functionality of extending the drug half-life and circulation time along with enhanced cancer targetability.

To activate Folic acid (Scheme 6), folic acid, dicyclohexylcarbodiimide, and *N*-hydroxysuccinimde are dissolved in DMSO (molar ratio of FA:DCC:NHS = 1:1.1:1.1) and stirred for 24 h in a dark room at room temperature. The white precipitate

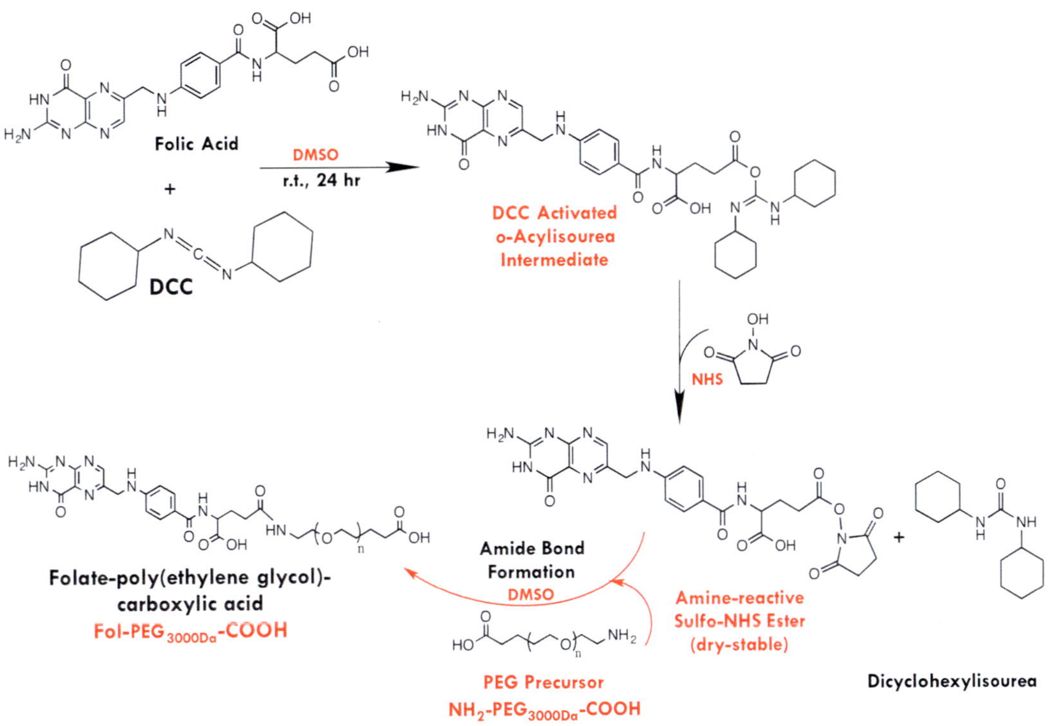

Scheme 6 Synthesis of folate-poly(ethylene glycol)-carboxylic acid conjugate (FA-PEG_{3000Da}-COOH)

(reaction by-product, dicyclohexyl urea) formed is removed via filtration.

Following the folic acid activation step, the esterification step takes place (Scheme 6). In the esterification step, PEG Precursor (NH_2-PEG_{3000Da} COOH) is dissolved in DMSO and added to the now activated folic acid and stirred for 12 h in a dark room at room temperature. After reaction completion, the reaction solution is evaporated via rotary evaporation and dialyzed in deionized water using a dialysis tube (2000 MWCO) to remove any low molecular weight contaminants (unconjugated activated folic acid and regenerated NHS) for 48 h. Finally, the folic acid conjugated poly(ethylene glycol) (FA-PEG_{3000Da}-COOH) is obtained on freeze-drying, yielding a yellow solid. Yield should be in the range of 40–70% over the two steps based on PEG content.

The chemical structure of synthesized FA-PEG_{3000Da}-COOH conjugate can be characterized by FT-IR spectrometer (Thermo-Mattson Satellite, Model 960M0017) using potassium bromide (KBr) pellets (Chart 9). Proton nuclear magnetic resonance spectrum (^{1}H NMR) of synthesized FA-PEG_{3000Da}-COOH conjugate can be obtained on Varian Inova 300 MHz instrument using deuterated chloroform ($CDCl_3$) as solvent (Chart 10). The molecular weight of the of synthesized FA-PEG_{3000Da}-COOH conjugate can be determined by DE Pro Workstation (ABI, Applied

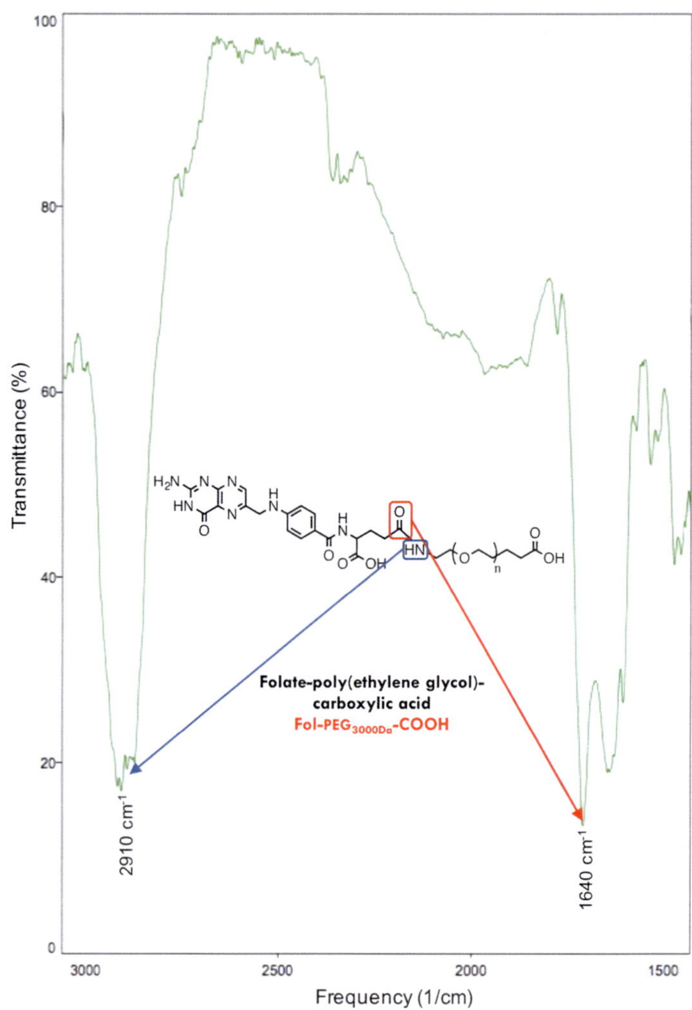

Chart 9 Amide IR bands of PEG_{3000Da}-COOH conjugate

Biosciences Inc.) mass spectrometer with MALDI ionization. The matrix used is saturated alpha-cyano-4-hydroxycinnamic acid dissolved in 50/50 acetonitrile/water with 0.1% TFA with conjugate concentration of 1 mg/ml. The conjugate is spotted with an equal volume of matrix and air-dried on the plate (Chart 11). For IR spectra, the characteristic amide C=O and NH stretching bands were shown at 1640 and 2910 cm^{-1}, respectively (Chart 9). The ^{1}H-NMR chart for FA-PEG_{3000Da}-COOH conjugate should show the aliphatic region overlapping multiplets due to multiple CH_2 groups in the ranges of δ 3.20–3.69 and 1.40–1.52 ppm characteristic to Polyethylene Glycol (PEG) (Chart 10). The peaks in the ^{1}H-NMR chart for FA-PEG_{3000Da}-COOH (Chart 10) are broad and not as sharp as the peaks in the ^{1}H-NMR charts reported for the previous examples (Indomethacin for example,

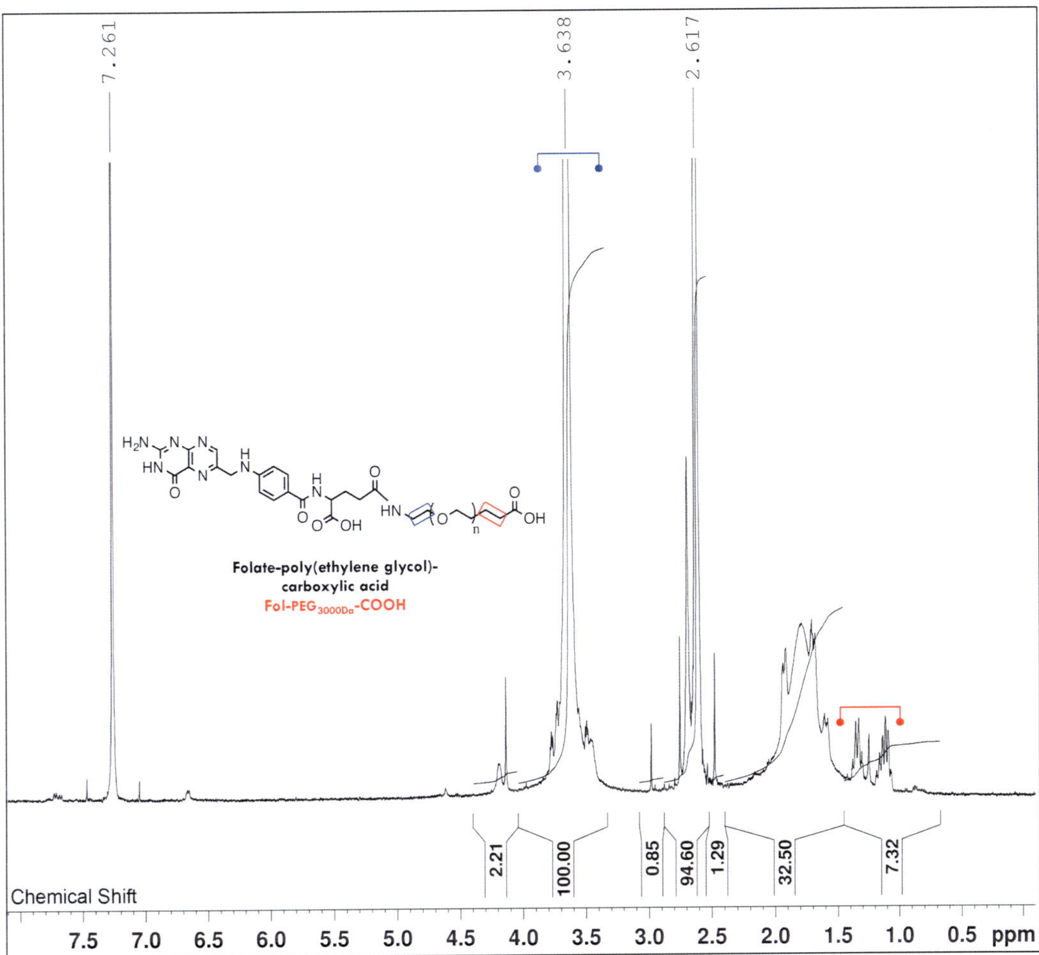

Chart 10 ^{1}H-NMR signals of PEG_{3000Da}-COOH conjugate

Chart 7c), due to the polymeric nature of PEG_{3000Da}-COOH [159]. In order to minimize the ^{1}H-NMR peaks broadening problem in polymers, it is recommended to increase the number of scans in each ^{1}H-NMR experiment/run. When undergoing a ^{1}H-NMR run, increasing the number of scans from 16 (~2 min experiment time) to 64 (about 5–6 min) would drastically increase the signal-to-noise ratio (S/N), subsequently generating sharper peaks for polymeric compounds [160]. Since the length of the experiment is directly proportional to the number of scans during the ^{1}H-NMR experiment, this approach to enhance the signal characteristics and decrease its broadening can become extremely time-prohibitive. The number average molecular weight (M_n) is found to be 3446 Da with a polydispersity index (PDI) of 1.06 as determined with MALDI-TOF (Chart 11). The polydispersity index (PDI) of 1.06 ensures the uniphasic distribution of the synthesized PEG_{3000Da}-COOH conjugate.

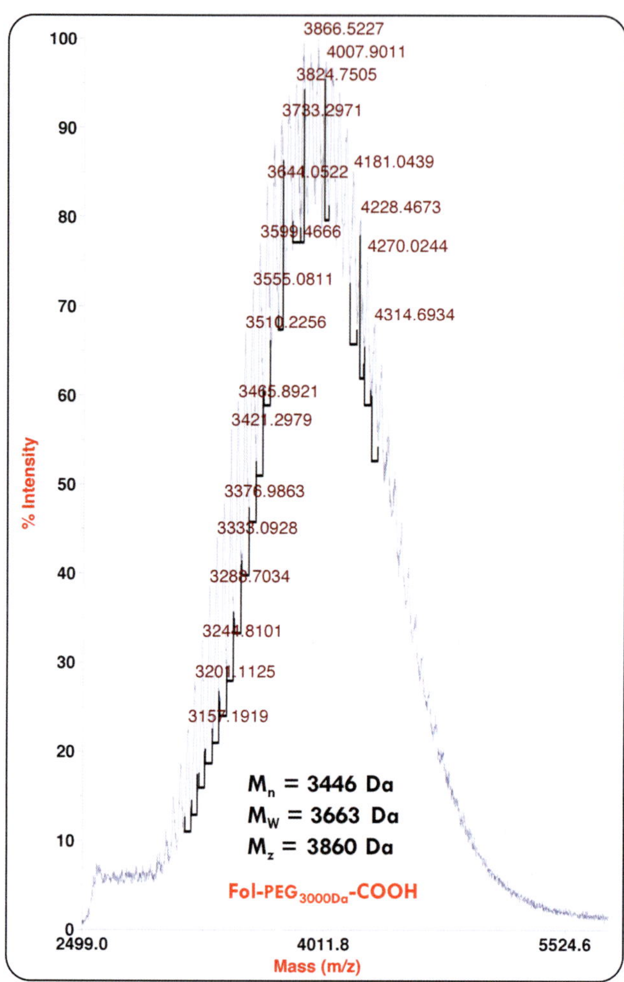

Chart 11 MALDI-TOF analysis of PEG_{3000Da}-COOH conjugate

5.4 (Strept) Avidin–Biotin System in Bioconjugation

Utilizing the natural strong binding of (strept)avidin for the small molecule biotin is one of the most widely used methods of non-covalent conjugation. The avidin–biotin interaction has been employed in nanoscale drug delivery systems for pharmaceutical agents, including small molecules, proteins, vaccines, monoclonal antibodies, and nucleic acids [161].

Furthermore, the derivatives of biotin and streptavidin can easily build dimeric, trimeric, or tetrameric complexes through modifying residues at four identical subunits of streptavidin each with a binding site for biotin. The specific characteristics and flexibility of avidin–biotin have been shown to be advantageous in the areas of sensing, labeling, and site-specific delivery (Fig. 22) [162].

Avidin is a basic tetrameric glycoprotein consisting of four identical subunits; each subunit binds to biotin with high specificity and affinity (K_D, the equilibrium dissociation constant between the antibody and its antigen, ~10^{-15} M) (Fig. 23) [163]. Avidin is

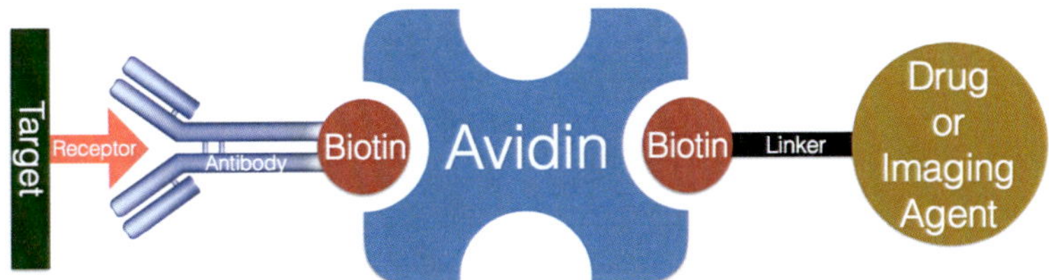

Fig. 22 Schematic diagram of (strept)avidin–biotin system

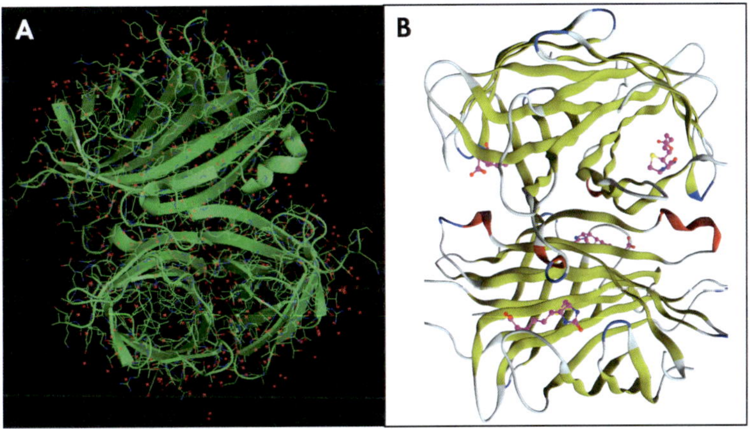

Fig. 23 3D view of streptavidin binding four biotin molecules generated by (**a**) MacPyMOL, PyMOL v1.7.4.5 Schrödinger, LLC and (**b**) MOE 2016.0802 (PDB ID 4YVB for structure of D128N streptavidin) [163]

originally derived from the eggs of aves, reptiles, and amphibians. Avidin–biotin interaction is considered one of the most specific and stable noncovalent interactions (Fig. 24), which is about 10^3–10^6 times higher than an antigen-antibody interaction [163, 164].

The biggest merit of such system is its high affinity interaction, which is robust and stable against manipulation, proteolytic enzymes, temperature, pH, harsh organic reagents, and other denaturing reagents. In addition, biotin-based conjugates are synthetically feasible and have less impact on the activity of the biomolecules. Biotin and avidin are also readily available with various functional groups for chemical conjugations. Moreover, the chemical conjugation on biotin or avidin avoids direct modification of the active biomolecule, thus maintaining their activity [161].

Despite its enormous advantages and wide applicability, avidin has several limitations including nonspecific binding and possible immunogenicity. To circumvent these limitations, tremendous efforts have been devoted to discovering and engineering superior variants of avidin by genetic modification or finding a completely new source, e.g., a different species.

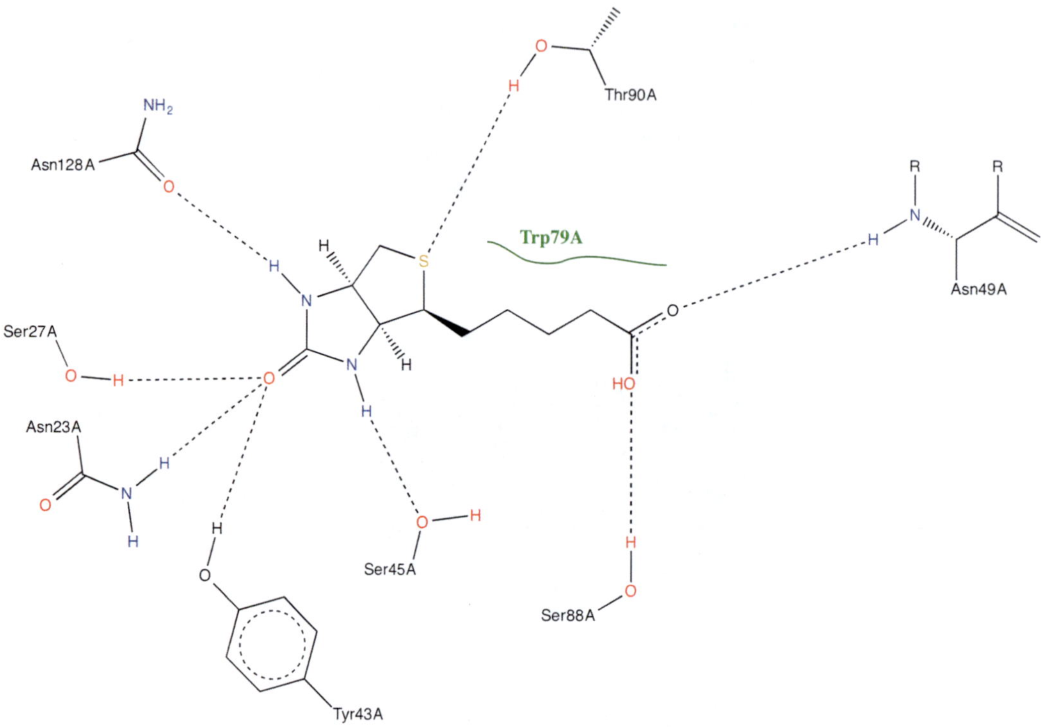

Fig. 24 2D streptavidin–biotin interactions from protein data bank (PDB ID 4YVB for structure of D128N streptavidin). Black dashed lines indicate hydrogen bonds and green solid line shows hydrophobic interactions [163]

The most widely used analogue of avidin is streptavidin. Derived from *Streptomyces avidinii*, streptavidin is a ~56 kDa nonglycosylated tetrameric protein that binds to four biotins with a K_D of ~10–14 M [165]. Homologs of streptavidin have been discovered from other species, including fungus, bacteria, chickens, and frogs [166]. Similar to avidin, streptavidin is also resistant to denaturing agents, temperature, pH, and proteolytic enzymes. Despite having a tertiary/quaternary structure and amino acid arrangement similar to those of avidin, streptavidin only shows a moderate sequence homology level of ~30% sequence identity and 40% similarity with avidin [167, 168]. Moreover, streptavidin is nonglycosylated and has a slightly acidic pH of ~5–6 [169, 170]. Due to its different physical-chemical properties, streptavidin shows an in vivo tissue distribution and clearance profile very different from those of avidin [169]. Furthermore, streptavidin protects the biotinyl esters from hydrolysis, whereas avidin augments this hydrolysis.

Biotin is a vitamin also known as vitamin H, vitamin B7, or coenzyme R. Biotin is composed of a tetrahydrothiophene ring fused to a tetrahydroimidizalone (ureido) ring. It plays a key role in cell signaling and acts as a cellular growth promoter. Biotin receptor (sodium-dependent multivitamin transporter and high-affinity biotin transporter) is widely expressed in nearly all living cells. Moreover, its expression in dividing cancer cells is higher than in normal cells, making biotin a potential targeting moiety for cancer therapeutics [171]. Extensive effort has therefore been made to develop biotin-based platforms for tumor targeting and diagnosis.

Li and coworkers [172] developed a multifunctional ternary complex-based cancer targeting drug (DOX) and gene codelivery system. This multifunctional vector was constructed based on polyion complex micelles (PIC) and the avidin–biotin system. As described by Li et al. [172], DOX-conjugated PIC (PIC-D) were prepared based on poly(L-aspartic acid) (PASP) and poly(2-(2-aminoethylamino)ethyl methacrylate) (PAEAEMA). pDNA was added to generate PIC-D–pDNA NPs with positively charged surfaces. Then, the negative macromolecule avidin–biotin–PEG-co-poly-(L-glutamate acid) (AB) was coated onto PIC-D–pDNA NPs to build AB–PIC-D–pDNA complexes. Finally, using avidin–biotin technology, the transferrin was functionalized on the surface of complexes as a targeting unit to generate TAB–PIC-D–pDNA ternary complexes. The complexes protected pDNA against nuclease degradation, minimized interference of blood proteins, facilitated tumor cell uptake, and delivered both DOX and the gene payload. In vitro cell tests indicated that TAB–PIC-D–pDNA complexes had increased transfection efficiency in the serum and enhanced luciferase expression in HeLa and HepG2 cells. This multifunctional ternary complex was an efficient carrier for the targeted release of anticancer drugs and genes.

The application of avidin–biotin system is more than just coupling biotinylated molecules to avidin-conjugated moieties. Properties such as higher relative tumor accumulation, immune modulation, and easy genetic engineering make them highly advantageous for a variety of applications in nanotechnology. Particularly in the pretargeting field, the avidin–biotin system is a leap ahead of the conventional radiolabeled antibody approach.

However, there is a concern regarding the application of avidin–biotin technology in vivo because of the potential immunogenic risk of modified streptavidin [173, 174]. Even though elevated anti-avidin antibodies in human cancer patients and mice did not significantly reduce the therapeutic properties or induce serious immune responses, possible biases related to an immunogenic response against avidin should be considered [175]. As a means of improving this issue, alternative biotin-binding proteins have been developed. NeutrAvidin was developed for reducing the immunogenic potential, and CaptAvidin was designed to have less nonspecific binding [176].

6 Future Directions

For the past couple of decades, academic research has been mainly focusing on novel carrier systems and nanoparticulate colloidal technologies for drug delivery, such as nanoparticles, nanospheres, vesicular systems, liposomes, and nanocapsules. Such efforts aided in the creation of newly marketed products such as Doxil® in the market [177, 178]. Such systems provide the tools to custom design a superior drug delivery system, impart novel functions to old drugs such as longer half-life and stealth properties (as in the case of Doxil®), and provide them with either passive or active targeting properties via grafting the carrier system with targeting moieties and/or imaging agents or another drug within the same carrier system [179]. Such technologies opened the gate towards more sophisticated and effective multi-acting platform(s) which can offer site-targeting, imaging, and treatment using a single multi-functional system [180]. Unfortunately, such technologies are faced with major problems including high cost, low stability profile, short shelf-life, and poor reproducibility across and within production batches leading to harsh bench-to-bedside transformation. Major process and formulation development concerns exist with respect to scale-up processes of complex nanoparticluate carriers. Most of the reagents and inactive moieties in the formulation of such novel therapeutic systems are not included in the FDA-approved inactive ingredient database (IID).

On the other hand, pharmaceutical industry invested heavily in bioconjugate structures. Bioconjugate technologies offered an attractive alternative to nanoparticulate carriers with all its flexible advantages when it comes to custom design and tailor grafting along with avoiding most of its disadvantages. Bioconjugates offer the flexibility of custom designing personalized products. Bioconjugates facilitate simple and easy drug (active pharmaceutical ingredient) conjugation, using various smart biocompatible, bioreducible, or biodegradable linkers, to targeting agents, PEG layer, or another drug. Such technology enables the construction of smart multifunctional platform(s) offered by nanoparticulate carriers. Furthermore, conjugates are still considered chemical compounds. This fact simply allows the use of traditional analytical and manufacturing technologies in the characterization and manufacturing of traditional active pharmaceutical ingredients offering high probability for their successful transition from bench to bedside (Fig. 25). Moreover, the final formulation could be a simple injectable or solid formulation, which offers long shelf-life and enhanced stability profile.

Subsequently, bioconjugation technologies can aid in creating safer, cheaper, stable, and effective novel therapeutics. It can also be a rate-limiting step in reinventing old drugs and imparting new

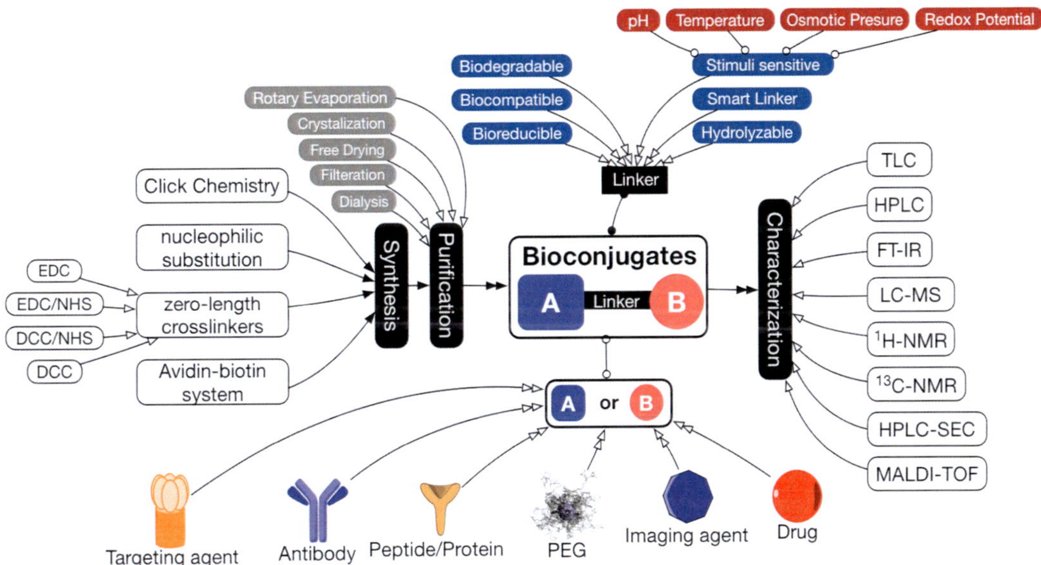

Fig. 25 Schematic diagram of bioconjugates' structure, design, synthesis, purification, and characterization

functions to them that would enhance their targetability, pharmacokinetic, and pharmacodynamic parameters, and their overall formulation patient compliance, easing their transition to market. A major focus should be the transformation of such novel bioconjugates' technologies from bench to bedside. The use of click chemistry, bioconjugation technologies, ligand post-insertion, and labeling techniques need to be extensively researched for ease of scale-up and proper bench-to-bedside transformation.

Consequently, a current focus is on simple bioconjugate structures, which can be easily synthesized with high yield, reduced cost, and high stability profile of the final formulation. This could provide a practical direction for the development of novel management tools and therapeutics, paving the road to affordable, scalable, stable, efficient, and safe disease-management strategies.

References

1. Hermanson GT (2013) Chapter 1 - introduction to bioconjugation. In: Bioconjugate techniques, 3rd edn. Academic Press, Boston, pp 1–125. https://doi.org/10.1016/B978-0-12-382239-0.00001-7
2. Kalia J, Raines RT (2010) Advances in bioconjugation. Curr Org Chem 14(2):138–147
3. Vicent MJ, Duncan R (2006) Polymer conjugates: nanosized medicines for treating cancer. Trends Biotechnol 24(1):39–47. https://doi.org/10.1016/j.tibtech.2005.11.006
4. Duncan R (1992) Drug-polymer conjugates: potential for improved chemotherapy. Anti-Cancer Drugs 3(3):175–210
5. Duncan R (2006) Polymer conjugates for drug targeting. From inspired to inspiration! J Drug Target 14(6):333–335. https://doi.org/10.1080/10611860600833880
6. Duncan R (2006) Polymer conjugates as anticancer nanomedicines. Nat Rev Cancer 6 (9):688–701. https://doi.org/10.1038/nrc1958
7. Faghihnejad A, Feldman KE, Yu J, Tirrell MV, Israelachvili JN, Hawker CJ, Kramer EJ, Zeng H (2014) Adhesion and surface interactions of a self-healing polymer with multiple hydrogen-bonding groups. Adv Funct Mater

24(16):2322–2333. https://doi.org/10.1002/adfm.201303013

8. Geyik C, Evran S, Timur S, Telefoncu A (2014) The covalent bioconjugate of multi-walled carbon nanotube and amino-modified linearized plasmid DNA for gene delivery. Biotechnol Prog 30(1):224–232. https://doi.org/10.1002/btpr.1836
9. Duncan R (2003) The dawning era of polymer therapeutics. Nat Rev Drug Discov 2 (5):347–360. https://doi.org/10.1038/nrd1088
10. Kroon J, Metselaar JM, Storm G, van der Pluijm G (2014) Liposomal nanomedicines in the treatment of prostate cancer. Cancer Treat Rev 40(4):578–584. https://doi.org/10.1016/j.ctrv.2013.10.005
11. Niraula S, Seruga B, Ocana A, Shao T, Goldstein R, Tannock IF, Amir E (2012) The price we pay for progress: a meta-analysis of harms of newly approved anticancer drugs. J Clin Oncol 30(24):3012–3019. https://doi.org/10.1200/JCO.2011.40.3824
12. Solyanik GI (2010) Multifactorial nature of tumor drug resistance. Exp Oncol 32 (3):181–185
13. Chari RVJ, Miller ML, Widdison WC (2014) Antibody–drug conjugates: an emerging concept in Cancer therapy. Angew Chem Int Ed 53(15):3796–3827. https://doi.org/10.1002/anie.201307628
14. Kim EG, Kim KM (2015) Strategies and advancement in antibody-drug conjugate optimization for targeted Cancer therapeutics. Biomol Ther (Seoul) 23(6):493–509. https://doi.org/10.4062/biomolther.2015.116
15. Diamantis N, Banerji U (2016) Antibody-drug conjugates—an emerging class of cancer treatment. Br J Cancer 114(4):362–367. https://doi.org/10.1038/bjc.2015.435
16. Ravin HA, Seligman AM, Fine J (1952) Polyvinyl pyrrolidone as a plasma expander; studies on its excretion, distribution and metabolism. N Engl J Med 247 (24):921–929. https://doi.org/10.1056/nejm195212112472403
17. Shelanski HA, Shelanski MV (1956) PVP-iodine: history, toxicity and therapeutic uses. J Int Coll Surg 25(6):727–734
18. Ringsdorf H (1975) Structure and properties of pharmacologically active polymers. J Polym Sci: Polym Symp 51(1):135–153. https://doi.org/10.1002/polc.5070510111
19. Matsumura Y, Maeda H (1986) A new concept for macromolecular therapeutics in cancer chemotherapy: mechanism of tumoritropic accumulation of proteins and the antitumor agent smancs. Cancer Res 46(12 Pt 1):6387–6392
20. Meerum Terwogt JM, ten Bokkel Huinink WW, Schellens JH, Schot M, Mandjes IA, Zurlo MG, Rocchetti M, Rosing H, Koopman FJ, Beijnen JH (2001) Phase I clinical and pharmacokinetic study of PNU166945, a novel water-soluble polymer-conjugated prodrug of paclitaxel. Anti-Cancer Drugs 12 (4):315–323
21. Wachters FM, Groen HJM, Maring JG, Gietema JA, Porro M, Dumez H, de Vries EGE, van Oosterom AT (2004) A phase I study with MAG-camptothecin intravenously administered weekly for 3 weeks in a 4-week cycle in adult patients with solid tumours. Br J Cancer 90(12):2261–2267
22. Duncan R, Vicent MJ (2010) Do HPMA copolymer conjugates have a future as clinically useful nanomedicines? A critical overview of current status and future opportunities. Adv Drug Deliv Rev 62 (2):272–282. https://doi.org/10.1016/j.addr.2009.12.005
23. Larson N, Ghandehari H (2012) Polymeric conjugates for drug delivery. Chem Mater 24 (5):840–853. https://doi.org/10.1021/cm2031569
24. Veronese FM, Harris JM (2002) Introduction and overview of peptide and protein pegylation. Adv Drug Deliv Rev 54(4):453–456
25. Pasut G, Sergi M, Veronese FM (2008) Anticancer PEG-enzymes: 30 years old, but still a current approach. Adv Drug Deliv Rev 60 (1):69–78. https://doi.org/10.1016/j.addr.2007.04.018
26. Bailon P, Palleroni A, Schaffer CA, Spence CL, Fung W-J, Porter JE, Ehrlich GK, Pan W, Xu Z-X, Modi MW, Farid A, Berthold W, Graves M (2001) Rational design of a potent, long-lasting form of interferon: a 40 kDa branched polyethylene glycol-conjugated interferon α-2a for the treatment of hepatitis C. Bioconjug Chem 12 (2):195–202. https://doi.org/10.1021/bc000082g
27. Wang YS, Youngster S, Grace M, Bausch J, Bordens R, Wyss DF (2002) Structural and biological characterization of pegylated recombinant interferon alpha-2b and its therapeutic implications. Adv Drug Deliv Rev 54 (4):547–570
28. Roelfsema F, Biermasz NR, Pereira AM, Romijn JM (2006) Nanomedicines in the treatment of acromegaly: focus on pegvisomant. Int J Nanomedicine 1(4):385–398

29. Trainer PJ, Drake WM, Katznelson L, Freda PU, Herman-Bonert V, van der Lely AJ, Dimaraki EV, Stewart PM, Friend KE, Vance ML, Besser GM, Scarlett JA, Thorner MO, Parkinson C, Klibanski A, Powell JS, Barkan AL, Sheppard MC, Malsonado M, Rose DR, Clemmons DR, Johannsson G, Bengtsson BA, Stavrou S, Kleinberg DL, Cook DM, Phillips LS, Bidlingmaier M, Strasburger CJ, Hackett S, Zib K, Bennett WF, Davis RJ (2000) Treatment of acromegaly with the growth hormone-receptor antagonist pegvisomant. N Engl J Med 342(16):1171–1177. https://doi.org/10.1056/nejm200004203421604
30. Rowinsky EK, Rizzo J, Ochoa L, Takimoto CH, Forouzesh B, Schwartz G, Hammond LA, Patnaik A, Kwiatek J, Goetz A, Denis L, McGuire J, Tolcher AW (2003) A phase I and pharmacokinetic study of pegylated camptothecin as a 1-hour infusion every 3 weeks in patients with advanced solid malignancies. J Clin Oncol 21(1):148–157. https://doi.org/10.1200/jco.2003.03.143
31. Fang J, Nakamura H, Maeda H (2011) The EPR effect: unique features of tumor blood vessels for drug delivery, factors involved, and limitations and augmentation of the effect. Adv Drug Deliv Rev 63(3):136–151. https://doi.org/10.1016/j.addr.2010.04.009
32. Duncan R (2009) Development of HPMA copolymer–anticancer conjugates: clinical experience and lessons learnt. Adv Drug Deliv Rev 61(13):1131–1148. https://doi.org/10.1016/j.addr.2009.05.007
33. Kopecek J, Kopeckova P (2010) HPMA copolymers: origins, early developments, present, and future. Adv Drug Deliv Rev 62 (2):122–149. https://doi.org/10.1016/j.addr.2009.10.004
34. Lammers T (2010) Improving the efficacy of combined modality anticancer therapy using HPMA copolymer-based nanomedicine formulations. Advanced Drug Deliv Rev 62 (2):203–230. https://doi.org/10.1016/j.addr.2009.11.028
35. Caiolfa VR, Zamai M, Fiorino A, Frigerio E, Pellizzoni C, d'Argy R, Ghiglieri A, Castelli MG, Farao M, Pesenti E, Gigli M, Angelucci F, Suarato A (2000) Polymer-bound camptothecin: initial biodistribution and antitumour activity studies. J Control Release 65(1–2):105–119. https://doi.org/10.1016/S0168-3659(99)00243-6
36. Rademaker-Lakhai JM, Terret C, Howell SB, Baud CM, de Boer RF, Pluim D, Beijnen JH, Schellens JHM, Droz J-P (2004) A phase I and pharmacological study of the platinum polymer AP5280 given as an intravenous infusion once every 3 weeks in patients with solid tumors. Clin Cancer Res 10(10):3386–3395. https://doi.org/10.1158/1078-0432.ccr-03-0315
37. L-d Q, Yuan F, X-m L, J-g H, Alnouti Y, Wang D (2010) Pharmacokinetic and biodistribution studies of N-(2-Hydroxypropyl) methacrylamide copolymer-dexamethasone conjugates in adjuvant-induced arthritis rat model. Mol Pharm 7(4):1041–1049. https://doi.org/10.1021/mp100132h
38. Lammers T, Subr V, Ulbrich K, Peschke P, Huber PE, Hennink WE, Storm G (2009) Simultaneous delivery of doxorubicin and gemcitabine to tumors in vivo using prototypic polymeric drug carriers. Biomaterials 30(20):3466–3475. https://doi.org/10.1016/j.biomaterials.2009.02.040
39. Kasuya Y, Lu ZR, Kopeckova P, Minko T, Tabibi SE, Kopecek J (2001) Synthesis and characterization of HPMA copolymer-aminopropylgeldanamycin conjugates. J Control Release 74(1–3):203–211
40. Tomalia D, Baker H, Dewald J, Hall M, Kallos G, Martin S, Roeck J, Ryder J, Smith P (1985) A new class of polymers: starburst-dendritic. Polym J 17(1):117–132
41. Tomalia DA, Baker H, Dewald J, Hall M, Kallos G, Martin S, Roeck J, Ryder J, Smith P (1986) Dendritic macromolecules: synthesis of starburst dendrimers. Macromolecules 19(9):2466–2468
42. Thiagarajan G, Ray A, Malugin A, Ghandehari H (2010) PAMAM-Camptothecin conjugate inhibits proliferation and induces nuclear fragmentation in colorectal carcinoma cells. Pharm Res 27(11):2307–2316. https://doi.org/10.1007/s11095-010-0179-6
43. Menjoge AR, Kannan RM, Tomalia DA (2010) Dendrimer-based drug and imaging conjugates: design considerations for nanomedical applications. Drug Discov Today 15 (5–6):171–185. https://doi.org/10.1016/j.drudis.2010.01.009
44. Majoros IJ, Williams CR, Becker A, Baker JR (2009) Methotrexate delivery via folate targeted dendrimer-based nanotherapeutic platform. Wiley Interdiscip Rev Nanomed Nanobiotechnol 1(5):502–510. https://doi.org/10.1002/wnan.37
45. Cheng Y, Zhao L, Li Y, Xu T (2011) Design of biocompatible dendrimers for cancer diagnosis and therapy: current status and future perspectives. Chem Soc Rev 40 (5):2673–2703. https://doi.org/10.1039/c0cs00097c

46. Sun H, Meng F, Dias AA, Hendriks M, Feijen J, Zhong Z (2011) α-Amino acid containing degradable polymers as functional biomaterials: rational design, synthetic pathway, and biomedical applications. Biomacromolecules 12(6):1937–1955. https://doi.org/10.1021/bm200043u
47. Couffin-Hoarau AC, Aubertin AM, Boustta M, Schmidt S, Fehrentz JA, Martinez J, Vert M (2009) Peptide-poly (L-lysine citramide) conjugates and their in vitro anti-HIV behavior. Biomacromolecules 10(4):865–876. https://doi.org/10.1021/bm801376v
48. Yang D, Van S, Liu J, Wang J, Jiang X, Wang Y, Yu L (2011) Physicochemical properties and biocompatibility of a polymer-paclitaxel conjugate for cancer treatment. Int J Nanomedicine 6:2557–2566. https://doi.org/10.2147/ijn.s25044
49. Metselaar JM, Bruin P, de Boer LW, de Vringer T, Snel C, Oussoren C, Wauben MH, Crommelin DJ, Storm G, Hennink WE (2003) A novel family of L-amino acid-based biodegradable polymer-lipid conjugates for the development of long-circulating liposomes with effective drug-targeting capacity. Bioconjug Chem 14(6):1156–1164. https://doi.org/10.1021/bc0340363
50. Baldwin AD, Kiick KL (2010) Polysaccharide-modified synthetic polymeric biomaterials. Biopolymers 94(1):128–140. https://doi.org/10.1002/bip.21334
51. Kemp JD, Cardillo T, Stewart BC, Kehrberg E, Weiner G, Hedlund B, Naumann PW (1995) Inhibition of lymphoma growth <em>in vivo</em> by combined treatment with Hydroxyethyl starch Deferoxamine conjugate and IgG monoclonal antibodies against the transferrin receptor. Cancer Res 55 (17):3817–3824
52. Pisal DS, Kosloski MP, Balu-Iyer SV (2010) Delivery of therapeutic proteins. J Pharm Sci 99(6):2557–2575. https://doi.org/10.1002/jps.22054
53. Yurkovetskiy AV, Fram RJ (2009) XMT-1001, a novel polymeric camptothecin pro-drug in clinical development for patients with advanced cancer. Adv Drug Deliv Rev 61 (13):1193–1202. https://doi.org/10.1016/j.addr.2009.01.007
54. Torchilin V (2009) Multifunctional and stimuli-sensitive pharmaceutical nanocarriers. Eur J Pharm Biopharm 71(3):431–444. https://doi.org/10.1016/j.ejpb.2008.09.026
55. Vaupel P, Kallinowski F, Okunieff P (1989) Blood flow, oxygen and nutrient supply, and metabolic microenvironment of human tumors: a review. Cancer Res 49 (23):6449–6465
56. Bawa P, Pillay V, Choonara YE, du Toit LC (2009) Stimuli-responsive polymers and their applications in drug delivery. Biomed Mater 4 (2):022001. https://doi.org/10.1088/1748-6041/4/2/022001
57. Ulbrich K, Etrych T, Chytil P, Jelınková M, Řıhová B (2003) HPMA copolymers with pH-controlled release of doxorubicin: in vitro cytotoxicity and in vivo antitumor activity. J Control Release 87(1–3):33–47. https://doi.org/10.1016/S0168-3659(02)00348-6
58. Chilkoti A, Dreher MR, Meyer DE, Raucher D (2002) Targeted drug delivery by thermally responsive polymers. Adv Drug Deliv Rev 54 (5):613–630. https://doi.org/10.1016/S0169-409X(02)00041-8
59. de Castro MDL, Capote FP, Ávila NS (2008) Is dialysis alive as a membrane-based separation technique? TrAC Trend Anal Chem 27 (4):315–326. https://doi.org/10.1016/j.trac.2008.01.015
60. Luo J, Wu C, Xu T, Wu Y (2011) Diffusion dialysis-concept, principle and applications. J Membr Sci 366(1–2):1–16. https://doi.org/10.1016/j.memsci.2010.10.028
61. Silva MM, Krug FJ, Oliveira PV, Nóbrega JA, Reis BF, Penteado DAG (1996) Separation and preconcentration by flow injection coupled to tungsten coil electrothermal atomic absorption spectrometry. Spectrochim Acta Part B: At Spectrosc 51(14):1925–1934. https://doi.org/10.1016/S0584-8547(96)01536-4
62. Sajid M, Kawde A-N, Daud M (2015) Designs, formats and applications of lateral flow assay: a literature review. J Saudi Chem Soc 19(6):689–705. https://doi.org/10.1016/j.jscs.2014.09.001
63. Yeh HM, Chen HY, Chen KT (2000) Membrane ultrafiltration in a tubular module with a steel rod inserted concentrically for improved performance. J Membr Sci 168 (1–2):121–133. https://doi.org/10.1016/S0376-7388(99)00315-4
64. Beckmann W (2013) Crystallization: introduction. In: Crystallization. Wiley-VCH Verlag GmbH & Co, KGaA, pp 1–5. https://doi.org/10.1002/9783527650323.ch1
65. Mersmann A (1995) Crystallization technology handbook. Dry Technol 13 (4):1037–1038. https://doi.org/10.1080/07373939508917003

66. Dhanaraj G, Byrappa K, Prasad V, Dudley M (2010) Crystal growth techniques and characterization: an overview. In: Dhanaraj G, Byrappa K, Prasad V, Dudley M (eds) Springer handbook of crystal growth. Springer, Berlin Heidelberg, Berlin, Heidelberg, pp 3–16. https://doi.org/10.1007/978-3-540-74761-1_1
67. Beckmann W (2013) Basics of industrial crystallization from solution. In: Crystallization. Wiley-VCH Verlag GmbH & Co, KGaA, pp 173–185. https://doi.org/10.1002/9783527650323.ch9
68. Pirrung MC (2017) 12 - Evaporation. In: Handbook of synthetic organic chemistry, 2nd edn. Academic Press, pp 139–142. https://doi.org/10.1016/B978-0-12-809504-1.00012-1
69. Adams G (2007) The principles of freeze-drying. Methods Mol Biol 368:15–38. https://doi.org/10.1007/978-1-59745-362-2_2
70. Perry SG, Amos R, Brewer PI (1972) The technique of thin-layer chromatography. In: Practical liquid chromatography. Springer US, Boston, MA, pp 115–164. https://doi.org/10.1007/978-1-4684-1935-1_6
71. Smith I, Ersser RS (1976) Introduction to paper and thin layer chromatography. In: Seakins ISWT (ed) Paper and thin layer chromatography, 4th edn. Butterworth-Heinemann, pp 5–11. https://doi.org/10.1016/B978-0-8151-7839-2.50007-6
72. Kaddi CD, Bennett RV, Paine MRL, Banks MD, Weber AL, Fernández FM, Wang MD (2016) DetectTLC: automated reaction mixture screening utilizing quantitative mass spectrometry image feature. J Am Soc Mass Spectrom 27(2):359–365. https://doi.org/10.1007/s13361-015-1293-9
73. Santiago M, Strobel S (2013) Thin layer chromatography. Methods Enzymol 533:303–324. https://doi.org/10.1016/b978-0-12-420067-8.00024-6
74. Wall PE (2006) Sample application. In: Thin-layer chromatography: a modern practical approach. The Royal Society of Chemistry, pp 65–85. https://doi.org/10.1039/9781847552464-00065
75. Ahadi A, Partoazar A, Abedi-Khorasgani M-H, Shetab-Boushehri SV (2011) Comparison of liquid-liquid extraction-thin layer chromatography with solid-phase extraction-high-performance thin layer chromatography in detection of urinary morphine. J Biomed Res 25(5):362–367. https://doi.org/10.1016/S1674-8301(11)60048-1
76. Wiemer B (1988) F. Geiss: fundamentals of thin layer chromatography planar chromatography. Dr. A. Hüthig Verlag., 1987, 482 S., 202 Abb., 40 tab., DM 192,–, ISBN 3-7785-0854-7. Acta Hydrochim Hydrobiol 16(6):653–653, Heidelberg. https://doi.org/10.1002/aheh.19880160622
77. Ettre LS, Sakodynskii KI (1993) M. S. Tswett and the discovery of chromatography I: Early work (1899–1903). Chromatographia 35 (3–4):223–231. https://doi.org/10.1007/BF02269707
78. Scott RPW (2012) Liquid chromatography (Chrom-Ed series). Reese-Scott Partnership, 1st edn, 23 Jan 2012
79. Martin AJP, Synge RLM (1941) A new form of chromatogram employing two liquid phases: a theory of chromatography. 2. Application to the micro-determination of the higher monoamino-acids in proteins. Biochem J 35(12):1358–1368
80. Lathe GH, Ruthven CRJ (1956) The separation of substances and estimation of their relative molecular sizes by the use of columns of starch in water. Biochem J 62(4):665–674
81. Porath J, Flodin P (1959) Gel filtration: a method for desalting and group separation. Nature 183(4676):1657–1659
82. Hjerten S, Mosbach R (1962) "Molecular-sieve" chromatography of proteins on columns of cross-linked polyacrylamide. Anal Biochem 3:109–118
83. Polson A (1961) Fractionation of protein mixtures on columns of granulated agar. Biochim Biophys Acta 50:565–567
84. Paul-Dauphin S, Karaca F, Morgan TJ, Millan-Agorio M, Herod AA, Kandiyoti R (2007) Probing size exclusion mechanisms of complex hydrocarbon mixtures: the effect of altering eluent compositions. Energy Fuel 21(6):3484–3489. https://doi.org/10.1021/ef700410e
85. Berek D (2010) Size exclusion chromatography – a blessing and a curse of science and technology of synthetic polymers. J Sep Sci 33 (3):315–335. https://doi.org/10.1002/jssc.200900709
86. Wang J, Huang X, Ruan L, Lan T, Ren J (2013) Size exclusion chromatography as a universal method for the purification of quantum dots bioconjugates. Electrophoresis 34 (12):1764–1771. https://doi.org/10.1002/elps.201200649
87. Douville V, Lodi A, Miller J, Nicolas A, Clarot I, Prilleux B, Megoulas N, Koupparis M (2006) Evaporative light scattering detection (ELSD): a tool for improved quality control of drug substances. Pharmeur Sci Notes 2006(1):9–15

88. Viegas A, Macedo AL, Cabrita EJ (2009) Ligand-based nuclear magnetic resonance screening techniques. Methods Mol Biol 572:81–100. https://doi.org/10.1007/978-1-60761-244-5_6
89. Balci M (2005) 1 - Introduction. In: Basic 1H- and 13C-NMR spectroscopy. Elsevier Science, Amsterdam, pp 3–8. https://doi.org/10.1016/B978-044451811-8.50001-2
90. Bharti SK, Roy R (2012) Quantitative 1H NMR spectroscopy. TrAC Trend Anal Chem 35:5–26. https://doi.org/10.1016/j.trac.2012.02.007
91. Jacobsen NE (2007) Interpretation of proton (1H) NMR spectra. In: NMR spectroscopy explained. John Wiley & Sons, pp 39–73. https://doi.org/10.1002/9780470173350.ch2
92. Jungnickel JL, Forbes JW (1963) Quantitative measurement of hydrogen types by Intergrated nuclear magnetic resonance intensities. Anal Chem 35(8):938–942. https://doi.org/10.1021/ac60201a005
93. Balci M (2005) 8 - dynamic NMR spectroscopy. In: Basic 1H- and 13C-NMR spectroscopy. Elsevier Science, Amsterdam, pp 213–231. https://doi.org/10.1016/B978-044451811-8.50008-5
94. Creary X, Anderson A, Brophy C, Crowell F, Funk Z (2012) Method for assigning structure of 1,2,3-triazoles. J Org Chem 77(19):8756–8761. https://doi.org/10.1021/jo301265t
95. Nishizawa K, Takai M, Ishihara K (2011) A bioconjugated phospholipid polymer biointerface with nanometer-scaled structure for highly sensitive immunoassays. Methods Mol Biol 751:491–502. https://doi.org/10.1007/978-1-61779-151-2_31
96. Balci M (2005) 13 - 13C chemical shifts of organic compounds. In: Basic 1H- and 13C-NMR spectroscopy. Elsevier Science, Amsterdam, pp 293–324. https://doi.org/10.1016/B978-044451811-8.50013-9
97. Sakata K, Uzawa J, Sakurai A (1977) Application of carbon-13 n.m.r. Spectroscopy to the structural investigation of ezomycins. Org Magn Reson 10(1):230–234. https://doi.org/10.1002/mrc.1270100152
98. Jacobsen NE (2007) Carbon-13 (13C) NMR spectroscopy. In: NMR spectroscopy explained. John Wiley & Sons, pp 135–154. https://doi.org/10.1002/9780470173350.ch4
99. Breitmaier E (2002) Short introduction to basic principles and methods. In: Structure elucidation by NMR in organic chemistry. John Wiley & Sons, pp 1–10. https://doi.org/10.1002/0470853069.ch1
100. Mancini L, Payne GS, Leach MO (2003) Comparison of polarization transfer sequences for enhancement of signals in clinical 31P MRS studies. Magn Reson Med 50(3):578–588. https://doi.org/10.1002/mrm.10551
101. Singhal N, Kumar M, Kanaujia PK, Virdi JS (2015) MALDI-TOF mass spectrometry: an emerging technology for microbial identification and diagnosis. Front Microbiol 6:791. https://doi.org/10.3389/fmicb.2015.00791
102. Lavigne JP, Espinal P, Dunyach-Remy C, Messad N, Pantel A, Sotto A (2013) Mass spectrometry: a revolution in clinical microbiology? Clin Chem Lab Med 51(2):257–270. https://doi.org/10.1515/cclm-2012-0291
103. Susnea I, Bernevic B, Wicke M, Ma L, Liu S, Schellander K, Przybylski M (2013) Application of MALDI-TOF-mass spectrometry to proteome analysis using stain-free gel electrophoresis. Top Curr Chem 331:37–54. https://doi.org/10.1007/128_2012_321
104. Calderaro A, Arcangeletti M-C, Rodighiero I, Buttrini M, Gorrini C, Motta F, Germini D, Medici M-C, Chezzi C, De Conto F (2014) Matrix-assisted laser desorption/ionization time-of-flight (MALDI-TOF) mass spectrometry applied to virus identification. Sci Rep 4:6803. https://doi.org/10.1038/srep06803. http://www.nature.com/articles/srep06803 - supplementary-information
105. Wang H, Zhao Z, Guo Y (2013) Chemical and Biochemical applications of MALDI TOF-MS based on analyzing the small organic compounds. In: Cai Z, Liu S (eds) Applications of MALDI-TOF spectroscopy. Springer, Berlin Heidelberg, pp 165–192. https://doi.org/10.1007/128_2012_364
106. Machado YJ, Rabasa Y, Montesinos R, Cremata J, Besada V, Fuentes D, Castillo A, de la Luz KR, Vázquez AM, Himly M (2011) Physicochemical and biological characterization of 1E10 anti-Idiotype vaccine. BMC Biotechnol 11:112–112. https://doi.org/10.1186/1472-6750-11-112
107. Stuart BH (2005) Introduction. In: Infrared spectroscopy: fundamentals and applications. John Wiley & Sons, pp 1–13. https://doi.org/10.1002/0470011149.ch1
108. Blinder SM (2004) Introduction to quantum mechanics: in chemistry, materials science, and biology. Elsevier
109. Ojeda JJ, Dittrich M (2012) Fourier transform infrared spectroscopy for molecular

analysis of microbial cells. Methods Mol Biol 881:187–211. https://doi.org/10.1007/978-1-61779-827-6_8

110. Skoog DA, Crouch SR, Holler FJ (2007) Principles of instrumental analysis. Thomson Brooks/Cole, Belmont, CA

111. Abbas O, Dardenne P, Baeten V (2012) Chapter 3 - near-infrared, mid-infrared, and Raman spectroscopy. In: Picó Y (ed) Chemical analysis of food: techniques and applications. Academic Press, Boston, pp 59–89. https://doi.org/10.1016/B978-0-12-384862-8.00003-0

112. Tanaka K, Kanazawa T, Horiuchi S, Ando T, Sugawara K, Takashima Y, Seta Y, Okada H (2013) Cytoplasm-responsive nanocarriers conjugated with a functional cell-penetrating peptide for systemic siRNA delivery. Int J Pharm 455(1–2):40–47. https://doi.org/10.1016/j.ijpharm.2013.07.069

113. Taranejoo S, Chandrasekaran R, Cheng W, Hourigan K (2016) Bioreducible PEI-functionalized glycol chitosan: a novel gene vector with reduced cytotoxicity and improved transfection efficiency. Carbohydr Polym 153:160–168. https://doi.org/10.1016/j.carbpol.2016.07.080

114. Knapinska AM, Tokmina-Roszyk D, Amar S, Tokmina-Roszyk M, Mochalin VN, Gogotsi Y, Cosme P, Terentis AC, Fields GB (2015) Solid-phase synthesis, characterization, and cellular activities of collagen-model nanodiamond-peptide conjugates. Biopolymers 104(3):186–195. https://doi.org/10.1002/bip.22636

115. Imani R, Emami SH, Faghihi S (2015) Synthesis and characterization of an octaarginine functionalized graphene oxide nano-carrier for gene delivery applications. Phys Chem Chem Phys 17(9):6328–6339. https://doi.org/10.1039/c4cp04301d

116. Kolb HC, Finn MG, Sharpless KB (2001) Click chemistry: diverse chemical function from a few good reactions. Angew Chem Int Ed Engl 40(11):2004–2021

117. Tornoe CW, Christensen C, Meldal M (2002) Peptidotriazoles on solid phase: [1,2,3]-triazoles by regiospecific copper(i)-catalyzed 1,3-dipolar cycloadditions of terminal alkynes to azides. J Org Chem 67(9):3057–3064

118. Wang X, Huang B, Liu X, Zhan P (2016) Discovery of bioactive molecules from CuAAC click-chemistry-based combinatorial libraries. Drug Discov Today 21(1):118–132. https://doi.org/10.1016/j.drudis.2015.08.004

119. Hermanson GT (2013) Chapter 17 - Chemoselective ligation; bioorthogonal reagents. In: Bioconjugate techniques, 3rd edn. Academic Press, Boston, pp 757–785. https://doi.org/10.1016/B978-0-12-382239-0.00017-0

120. Wang C, Ikhlef D, Kahlal S, Saillard J-Y, Astruc D (2016) Metal-catalyzed azide-alkyne “click” reactions: mechanistic overview and recent trends. Coordin Chem Rev 316:1–20. https://doi.org/10.1016/j.ccr.2016.02.010

121. Worrell BT, Malik JA, Fokin VV (2013) Direct evidence of a dinuclear copper intermediate in Cu(I)-catalyzed azide-alkyne cycloadditions. Science (New York, NY) 340 (6131):457–460. https://doi.org/10.1126/science.1229506

122. Massarotti A, Aprile S, Mercalli V, Del Grosso E, Grosa G, Sorba G, Tron GC (2014) Are 1,4- and 1,5-disubstituted 1,2,3-triazoles good pharmacophoric groups? ChemMedChem 9(11):2497–2508. https://doi.org/10.1002/cmdc.201402233

123. Kolb HC, Sharpless KB (2003) The growing impact of click chemistry on drug discovery. Drug Discov Today 8(24):1128–1137. https://doi.org/10.1016/S1359-6446(03)02933-7

124. Jewett JC, Bertozzi CR (2010) Cu-free click cycloaddition reactions in chemical biology. Chem Soc Rev 39(4):1272–1279

125. Prescher JA, Bertozzi CR (2005) Chemistry in living systems. Nat Chem Biol 1(1):13–21. https://doi.org/10.1038/nchembio0605-13

126. Jarrad AM, Karoli T, Debnath A, Tay CY, Huang JX, Kaeslin G, Elliott AG, Miyamoto Y, Ramu S, Kavanagh AM, Zuegg J, Eckmann L, Blaskovich MAT, Cooper MA (2015) Metronidazole-triazole conjugates: activity against Clostridium difficile and parasites. Eur J Med Chem 101:96–102. https://doi.org/10.1016/j.ejmech.2015.06.019

127. Negi B, Kumar D, Kumbukgolla W, Jayaweera S, Ponnan P, Singh R, Agarwal S, Rawat DS (2016) Anti-methicillin resistant *Staphylococcus aureus* activity, synergism with oxacillin and molecular docking studies of metronidazole-triazole hybrids. Eur J Med Chem 115:426–437. https://doi.org/10.1016/j.ejmech.2016.03.041

128. Link AJ, Vink MK, Tirrell DA (2004) Presentation and detection of azide functionality in bacterial cell surface proteins. J Am Chem Soc 126(34):10598–10602. https://doi.org/10.1021/ja047629c

129. Lo Conte M, Pacifico S, Chambery A, Marra A, Dondoni A (2010) Photoinduced addition of glycosyl thiols to alkynyl peptides: use of free-radical thiol-yne coupling for post-translational double-glycosylation of peptides. J Org Chem 75(13):4644–4647. https://doi.org/10.1021/jo1008178
130. Clayton R, Ramsden CA (2005) N-vinyl-Nitroimidazole Cycloadditions: potential routes to nucleoside analogues. Synthesis 2005(16):2695–2700. https://doi.org/10.1055/s-2005-872083
131. Zhang W, Li Z, Zhou M, Wu F, Hou X, Luo H, Liu H, Han X, Yan G, Ding Z, Li R (2014) Synthesis and biological evaluation of 4-(1,2,3-triazol-1-yl)coumarin derivatives as potential antitumor agents. Bioorg Med Chem Lett 24(3):799–807. https://doi.org/10.1016/j.bmcl.2013.12.095
132. Hermanson GT (2013) Chapter 4 - zero-length crosslinkers. In: Bioconjugate techniques, 3rd edn. Academic Press, Boston, pp 259–273. https://doi.org/10.1016/B978-0-12-382239-0.00004-2
133. Williams A, Ibrahim IT (1981) A new mechanism involving cyclic tautomers for the reaction with nucleophiles of the water-soluble peptide coupling reagent 1-ethyl-3-(3-′-(dimethylamino)propyl)carbodiimide (EDC). J Am Chem Soc 103 (24):7090–7095. https://doi.org/10.1021/ja00414a011
134. Sheehan J, Cruickshank P, Boshart G (1961) Notes- a convenient synthesis of water-soluble carbodiimides. J Org Chem 26 (7):2525–2528. https://doi.org/10.1021/jo01351a600
135. Nakajima N, Ikada Y (1995) Mechanism of amide formation by carbodiimide for bioconjugation in aqueous media. Bioconjug Chem 6(1):123–130. https://doi.org/10.1021/bc00031a015
136. Young J-J, Cheng K-M, Tsou T-L, Liu H-W, Wang H-J (2004) Preparation of cross-linked hyaluronic acid film using 2-chloro-1-methylpyridinium iodide or water-soluble 1-ethyl-(3,3-dimethylaminopropyl)carbodiimide. J Biomater Sci Polym Ed 15(6):767–780. https://doi.org/10.1163/156856204774196153
137. Wang C, Yan Q, Liu H-B, Zhou X-H, Xiao S-J (2011) Different EDC/NHS activation mechanisms between PAA and PMAA brushes and the following Amidation reactions. Langmuir 27(19):12058–12068. https://doi.org/10.1021/la202267p
138. Hoare DG, Koshland DE (1967) A method for the quantitative modification and estimation of carboxylic acid groups in proteins. J Biol Chem 242(10):2447–2453
139. Larsson P-O, Mosbach K (1971) Preparation of a NAD(H)-polymer matrix showing coenzymic function of the bound pyridine nucleotide. Biotechnol Bioeng 13(3):393–398. https://doi.org/10.1002/bit.260130306
140. Lowe CR, Harvey MJ, Craven DB, Dean PDG (1973) Some parameters relevant to affinity chromatography on immobilized nucleotides. Biochem J 133(3):499–506. https://doi.org/10.1042/bj1330499
141. Staros JV (1982) N-hydroxysulfosuccinimide active esters: bis(N-hydroxysulfosuccinimide) esters of two dicarboxylic acids are hydrophilic, membrane-impermeant, protein cross-linkers. Biochemistry 21 (17):3950–3955. https://doi.org/10.1021/bi00260a008
142. Khalfan H, Abuknesha R, Rand-Weaver M, Price RG, Robinson D (1986) Aminomethyl coumarin acetic acid: a new fluorescent labelling agent for proteins. Histochem J 18 (9):497–499. https://doi.org/10.1007/bf01675617
143. Hermanson GT (2013) Chapter 10 - fluorescent probes. In: Bioconjugate techniques, 3rd edn. Academic Press, Boston, pp 395–463. https://doi.org/10.1016/B978-0-12-382239-0.00010-8
144. Harvey DJ (2011) Derivatization of carbohydrates for analysis by chromatography; electrophoresis and mass spectrometry. J Chromatogr B 879(17–18):1196–1225. https://doi.org/10.1016/j.jchromb.2010.11.010
145. Nakano M, Kakehi K, Taniguchi N, Kondo A (2011) Capillary electrophoresis and capillary electrophoresis–mass spectrometry for structural analysis of N-Glycans derived from glycoproteins. In: Volpi N (ed) Capillary electrophoresis of carbohydrates: from Monosaccharides to complex polysaccharides. Humana Press, Totowa, NJ, pp 205–235. https://doi.org/10.1007/978-1-60761-875-1_9
146. Gibrat C, Cicchetti F (2011) Potential of cystamine and cysteamine in the treatment of neurodegenerative diseases. Prog Neuro-Psychopharmacol Biol Psychiatry 35 (2):380–389. https://doi.org/10.1016/j.pnpbp.2010.11.023
147. Fujisawa T, Rubin B, Suzuki A, Patel PS, Gahl WA, Joshi BH, Puri RK (2012) Cysteamine suppresses invasion, metastasis and prolongs survival by inhibiting matrix Metalloproteinases in a mouse model of human pancreatic

Cancer. PLoS One 7(4):e34437. https://doi.org/10.1371/journal.pone.0034437

148. Brunelli C, Amici C, Angelini M, Fracassi C, Belardo G, Santoro MG (2012) The non-steroidal anti-inflammatory drug indomethacin activates the eIF2alpha kinase PKR, causing a translational block in human colorectal cancer cells. Biochem J 443 (2):379–386. https://doi.org/10.1042/bj20111236
149. Suh W, Chung JK, Park SH, Kim SW (2001) Anti-JL1 antibody-conjugated poly (L-lysine) for targeted gene delivery to leukemia T cells. J Control Release 72(1–3):171–178
150. Toni R, Mirandola P, Gobbi G, Vitale M (2007) Neuroendocrine regulation and tumor immunity. Eur J Histochem 51(Suppl 1):133–138
151. Shadidi M, Sioud M (2003) Selective targeting of cancer cells using synthetic peptides. Drug Resist Updat 6(6):363–371
152. Benns JM, Maheshwari A, Furgeson DY, Mahato RI, Kim SW (2001) Folate-PEG-folate-graft-polyethylenimine-based gene delivery. J Drug Target 9(2):123–139
153. Hong G, Yuan R, Liang B, Shen J, Yang X, Shuai X (2008) Folate-functionalized polymeric micelle as hepatic carcinoma-targeted, MRI-ultrasensitive delivery system of antitumor drugs. Biomed Microdevices 10 (5):693–700. https://doi.org/10.1007/s10544-008-9180-9
154. Low PS, Antony AC (2004) Folate receptor-targeted drugs for cancer and inflammatory diseases. Adv Drug Deliv Rev 56 (8):1055–1058. https://doi.org/10.1016/j.addr.2004.02.003
155. Liang B, He ML, Xiao ZP, Li Y, Chan CY, Kung HF, Shuai XT, Peng Y (2008) Synthesis and characterization of folate-PEG-grafted-hyperbranched-PEI for tumor-targeted gene delivery. Biochem Biophys Res Commun 367 (4):874–880. https://doi.org/10.1016/j.bbrc.2008.01.024
156. Gabizon A, Horowitz AT, Goren D, Tzemach D, Mandelbaum-Shavit F, Qazen MM, Zalipsky S (1999) Targeting folate receptor with folate linked to extremities of poly(ethylene glycol)-grafted liposomes: in vitro studies. Bioconjug Chem 10 (2):289–298. https://doi.org/10.1021/bc9801124
157. Yoo HS, Park TG (2004) Folate receptor targeted biodegradable polymeric doxorubicin micelles. J Control Release 96(2):273–283. https://doi.org/10.1016/j.jconrel.2004.02.003
158. Hwa Kim S, Hoon Jeong J, Chul Cho K, Wan Kim S, Gwan Park T (2005) Target-specific gene silencing by siRNA plasmid DNA complexed with folate-modified poly(ethylenimine). J Control Release 104(1):223–232. https://doi.org/10.1016/j.jconrel.2005.02.006
159. Cudaj M, Cudaj J, Hofe T, Luy B, Wilhelm M, Guthausen G (2012) Polystyrene solutions: characterization of molecular motional modes by spectrally resolved low- and high-field NMR relaxation. Macromol Chem Phys 213(17):1833–1840. https://doi.org/10.1002/macp.201200092
160. Rule GS, Hitchens TK (2006) Fundamentals of protein NMR spectroscopy. In: Focus on Structural Biology, vol 5. Springer Netherlands, p 532. https://doi.org/10.1007/1-4020-3500-4
161. Jain A, Cheng K (2017) The principles and applications of avidin-based nanoparticles in drug delivery and diagnosis. J Control Release 245:27–40. https://doi.org/10.1016/j.jconrel.2016.11.016
162. Wilchek M, Bayer EA, Livnah O (2006) Essentials of biorecognition: the (strept)avidin-biotin system as a model for protein-protein and protein-ligand interaction. Immunol Lett 103(1):27–32. https://doi.org/10.1016/j.imlet.2005.10.022
163. Berman HM, Westbrook J, Feng Z, Gilliland G, Bhat TN, Weissig H, Shindyalov IN, Bourne PE (2000) The protein data bank. Nucleic Acids Res 28(1):235–242. https://doi.org/10.1093/nar/28.1.235
164. Diamandis EP, Christopoulos TK (1991) The biotin-(strept)avidin system: principles and applications in biotechnology. Clin Chem 37 (5):625–636
165. Tausig F, Wolf FJ (1964) Streptavidin—A substance with avidin-like properties produced by microorganisms. Biochem Biophys Res Commun 14(3):205–209. https://doi.org/10.1016/0006-291X(64)90436-X
166. Dundas CM, Demonte D, Park S (2013) Streptavidin-biotin technology: improvements and innovations in chemical and biological applications. Appl Microbiol Biotechnol 97(21):9343–9353. https://doi.org/10.1007/s00253-013-5232-z
167. Hendrickson WA, Pahler A, Smith JL, Satow Y, Merritt EA, Phizackerley RP (1989) Crystal structure of core streptavidin determined from multiwavelength anomalous diffraction of synchrotron radiation. Proc Natl Acad Sci U S A 86(7):2190–2194

168. Huberman T, Eisenberg-Domovich Y, Gitlin G, Kulik T, Bayer EA, Wilchek M, Livnah O (2001) Chicken avidin exhibits pseudo-catalytic properties: biochemical, structural, and electrostatic consequences. J Biol Chem 276(34):32031–32039. https://doi.org/10.1074/jbc.M102018200
169. Schechter B, Silberman R, Arnon R, Wilchek M (1990) Tissue distribution of avidin and streptavidin injected to mice. Effect of avidin carbohydrate, streptavidin truncation and exogenous biotin. Eur J Biochem 189 (2):327–331
170. Nguyen TT, Sly KL, Conboy JC (2012) Comparison of the energetics of avidin, streptavidin, neutrAvidin, and anti-biotin antibody binding to biotinylated lipid bilayer examined by second-harmonic generation. Anal Chem 84(1):201–208. https://doi.org/10.1021/ac202375n
171. Chen S, Zhao X, Chen J, Chen J, Kuznetsova L, Wong SS, Ojima I (2010) Mechanism-based tumor-targeting drug delivery system. Validation of efficient vitamin receptor-mediated endocytosis and drug release. Bioconjug Chem 21(5):979–987. https://doi.org/10.1021/bc9005656
172. Ma M, Yuan ZF, Chen XJ, Li F, Zhuo RX (2012) A facile preparation of novel multifunctional vectors by non-covalent bonds for co-delivery of doxorubicin and gene. Acta Biomater 8(2):599–607. https://doi.org/10.1016/j.actbio.2011.11.006
173. Scott Wilbur D, Pathare PM, Hamlin DK, Stayton PS, To R, Klumb LA, Buhler KR, Vessella RL (1999) Development of new biotin/streptavidin reagents for pretargeting. Biomol Eng 16(1–4):113–118. https://doi.org/10.1016/S1050-3862(99)00044-3
174. Arribillaga L, Durantez M, Lozano T, Rudilla F, Rehberger F, Casares N, Villanueva L, Martinez M, Gorraiz M, Borras-Cuesta F, Sarobe P, Prieto J, Lasarte JJ (2013) A fusion protein between streptavidin and the endogenous TLR4 ligand EDA targets Biotinylated antigens to dendritic cells and induces T cell responses in vivo. Biomed Res Int 2013:9. https://doi.org/10.1155/2013/864720
175. Petronzelli F, Pelliccia A, Anastasi AM, Lindstedt R, Manganello S, Ferrari LE, Albertoni C, Leoni B, Rosi A, D'Alessio V, Deiana K, Paganelli G, De Santis R (2010) Therapeutic use of avidin is not hampered by antiavidin antibodies in humans. Cancer Biother Radiopharm 25(5):563–570. https://doi.org/10.1089/cbr.2010.0797
176. Ren WX, Han J, Uhm S, Jang YJ, Kang C, Kim J-H, Kim JS (2015) Recent development of biotin conjugation in biological imaging, sensing, and target delivery. Chem Commun 51(52):10403–10418. https://doi.org/10.1039/c5cc03075g
177. Muggia FM (1998) Doxil in breast cancer. J Clin Oncol 16(2):811–812. https://doi.org/10.1200/JCO.1998.16.2.811
178. Porche DJ (1996) Liposomal doxorubicin (Doxil). J Assoc Nurses AIDS Care 7 (2):55–59. https://doi.org/10.1016/S1055-3290(96)80016-1
179. Tagami T, Ozeki T (2017) Recent trends in clinical trials related to carrier-based drugs. J Pharm Sci 106(9):2219–2226. https://doi.org/10.1016/j.xphs.2017.02.026
180. Cagel M, Tesan FC, Bernabeu E, Salgueiro MJ, Zubillaga MB, Moretton MA, Chiappetta DA (2017) Polymeric mixed micelles as nanomedicines: achievements and perspectives. Eur J Pharm Biopharm 113:211–228. https://doi.org/10.1016/j.ejpb.2016.12.019

Chapter 12

Conjugation of Triphenylphosphonium Cation to Hydrophobic Moieties to Prepare Mitochondria-Targeting Nanocarriers

Diana Guzman-Villanueva, Mark R. Mendiola, Huy X. Nguyen, Francis Yambao, Nusem Yu, and Volkmar Weissig

Abstract

The contribution of mitochondrial dysfunctions to diseases such as cancer, diabetes, cardiovascular, and neurodegenerative diseases has made mitochondria an attractive pharmacological target. To deliver biologically active molecules to mitochondria, however, cellular and mitochondrial barriers must be first overcome. The mitochondrial transmembrane electric potential (negative inside) is among the most commonly used strategies to deliver molecules to mitochondria as it allows the accumulation of positively charged molecules. Thus, therapeutic molecules are either covalently conjugated to lipophilic cations like triphenylphosphonium (TPP) or loaded into nanocarriers conjugated to TPP.

Key words TPP, Triphenylphosphonium, Mitochondria-targeting, Nanocarriers, Mitochondrial diseases

1 Introduction

In the last decade, mitochondria have gained attention as a potential therapeutic target due to the evidence of their implication in diseases such as cancer, diabetes, ischemia-reperfusion injury, and neurodegenerative conditions like Alzheimer's and Parkinson's disease [1–3].

Despite the incidence of mitochondrial dysfunction-related diseases and the multiple pharmacological targets that mitochondria possess (electron transport chain, voltage-dependent anion channel-VDAC, and permeability transition pore-PTP) [3–6], the delivery of biologically active molecules to these organelles still represents a challenge. To reach mitochondria, for example, drugs and molecules need to overcome extracellular and intracellular barriers, including the plasma membrane and the outer and inner mitochondrial membranes [7, 8]. Although the outer

Volkmar Weissig and Tamer Elbayoumi (eds.), *Pharmaceutical Nanotechnology: Basic Protocols*, Methods in Molecular Biology, vol. 2000, https://doi.org/10.1007/978-1-4939-9516-5_12, © Springer Science+Business Media, LLC, part of Springer Nature 2019

mitochondrial membrane (OMM) is relatively permeable to molecules up to 5 kDa, the inner mitochondrial membrane (IMM) is highly impermeable to most solutes and ions. Their large mitochondrial transmembrane electric potential (−180 mV, inside), however, facilitates the accumulation of lipophilic and positively charged molecules within mitochondria [9–11]. Therefore, a common strategy to delivery molecules to mitochondria is the use of the lipophilic cations [9, 12, 13]. Lipophilic cations easily penetrate the phospholipid bilayers and are taken up by mitochondria without requiring a transporter [9, 14].

Triphenylphosphonium (TPP), is among the most extensively studied lipophilic cations used for mitochondrial delivery. TPP efficiently accumulates within mitochondria due to its charge is delocalized and dispersed over a large hydrophilic surface area, which reduces its activation energy and facilitates its movement across the phospholipid bilayers [2, 10, 15].

To exert a pharmacological effect within mitochondria, drugs or molecules can be directly attached to TPP via a covalent conjugation reaction or loaded into a nanocarrier conjugated to TPP. Although the delivery of molecules directly conjugated to TPP (vitamin E, coenzyme-Q, lipoid acid, superoxide dismutase, and α-phenyl-*N*-*tert*-butylnitrone-PBN) has been successful [10, 13, 16], the modifications resulting from the conjugation reaction could alter the chemical properties of the drugs or molecules, potentially leading to loss of their therapeutic activity [17]. An alternative to these modifications is the design and preparation of mitochondria-targeting nanocarriers that incorporate the drug without requiring any changes. To facilitate the incorporation into the carrier, TPP is covalently conjugated to a moiety that is subsequently anchored/incorporated into the carrier [18–21]. Examples of nanocarriers used for mitochondria targeting include liposomes, emulsions, and polymeric systems like dendrimers [7, 22].

Our research group conjugated TPP cation to the two-tailed phospholipid phosphatidylethanolamine (PE) [23], prepared nanoliposomes, and compared them with the previously synthesized STPP conjugate [21].

This conjugation took place via a carboxy-to-amine reaction using the crosslinker EDC and Sulfo-NHS (*see* Fig. 1). The purified TPP-PE conjugate was subsequently incorporated as an anchor in the phospholipid bilayer of nanoliposomes along with phosphatidylcholine and cholesterol at 70:25:5 (PC:Chol:TPP-PE) mol% ratios.

We evaluated the cytotoxicity of TPP-PE nanoliposomes in transformed (4T1) and nontransformed (H9c2) cell lines. As seen in Fig. 2, our data suggested that TPP-PE nanoliposomes showed low cytotoxicity, as the cell viability remained up to 90% at a concentration of 300 μg/mL. In addition, no alterations in the

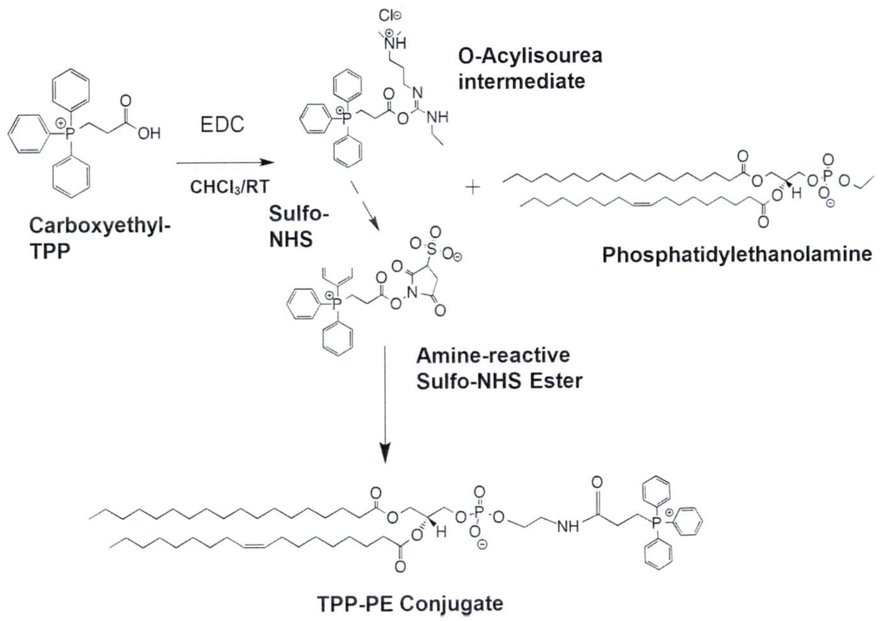

Fig. 1 Conjugation reaction between the TPP cation and the two-tail phospholipid L-α-phosphatidylethanolamine (PE)

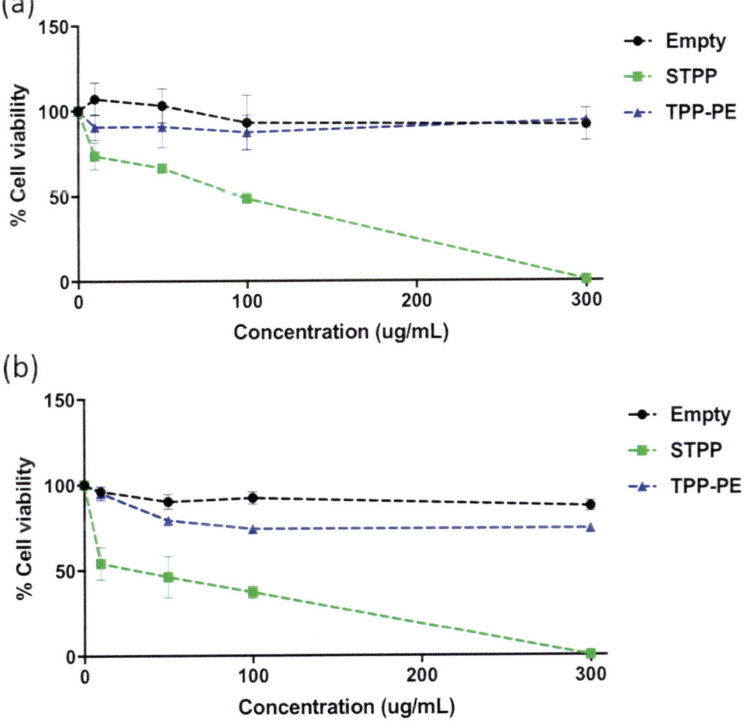

Fig. 2 Evaluation of the toxicity of TPP-PE conjugate anchored into nanoliposomes in (**a**) nontransformed (H9c2) and (**b**) transformed (4T1) cell lines. TPP-PE nanoliposomes were compared with plain liposomes and STPP

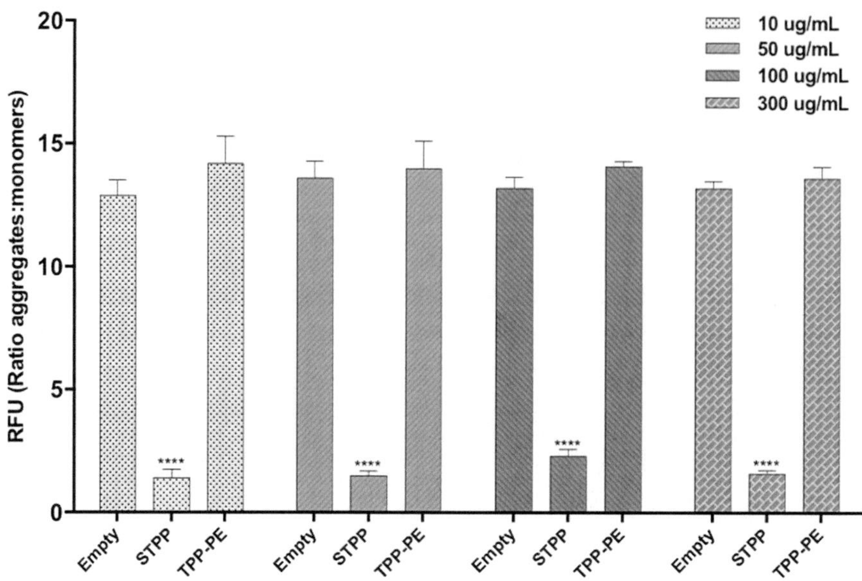

Fig. 3 Mitochondrial membrane potential determination in cells treated with TPP-PE nanoliposomes for 24 h. TPP-PE conjugate did not disrupt mitochondria compared to STPP

mitochondrial membrane potential nor changes in the mitochondrial selectivity of TPP occurred after the conjugation reaction with PE and liposome preparation, as illustrated in Figs. 3 and 4.

In this chapter, we describe the conjugation of TPP to a two-tail hydrophobic moiety as an anchor suitable to prepare mitochondria-targeting lipid-based nanocarriers.

2 Materials

2.1 Synthesis of TPP Derivative

(2-Carboxyethyl)-triphenylphosphonium chloride (TPP).

L-α-phosphatidylethanolamine (Egg, Chicken, PE).

N-(3-Dimethylaminopropyl)-*N*′-ethylcarbodiimide hydrochloride (EDC).

N-Hydroxysulfosuccinimide sodium salt (Sulfo-NHS).

Chloroform.

MilliQ water.

10 mL round-bottom flask.

1 mL pipette tips.

1 mL pipette.

Magnetic bar.

Magnetic stirrer.

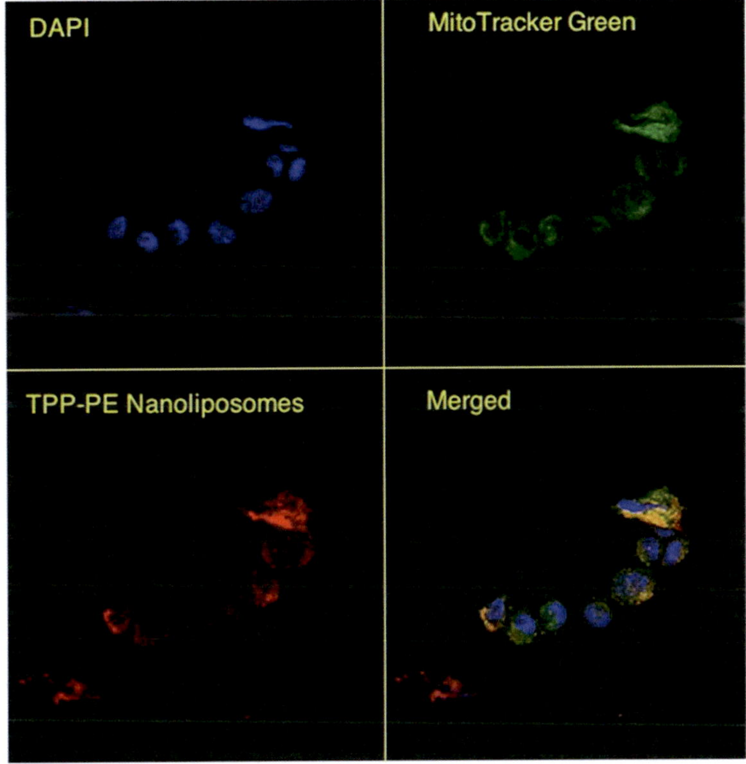

Fig. 4 Mitochondrial accumulation of TPP-PE nanoliposomes. TPP-PE conjugate maintained its selectivity for mitochondria after its incorporation into nanoliposomes

2.2 Purification of TPP Derivative Product

Dialysis tubing (2000 Da MWCO).

Dialysis tubing closures.

Glass vials.

Lyophilizer.

3 Methods

3.1 Synthesis of TPP Derivative

1. Weight 8 mg of CTPP and transfer it to a 10 mL round-bottom flask (*see* Fig. 5).
2. Add 1 mL of chloroform and gently stir. Immediately cover the flask to avoid chloroform evaporation.
3. Transfer 65.5 μM EDC and 65.5 μM Sulfo-NHS to the round-bottom flask and incubate at room temperature for 2 h under stirring.
4. Add 12 mg of PE and incubate at room temperature overnight.

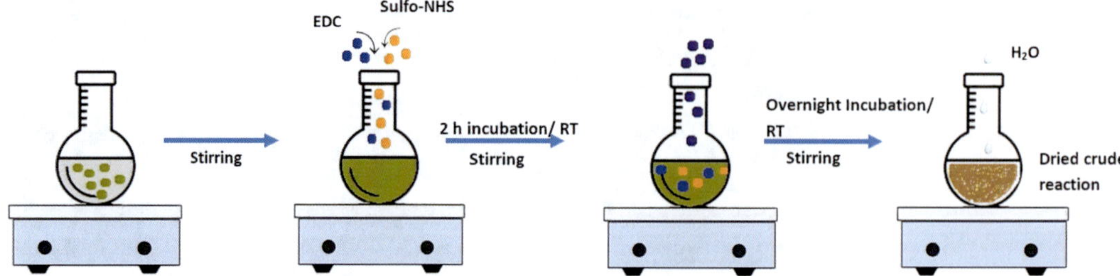

Fig. 5 Schematic representation of the covalent conjugation reaction between TPP cation and the two-tail lipid L-α-phosphatidylethanolamine (PE). The TPP-PE conjugate is incorporated as an anchor in the phospholipid bilayer of nanoliposomes to deliver cargoes to mitochondria

5. Next day, remove the chloroform under a Nitrogen stream for 30 min.
6. Resuspend the product in 3 mL of MilliQ water (*see* **Note 1**).

3.2 TPP Derivative Purification

1. Wash or soak a portion of membrane tubing in a 2 L beaker containing MilliQ water (*see* **Note 2**).
2. Hold the corner of the membrane, twist or fold, and secure it with one of the dialysis tubing closures.
3. Carefully transfer the product to the dialysis tubing and secure the opposite end of the membrane with a tubing closure.
4. Transfer the dialysis bag to a 2 L beaker and dialyze it against MilliQ water under gentle stirring for 24 h at room temperature.
5. Change the water from the beaker every 3–4 h.
6. Transfer the purified product to a glass vial and freeze it for 1 h at −80 °C.
7. Lyophilize the product for 24 h or until obtaining a white, pale yellow powder (*see* **Note 3**).
8. Weigh the TPP derivative and dissolve it in chloroform to the desired concentration.
9. Store the TPP derivative at −20 °C.

4 Notes

1. A sticky mass will be formed after removing the solvent from the crude reaction in **step 5** in Sect. 3.1 (Synthesis of TPP derivative). Because of the stickiness of the mass, resuspending the crude in water will require stirring at moderate to high speed. Start by adding 1 mL of MilliQ water and stir the flask until the mass disintegrates. Transfer the product to the dialysis

membrane. Repeat this step two more times until the product is completely removed from the flask.

2. Do not let the dialysis tubing to dry, otherwise it will turn brittle. After transferring the product, return the membrane to the 2 L beaker. Be careful not to perforate the membrane while transferring the product.

3. Due to the hygroscopic nature of the TPP derivative, store it at −20 °C immediately after lyophilization.

References

1. Frantz MC, Wipf P (2010) Mitochondria as a target in treatment. Environ Mol Mutagen 51:462–475
2. Reily C, Mitchell T, Chacko BK et al (2013) Mitochondrially targeted compounds and their impact on cellular bioenergetics. Redox Biol 1:86–93
3. Olszewska A, Szewczyk A (2013) Mitochondria as a pharmacological target: magnum overview. IUBMB Life 65:273–281
4. Szewczyk A, Wojtczak L (2002) Mitochondria as a pharmacological target. Pharmacol Rev 54:101–127
5. Fulda S, Galluzzi L, Kroemer G (2010) Targeting mitochondria for cancer therapy. Nat Rev Drug Discov 9:447–464
6. Malty RH, Jessulat M, Jin K et al (2015) Mitochondrial targets for pharmacological intervention in human disease. Proteome Res 14:5–21
7. Wongrakpanich A, Geary SM, Joiner MA et al (2014) Mitochondria-targeting particles. Nanomed 9:2531–2543
8. Agrawal U, Sharma R, Vyas SP (2015) Targeted drug delivery to the mitochondria. In: Devarajan PV, Jain S (eds) Targeted drug delivery: concepts and design. Advances in delivery sciences and technology. Springer
9. Murphy MP (2008) Targeting lipophilic cations to mitochondria. Biochim Biophys Acta 1777:1028–1031
10. Smith RAJ, Porteous CM, Gane AM et al (2003) Delivery of bioactive molecules to mitochondria in vivo. Proc Natl Acad Sci 100:5407–5412
11. Liberman EA, Topaly VP, Tsofina LM et al (1969) Mechanism of coupling of oxidative phosphorylation and the membrane potential of mitochondria. Nature 222:1076–1078
12. Murphy MP (2001) Development of lipophilic cations as therapies for disorders due to mitochondrial dysfunction. Expert Opin Biol Ther 1:753–764
13. Murphy MP, Smith RAJ (2007) Targeting antioxidants to mitochondria by conjugation to lipophilic cations. Annu Rev Pharmacol Toxicol 47:629–656
14. Murphy MP (1997) Selective targeting of bioactive compounds to mitochondria. Trends Biotechnol 15:326–330
15. Smith RAJ, Hartley RC, Cochemé HM et al (2012) Mitochondrial pharmacology. Trends Pharmacol Sci 3:341–352
16. Kelso GF, Porteous CM, Hughes G et al (2002) Prevention of mitochondrial oxidative damage using targeted antioxidants. Ann N Y Acad Sci 959:263–274
17. Zhang XY, Zhang PY (2016) Mitochondria targeting nano agents in cancer therapeutics. Oncol Lett 12:4887–4890
18. Boddapati SV, Tongcharoensirikul P, Hanson RN et al (2005) Mitochondriotropic liposomes. J Liposome Res 15:49–58
19. Weissig V (2003) Mitochondrial-targeted drug and DNA delivery. Crit Rev Ther Drug Carrier Syst 20:1–62
20. Yamada Y, Harashima H (2008) Mitochondrial drug delivery systems for macromolecule and their therapeutic application to mitochondrial diseases. Adv Drug Deliv Rev 60:1439–1462
21. Boddapati SV, D'Souza GGM, Weissig V (2010) Liposomes for drug delivery to mitochondria. Methods Mol Biol 605:295–303
22. Biswas S, Dodwadkar NS, Piroyan A et al (2012) Surface conjugation of triphenylphosphonium to target poly(amidoamine) dendrimers to mitochondria. Biomaterials 33:4773–4782
23. Guzman-Villanueva D, Mendiola MR, Nguyen HX et al (2015) Influence of triphenylphosphonium (TPP) cation hydrophobization with phospholipids on cellular toxicity and mitochondrial selectivity. SOJ Pharma Pharm Sci 2:1–9

Chapter 13

Surface Modification of Biomedically Essential Nanoparticles Employing Polymer Coating

Rahul Maheshwari, Nidhi Raval, and Rakesh Kumar Tekade

Abstract

Colloidal nanoparticles offering multiple biological applications carry tremendous potential to be developed as future medicines or nanomedicines. However, to decrease the particle agglomeration and enhance the stability of nanoparticles, functionalization could be of great interest. Functionalization is also capable of molding the delivery system for targeting and selective delivery of drugs and other biomolecules. In particular, the control over the size and the surface chemistry is crucial, since the successful applications in the prevention of diseases required biocompatibility at biological interfaces. Regardless of the advancements noted in nanotechnology-based nanoparticles, the development of nontoxic/biocompatible multifunctionalized nanoparticles is still a critical problem for researchers and requires urgent attention. In this chapter, an overview of nanoparticle functionalization with particular emphasis on its principle, needs, and formulation strategies has been discussed. Moreover, various applications of different surface-functionalized nanoparticles such as gold, silver, silicon, magnetic, liposomes, dendrimers, poly-lactic-co-glycolic acid, and solid lipid nanoparticles have also been presented.

Key words Nanoparticles, Surface-modified nanoparticles, Gold nanoparticles, Silver nanoparticles, Functionalization methods, Polymer coating, Dendrimers, Liposomes, Solid lipid nanoparticles

1 Introduction

One of the advancements of nanotechnology in medicine is the creation of carrier system, in particular, "NPs" which are studied extensively and possess the potential to become a key component of the commercial market shortly [1, 2]. NPs were formerly known as "ultrafine particles," ranging from 1 nm to few based on the method of production or type of particles used, and behave as a whole unit on its physicochemical and biological attributes [3, 4]. The most challenging perspective is to develop NPs, mainly inorganic, as a delivery system for the efficient transportation of drugs, gene, amino acids, and peptides [5–7].

Despite their increasing use, specific factors limit the application of NPs because of their restricted behavior in different

Volkmar Weissig and Tamer Elbayoumi (eds.), *Pharmaceutical Nanotechnology: Basic Protocols*, Methods in Molecular Biology, vol. 2000, https://doi.org/10.1007/978-1-4939-9516-5_13,

solvents. However this can be overcome by functionalizing NP surface [8, 9]. The need of surface modification of NPs is observed to help in tuning their properties to suit different applications in pharmaceutical science. These involve the changes in hydrophilicity, hydrophobicity, solubility, conductivity, anticorrosive property, bioavailability, and site specificity [10–12].

The surface properties of NPs differ to those of the material in bulk, and further modification/functionalization of the surface of NPs determines their interaction with the biological system. As a principle, the interaction between reactive monomers with aggregated NPs involves the penetration of the triggered sites on the NP periphery, and therefore the interstitial volume at the core of NPs becomes occupied to some degree [13]. The low molecular mass of monomer by their nature plays an essential role in this chemistry. This interaction finally alters the stability of the NPs and participates in site-specific/targeted delivery through the functional molecules on the particle surface [14, 15].

Functionalization is made possible using various principle techniques such as antibody conjugation, folic acid conjugation, aptamer linkage, glycine, and PEG linkage with/to NPs (Fig. 1). Wang et al. demonstrated doxorubicin (DOX)-entrapped folate-targeted phytosterol-alginate NPs for site-specific delivery. Their findings of cellular uptake and internalization of conjugated NPs revealed better than that of the control one. Similarly, higher intracellular uptake was also reported with conjugated NPs exhibiting potential as an emerging nanocarrier for site-specific delivery of drugs with no significant cellular toxicity to noncancerous cells [16]. Many investigators used this technique of folate conjugation and got better results in comparison to control or non-functionalized NPs, clearly indicating how functionalization increases the medicinal value of NPs. The different agents for functionalization, functionalization strategies, and their advantages and disadvantages are represented in Fig. 1.

With regard to the methods used for the functionalization, "grafting to" method offers an advantage to those polymers which are more sensitive toward polymerization conditions. However, "grafting from" method produces a polymeric shell of better thickness and homogeneity. However, the initiation of NP coating at the initial stage is important to stimulate reaction and homogeneous polymerization in both the cases [17]. Some of the commonly used formulation strategies for polymeric nanoparticles are depicted in Fig. 2.

The fundamental objective of tailoring the exterior of NP surface is to eliminate the difficulty of particle aggregation and prolonging the stability of the formulation. Also, it is a well-established fact that NPs may produce diffrent degrees of toxicity depending upon their type and mode of interaction with biological

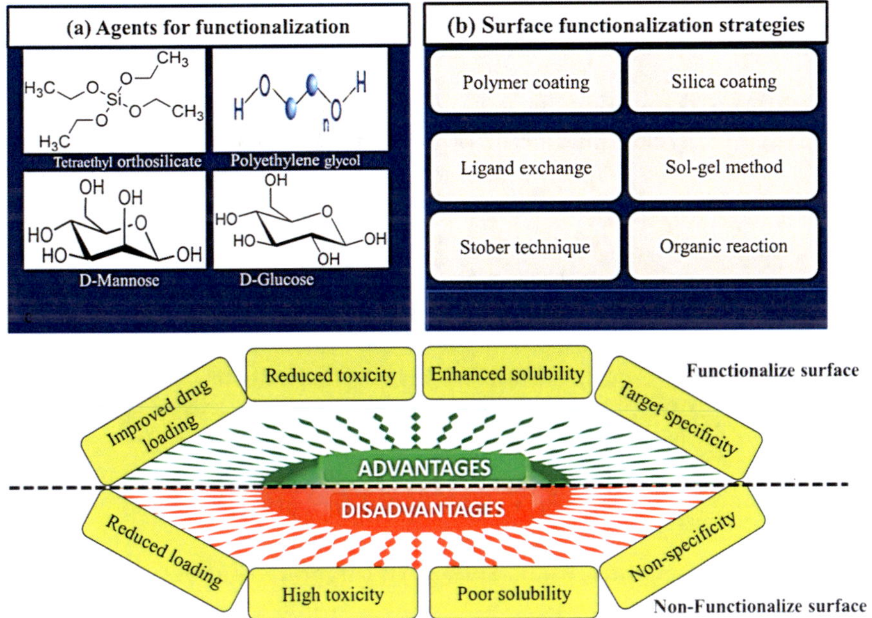

Fig. 1 Depicting the concept and components of surface functionalization of NPs. (**a**) Various agents employed for the modification of the NP surface. (**b**) Several methodologies utilized for the synthesis/formulation of NPs. (**c**) Conceptual figure representing the various advantages of NP surface modification and disadvantages associated with the non-functionalized surface

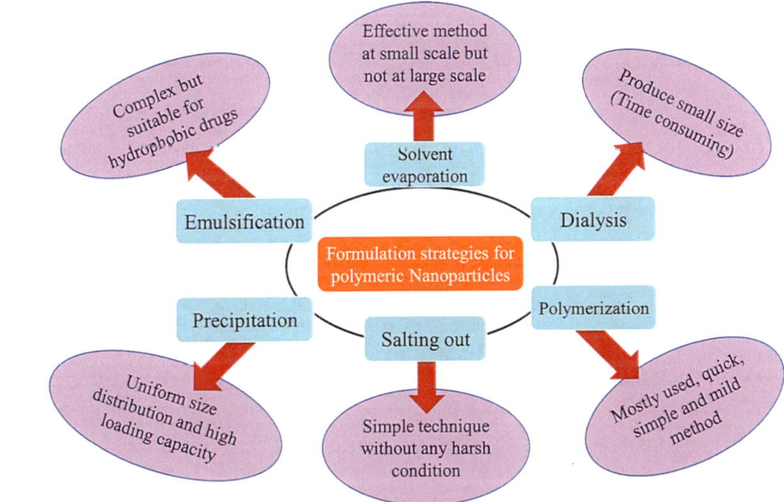

Fig. 2 Some of the commonly used formulation strategies for polymeric nanoparticles

membranes effects functionalization also useful in delivering the drugs to a specific site or can be targeted to particular tissue or cells which further reduces the chances of toxicity due to NPs.

Maisel et al. demonstrated the biological interaction of mucoadhesive NPs and mucus-penetrating NPs with either mucoadhesive surface or non-mucoadhesive surfaces in GI tract of normal mice. Authors studied nanoparticle absorption administering fluid orally or by injection into ligated intestinal loops, and found that both mucoadhesive NPs and mucus-penetrating NPs accumulated in the intestine, confirming that hurdles of GI barriers were compromised [18].

Moreover, Nasir et al. showed that their respective characters could govern the nature and level of interactions between a biological protein and NPs. A relevant NP characteristic involves the size, shape, and presence of functional groups whereas the morphology and configuration are critical parameters for a biological protein. They investigated the adsorption-provoked structural alterations of human carbonic anhydrase I (metalloenzymes having a role in acid-base balance) in the presence of NPs of varying dimensions and polarities. Further evaluating binding potential using isothermal titration calorimetry exhibited that the coupling to apolar surfaces is influenced via the NP dimension together with the inherent protein stability [19].

Furthermore, fluorescence study using 8-anilino-1-naphthalene sulfonic acid fluorescence demonstrated that human carbonic anhydrase adsorbs to both glasses of water-soluble and water-insoluble surfaces. However, the dynamics of the unfolding at the NP periphery is significantly different with the differences in polarity.

Recently, Podgorna used a new layer-by-layer deposition of chitosan-based polyelectrolytes to form calcium alginate and zinc pectin gel NPs with a diameter of approximately 100 nm size using reverse microemulsion technique. The chemical cross-linking was the principle behind the approach. Upon evaluating the results, authors concluded that the formed NPs were stable for up to 2 months and could be able to modify further using a layer-by-layer adsorption of polyelectrolytes. Moreover, they were nontoxic to human neuroblastoma cell line SH-SY5Y. However, in many different studies, the nanoparticle-nanoparticle combination was used to embed different functionalities into a single nanostructure using various approaches [20]. In this context, our next section will describe the various strategies used to coat or modify the surface of NPs.

2 Materials

2.1 Polymer Coating

1. Methoxy-terminated poly(ethylene glycol).
2. Poly(lactic acid) (PEG–PLA) block copolymers.
3. 2-Methoxyethanol.

4. Potassium naphthalene (1 M equivalent of 2-methoxyethanol).
5. Tetrahydrofuran.
6. Ethylene oxide.
7. Potassium 2-methoxyethoxide.
8. 2-Propanol
9. Benzene.
10. Dimethylformamide.
11. Lithium bromide.
12. Polylactic acid.
13. Dichloromethane-toluene.
14. Polyvinyl alcohol.
15. Chitosan 0.1% w/v solution: Prepare a stock chitosan solution of 0.1% (w/v) by solubilizing in acetic acid solution at 20 °C under continuous stirring for 15 min approximately, followed by a vacuum filtration to eliminate insoluble materials.
16. Gallic acid.
17. Sodium tripolyphosphate (TPP) 0.1% v/v solution: Prepare a stock TPP solution of 0.1% (v/v) by solubilizing in Milli-Q water under continuous stirring for 5 min approximately.
18. Nisin (polycyclic antimicrobial-peptide).
19. Poly(3-dimethylammonium-1-propyne hydrochlori de).

2.2 Silica Coating

1. Cobalt ferrite.
2. Tetraethyl orthosilicate.
3. Sodium hydroxide.
4. Hydrochloric acid.

2.3 Ligand Exchange

1. Platinum(II) acetylacetonate.
2. 1,2-Hexadecanediol.
3. Dioctyl ether.
4. Oleic acid.
5. Oleylamine.
6. Iron(0) pentacarbonyl.
7. Trioctylphosphine.
8. Iron(II) acetylacetonate (0.26 g).
9. Oleylamine (1 mL).
10. Trioctylphosphine (5 mL).
11. Ethanol.
12. Methanol.

13. Tetrahydrofuran.
14. *N*,*N*-dimethylformamide.
15. Hexanes.
16. *N*,*N*-dimethylformamide phase.

3 Methods

3.1 Synthesis of Methoxy-Terminated Poly(Ethylene Glycol)–Poly(Lactic Acid) (PEG–PLA) Block Copolymers

1. Mix the appropriate amount of 2-methoxyethanol (0.7 mmol) and potassium naphthalene (1 M equivalent of 2-methoxyethanol) in tetrahydrofuran for 1 h.
2. Add the purified ethylene oxide (80–131 mmol) to the obtained potassium 2-methoxyethoxide solution (total volume: 50 mL) and then stir the solution for 48 h.
3. Add purified DL-lactide to this solution. After the reaction precipitate the resulting block copolymer using cold 2-propanol, store in a freezer for 12 h, centrifuge at 17722 × *g*, and lyophilize in benzene.
4. Determine the average molecular weight of the resultant block copolymer by the use of gel permeation chromatography (GPC). GPC conditions: eluent; DMF in the presence of 10 mM LiBr, flow; 1 mL/min; column temperature: 40 °C.
5. A recent finding on the example of polymer coating and experiment of developed formulation in animal model is represented in Fig. 3 (*see* **Note 1**).

3.2 Preparation of Surface-Modified and Surface-Unmodified PLGA Particles

1. Dissolve PLGA7510 (0.06–1.25 w/v%) and PEG–PLA block copolymers in a dichloromethane-toluene mixed solvent and keep the molar ratio of PEG–PLA/PLGA at 1:5.
2. Dissolve PVA as an emulsion stabilizer in pure Milli-Q water (0.2–2 w/v %).
3. Add the organic solvent to the PVA solution by keeping the volume ratio of the organic solvent/PVA solution at 0.25–2.
4. Homogenize the mixture to form an o/w emulsion at 3836–43456 × *g* for 2–10 min.
5. Stir the emulsion at 150 rpm overnight at room temperature with a propeller-type impeller so that the organic solvent from the o/w emulsion droplets would evaporate.
6. Centrifuge the solution at 5703 × *g*.
7. Wash the particles with pure Milli-Q water three times to get purified surface-modified particles.
8. Similarly, surface-unmodified PLGA particles can be prepared using the same method except that PEG will be absent (*see* **Note 2**).

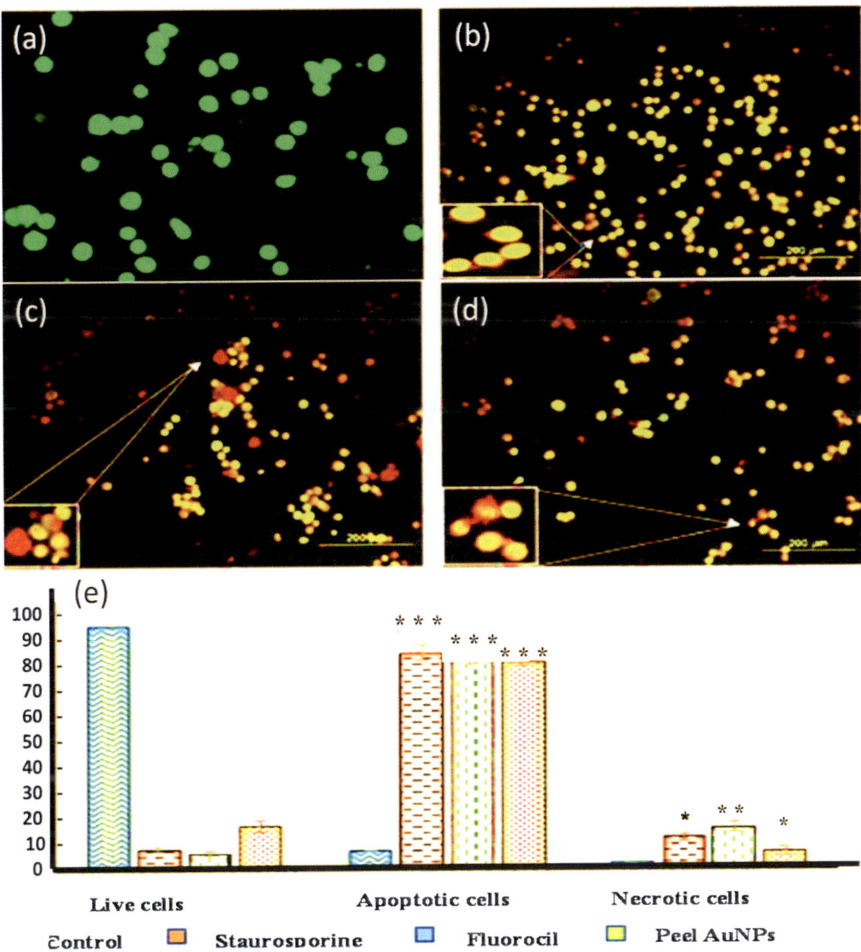

Fig. 3 Influence of *Vitis vinifera* peel AuNPs on A431 lung carcinoma cells by promoting structural changes in cell morphology. A431 lung carcinoma cells were exposed to AO/EtBr and seen using fluorescent microscope. (**a**) Control cells, (**b**) exposed to staurosporine (0.1 nM), (**c**) exposed to fluorouracil (23.43 μM), (**d**) exposed to grape peel AuNPs (23.6 μM), (**e**) histogram showing percentage structural alterations in the cells exposed to peel NPs. Investigation was done in triplicates. Data taken from [21] and used without modification

3.3 Chitosan and Gallic Acid-Based Coating

1. Dropwise add 5 mL sodium tripolyphosphate (TPP; cross-linking agent) 0.1% v/v solution in chitosan stock solution (5 mL) admixed with GA (*see* **Note 3**).
2. Homogenize the obtained suspensions at 21293 × *g* for 12 min.

3.4 PLGA-Based Polymeric Coating

1. Dissolve 20 mg of the PLGA in 2 mL of an organic solvent that is miscible with water.
2. Add organic phase dropwise into 4 mL of an aqueous phase under moderate magnetic stirring 104 (360 rpm).
3. Stir the solution to remove organic phase or uncover it for 16 h at room temperature (*see* **Note 4**).

4. Isolate the particles using centrifugation for 15 min at 3836 × *g*.
5. Collect dry particles by freeze-drying or by drying under vacuum over phosphorus pentoxide (*see* **Note 5**).
6. The dry powder can be stored at 4 °C until further use (*see* **Note 6**).

3.5 Ligand Exchange Coating Method for Magnetic Nanoparticles (Method I)

1. Add platinum(II) acetylacetonate (197 mg, 0.5 mmol), 1,2-hexadecanediol (390 mg, 1.5 mmol), and dioctyl ether (20 mL) into a three-necked round-bottom flask and stir using a magnetic stirring bar (*see* **Note 7**).
2. Heat the mixture to 100 °C under a gentle flow of N_2 to remove oxygen and moisture (*see* **Note 8**).
3. Add oleic acid (0.16 mL, 0.5 mmol) and oleylamine (0.17 mL, 0.5 mmol) to the mixture and set to equilibrate for 10 min before injecting iron(0) pentacarbonyl (0.13 mL, 1 mmol) (*see* **Note 9**).
4. Add trioctylphosphine (0.22 mL, 0.5 mmol) as a third protecting ligand. Heat the mixture to reflux at 297 °C for 30 min (ramp rate as a variable, 5–15 °C/min).
5. Stop the reaction by removing the heat source and allowing cooling naturally to room temperature. Collect the final NPs by centrifugation using ethanol as the antisolvent (*see* **Note 10**).
6. Wash the particles with hexanes and ethanol two to three times to remove excessive ligands.
7. Store the obtained particles (*see* **Note 11**).

3.6 Ligand Exchange Coating Method for Magnetic Nanoparticles (Method II)

1. In three-necked round-bottom flask combine the iron (II) acetylacetonate (0.26 g), oleylamine (1 mL), and trioctylphosphine (5 mL) followed by the complete removal of residue oxygen using the suitable technique.
2. Heat the reaction to 100 °C under the protection of N_2 and equilibrate for 10 min to ensure the even distribution of reagents.
3. Increase the reaction temperature to 200 °C for 30 min.
4. Store the obtained particles.

3.7 Ligand Exchange Coating Method for Magnetic Nanoparticles (Method III)

1. Mix 50 mg of hydrophobic NP powder and 500 mg of the hydrophilic ligand with 20 mL of any of the solvent (ethanol, methanol, tetrahydrofuran, or *N*,*N*-dimethylformamide) in a scintillation vial.
2. Sonicate for 10 min.
3. Perform magnetic stirring (300 rpm; 10 min), to ensure a satisfying conversion rate.

4. Keep the obtained solution aside and observe for color change (dark to transparent).
5. Precipitate the exchanged NPs with the addition of excessive hexanes as the antisolvent and collect by centrifugation at 3836 × g for 5 min.
6. Wash the resulting NPs with ethanol/hexane combination two or three times to remove any excessive ligands.

4 Notes

1. Specific to cancer, Nirmala et al. reported that *Vitis vinifera* peel polyphenol-functionalized AuNPs induced cellular toxicity and cell death in A431 skin metastasis cell cultures [21]. In this case, the multifunctional phytochemicals from *Vitis vinifera* peel polyphenols facilitated "green" biotechnical process exhibited to be advantageous in both the reduction and stabilization through capping of AuNPs for applications in several biomedical and therapeutic uses, including skin cancer. The development of green chemistry in synthesizing and functionalizing NPs for a required size, shape, and dispersivity is also getting huge attention as evident by the investigators. Influence of *Vitis vinifera* peel AuNPs on A431 cells by promoting morphological alterations has been presented in Fig. 3.
2. Along with functionalized formulation, the non-functionalized formulation is also required for appropriate evaluation of the efficiency of the coated system.
3. Sodium tripolyphosphate is an anionic counter ion which helps in cross-linking with cationic polymers such as chitosan.
4. Stirring speed may vary as per the need.
5. Apart from freeze-drying particles can also be lyophilized based on the intended application.
6. Freeze-drying or lyophilization is exclusively used to protect the nanoformulations from oxidation or degradation and enhance their stability.
7. A three-necked round-bottom flask is used so that the evaporating gases can be suitably eliminated out of the reaction.
8. Nitrogen environment is required to maintain the necessary conditions for reaction.
9. The solution can be equilibrating by keeping the mixture aside for some time.
10. Antisolvent is used to make the particles harden so that upon dispersion into the water they must be in colloidal phase.
11. Particles can be stored at 4 °C.

Acknowledgments

The authors would like to acknowledge Science and Engineering Research Board (Statutory Body Established Through an Act of Parliament: SERB Act 2008), Department of Science and Technology, Government of India, for the award of early carrier research grant (File Number: ECR/2016/001964) and DST-NPDF to Dr. Maheshwari (PDF/2016/003329) in Dr. Tekades's lab. Authors would also like to thank NIPER Ahmedabad for providing research support for research on cancer and arthritis.

References

1. Maheshwari R, Tekade M, Sharma PA et al (2015) Nanocarriers assisted siRNA gene therapy for the management of cardiovascular disorders. Curr Pharm Des 21(30):4427–4440
2. Sharma PA, Maheshwari R, Tekade M et al (2015) Nanomaterial based approaches for the diagnosis and therapy of cardiovascular diseases. Curr Pharm Des 21(30):4465–4478
3. Lalu L, Tambe V, Pradhan D et al (2017) Novel nanosystems for the treatment of ocular inflammation: current paradigms and future research directions. J Control Release 268:19–39
4. Maheshwari RG, Tekade RK, Sharma PA et al (2012) Ethosomes and ultradeformable liposomes for transdermal delivery of clotrimazole: a comparative assessment. Saudi Pharm J 20 (2):161–170
5. Tekade RK, Maheshwari R, Soni N et al (2017) Chapter 1—Nanotechnology for the development of nanomedicine A2—Mishra, Vijay. In: Kesharwani P, Amin MCIM, Iyer A (eds) Nanotechnology-based approaches for targeting and delivery of drugs and genes. Academic, New York, pp 3–61
6. Tekade RK, Maheshwari R, Tekade M (2017) 4—Biopolymer-based nanocomposites for transdermal drug delivery. In: Jana S, Maiti S, Subrata BT (eds) Biopolymer-based composites. Woodhead Publishing, Cambridge, pp 81–106
7. Tekade RK, Maheshwari R, Tekade M et al (2017) Chapter 8—Solid lipid nanoparticles for targeting and delivery of drugs and genes A2—Mishra, Vijay. In: Kesharwani P, Amin MCIM, Iyer A (eds) Nanotechnology-based approaches for targeting and delivery of drugs and genes. Academic, New York, pp 256–286
8. Maheshwari RG, Thakur S, Singhal S et al (2015) Chitosan encrusted nonionic surfactant based vesicular formulation for topical administration of ofloxacin. Sci Adv Mater 7 (6):1163–1176
9. Tekade RK, Maheshwari R, Jain NK (2018) 9—Toxicity of nanostructured biomaterials A2—Narayan, Roger. Nanobiomaterials. Woodhead Publishing, p 231–256
10. Kumar Tekade R, Maheshwari RGS, Sharma PA et al (2015) siRNA therapy, challenges and underlying perspectives of dendrimer as delivery vector. Curr Pharm Des 21 (31):4614–4636
11. Maheshwari R, Tekade M, Gondaliya P et al (2017) Recent advances in exosome-based nanovehicles as RNA interference therapeutic carriers. Nanomedicine (Lond) 12 (21):2653–2675
12. Soni N, Soni N, Pandey H et al (2016) Augmented delivery of gemcitabine in lung cancer cells exploring mannose anchored solid lipid nanoparticles. J Colloid Interface Sci 481:107–116
13. Tekade RK, Maheshwari R, Soni N et al (2017) Chapter 12—carbon nanotubes in targeting and delivery of drugs A2—Mishra, Vijay. In: Kesharwani P, Amin MCIM, Iyer A (eds) Nanotechnology-based approaches for targeting and delivery of drugs and genes. Academic, New York, pp 389–426
14. Yan J, Li Z, Wang L et al (2008) Preparation and characterization of BaLiF3:Er3+ nanoparticles by the hydrothermal microemulsion synthesized method. J Rare Earths 26 (1):48–50
15. Fang L, Wang F, Chen Z et al (2017) Flower-like MoS2 decorated with Cu2O nanoparticles for non-enzymatic amperometric sensing of glucose. Talanta 167:593–599
16. Wang J, Wang M, Zheng M et al (2015) Folate mediated self-assembled phytosterol-alginate nanoparticles for targeted intracellular anticancer drug delivery. Colloids Surf B: Biointerfaces 129:63–70
17. Bennet D, Kim S (2014) Polymer nanoparticles for smart drug delivery. In: Sezer AD

(ed) Application of nanotechnology in drug delivery. InTech, Rijeka

18. Maisel K, Ensign L, Reddy M et al (2015) Effect of surface chemistry on nanoparticle interaction with gastrointestinal mucus and distribution in the gastrointestinal tract following oral and rectal administration in the mouse. J Control Release 197:48–57
19. Nasir I, Lundqvist M, Cabaleiro-Lago C (2015) Size and surface chemistry of nanoparticles lead to a variant behavior in the unfolding dynamics of human carbonic anhydrase. Nanoscale 7(41):17504–17515
20. Podgórna K, Jankowska K, Szczepanowicz K (2017) Polysaccharide gel nanoparticles modified by the Layer-by-Layer technique for biomedical applications. Colloids Surf A Physicochem Eng Asp 519:192–198
21. Nirmala JG, Akila S, Narendhirakannan R et al (2017) *Vitis vinifera* peel polyphenols stabilized gold nanoparticles induce cytotoxicity and apoptotic cell death in A431 skin cancer cell lines. Adv Powder Technol 28 (4):1170–1184

Chapter 14

Peptide-Modified Gemini Surfactants: Preparation and Characterization for Gene Delivery

Mays Al-Dulaymi, Waleed Mohammed-Saeid, Anas El-Aneed, and Ildiko Badea

Abstract

Diquaternary ammonium-based gemini surfactants have been investigated widely as nonviral gene delivery systems. These unique cationic lipids have versatility in their chemical structure, show relatively low toxicity, are able to compact genetic material (pDNA, RNA) into nano-sized lipoplexes, and can be easily produced. In addition, the gemini surfactants show significant improvement in the transfection activity and biocompatibility compared to other cationic lipids used as nonviral gene delivery agents. The successful applications of gemini surfactant-based lipoplexes as topical gene delivery systems in animal models indicate their potential as noninvasive carriers for genetic immunization, theranostic agents, and in other gene therapy treatments. Detailed physicochemical characterization of gemini surfactant lipoplexes is a key factor in terms of formulation optimization and elucidation of the cellular uptake and stability of the lipoplexes system. In this chapter, we describe in detail different formulation methods to prepare gemini surfactant lipoplexes and comprehensive physicochemical characterization. In addition, we illustrate general protocols for in vitro evaluations.

Key words Gene therapy, Gemini surfactants, Formulation development, Small-angle X-ray scattering, Lipid packing parameter, pDNA binding, Circular dichroism, In vitro transfection

1 Introduction

Advancements in the discovery of the genetic bases of many diseases and the development of the corresponding means of interventions have made gene therapy a promising therapeutic approach that can revolutionize the world of medicine. The last few decades have witnessed unprecedented interest in developing efficient vectors capable of delivering the genetic material into the targeted site, promoting gene expression—the ultimate goal of gene therapy [1, 2]. Diquaternary ammonium gemini surfactants have emerged as a promising class of nonviral gene delivery vectors. They are

Mays Al-Dulaymi and Waleed Mohammed-Saeid contributed equally to this work.

Volkmar Weissig and Tamer Elbayoumi (eds.), *Pharmaceutical Nanotechnology: Basic Protocols*, Methods in Molecular Biology, vol. 2000, https://doi.org/10.1007/978-1-4939-9516-5_14,

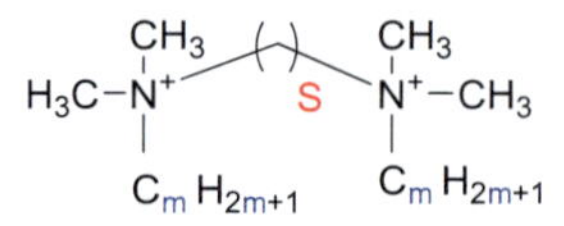

S = 2, 3, 4, 5, 6, 8, 10, 12, 16
m = 12, 16

Fig. 1 Schematic representation of the traditional diquaternary ammonium gemini surfactants

composed of two monomeric surfactant molecules connected by a spacer (Fig. 1) [3]. Gemini surfactants have the ability to electrostatically interact with the negatively charged nucleic acids forming nano-sized lipoplexes and favoring cellular internalization. In comparison to their corresponding monomeric counterparts, gemini surfactants display a number of superior characteristics such as higher efficiency in reducing surface tension, enhanced wetting properties, one to two orders of magnitude lower critical micelle concentration (CMC) and lower Krafft temperature [4]. In addition, the unique structure of gemini surfactants offers endless possibilities for structural modification allowing for the emergence of compounds specifically tailored to overcome delivery barriers.

The most extensively studied group of gemini surfactants is the traditional cationic *N*,*N*-bis(dimethylalkyl)-α,ω-alkane-diammonium surfactants (Fig. 1). These surfactants exhibit promising results in delivering genetic materials both in vitro and in vivo [5, 6]. In fact, lipoplexes composed of *N*,*N*′-bis(dimethylhexadecyl)-1,3-propanediammonium dibromide gemini surfactant, designated as 16-3-16, complexed with pDNA encoding for interferon gamma (INF-γ) were tested for the treatment of localized scleroderma in normal, knockout, and diseased mice [5–7]. Results revealed significant elevation in the level of INF-γ expression in animals treated with gemini surfactant-based lipoplexes compared to animals treated with naked DNA both in normal animals and INF-γ-deficient mice [5, 6]. Furthermore, topical application of gemini surfactant-based nanoparticles into Tsk/+ (tight-skin scleroderma) mouse model resulted in significant decrease in collagen production demonstrating the efficiency of the noninvasive delivery system [7].

Despite the success of the traditional gemini surfactants, concerns regarding the toxicity profile of gemini surfactants have arisen, impeding their progress toward a clinical success [8]. Investigations continue to explore new chemical modifications to address this problem, including the introduction of compounds with biocompatible and biodegradable moieties such as sugars, lipids, and amino acids [9–11]. A new series of gemini surfactants was introduced by coupling short chain peptides into the spacer

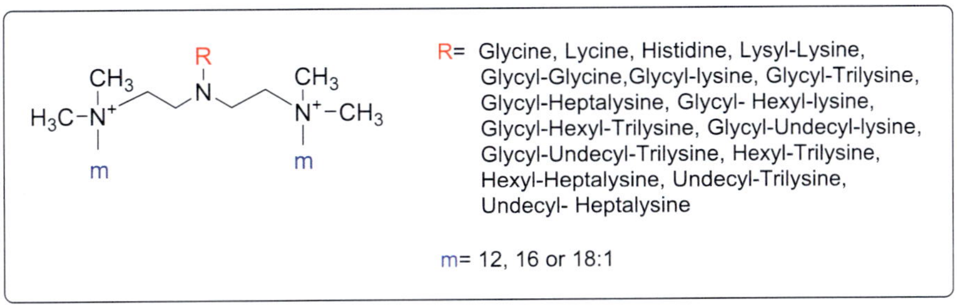

R= Glycine, Lycine, Histidine, Lysyl-Lysine, Glycyl-Glycine,Glycyl-lysine, Glycyl-Trilysine, Glycyl-Heptalysine, Glycyl- Hexyl-lysine, Glycyl-Hexyl-Trilysine, Glycyl-Undecyl-lysine, Glycyl-Undecyl-Trilysine, Hexyl-Trilysine, Hexyl-Heptalysine, Undecyl-Trilysine, Undecyl- Heptalysine

m= 12, 16 or 18:1

Fig. 2 Schematic representation of the peptide-modified gemini surfactants

region (Fig. 2) [12–15]. This generation of gemini surfactants demonstrated superior in vitro transfection efficiency compared to the traditional gemini surfactants with minimum cytotoxicity [12–15]. Furthermore, topical application of the peptide-modified gemini surfactants nanoparticles into rabbit vaginal cavities showed higher gene expression compared to the unsubstituted parent compound without visible toxicity [16]. The incorporation of the peptide chain gave rise to balanced binding properties with the nucleic acid mediating for both DNA compaction and subsequent release [13]. The addition of peptides demonstrate an enhanced buffering capacity that induced a pH-dependent increase in particle size and zeta potential, resulting in the production of "intelligent" nanoparticles that respond to endosomal acidification, facilitating endosomal escape [14, 17]. Finally, the presence of terminal amino groups in the peptide backbone imparted a higher positive charge to the modified compounds, hence a decreased number of gemini surfactants are required to neutralize and compact the DNA [14].

In addition to structural modifications, optimizing the physicochemical properties of the nonviral gene delivery systems is an essential aspect in the design of efficient and safe nucleic acid delivery vectors. Parameters such as particle size, surface charge, morphological characteristics, and lipid organization can greatly impact the route of cellular entry, efficiency of cellular uptake, and potential cytotoxicity [18–20]. Here, we present materials and methods for the preparation and characterization of peptide-modified gemini surfactants-based gene delivery system. Specifically, this chapter will discuss protocols for lipoplex preparation and modalities for enhancing their physicochemical stability in pharmaceutical formulations. In addition, we elaborate on various physicochemical characterization techniques including particle size, surface charge evaluation, and the impact of pH on their measurements. Morphological characteristics and lipid organization assessment techniques, such as the employment of small-angle X-ray scattering (SAXS) and Langmuir-Blodgett technique, are highlighted. Finally, protocols used to evaluate the interaction of the gemini surfactants with the DNA are described.

2 Materials

2.1 Formulation Preparation

1. pDNA that express a model protein (*see* **Note 1**).
2. Peptide-modified gemini surfactants.
3. Helper lipid 1,2 dioleyl-*sn*-glycero-phosphatidylethanolamine (DOPE) (*see* **Note 2**).
4. α-Tocopherol.
5. Anhydrous ethanol.

2.2 Determination of Particle Size and ζ-Potential

1. Zetasizer Nano ZS instrument (Malvern Instruments, Worcestershire, UK) for particle size determination by dynamic light scattering (DLS) and ζ-potential evaluation by laser Doppler velocimetry.
2. MPT-2 multipurpose titrator accessory (Malvern Instruments, Worcestershire, UK) connected to the Zetasizer Nano ZS apparatus to determine size and ζ-potential as a function of pH.
3. Disposable folded capillary cell (DTS1060) (Malvern Instruments, Worcestershire, UK).
4. Hydrochloric acid (HCl) and sodium hydroxide (NaOH) for pH adjustment.
5. Disposable titration cell (Malvern Instruments, Worcestershire, UK).
6. Magnetic stirring bar.

2.3 Determination of the Supramolecular Organization

1. Utilize Beamline BL4-2 at Stanford Synchrotron Radiation Lightsource (SSRL, Stanford, CA, USA) for SAXS data collection (*see* **Note 3**).
2. Detect the scattered X-ray is on MAR225-HE (225 mm × 225 mm (3072 × 3072 pixels, pixel size 73.24 μm).
3. Silver behenate (TCI, Mississauga, ON, Canada).
4. In-house 96-well polyethylene and aluminum sample plate of 2 mm thickness with wells of 4 mm diameter (Fig. 3) (*see* **Note 4**).
5. Kapton tape.
6. Eppendorf concentrator 5301 (Eppendorf, Hamburg, Germany).
7. GSASII software [21].

2.4 Circular Dichroism: pDNA Conformational Changes

1. Pi-star-180 circular dichroism spectroscopy (Applied Photophysics, Leatherehead, UK).
2. Quartz cuvette.

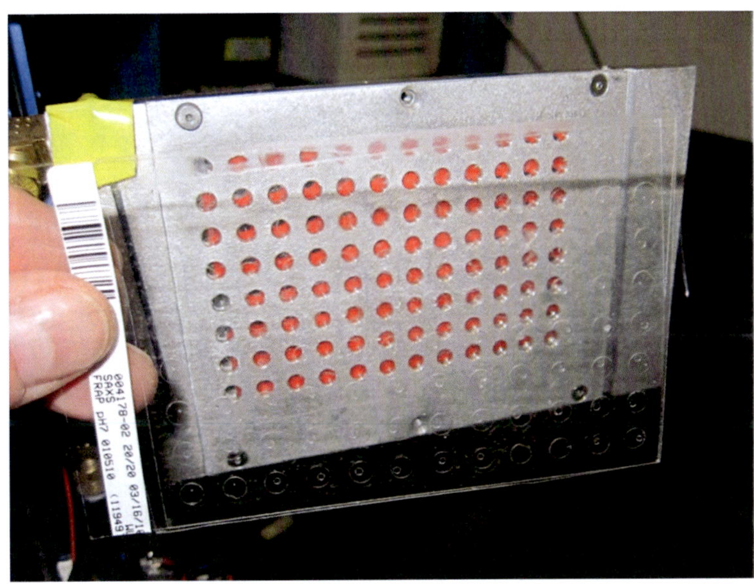

Fig. 3 Picture of the 96-well polyethylene sample plate

3. Freeze dryer Labconco® Freezone Plus 6 L cascade freeze dryer (Labconco, Kansas City, MO, USA).
4. Eppendorf concentrator 5301(Eppendorf, Hamburg, Germany) (*see* **Note 5**).

2.5 DNA Binding Assay

1. Bio-Rad PowerPac HC electrophoresis apparatus (Biorad, Mississauga, ON, Canada).
2. AlphaImager™ UV imager (Alpha Innotech, San Leandro, CA, USA) to visualize the gel.
3. Agarose for biological applications.
4. Tris-acetate-EDTA (TAE) buffer: 4 mM Tris–acetate, 1 mM EDTA, pH 8.

2.6 Determination of the Critical Micelle Concentration (CMC)

1. Kruss (Model K10T) tensiometer for determining the surface tension of peptide-modified gemini surfactants.
2. Weyne-Kerr precision component analyzer (Model 6425) operated at 1.5 kHz and equipped with a Tacussel electrode with cell constant of 1.52 cm^{-1} for assessing the gemini surfactants' specific conductivities.
3. VWR Scientific (Model 1160A) circulating water bath for maintaining the temperature.

2.7 Determination of the Molecular Packaging Parameter of the Lipids

1. Langmuir minitrough (KSV, Helsinki, Finland) equipped with a Wilhelmy plate balance for evaluating the surface pressure-mean molecular area isotherms.
2. 100 μL Hamilton syringe.

3. Chloroform.
4. Avogadro software for estimating the length of the hydrocarbon tails [22].
5. Gaussian 09 software, revision C.01, for calculating the volume of the hydrocarbon tails [23].

2.8 Cell Culture

1. COS-7 African green monkey kidney fibroblasts cell line (ATCC, CRL-1651).
2. PAM 212 murine keratinocytes kindly provided in 2000 by Dr. S. Yuspa, National Cancer Institute, Bethesda, MD, USA.
3. Minimum Essential Medium (MEM) and Dulbecco's Modified Eagle's Medium (DMEM).
4. Antibiotic Antimycotic Solution.
5. Fetal Bovine Serum Albumin (FBS).
6. 75-cm^2 tissue culture flasks.
7. 96-well tissue culture plates.
8. Lipofectamine Plus™ reagent (Invitrogen, CA, USA).

2.9 Transfection Efficiency Assessment

For INF-γ enzyme-linked immunosorbent assay (ELISA)

1. Clear Flat-Bottom ELISA 96-Well Plates.
2. BD Pharmingen™ Purified Rat Anti-Mouse IFN-γ (BD Pharmingen, BD Biosciences).
3. BD Pharmingen™ Biotin Rat Anti-Mouse IFN-γ (BD Pharmingen, BD Biosciences).
4. BD Pharmingen™ AKP Streptavidin (BD Pharmingen, BD Biosciences).
5. BD Pharmingen™ Recombinant Mouse IFN-γ (BD Pharmingen, BD Biosciences).
6. Phosphatase substrate.
7. Bovine Serum Albumin.
8. BioTek microplate reader (Bio-Tek Instruments, VT, USA).
9. TWEEN® 20.

2.10 In Vitro Cytotoxicity Evaluation

1. MTT (3-(4,5-Dimethylthiazol-2-yl)-2,5-Diphenyltetrazolium Bromide) Invitrogen Corporation.
2. Dimethyl sulfoxide (DMSO) ACS spectrophotometric grade.
3. Microplate reader: BioTek microplate reader (Bio-Tek Instruments, VT, USA).

2.11 Buffer Solutions

1. 1× PBS buffer (without Ca^{2+}/Mg^{2+}).
   137 mM NaCl, 2.7 mM KCl, 10 mM Na_2HPO_4 anhydrous and 1.8 mM KH_2PO_4 (final pH 7.3).

2. pNPP buffer.
   10 mM Diethanolamine and 0.5 mM $MgCl_2$.
   Adjust pH to 9.8 with conc. HCl.
3. Coating buffer ($NaHCO_3$ 0.05 M).
   (final pH should be 9.6).
4. Stop Solution (to stop further color production).
   0.3 M EDTA pH 8 with 1 N NaOH.
   Or
   50 μL of (1 N NaOH with 0.25 M EDTA).
5. PBST buffer (0.05% Tween 20 in 1× PBS).

3 Methods

3.1 Formulation Preparation

The basic components of gemini surfactant-based lipoplexes are pDNA, that express a model protein (e.g., GFP, Interferon-γ, Luciferase) (denoted *P* in formulation), helper lipid DOPE (denoted G in formulation), gemini surfactant (denoted *L* in formulation)

1. Prepare pDNA in sterile, biological-grade water at final concentration of 200 μg/mL.
2. Prepare gemini surfactant in Millipore water at 3 mM concentration, sterile filtration required for in vitro evaluation using 0.45 μM syringe filter.
3. Prepare 1 mM DOPE vesicles as follows.

3.1.1 Day 1: Preparation of DOPE Film

1. Weight/measure into a 50-mL round bottom flask:

DOPE	37.2 mg
α-Tocopherol	2.48 mg
Ethanol (100%)	3 mL

2. Dissolve the content of the flask by sonication for 10 min, or until clear dispersion is obtained.
3. Evaporate the ethanol by using rotary evaporators at 55 °C, under vacuum and using water cooling coil until thin uniform film is formed on the wall of the flask.
4. Freeze-dry overnight to remove solvent trace (*see* **Note 6**).

3.1.2 Day 2: Day of Lipoplexes Preparation

1. Prepare 9.25% Sucrose solution using fresh dd·H_2O, adjust the pH to 9 with conc. NaOH solution (*see* **Note 7**).
2. Add 5 g of sterile glass beads to the DOPE film (in the 100 round bottom flask), then add 50 mL of 9.25 sucrose solution.

3. Sonicate for 3 h at 55 °C or until clear solution is obtained (*see* **Note 8**).
4. In biosafety cabinet, filter the obtained solution to sterile container using sterile Acrodisc syringe filter 0.45 μM (*see* **Note 9**).

3.1.3 Prepare [P/G/L] Lipoplexes (See Note 10)

The first step in preparing the P/G/L lipoplex is the preparing of pDNA/gemini surfactant complex (in the following case at 1:10 ± charge ratio). This complex will form as a result of electrostatic interaction between the phosphate backbone of the pDNA and the cationic head group of the gemini surfactant.

1. In sterile 1.5 mL Eppendorf tube, add 37 μL of the 200 μg/mL pDNA and 37 μL of 3 mM gemini surfactant dispersion.
2. Incubate at room temperature for 15 min. The result of the incubation is the P/G complex (*see* **Notes 11–13**).
3. Upon the preparation of the DOPE vesicles and P/G complex: add 926 μL of 1 mM molar DOPE dispersion to the P/G complex.
4. Incubate at room temperature for 15 min.
5. The result of the incubation is the P/G/L lipoplexes at 1:10 ± charge ratio and 1:10 gemini surfactant/DOPE molar ratio (*see* **Notes 14–16**).

3.2 Determination of Particle Size and ζ-Potential

The particle size and the overall charge of nonviral lipoplex systems are essential physicochemical properties that need to be evaluated and optimized. These properties of gemini surfactant-based lipoplexes significantly influence formulation stability, cellular uptake, biodistribution, and clearance [24–26].

1. Calibrate the Zetasizer Nano ZS instrument for size and ζ-potential measurements (*see* **Notes 17** and **18**).
2. Transfer 750 μL of the peptide-modified gemini surfactants lipoplex formulations into the disposable folded capillary cell (*see* **Note 19**).
3. Place the cell into the Zetasizer Nano ZS apparatus.
4. Measure the size and ζ-potential according to the equipment protocol (*see* **Note 20**).
5. Report data as a volume distribution (*see* **Note 21**).

3.3 Determination of Particle Size and ζ-Potential as Function of pH

1. Calibrate the MPT-2 titrator pH probe according to the equipment protocol.
2. Fill one titrant container with 0.1 M HCl and the other with 0.05 M NaOH (*see* **Note 22**).
3. Prime the titrant syringe pumps and tubes (*see* **Note 23**).

4. Transfer 10 mL of the peptide-modified gemini surfactants lipoplex formulations into the disposable titration cell of the autotitrator (*see* **Note 24**).
5. Place magnetic stirring bar inside the titration cell (*see* **Note 25**).
6. Connect the titration cell to the DTS1060 measurement cell using Luer lock connectors.
7. Fill the measurement cell to ensure that the sample path is free of air.
8. Measure the size and ζ-potential according to the equipment protocol.
9. Titrate over basic to acidic pH range at 0.2–0.5 pH unit increments (*see* **Note 26**).

3.4 Determination of the Supramolecular Organization

Determination of supramolecular arrangement of the lipid phase of gemini surfactant-based lipoplexes is an essential factor that can influence the transfection activity of the lipoplex system. Lipoplexes can form in several lipid phases which include lamellar, cubic, hexagonal, and inverted hexagonal phases. Synchrotron-based small-angle X-ray technique is used for this purpose.

1. Prepare peptide-modified gemini surfactants lipoplexes as discussed in the formulation section using ten times higher concentrations (*see* **Note 27**).
2. Set incident beam wavelength to 1.1271 Å (11 KeV energy).
3. Position detector 1.1 m from the sample.
4. Calibrate the SAXS detector with silver behenate.
5. Seal one side of the 96 well samples holder with Kapton tape as demonstrated in Fig. 4.
6. Load 50 μL of the sample into each well on the plate.
7. Seal the top of the samples with Kapton tape as demonstrated in Fig. 4.
8. Collect the scattering data of the blank solutions for background subtraction (*see* **Note 28**).
9. Collect scattering data of the individual constituent of the peptide-modified gemini surfactant lipoplex formulation (i.e. DNA, gemini surfactants, and helper lipid).
10. Collect the scattering profile of the peptide-modified gemini surfactants lipoplexes (Plasmid/Gemini/Lipid complexes, P/G/L).
11. Expose the samples to X-ray for 20 s (*see* **Note 29**).
12. Plot diffraction intensity versus 2θ (where θ is the diffraction angle) or the scattering vector $\left(q = \frac{4\pi}{\lambda}\sin\theta\right)$ by radial integration of the 2D patterns using GSASII software (*see* **Notes 30** and **31**).

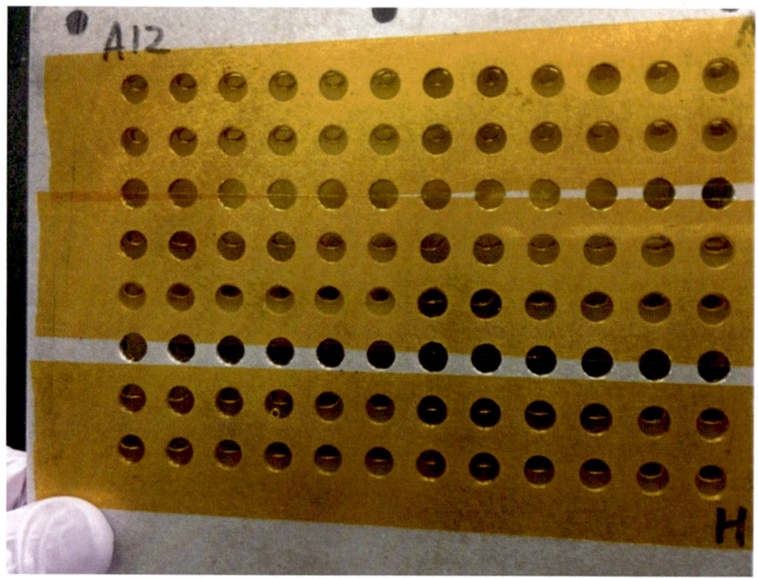

Fig. 4 96-well aluminum sample plate sealed with Kapton tape

3.5 Circular Dichroism: pDNA Conformational Changes

Circular dichroism spectroscopy (CD) provides information about the helical structure of the pDNA in solution and the possible conformational changes in pDNA native structure resulting from ligand binding of noncovalent interaction (as in case of pDNA-gemini surfactant interaction). The sensitivity of the CD to detect the changes in pDNA conformational changes depends on several factors including the pDNA concentration, the presence of metals or optically active chiral molecules, and the type of the instrument.

1. Utilize Pi-star-180 instrument CD with 2 nm slit at 37 °C under a N2 atmosphere.
2. Prepare the pDNA, PG, and P/L/G in pDNA final concentration of 10–15 μg/mL. The high concentration lipoplex can be obtained by the following methods:
   (a) Use high concentration pDNA, gemini surfactant, and DOPE and follow the method of preparing P/G or P/G/L as described in the formulation preparation section (*see* **Note 32**).
   (b) Prepare the P/G and P/L/G formulation at the same concentrations as described in the formulation preparation section then concentrate the sample either by using lyophilization or speed vacuum techniques (*see* **Note 33**).
3. Record CD spectrum from 340 to 200 nm every 2 nm using standard 1 cm path length quartz cuvette.
4. Analyze CD spectrum by recoding any peak shifts resulting from pDNA conformational transitions.

3.6 DNA Binding Assay

Ethidium bromide (EtBr) binding assay can be used to evaluate the ability of gemini surfactants to compact and protect the pDNA, in addition, to study the release of the pDNA from the lipoplexes as a function of pH changes.

3.6.1 1% Agarose Gel for Electrophoresis

1. Weigh 1 g agarose.
2. Add the agarose in 200 mL microwavable conical flask and add 100 mL 1× tris-acetate-EDTA (TAE) buffer.
3. Dissolve the agarose using microwave for 1–3 min (depending on the power of the microwave) and mix the beaker during the heating by stopping the microwave and shaking the beaker every 30 s (*see* **Note 34**).
4. Cool down the agarose to around 50 °C.
5. Add EtBr to final concentration of 0.5 μg/mL (*see* **Note 35**).
6. Pour the agarose gel solution to gel tray using proper size comb, cover with aluminum foil, and leave it at room temperature (20–30 min) to solidify (*see* **Note 36**).
7. Fill the electrophoresis chamber with 1× TEA buffer with 0.5 μg/mL EtBr (around 1.5 L).
8. After the gel becomes solid, remove the comb and transfer the tray containing the gel to the electrophoresis chamber (*see* **Note 37**).

3.6.2 Loading Samples and Running Gel

1. In Eppendorf tube, mix the naked pDNA and previously prepared P/G or P/G/L samples, respectively, with loading buffer (composed of bromophenol blue and xylene cyanol FF in glycerol):

Naked pDNA or sample	10 μL (formulation P/G or P/G/L)
Loading buffer	2 μL

2. Add the ladder (10 μL) to the first well.
3. Add your samples (10 μL/well) into the other wells (*see* **Note 38**).
4. Run the electrophoresis at 100 V for 45 min or until the dye lines are two-third away from the starting wells (*see* **Note 39**).
5. Visualize the gel by using AlphaImager™. No pDNA band will be observed for P/G or P/G/L if the gemini surfactant lipoplexes are able to completely protect the pDNA from migration toward the positive side (*see* **Note 40**).

3.7 Determination of the Critical Micelle Concentration

Critical micelle concentration (CMC) is defined as the concentration of amphiphiles above which aggregates, so-called micelles, start to form. It is an important structural parameter that quantitatively describes the propensity of an amphiphile to assemble. As such, provides an indication about the ability of the surfactants to form stable complexes with the nucleic acid [27].

3.7.1 Surface Tension Method

1. Measure CMC values using the du Noüy ring method (*see* **Note 41**).
2. Calibrate the tensiometer using a set of calibration weights.
3. Clean the ring and sample vessel before each measurement in accordance to the manufacturer's protocol.
4. Insert the sample vessel and fill it with 15 mL of ultrapure water (Millipore, resistivity 18 MΩ·cm).
5. Install the clean ring and lift the sample vessel until the ring dips into the liquid.
6. Set and maintain the temperature at 45.0 ± 0.1 °C using a circulating water bath.
7. Zero the instrument according to the manufacturer's protocol.
8. Titrate the concentrated aqueous solution of gemini surfactants into the sample vessel.
9. Measure the surface tension after each addition (*see* **Note 42**).
10. Correct the measured surface tension values (γ) using the method of Harkins and Jordan [28].
11. Determine the CMC values through linear fitting of the premicellar and postmicellar regions of the surface tension vs. logarithmic concentration plot.

3.7.2 Specific Conductance Method

1. Maintain the conductance cell temperature at 45.0 ± 0.1 °C using a Haake (Model F3) circulating water bath.
2. Titrate the gemini surfactant concentrated aqueous solution into 15 mL of water under nitrogen atmosphere, allowing the solution to equilibrate.
3. Measure the specific conductance after each addition using the Weyne-Kerr precision component analyzer.
4. Determine the CMC values through fitting the specific conductivity data according to Carpena et al. method [29], using the relation below:

$$k = k^{\circ} + A_1 c + d(A_2 - A_1) \ln \frac{1 + e^{(c - \mathrm{CMC})/d}}{1 - e^{-\mathrm{CMC}/d}} \quad (1)$$

where k is specific conductivity, k° is the specific conductivity of the surfactant-free solvent, c is the total concentration of surfactant, A_1 and A_2 are the premicellar and postmicellar slopes of the k vs. concentration curve, and d is the width of the cmc transition region (*see* **Note 43**).

3.8 Determination of the Molecular Packaging Parameter of the Lipids

The molecular packing parameter (P) is a concept that describes the shape of aggregates formed by amphiphilic compounds in aqueous solution [30]. A specific value P can be translated into a particular geometrical shape which can be correlated to the lipoplexes transfection efficiency [30–32].

1. Determine the molecular packing parameter using values of parameters based on the behavior of the gemini surfactant at the air-water interface using the relation below:

$$P = V/a_0 l \tag{2}$$

   where V is the volume of the hydrophobic tails, l is the length of the hydrocarbon tails, and a_0 is the head group area per molecule at the aggregate surface [30].
2. Utilize Langmuir-Blodgett technique to evaluate the surface area occupied by the gemini surfactant head group (*see* **Note 44**).
3. Calibrate the Wilhelmy balance using calibration weights.
4. Calibrate the Langmuir minitrough using calibration standard (*see* **Note 45**).
5. Fill the trough with ultrapure water (Millipore, resistivity 18 MΩ·cm) as a subphase.
6. Maintain the subphase temperature at 22 °C.
7. Prepare stock solutions of gemini lipids at a concentration of 1 mM in chloroform.
8. Spread 30 μL of the stock solution dropwise on the surface of the subphase using Hamilton syringe.
9. Allow evaporation of the chloroform for 10 min.
10. Compress the trough monolayer at a speed of 20 mm/min.
11. Generate a surface pressure-mean molecular area isotherm.
12. Determine the surface area occupied by the gemini surfactant head group by extrapolating a tangent line of the solid phase to the zero pressure as shown in Fig. 5.
13. Utilize Gaussian 09 software to calculate the volume of the hydrocarbon tails (*see* **Note 46**).
14. Optimize the geometry according to the B3LYP level of theory with 6-311+G(d,p) basis set.
15. Confirm the optimized structures using harmonic frequency calculations.
16. Calculate volume for the optimized structures using united atom radii.
17. Estimate the length of the alkyl tails by measuring the distance between the first and last carbon on in the aliphatic tail using Avogadro software (*see* **Notes 47** and **48**).

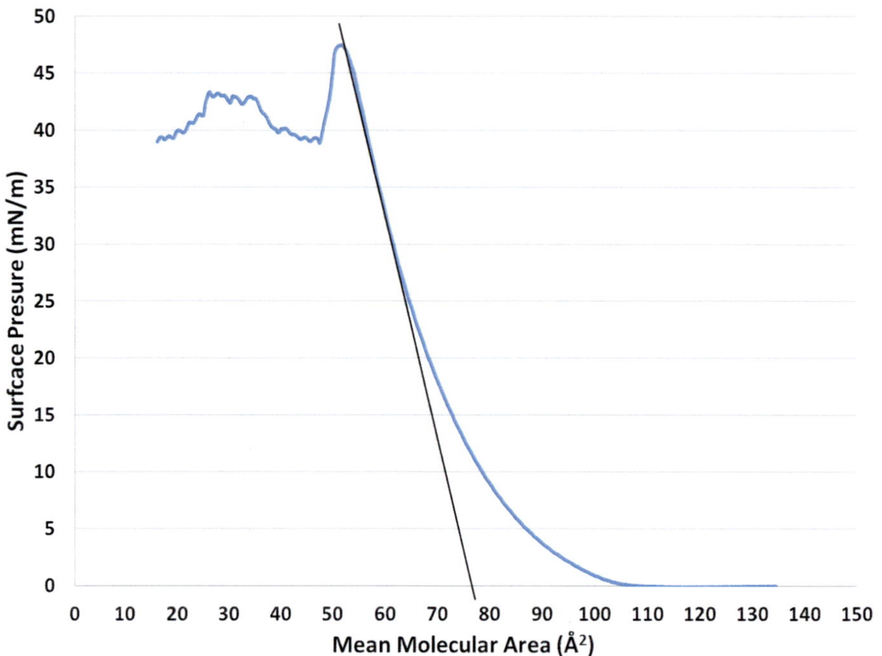

Fig. 5 Pressure-area isotherms of a representative example of the peptide-modified gemini surfactants

3.9 In Vitro Transfection Protocol

1. Grow cells to 80% confluency in 75-cm^2 tissue culture flasks using cell medium according to the supplier's recommendations. Supplement media with 10% (vol/vol) FBS and 1% (vol/vol) antibiotic antimycotic agents (*see* **Notes 49–51**).
2. One day prior to transfection, seed cells at a density of 15×10^3 for PAM 212 and 1×10^4 cells/well for COS-7 in 96-well tissue culture plates (*see* **Note 52**).
3. Replace the supplemented medium with un-supplemented medium 1 h prior to transfection.
4. Transfect cells with 0.2 μg/well pDNA using the peptide-modified gemini surfactants lipoplexes (*see* **Notes 53–56**).
5. Incubate cells at 37 °C in with 5% CO_2 for 5 h (*see* **Note 57**).
6. Replace the transfection mixture with supplemented medium after 5 h.
7. Collect the supernatants (which contain the expressed IFN-γ) at 24, 48, and 72 h and replace it with fresh medium (*see* **Note 58**).
8. Store the collected supernatants at −80 °C for further analysis.

3.10 Transfection Efficiency Assessment

It is important to use a sensitive and selective method to determine the level of expressed gene. The pDNA used is (pGThCMV.IFN-GFP), encoding for murine interferon gamma (IFN-γ) and green fluorescent protein (GFP) which both can be quantified (*see* **Note 59**). Interferon gamma (IFN-γ) is a secreted protein that will be expressed by transfected cells in the surrounding media. Different ELISA kits are available for accurate measurement of expressed IFN, from different suppliers (*see* **Note 60**).

3.10.1 Day 1

1. Coat ELISA plate with purified anti-mouse IFN-γ (capture).
   For each 96-well ELISA plate we need 10 mL purified anti-mouse IFN-γ (capture) and 5 mL coating buffer (*see* **Note 61**).
2. Coat ELISA plate with 50 μL/well of above solution and cover it with aluminum foil.
3. Incubate at 4 °C overnight.
4. Prepare 1% BSA buffer, 40 mL/plate (*see* **Note 62**).

3.10.2 Day 2

1. Wash ELISA plate four times with 1× PBST and add 100 μL of 1% BSA in each well.
2. Cover the plate with aluminum foil and incubate at 4 °C 1 h.
3. Wash the plate six times with 1× PBST and then add 70 μL of 1% BSA in each well (where collected supernatant or blank will be added) (*see* **Note 63**).
4. For standard curve, add standard IFN (BD Pharmingen™ Recombinant Mouse IFN-γ) as described below in **steps 5–11**.
5. Prepare 4 ng/ mL IFN-γ standard solution in 1% BSA.
6. Leave row E10 to E12 empty. Add 100 μL 1% BSA to rows F10-F12, G10-G12, and H10-H12.
7. Add 133 μL/well of IFN-γ standard solution (4 ng/mL) to row E10-E12.
8. Take 33 μL from E10-E12 and transfer to F10-F12, mix (1 ng/mL).
9. Take 33 μL from F10-F12 and transfer to G10-G12, mix (0.25 ng/mL).
10. Take 33 μL from G10-G12 and transfer to H10-H12, mix (0.06 ng/mL).
11. Take 33 μL from H10-H12 and discard.
12. For sample (collected supernatant) add 30 μL/well of collected supernatant (final volume is 100 μL/well – sample + 1% BSA) (*see* **Notes 64** and **65**).
13. Cover the plate with aluminum foil and incubate at 4 °C overnight.

3.10.3 Day 3

1. Wash the plate six times with 1× PSBT, and 100 μL of secondary antibody (Biotin Rat anti-mouse IFN-γ). For each 96-well ELISA plate add 10 μL of secondary antibody in 10 mL 1% BSA.

2. Cover the plate with aluminum foil and incubate at room temperature for 2 h.
3. Wash the plate six times with 1× PSBT and then add 100 μL/well of AKP Streptavidin. For each 96-well ELISA plate add 10 μL of AKP Streptavidin in 10 mL 1% BSA.
4. Cover the plate with aluminum foil and incubate at room temperature for 1 h.
5. Wash the plate six times with 1× PSBT and set up the Biotek® microplate reader.
6. Once the instrument ready, add 100 μL/well of pNPP solution (1 mg/mL); for each 96-well ELISA plate; 10 mg pNPP (two 5 mg tablets) in 10 mL pNPP buffer (*see* **Note 66**).
7. Allow yellow color to develop in wells (30–45 min) containing the IFN-γ standard and read the absorbance at 450 using the Biotek® microplate reader (*see* **Notes 67–70**).

3.10.4 Calculations

1. Calculate the mean absorbance of standard and subtract the standard optical density.
2. Calculate the mean absorbance of samples and subtract the negative control absorbance and standard optical density values.
3. Plot the IFN-γ standard curve with standard concentration on the *x*-axis and absorbance on the *y*-axis.
4. Use the linear equation from the standard curve to calculate sample concentration.

3.11 Assessment of Cell Viability/Toxicity

Different cytotoxicity assays can be used to assess cell viability/toxicity of gemini surfactant-based lipoplexes. In our work we used MTT (3-(4,5-Dimethylthiazol-2-yl)-2,5-Diphenyltetrazolium Bromide) assay as a simple and reliable assay to determine the cellular toxicity. Basic protocol of MTT can be obtained from the product provider (*see* **Note 71**).

4 Notes

1. The pDNA used is (pGThCMV.IFN-GFP). Encoding for murine interferon gamma (IFN-γ) and green fluorescent protein (GFP) with a total size of 5588 bps was created to insert the protein of interest, interferon-γ, and a fluorescent protein (driven by an internal ribosomal entry site) for monitoring gene expression during the experiment described previously [5].
2. Other helper lipids such as cholesterol can be used in preparing lipoplexes.

3. SAXS experiments can also be performed on other synchrotron radiation facilities such as the Advanced Photon Source Laboratory (APS) at Argonne National Laboratory (Argonne, Illinois, USA).
4. It is recommended to a make the plates with a rigid material, such as aluminum, as it is easier to load them without cross-contamination.
5. Freeze dryer or speed vacuum can be used to concentrate the sample for CD analysis.
6. The film can be stored at −20 °C until utilization.
7. It is important to adjust the pH of the sucrose solution to 9 to form DOPE vesicles.
8. The water-bath temperature must be monitored during this step to avoid excessive heating which can cause the disruption of DOPE vesicles and degradation of the phospholipid.
9. Avoid fast filtration, as it can cause disruption of the formed vesicles. It is recommended to measure the pH, particle size, and zeta potential of the resulted DOPE vesicles every time to ensure the reproducibility.
10. The following method is to prepare [P/G/L] lipoplexes using 1:10 ± charge ratio and 1:10 gemini surfactant-to-DOPE molar ratio. Necessary calculations can be done to adjust the charge ratio or helper lipid amount.
11. It is important to add the pDNA first to tube then followed by the addition of the gemini surfactant to ensure the formation of the complex.
12. Avoid mixing by pipetting or vortexing as these processes could disrupt the complex formation.
13. In the case of downscaling the volume of the formulation, a quick pulse spin in a centrifuge can be performed to ensure that all the liquid aliquots are at the bottom of the tube.
14. If the lipoplexes are prepared for in vitro evaluation, all the steps must be performed under sterile conditions.
15. It is recommended to prepare the formulation fresh just prior to the experimental evaluation.
16. To ensure the quality of the formulation and reproducibility of the data, it is recommended to measure the pH, particle size, and zeta potential of the produced P/G/L lipoplexes.
17. To calibrate for size measurements, use the 3000 Series Nanosphere™ size standards of 60 nm ± 4 nm.
18. To calibrate for ζ-potential measurements, utilize zeta potential transfer standard (DTS1235) of −42 mV ± 4.2 mV.

19. Fill the cell slowly avoiding the creation of air bubbles. Alternatively, the cells can be sonicated for 15–20 s.
20. Repeat each measurement at least three times.
21. Volume distribution representation of the particle size is considered to be the most preferred way for reporting particle size in the pharmaceutical industry for most applications [33].
22. The titrant volume should be more than 5 mL and should not exceed 25 mL.
23. It is recommended to prime all the tubing at the beginning of a new measurement session to ensure the absence of air bubbles.
24. Although it is ideal to conduct measurements with a sample volume of 10 mL, smaller volume samples can also be used for scarce samples. However, sample volume should not be less than 8 mL.
25. Optimize the stirring speed to ensure adequate distribution of the titrant within the sample without introducing air bubbles into the sample.
26. For more reliable measurements, it is recommended to start the measurements from the intrinsic pH of the sample then titrate towards either the basic or acidic range.
27. To prepare 10 mM DOPE, it is recommended to start with 2 mM solution then concentrate it using centrifugal evaporator.
28. The blank solution for DNA and gemini surfactants samples is water, while the blank solution of DOPE and peptide-modified gemini surfactants lipoplexes is water with sucrose as discussed in the formulation section.
29. Exposure time of 20 s was ideal for peptide-modified gemini surfactants nanosystem. Using shorter exposure time resulted in weak structural information while long exposure time resulted in beam damage.
30. Make sure to mask the beam stopper before integrating the data.
31. Make sure to subtract the background scattering for each sample.
32. Preparing the DOPE vesicles in high concentration could be an obstacle, as obtaining clear DOPE dispersion is difficult.
33. Concentrating the lipoplexes could cause changes in pDNA conformation as a damage to pDNA structure can be occurred as a result of freezing (in case of lyophilization without proper cryoprotectant) or due to excessive heating (in the case of speed vacuum).

34. Avoid excessive boiling of the agarose solution as it can cause evaporation which can affect the quality of the gel. Use appropriate protective gloves while handling the hot beaker.
35. As EtBr is mutagenetic, use appropriate protective equipment (PPE) (gloves, eye goggles, lab coat) and follow the proper disposal policy.
36. Pour the agarose solution slowly in the corner of the tray and avoid creating air bubbles. If air bubbles are formed, it can be moved to the sides using pipette tip.
37. Place the gel wells near the negative side of chamber (Black) away from the positive (red). pDNA is negatively charged and will run to the positive electrode.
38. Carefully add your sample to gel wells and avoid introducing air bubbles. Avoid loading more that well capacity.
39. The dyes migrate at the 3000- and 500-basepair fragments level, thus the migration of the DNA can be visually monitored.
40. Again EtBr is a known mutagenic. Use proper PPE and dispose the gel and running after according to the mutagen waste disposal policy. EtBr can be replaced with other DNA-staining dye such as SYBR Green I or GelRed.
41. Several alternative methods can be used to measure the surface tension such as Wilhelmy plate and Pendant drop methods.
42. Repeat each measurement at least three times.
43. Several alternative mathematical models are available for determining the CMC value from the specific conductance vs. concentration curve such as Pe'rez-Rodrı'guez et al. method, Garcı'a-Mateos et al., and others [34, 35].
44. The head group area of the gemini surfactant can also be calculated from the surface excess concentration according to the relation:

$$a_0 = (N_A \Gamma)^{-1} \tag{3}$$

where N_A is the Avogadro number, and Γ is surface excess concentration derived from the Gibbs adsorption isotherm as shown in the equation below:

$$\Gamma = -\frac{1}{2.303n\,RT}\left(\frac{d\gamma}{d\log C}\right)_T \tag{4}$$

where R is the gas constant, T is the absolute temperature, and n is the number of species at the interface resulting from the dissociation of the surfactants, which for gemini surfactants is equal to 3. The value for the premicellar slop, $(d\gamma/d\log C)_T$, is obtained from the surface tension measurements.

45. An example of calibration standard is the use of 1 mM stearic acid in hexane.
46. An alternative method to estimate the volume of the hydrocarbon tails is simply from the known values for the volume of methylene, methyl, and methine groups using the relation below:

$$v = n\left(V_{\text{methylene}}\right) + m\left(V_{\text{methine}}\right) + h\left(V_{\text{methyl}}\right) \quad (5)$$

where $V_{\text{methylene}} = 27\ \text{Å}^3$, $V_{\text{methine}} = 20.5\ \text{Å}^3$ and $V_{\text{methyl}} = 54\ \text{Å}^3$ [36].

However, this method does not account for the stereochemical arrangement in the presence of a double bond (cis or trans configuration). Thus, it is recommended to use this method only for saturated aliphatic tail.
47. If you have a double bond, make sure to optimize for the stereochemical configuration.
48. Alternatively, the length of the hydrocarbon tails can be calculated according to Tanford equation [37]:

$$l = 1.5 + 1.265 \times n_C \quad (6)$$

where n_C is the number of carbons in the alkyl tail. This method also does not account for the stereochemical arrangement, hence it is recommended to be used only for saturated aliphatic chains.
49. Use fresh cells from stock for each experiment, keeping the cellular passage number constant.
50. Before any transfection experiment make sure to subculture the cells once.
51. It is recommended not to use media that has been supplemented for more than a month.
52. The density of the seeded cells needs to be optimized for each cell line to achieve best response to the transfection experiment.
53. Make sure to have all the necessary positive and negative controls.
54. Utilize Lipofectamine Plus™ as a positive control according to the manufacturer's protocol.
55. Nontreated cells are used as a negative control.
56. In order to evaluate the role of each component of the P/G/L nanoparticle in the delivery of the genetic material, make sure to treat cells with P, P/G and P/L formulations.
57. The treatment incubation time needs to be determined based on optimization studies.

58. The duration of the treatment needs to be determined based on time-course studies.
59. Green fluorescent protein (GFP) can be also quantified using appropriate microplate reader with fluorometer ability or by using fluorescent microscopy. However, assessment of GFP will be relative quantification.
60. Different IFN- ELISA kits are available with some modifications to standard protocols, if ELISA kit is used, follow provider's protocol.
61. Coating buffer must store in 4 °C until utilization.
62. BSA buffer is light sensitive and can be subjected to hydrolysis and microbial contamination. So cover the bottle containing the 1% BSA buffer and store at 4 °C.
63. Washing step is important. Improper or inefficient washing can lead to false results.
64. Collected supernatant can be centrifuged by plate centrifugation to ensure the removal of any suspended materials or cells debris in the media which can interfere with the measurements.
65. Use fresh pipette tips for different sample/concentration to avoid cross contamination.
66. pNPP buffer must store in 4 °C until utilization.
67. All reagents used in Day 3 must be equilibrated to room temperature before adding to plates.
68. All incubation periods in this day are at room temperature.
69. Stop solution can be used to stop the excessive development of yellow color which can cause overflow readings. If stop solution is used, reading must be taken immediately.
70. The time for color development depends on the amount of IFN-captured and the sensitivity of the microplate reader.
71. MTT assay estimates the viable cells in the tissue culture, as viable cells have the ability to metabolize soluble MTT into insoluble purple formazan crystals by redox activity. To achieve reliable data, a well-controlled study is required. This can be accomplished by using the following controls in the 96-well plate: three wells with nontransfected cells (negative control), three wells with only the supplemented cell culture medium (blank) but no cells, and three wells with medium and lipoplexes and no cells. All these control wells need to be treated with MTT, and any reading from these well must be subtracted from nontreated and treated wells.

References

1. Jin L, Zeng X, Liu M, Deng Y, He N (2014) Current progress in gene delivery technology based on chemical methods and nano-carriers. Theranostics 4:240–255
2. Singh J, Mohammed-Saied W, Kaur R, Badea I (2013) Nanoparticles in gene therapy: from design to clinical applications. Rev Nanosci Nanotechnol 2:275–299
3. Menger FM, Littau C (1991) Gemini-surfactants: synthesis and properties. J Am Chem Soc 113:1451–1452
4. Rosen MJ, Tracy DJ (1998) Gemini surfactants. J Surfactant Deterg 1:547–554
5. Badea I, Verrall R, Baca-Estrada M et al (2005) In vivo cutaneous interferon-γ gene delivery using novel dicationic (gemini) surfactant--plasmid complexes. J Gene Med 7:1200–1214
6. Badea I, Wettig S, Verrall R, Foldvari M (2007) Topical non-invasive gene delivery using gemini nanoparticles in interferon-γ-deficient mice. Eur J Pharm Biopharm 65:414–422
7. Badea I, Virtanen C, Verrall R, Rosenberg A, Foldvari M (2011) Effect of topical interferon-γ gene therapy using gemini nanoparticles on pathophysiological markers of cutaneous scleroderma in Tsk/+ mice. Gene Ther 19:978–987
8. Garcia MT, Kaczerewska O, Ribosa I, Brycki B, Materna P, Drgas M (2016) Biodegradability and aquatic toxicity of quaternary ammonium-based gemini surfactants: effect of the spacer on their ecological properties. Chemosphere 154:155–160
9. Wasungu L, Scarzello M, van Dam G et al (2006) Transfection mediated by pH-sensitive sugar-based gemini surfactants; potential for in vivo gene therapy applications. J Mol Med 84:774–784
10. Kim B-K, Doh K-O, Bae Y-U, Seu Y-B (2011) Synthesis and optimization of cholesterol-based diquaternary ammonium gemini surfactant (chol-gs) as a new gene delivery vector. J Microbiol Biotechnol 21:93–99
11. Pérez L, Pinazo A, Pons R, Infante M (2014) Gemini surfactants from natural amino acids. Adv Colloid Interf Sci 205:134–155
12. Yang P, Singh J, Wettig S, Foldvari M, Verrall RE, Badea I (2010) Enhanced gene expression in epithelial cells transfected with amino acid-substituted gemini nanoparticles. Eur J Pharm Biopharm 75:311–320
13. Singh J, Yang P, Michel D, E Verrall R, Foldvari M, Badea I (2011) Amino acid-substituted gemini surfactant-based nanoparticles as safe and versatile gene delivery agents. Curr Drug Deliv 8:299–306
14. Al-Dulaymi MA, Chitanda JM, Mohammed-Saeid W et al (2016) Di-peptide-modified gemini surfactants as gene delivery vectors: exploring the role of the alkyl tail in their physicochemical behavior and biological activity. Am Assoc Pharm Scient J 18:1168–1181
15. Al-Dulaymi M, Michel D, Mohammed Saeid W et al (2016) Novel peptide-modified gemini surfactants as gene carriers structure activity relationship, physicochemical characterizations and mass spectrometric dissociation behaviour. American Association of Pharmaceutical Scientists Annual Meeting and Exhibition, Denver, CO
16. Singh J, Michel D, Getson HM, Chitanda JM, Verrall RE, Badea I (2015) Development of amino acid substituted gemini surfactant-based mucoadhesive gene delivery systems for potential use as noninvasive vaginal genetic vaccination. Nanomedicine 10:405–417
17. Singh J, Michel D, Chitanda JM, Verrall RE, Badea I (2012) Evaluation of cellular uptake and intracellular trafficking as determining factors of gene expression for amino acid-substituted gemini surfactant-based DNA nanoparticles. J Nanobiotechnol 10:7
18. Prabha S, Arya G, Chandra R, Ahmed B, Nimesh S (2014) Effect of size on biological properties of nanoparticles employed in gene delivery. Artif Cell Nanomed Biotechnol 44:1–9
19. Fröhlich E (2012) The role of surface charge in cellular uptake and cytotoxicity of medical nanoparticles. Int J Nanomedicine 7:5577
20. Ma B, Zhang S, Jiang H, Zhao B, Lv H (2007) Lipoplex morphologies and their influences on transfection efficiency in gene delivery. J Control Release 123:184–194
21. Toby BH, Von Dreele RB (2013) GSAS-II: the genesis of a modern open-source all purpose crystallography software package. J Appl Crystallogr 46:544–549
22. Hanwell MD, Curtis DE, Lonie DC, Vandermeersch T, Zurek E, Hutchison GR (2012) Avogadro: an advanced semantic chemical editor, visualization, and analysis platform. J Chem 4:17
23. Frisch MJ, Trucks GW, Schlegel HB, Scuseria GE, Robb MA, Cheeseman JR, Scalmani G, Barone V, Mennucci B, Petersson GA et al (2013) Gaussian 09, revision D.01. Gaussian, Wallingford, CT
24. Gratton SE, Ropp PA, Pohlhaus PD et al (2008) The effect of particle design on cellular internalization pathways. Proc Natl Acad Sci 105:11613–11618

25. Foged C, Brodin B, Frokjaer S, Sundblad A (2005) Particle size and surface charge affect particle uptake by human dendritic cells in an in vitro model. Int J Pharm 298:315–322
26. Zantl R, Baicu L, Artzner F, Sprenger I, Rapp G, Rädler JO (1999) Thermotropic phase behavior of cationic lipid-DNA complexes compared to binary lipid mixtures. J Phys Chem B 103:10300–10310
27. Dauty E, Remy J-S, Blessing T, Behr J-P (2001) Dimerizable cationic detergents with a low cmc condense plasmid DNA into nanometric particles and transfect cells in culture. J Am Chem Soc 123:9227–9234
28. Harkins WD, Jordan HF (1930) Surface tension by the ring method. Science 72:73–75
29. Carpena P, Aguiar J, Bernaola-Galván P, Carnero Ruiz C (2002) Problems associated with the treatment of conductivity–concentration data in surfactant solutions: simulations and experiments. Langmuir 18:6054–6058
30. Israelachvili JN, Mitchell DJ, Ninham BW (1976) Theory of self-assembly of hydrocarbon amphiphiles into micelles and bilayers. J Chem Soc Faraday Trans 72:1525–1568
31. Kumar V (1991) Complementary molecular shapes and additivity of the packing parameter of lipids. Proc Natl Acad Sci 88:444–448
32. Moghaddam B, Mcneil SE, Zheng Q, Mohammed AR, Perrie Y (2011) Exploring the correlation between lipid packaging in lipoplexes and their transfection efficacy. Pharmaceutics 3:848–864
33. Burgess DJ, Duffy E, Etzler F, Hickey AJ (2004) Particle size analysis: American Association of Pharmaceutical Scientists workshop report, cosponsored by the Food and Drug Administration and the United States Pharmacopeia. Am Assoc Pharm Scient J 6:23–34
34. Pérez-Rodríguez M, Prieto G, Rega C, Varela LM, Sarmiento F, Mosquera V (1998) A comparative study of the determination of the critical micelle concentration by conductivity and dielectric constant measurements. Langmuir 14:4422–4426
35. Garcia-Mateos I, Mercedes Velazquez M, Rodriguez LJ (1990) Critical micelle concentration determination in binary mixtures of ionic surfactants by deconvolution of conductivity/concentration curves. Langmuir 6:1078–1083
36. Koenig BW, Gawrisch K (2005) Specific volumes of unsaturated phosphatidylcholines in the liquid crystalline lamellar phase. Biochim Biophys Acta 1715:65–70
37. Tanford C (1980) The hydrophobic effect: formation of micelles and biological membranes, 2nd edn. Wiley, New York

Chapter 15

Preparation of Responsive Carbon Dots for Anticancer Drug Delivery

Tao Feng and Yanli Zhao

Abstract

Fluorescent carbon dots (CDs) have been extensively utilized as responsive drug nanocarriers to deliver anticancer agents, owing to their facile preparation, excellent water solubility, good photostability, and high quantum yield. Herein, we summarize the protocols for the synthesis and application of responsive CDs toward anticancer drug delivery both in vitro and in vivo. Specially, this chapter includes the preparation and structural characterization of CDs and anticancer prodrug-loaded CDs, in vitro anticancer drug release, in vitro and in vivo fluorescence imaging, and toxicity studies.

Key words Anticancer prodrug, Carbon dots, Cytotoxicity, Drug release, Fluorescence imaging, In vitro study, In vivo study

1 Introduction

Carbon dots (CDs), one kind of fluorescent carbon nanomaterials with size less than 10 nm and emissive wavelength from blue to near-infrared (NIR) region, have attracted increasing attention worldwide in biomedical fields [1–3]. On the one hand, CDs can be facilely prepared with high water solubility via different techniques, such as top-down and bottom-up methods [4]. The top-down synthesis is to break down the carbon structures of regular sp^2 carbon layers (graphite rod and carbon nanotubes) and amorphous carbon (candle soot and carbon black) through diverse approaches including sulfuric acid/nitric acid oxidation, laser ablation, electrochemical oxidation, and arc discharge [5, 6]. The bottom-up preparation is to use small molecules containing various hydroxyl, carboxyl, and amine groups, including ascorbic acid, citric acid, amino acid, and glycerol, as the carbon precursors by calcinations, microwave-assisted pyrolysis, and hydrothermal methods [7, 8]. On the other hand, CDs have good photostability and high quantum yield, which can surmount the disadvantages of commonly used fluorescent materials, such as photoblinking for

Volkmar Weissig and Tamer Elbayoumi (eds.), *Pharmaceutical Nanotechnology: Basic Protocols*, Methods in Molecular Biology, vol. 2000, https://doi.org/10.1007/978-1-4939-9516-5_15, © Springer Science+Business Media, LLC, part of Springer Nature 2019

quantum dots and photobleaching for organic dyes [9, 10]. Benefited from these excellent properties, CDs have not only been applied in bioimaging in vitro and in vivo [11–13], but also as drug nanocarriers to deliver chemotherapeutic drugs, therapeutic genes, and photosensitizers to treat cancer [14–16]. Herein, two representative examples are introduced. In 2012, Tao et al. synthesized CDs through oxidizing carbon nanotubes or graphite with mixed acids, and applied the obtained CDs for in vivo NIR fluorescence imaging [11]. Their results showed that CDs could be cleared via renal and fecal excretion without noticeable in vivo toxicity, demonstrating its great potential as nontoxic fluorescent nanoprobes for biomedical imaging. In 2013, Tang et al. fabricated a CD-based drug delivery system by loading fluorescent anticancer drug doxorubicin (DOX) on the surface through π–π stacking and electrostatic interactions [14]. Meanwhile, CDs were Förster resonance energy transfer (FRET) donor and DOX was FRET acceptor. By changing the environment pH, the FRET signal change could be used to monitor the release of DOX in real time. Under 810 nm excitation, two-photon imaging of this drug nanocarrier could also be utilized to monitor the drug release at deep tumor tissues of 65–300 μm.

In view of the wide application of fluorescent CDs as responsive drug nanocarriers to deliver anticancer agents, in this chapter, we summarize the protocols for the design and application of responsive CDs for anticancer drug delivery both in vitro and in vivo (Fig. 1), including the synthesis of amino group-functionalized CDs and anticancer cisplatin(IV) prodrug, structural investigation of CDs and anticancer prodrug, loading of anticancer prodrug and stealthy polyethylene glycol (PEG) polymer on CDs, in vitro anticancer drug release, investigation of cellular uptake by confocal laser scanning microscopy (CLSM) and flow cytometry, cytotoxicity study, in vivo fluorescence imaging and toxicity, and histology examination [17, 18].

2 Materials

1. Carboxylic acid-functionalized methoxyl polyethylene glycol (mPEG-COOH, 5 kDa).
2. Citric acid.
3. Dialysis membranes with molecular weight cutoff (MWCO) of 1, 3.5, and 12 kDa.
4. Diethylenetriamine.
5. 1-(3-(Dimethylamino)propyl)-3-ethylcarbodiimide hydrochloride (EDC·HCl).
6. Fetal bovine serum (FBS).

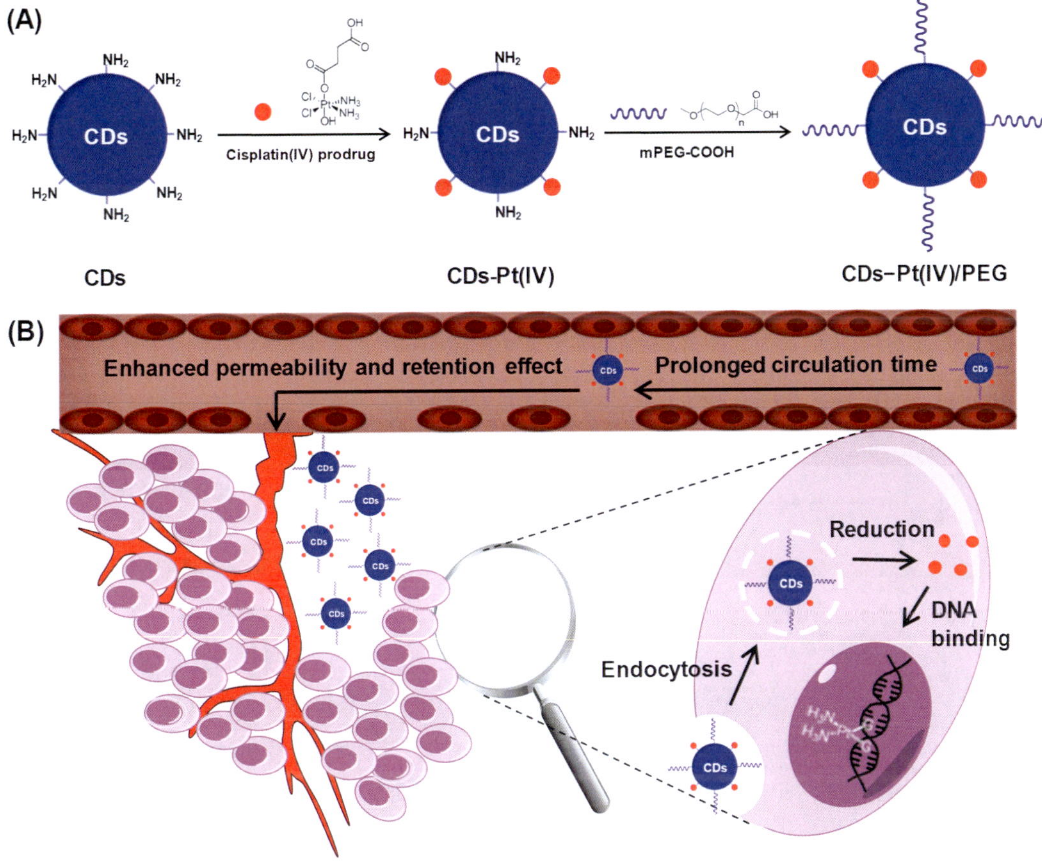

Fig. 1 Schematic illustration for (**a**) the preparation and (**b**) the drug delivery process of CD-based nanocarrier CDs–Pt(IV)/PEG. Reproduced with permission from ref. 17

7. Human ovarian carcinoma A2780 cells (ECACC).
8. Hydrogen peroxide (30 wt%).
9. *N*-hydroxysuccinimide (NHS).
10. *cis*-[$PtCl_2(NH_3)_2$].
11. Roswell Park Memorial Institute (RPMI) 1640 medium.
12. Succinic anhydride.

3 Methods

3.1 Synthesis of Amino Group-Functionalized CDs

1. Citric acid (2.1 g) and diethylenetriamine (3.5 g) are reacted in nitrogen atmosphere at 170 °C for 3 h (*See* **Note 1**).
2. The mixture is cooled down to room temperature, followed by adding a certain amount of water.
3. Upon dialysis against water for 48 h with a dialysis membrane (MWCO = 3.5 kDa), CDs are obtained through freeze-drying.

4. CDs are stored at room temperature for future use.
5. All CDs could be characterized by transmission electron microscopy, atomic force microscopy, dynamic light scattering (CD size), powder X-ray diffraction (crystalline structure), Fourier-transform infrared spectroscopy and X-ray photoelectron spectroscopy (composition and surface functional groups), zeta potential (surface charge), ultraviolet–visible and photoluminescence spectra (optical properties), and ninhydrin colorimetric analysis with alanine as the reference (amount of amino groups on the surface).

3.2 Synthesis of Anticancer Cisplatin (IV) Prodrug (c,c,t-[$PtCl_2(OH)(NH_3)_2(O_2CCH_2CH_2CO_2H$])

1. *cis*-[$PtCl_2(NH_3)_2$] (300.0 mg) is dispersed in water (7.5 mL), and then hydrogen peroxide (10.5 mL, 30 wt%) is added dropwise at 50 °C.
2. The mixture is stirred for 2 h before cooling down to room temperature.
3. The solvent is reduced to about 2 mL using rotary evaporator, and the remaining solution is stored at 0 °C overnight.
4. Upon the collection by filtration, pale yellow crystals are washed with cold water and cold diethyl ether, and then dried in vacuo to obtain *c,t,c*-[$PtCl_2(OH)_2(NH_3)_2$] (*See* **Note 2**).
5. *c,t,c*-[$PtCl_2(OH)_2(NH_3)_2$] (120.0 mg) and succinic anhydride (36.0 mg) are stirred in dimethyl sulfoxide (DMSO, 2.0 mL) at room temperature for 24 h.
6. The mixture is freeze-dried, and the obtained solid is rinsed with cold acetone and cold diethyl ether to afford *c,c,t*-[$PtCl_2(OH)(NH_3)_2(O_2CCH_2CH_2CO_2H$] (*See* **Note 3**).
7. *c,c,t*-[$PtCl_2(OH)(NH_3)_2(O_2CCH_2CH_2CO_2H$] is characterized by proton nuclear magnetic resonance spectroscopy (DMSO-d_6) and electrospray ionization mass spectrometry.

3.3 Loading Cisplatin (IV) Prodrug on CDs

1. Cisplatin(IV) prodrug (8.7 mg), EDC·HCl (3.8 mg), and NHS (2.3 mg) are stirred in water at room temperature for 30 min.
2. CDs (40.0 mg) are added to the above solution, and the obtained mixture is stirred for another 24 h.
3. The mixture is dialyzed against water for 48 h with a dialysis membrane (MWCO = 1 kDa) to discard unreacted compounds.
4. After being frozen dried, cisplatin(IV) prodrug-loaded CDs (CDs–Pt(IV)) are stored at room temperature for future use.
5. The characterization methods of CDs–Pt(IV) are similar to those of CDs.
6. The loading percentage of platinum is measured by inductively coupled plasma mass spectrometer (ICP–MS).

3.4 Synthesis of CDs– Pt(IV)/PEG

1. mPEG-COOH (5 kDa, 29.8 mg), EDC·HCl (3.4 mg), and NHS (1.2 mg) are stirred in DMSO at room temperature for 30 min.
2. CDs–Pt(IV) (20.0 mg) is added to the above solution, and the obtained mixture is stirred for another 24 h.
3. The mixture is dialyzed against water for 48 h with a dialysis membrane (MWCO = 12 kDa) to discard unreacted compounds.
4. After being freeze-dried, PEG covalently conjugated CDs–Pt (IV) (CDs–Pt(IV)/PEG) is stored in the dark at −20 °C for future use (*See* **Note 4**).

3.5 In Vitro Drug Release from CDs–Pt (IV)/PEG

1. CDs–Pt(IV)/PEG (30.0 mg) is dissolved in phosphate buffer saline (PBS, 1.0 mL, 10 mM, pH 7.4) containing 0 or 10 mM glutathione (GSH), and transferred into a dialysis tube (MWCO = 1 kDa), followed by dialysis against PBS buffer (69 mL, containing 0 or 10 mM GSH) at 37 °C in the dark (*See* **Note 5**).
2. At each designated time point, a certain volume of the release medium is withdrawn and the same amount of new buffer solution is added back to the original release medium.
3. The collected release medium is used for ICP–MS measurements to determine the amount of released platinum.

3.6 Cellular Uptake Measured by CLSM

1. A2780 cells are cultured in RPMI 1640 medium with supplement of 10% FBS at 37 °C in a humidified incubator with 5% CO_2.
2. A2780 cells at a density of 1×10^5 are cultured in a confocal dish (ibidi GmbH, Germany) at 37 °C for 24 h.
3. The medium is replaced by RPMI 1640 medium containing CDs–Pt(IV)/PEG before the incubation at 37 °C for another 2 h.
4. A2780 cells are washed by PBS twice and used for CLSM imaging (*See* **Note 6**).

3.7 Cellular Uptake Measured by Flow Cytometry

1. A2780 cells are cultured in 6-well plates at a density of 5×10^5 per well at 37 °C for 24 h.
2. The medium is replaced by RPMI 1640 medium containing CDs–Pt(IV)/PEG, followed by the incubation at 37 °C for another 2 h.
3. After being washed and trypsinized, A2780 cells are measured by a flow cytometer in Indo-1 violet channel (excitation wavelength: 355 nm, emission wavelength range: 450/50 nm) (*see* **Note 6**).

3.8 In Vitro Cytotoxicity

1. A2780 cells are seeded in 96-well plates at a density of 1×10^4 cells per well within RPMI 1640 medium (100 μL) at 37 °C for 24 h.
2. The medium is replaced by RPMI 1640 medium containing CDs–Pt(IV)/PEG with different platinum concentrations before the incubation at 37 °C for another 2 h.
3. After replacing with fresh complete RPMI 1640 medium, A2780 cells are incubated at 37 °C for another 70 h.
4. A2780 cells are washed with PBS, and incubated with 20% 3-(4,5-dimethylthiazol-2-yl)-2,5-diphenyltetrazolium bromide in RPMI 1640 medium at 37 °C for 5 h.
5. After removing the medium, DMSO (100 μL) is added to lyse A2780 cells.
6. The absorbance at 490 nm is measured using a microplate reader.
7. Every experiment is performed three times, and the cell viability is determined with reference to the control group without any treatment.

3.9 In Vivo Fluorescence Imaging

1. All xenograft animal studies are performed according to guidelines from the MIT Committee on Animal Care with approved protocols.
2. The tumor model is established by injecting a suspension of A2780 cells in PBS into the buttock of each female nude mouse, and the in vivo experiments are studied until the tumor size is about 6–8 mm.
3. Upon intravenous injection of CDs–Pt(IV)/PEG, fluorescence imaging is performed at different time points with a small-animal imaging system (*See* **Note 7**).

3.10 In Vivo Toxicity

1. The tumor-bearing mice are randomly divided into two groups ($n = 5$, each group) and, respectively, administrated with saline and CDs–Pt(IV)/PEG through intravenous injection.
2. The injected amount for CDs–Pt(IV)/PEG is 100 μL saline with platinum dose of 1.5 mg/kg body weight, and the dates of the treatment are 0, 4, and 10 days.
3. The tumor size and body weights are measured every day after the treatment.

3.11 Histology Examination

1. Histology analysis is carried out at the 14th day after the treatment.
2. The tissues of the mouse organs, including heart, liver, spleen, lung, and kidney, are isolated.

3. The organs are dehydrated with buffered formalin, ethanol of different concentrations, and xylene.
4. The organs are embedded in liquid paraffin.
5. The sliced organs and tumor tissues (3–5 mm) are stained with hematoxylin and eosin, and imaged on a microscope.

4 Notes

1. To prepare amino-group-functionalized CDs, parameters including the molar ratio between citric acid and diethylenetriamine, reaction time, and reaction temperature should be optimized.
2. To synthesize *c,t,c*-[$PtCl_2(OH)_2(NH_3)_2$], cold water and cold diethyl ether should be used to wash the product in order to increase the yield.
3. To synthesize *c,c,t*-[$PtCl_2(OH)(NH_3)_2(O_2CCH_2CH_2CO_2H$], cold acetone and cold diethyl ether should be used to wash the product in order to increase the yield.
4. The CD-based drug nanocarriers (CDs–Pt(IV)/PEG) should be stored at −20 °C in the dark for future use in order to avoid any potential degradation.
5. To study the in vitro drug release, the dialysis should be processed in a sealed container to avoid the water evaporation that will lead to experimental errors.
6. For CLSM imaging and flow cytometry studies, the incubation time of CDs–Pt(IV)/PEG should be optimized.
7. For in vivo fluorescence imaging, the concentration of CDs–Pt(IV)/PEG and the time point for imaging upon CDs–Pt(IV)/PEG administration should be optimized.

Acknowledgments

The authors thank the financial support from the Singapore National Research Foundation Investigatorship (NRF-NRFI2018-03).

References

1. Wang Y, Hu A (2014) Carbon quantum dots: synthesis, properties and applications. J Mater Chem C 2(34):6921–6939
2. Luo P, Yang F, Yang S et al (2014) Carbon-based quantum dots for fluorescence imaging of cells and tissues. RSC Adv 4 (21):10791–10807
3. Lim S, Shen W, Gao Z (2015) Carbon quantum dots and their applications. Chem Soc Rev 44(1):362–381

4. Song Y, Zhu S, Yang B (2014) Bioimaging based on fluorescent carbon dots. RSC Adv 4 (52):27184–27200
5. Xu X, Ray R, Gu Y et al (2004) Electrophoretic analysis and purification of fluorescent single-walled carbon nanotube fragments. J Am Chem Soc 126(40):12736–12737
6. Hu S, Niu K, Sun J et al (2009) One-step synthesis of fluorescent carbon nanoparticles by laser irradiation. J Mater Chem 19 (4):484–488
7. Jia X, Li J, Wang E (2012) One-pot green synthesis of optically pH-sensitive carbon dots with upconversion luminescence. Nanoscale 4 (18):5572–5575
8. Zhai X, Zhang P, Liu C et al (2012) Highly luminescent carbon nanodots by microwave-assisted pyrolysis. Chem Commun 48 (64):7955–7957
9. Zhu S, Meng Q, Wang L et al (2013) Highly photoluminescent carbon dots for multicolor patterning, sensors, and bioimaging. Angew Chem Int Ed 52(14):3953–3957
10. Yang C, Thomsen RP, Ogaki R et al (2015) Ultrastable green fluorescence carbon dots with a high quantum yield for bioimaging and use as theranostic carriers. J Mater Chem B 3 (22):4577–4584
11. Tao H, Yang K, Ma Z et al (2012) In vivo NIR fluorescence imaging, biodistribution, and toxicology of photoluminescent carbon dots produced from carbon nanotubes and graphite. Small 8(2):281–290
12. Huang X, Zhang F, Zhu L et al (2013) Effect of injection routes on the biodistribution, clearance, and tumor uptake of carbon dots. ACS Nano 7(7):5684–5693
13. Shi Q, Li Y, Xu Y et al (2014) High-yield and high-solubility nitrogen-doped carbon dots: formation, fluorescence mechanism and imaging application. RSC Adv 4(4):1563–1566
14. Tang J, Kong B, Wu H et al (2013) Carbon nanodots featuring efficient FRET for real-time monitoring of drug delivery and two-photon imaging. Adv Mater 25(45):6569–6574
15. Wang L, Wang X, Bhirde A et al (2014) Carbon-dot-based two-photon visible nanocarriers for safe and highly efficient delivery of SiRNA and DNA. Adv Healthc Mater 3 (8):1203–1209
16. Huang P, Lin J, Wang X et al (2012) Light-triggered theranostics based on photosensitizer-conjugated carbon dots for simultaneous enhanced-fluorescence imaging and photodynamic therapy. Adv Mater 24 (37):5104–5110
17. Feng T, Ai X, An G et al (2016) Charge-convertible carbon dots for imaging-guided drug delivery with enhanced in vivo cancer therapeutic efficiency. ACS Nano 10 (4):4410–4420
18. Feng T, Ai X, Ong H et al (2016) Dual-responsive carbon dots for tumor extracellular microenvironment triggered targeting and enhanced anticancer drug delivery. ACS Appl Mater Interfaces 8(29):18732–18740

Chapter 16

Surface Modification of Nanoparticles and Nanovesicles via Click-Chemistry

Matthias Voigt, Thomas Fritz, Matthias Worm, Holger Frey, and Mark Helm

Abstract

Surface modification of nanocarriers offers the possibility of targeted drug delivery, which is of major interest in modern pharmaceutical science. Click-chemistry affords an easy and fast way to modify the surface with targeting structures under mild reaction conditions. Here we describe our current method for the post-preparational surface modification of multifunctional sterically stabilized (stealth) liposomes via copper-catalyzed azide–alkyne cycloaddition (CuAAC) and inverse electron demand Diels-Alder norbornene–tetrazine cycloaddition (IEDDA). We emphasize the use of these in a one-pot orthogonal reaction for deep investigation on stability and targeting of nanocarriers. As the production of clickable amphiphilic polymers is a limiting factor in most cases, we also describe our nanocarrier preparation technique called dual centrifugation, which enables the formulation of liposomes on a single-digit milligram scale of total lipid mass.

Key words Nanoparticles, Nanovesicles, Nanocarriers, Liposomes, Click-chemistry, Surface modification, Post-preparation, Orthogonal click, Multifunctional, Dual centrifugation

1 Introduction

Targeted delivery of nanoscale drug carriers like liposomes via ligand–receptor interaction to specific cells is one of the most promising concepts in pharmaceutical science [1]. It can be realized by surface modification of the nanocarrier with targeting structures pre-, intra-, or post-preparational. Certain limitations are connected to pre- and intra-preparational modification, as targeting structures can also occur on the inner membrane surface making them inaccessible to the target cells, as well as changed nanocarrier characteristics due to an alteration of the formulation process [2]. The so-called post-insertion technique overcomes these limitations, as ligand-functionalized amphiphiles are inserted into the preformed nanocarrier [3, 4]. However, certain challenges are connected with this method, like cargo-leakage during insertion,

Volkmar Weissig and Tamer Elbayoumi (eds.), *Pharmaceutical Nanotechnology: Basic Protocols*, Methods in Molecular Biology, vol. 2000, https://doi.org/10.1007/978-1-4939-9516-5_16,

reproducibility and equality in distribution of the targeting vector, and insertion into a saturated, polymer stabilized surface [5]. Therefore, we use a post-preparational strategy which is based on click-chemistry on the nanocarrier surface, ensuring reproducibility of the formulation, cargo protection and equally distributed ligand presentation only on the outer membrane surface at accessible acceptor sites [6].

We use highly biocompatible and biodegradable liposomes as nanocarriers, which mainly consist of naturally occurring phosphatidylcholine and cholesterol forming a lipid bilayer around an aqueous core [7, 8], as well as amphiphilic polymers which provide a clickable moiety and stealth properties. Polyethylene glycol (PEG) and hyperbranched polyglycerol (*hb*PG) are used as biocompatible polymers, attached to a cholesterol or dialkyl-moiety which serves as lipid membrane anchor [9–11]. The occurring stealth effect caused by these amphiphilic polymers leads to a prolonged circulation time of the liposomes in vivo due to a lowered uptake by the reticuloendothelial system, lowered protein adsorption, and higher colloidal stability [12–16]. However, we were able to show that cholesterol as lipid anchor is less stable compared to dialkyl-anchors, resulting in an exchange with cellular membranes when applied in vitro and therefore in less effective targeting of the liposomes [17]. To achieve targeting, PEG and *hb*PG were functionalized at their terminal hydroxyl-groups with alkyne- or norbornene-moieties, whereas *hb*PG has the advantage of providing a larger number of these possible derivatization sites [9–11]. Azide- or tetrazine-bearing ligands can then be clicked via the copper-catalyzed azide–alkyne cycloaddition (CuAAC) [18–21] or the inverse electron demand Diels-Alder norbornene-–tetrazine cycloaddition (IEDDA) [22–24] to the liposomal surface post-preparation, respectively [6, 17]. The use of fluorescent dyes as model ligands clicked to the liposomal surface facilitated a detailed investigation on their intra- and extracellular fate.

Our current approach for the preparation of liposomes relies on a rather new technique called dual centrifugation [25, 26], which allows us to quickly and easily prepare liposomes with the valuable amphiphilic polymers down to a single-digit milligram scale of total lipid mass. This technique overcomes limitations of typically used formulation techniques as extrusion, sonication, or high-pressure homogenization, which are very material-intensive. It provides high encapsulation efficiencies of hydrophilic cargo up to 80%, reproducible small liposomes in a size range of 100–300 nm, and small polydispersities of below 0.3 [6, 17]. Subsequent liposome purification by size exclusion chromatography assures removal of unclicked ligands, non-encapsulated cargo, and click reactants and therefore enables the determination of encapsulation- and click-efficiencies.

2 Materials

2.1 Liposome Preparation

1. Phosphate buffered saline (PBS) 10×: 1.4 M NaCl, 27 mM KCl, 15 mM KH_2PO_4, 80.6 mM Na_3HPO_4 in sterile water. The resulting solution has a pH of 6.8. Sterile filtrate the PBS through 0.2 μm pores and store at room temperature (*see* **Note 1**).
2. PBS 1×: Dilution of PBS 10× with sterile water results in a pH of 7.4 with final concentrations of 140 mM NaCl, 2.7 mM KCl, 1.5 mM KH_2PO_4, 8.06 mM Na_2HPO_4. Sterile filtrate the PBS through 0.2 μm membrane and store at room temperature (*see* **Note 2**).
3. Ethanol, 99.5%. Used for dissolving of lipids.
4. 20 mg/mL cholesterol in absolute ethanol (*see* **Note 3**)
5. 50 mg/mL egg phosphatidyl choline (EPC3) (Lipoid) in absolute ethanol (*see* **Note 3**).
6. 20 mg/mL functional amphiphilic polymers in absolute ethanol: All dialkyl- or cholesterol-based polyethylene glycol (PEG) or hyperbranched polyglycerol (*hb*PG) amphiphiles were provided by the Frey group, Johannes Gutenberg University Mainz, Germany. They are synthesized via oxyanionic ring-opening polymerization and subsequent derivatization with propargyl bromide to yield terminal alkynes, or by esterification with dicarboxy-norbornene anhydride to yield norbornene residues as previously reported [9, 10]. Functionalization degrees are determined by ^{1}H-NMR analysis and diffusion-ordered NMR spectroscopy and verified 1 alkyne or norbornene group for PEG and up to 4 for *hb*PG. Molecular weights range from 2000 to 8000 g/mol (*see* **Note 3**).
7. Fluorophores: 1 mM calcein or sulforhodamine B as model drugs for encapsulation (*see* Table 1) in sterile water (*see* **Note 4**).
8. Ceramic beads: SiLiBeads® ZY, 0.3–0.4 mm (Sigmund Lindner, Warmensteinach, Germany). These beads can be used for preparations up to 20 mg of total lipid mass (*see* **Note 5**).
9. PCR vials with a volume of 200 μL for dual centrifugation (Kisker Biotech, Steinfurt, Germany) (*see* **Note 6**).
10. Dual centrifuge (Hettich, Tuttlingen, Germany): A Rotanta 400 centrifuge with a prototype DC-rotor and a custom-made, 3D-printed inset for PCR tubes (Helm Group, Johannes Gutenberg University Mainz, Germany).
11. SpeedVac vacuum centrifuge or a comparable instrument.
12. Lyophilization unit Alpha 2–4 LD (Christ, Osterode am Harz, Germany).

Table 1
Fluorophores used as model drugs for encapsulation or model ligands for click reactions

Name	Supplier	λ_{exc}/nm	λ_{em}/nm
Calcein	Sigma-Aldrich	393	517
Atto488 Azide	Atto-Tec (Siegen, Germany)	500	520
Atto488 Tetrazine	Jena Biosciences (Jena, Germany)	500	520
DiI	Sigma-Aldrich	549	565
Sulforhodamine	Sigma-Aldrich	565	586
Alexa Fluor 594 Azide	Thermo-Fisher Scientific	590	617
SulfoCy5 Azide	Sigma-Aldrich	647	663
SulfoCy5 Tetrazine	Jena Biosciences	647	663

2.2 Click-Reactions

1. 2 mM azide or tetrazine-functionalized fluorophores (*see* Table 1) in sterile water.
2. Phosphate buffer (PB): 53 mM NaH_2PO_4 and 947 mM Na_2HPO_4 in sterile water. The resulting solution has a pH of 8. Sterile filtrate the PB through 0.2 μm membrane and store at room temperature.
3. 5 mM $CuSO_4$ ($5 \times H_2O$) in sterile water.
4. 50 mM Tris(hydroxypropyltriazolylmethyl)amine (THPTA) (Helm Group, Johannes Gutenberg University Mainz, Germany) in sterile water.
5. 50 mM Sodium ascorbate in sterile water.
6. 20 mM Ethylenediaminetetraacetic acid (EDTA) in sterile water.
7. Nanocarriers bearing functional amphiphilic polymers (*see* Subheading 2.1, **item 6**).

2.3 Size Exclusion Chromatography (SEC)

1. Sepharose 2B-CL as matrix for the column.
2. Set of empty columns for size exclusion chromatography with 10 μm pore size lower filter order #S10011 and upper filters with 10 μm pore size order #S10031 (MoBiTec, Goettingen, Germany) (*see* **Note 7**).
3. 10 mL Luer solo syringes.
4. PBS 1× and absolute ethanol (*see* Subheading 2.1, **items 2** and **3**).

2.4 Characterization

1. Malvern Zetasizer Nano ZS or a comparable instrument (*see* **Note 8**).

2. Polystyrol cuvettes 10 × 4 × 45 mm (path length 1 cm). These are used for size and polydispersity determination only.
3. Disposable cuvettes DTS1070 for measuring the zetapotential, size, and polydispersity (Malvern, Worcestershire, UK).
4. Pure water and PBS 1× (*see* Subheading 2.1, **item 2**).
5. Sterile 96-well black microplates with lid, PS, and F-bottom.
6. 10 vol% Triton X 100 in PBS 1×, used to disintegrate the liposomal membrane and release the encapsulated cargo.
7. Infinite M200 Pro microplate reader (Tecan, Crailsheim, Germany).

3 Methods

3.1 Liposome Formulation by Dual Centrifugation

1. Dissolve all lipids and functionalized amphiphilic polymers in absolute ethanol (*see* **Note 3**).
2. Combine stock solutions of 20 mg/mL Cholesterol, 50 mg/mL EPC and 20 mg/mL amphiphilic polymer in a PCR tube to yield the intended compositions (*see* **Notes 6**, **9**, and **10**).
3. Dry the combined lipid solutions with a total lipid mass up to 20 mg in a vacuum centrifuge at 30 °C for at least 6 h.
4. Deep-freeze the samples at −80 °C for at least 1 h.
5. Lyophilize the samples for at least 48 h (*see* **Note 11**).

   The following values for liposome formulation represent an example for a 5 mg total lipid batch and scale linearly with the batch size.
6. Add 9.3 μL of PBS 1× or a solution of designated cargo in PBS 1× to the dry lipids and incubate them for 10 min at room temperature (*see* **Note 12**).
7. Add 71 mg of ceramic beads to each sample (*see* **Note 5**).
8. Subject the PCR tube to the dual centrifuge for 20 min at 2500 RPM (1048 × *g*) (Fig. 1a).
9. Dilute the obtained vesicular phospholipid gel with 28.5 μL PBS 1×.
10. Subject the PCR tubes again to dual centrifugation for 2 × 2 min at 2500 RPM, while turning the tube by 180° in between (*see* **Note 13**).

3.2 Post-Preparational Click Reactions on the Nanocarrier Surface

1. CuAAC.
   (a) Add components in the stated order with the final concentrations shown in Table 2 (*see* **Note 14**).
   (b) Mix the combined components and incubate them for 2 h at room temperature.
   (c) Add 1 μL of 20 mM EDTA to fix the reaction endpoint.

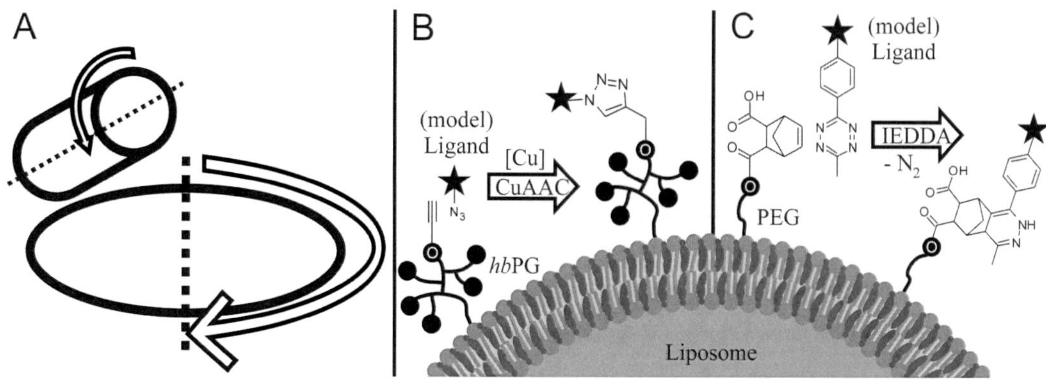

Fig. 1 (**a**) Dual centrifugation. The main axis rotates contrary to the sample container axis. (**b**) CuAAC: Copper-catalyzed azide–alkyne cycloaddition. (**c**) IEDDA: Inverse electron demand Diels-Alder norbornene–tetrazine cycloaddition

Table 2
Components in their order of addition for post-preparational CuAAC and in combination with IEDDA

Components in order of addition	Stock concentration	Final concentration
MilliQ water	To a final volume of 40 μL	
Phosphate buffer pH 8	53 mM NaH_2PO_4 947 mM Na_2HPO_4	5.3 mM NaH_2PO_4 94.7 mM Na_2HPO_4
THPTA	50 mM	0.5 mM
$CuSO_4 \times 5H_2O$	5 mM	0.1 mM
Sodium ascorbate	50 mM	2.5 mM
Azide (+Tetrazine)	2 mM	0.1 mM (0.05 mM each)
Alkyne (+Norbornene) functionalized Liposome stock		Add 10 μL

2. IEDDA.
   (a) Add tetrazine fluorophore to a final concentration of 0.1 mM and 10 μL of liposome suspension to PBS 1× to yield a final volume of 40 μL.
   (b) Incubate the PCR vials at 45 °C in a PCR-Thermocycler for 4 h.
3. CuAAC + IEDDA.
   (a) Add components as given in Table 2, with azide and tetrazine functionalized fluorophores and multifunctional liposomes bearing alkyne and norbornene moieties (Fig. 1a, b).
   (b) Carry out the orthogonal reaction in PCR vials at 45 °C in a PCR-Thermocycler for 4 h (*see* **Note 15**).
   (c) Add 1 μL of 20 mM EDTA to fix the reaction endpoint.

3.3 Size Exclusion Chromatography

1. Pack the MoBiTec columns with Sepharose 2 B-CL under a continuous flow of absolute ethanol and add a filter with 10 μm pore size on top.
2. Rinse the column multiple times with PBS 1× to remove the ethanol.
3. Add the nanocarrier solution on top of the column, let it sink in the matrix and then add PBS 1× on top (*see* **Note 16**).
4. Collect drop-sized fractions in a 96-well plate and measure it in a plate reader to receive a full chromatogram. The obtained elution volumes can then be used column-specifically to collect purified nanocarrier fractions and free cargo/ligand fractions of defined volumes (*see* **Note 17**).
5. Rinse the column with PBS 1× at least 3× the column volume in between different nanocarrier purifications.

3.4 Characterization

1. Determine the encapsulation and click efficiencies with a microwell plate reader.
   (a) Add a 20 μL sample (V_{S_n}) of the nanocarrier fraction (V_n) and a 50 μL sample ($V_{S_c/1}$) of the free cargo/ligand fraction ($V_{c/1}$) in a black 96-well plate.
   (b) Supplement each sample with Triton X-100 and PBS 1× to yield a final concentration of 5 vol% Triton-X 100 in 100 μL.
   (c) Quantify the samples on a microwell plate reader fluorometrically to obtain intensities for the nanocarrier sample fraction I_n and the free cargo/ligand sample fraction $I_{c/1}$.
   (d) Calculate the encapsulation (EE) and click efficiency (CE) with formula 1.
   (e) Carry out absolute quantification via external calibration with free fluorophores dissolved in PBS 1× to a final volume of 100 μL and addition of Triton-X 100 to a final concentration of 5 vol% (*see* **Note 18**).

$$\mathrm{EE/CE} = \frac{\frac{V_n}{V_{S_n}} \times I_n}{\frac{V_n}{V_{S_n}} \times I_n + \frac{V_{c/l}}{V_{S_c/l}} \times I_{c/l}} \tag{1}$$

2. Determine the size, polydispersity, and zetapotential of the nanocarrier by dynamic light scattering on a Malvern Zetasizer Nano ZS.
   (a) Dilute 10 μL of purified or 1 μL of unpurified nanocarrier suspension in 1 mL of freshly filtered (0.2 μm) PBS 1× in a polystyrol cuvette and then subject it to the instrument, resulting in count rates of 300 ± 50 kcounts per second for size measurement.

(b) Dilute the nanocarrier suspension in freshly filtered pure water in the cuvettes DTS1070 from Malvern for determination of the zetapotential.

(c) Set the viscosity of water/PBS to 0.8872 cP, the refractive index of water/PBS to 1.330 and of the liposomes to 1.59, and the liposome absorption to 0.01.

(d) Perform three measurements at 25 °C with a scattering angle of 173°, while attenuator and measurement positions are controlled by the instrument.

4 Notes

1. We use filter top vacuum bottles, PES, 0.2 μm pores, 500 mL from Sarstedt (Nuembrecht, Germany).
2. If sterile conditions are needed, you can alternatively use sterile D-PBS #14190 from Thermo-Fisher Scientific. Filtration is necessary to remove any contaminants. For long-term storage of the PBS 10× we suggest to use brown glass to protect it from light.
3. Lipid-ethanol stock solutions should be stored at −20 °C until usage. Allow the stock solutions to thaw at room temperature before use, which can take a considerable amount of time.
4. You can also use another desired concentration. Beware that the encapsulation efficiencies of your nanocarrier may vary and that this value is not the final cargo concentration after formulation.
5. For preparations above 20 mg of total lipid mass, bigger beads should be used, like SiLiBeads® ZY Ceramic beads, 0.6–0.8 mm #9607-53 from Sigmund Lindner (Warmensteinach, Germany).
6. Other PCR tubes (soft) burst during centrifugation. If your preparation exceeds the 200 μL volume of the PCR tubes (typically >10 mg of total lipid mass), use screw cap vials with 0.5, 0.65, or 2 mL volume from Carl Roth. Be sure to use the smallest vial volume possible for your preparation, because it results in lower polydispersities for your nanocarrier. Different insets for the Dual Centrifuge have to be used (Andreas Hettich GmbH, Tuttlingen, Germany).
7. Depending on the resolution you need to separate your unclicked ligands or non-encapsulated cargo from your nanocarrier, the volume of your preparation and the size of your nanocarrier, you will have to try different column volumes ranging from 2.5 to 10 mL. We use 2.5 mL, which is sufficient to separate the fluorophores we use from our nanocarrier with

diameters between 80 and 300 nm. NAP-5-based columns (GE Healthcare, Solingen, Germany) can also be used and packed with Sepharose 2 B-CL.

8. For highly polydisperse samples, we suggest the use of Nanoparticle Tracking Analysis (NTA). This method tracks each particle individually which results in more reasonable results. We conduct NTA on a Malvern NanoSight LM10 equipped with a sCMOS camera, a tempered chamber unit and a 532 nm laser. Beware that your nanocarrier solution has to be diluted around 10× more compared to measurements with the Zetasizer.
9. Conventional liposomes typically consist of 55:45 mol% for EPC:Cholesterol. For monofunctional liposomes with alkyne or norbornene-functionalized amphiphilic polymers anchored via cholesterol into the liposomal membrane, the equivalent amount of cholesterol is substituted, typically 5 mol%, resulting in a composition of 55:40:5 mol% for EPC:Cholesterol:Polymer. If the polymer bears a dialkyl-moiety as anchor, the equivalent amount of EPC is substituted. As an example for bifunctional liposomes bearing two dialkyl-anchored functionalized polymers, the composition is 50:45:2.5:2.5 mol% for EPC:Cholesterol:Polymer1:Polymer2.
10. For membrane labeling with DiI, add the DiI dissolved in ethanol in a concentration of 3 mg/mL and substitute the equal amount of Cholesterol, which was typically 0.2 mol%.
11. Depending on the batch size, more time could be necessary to completely dry your lipid solution. The dried lipids can be stored up to 3 days at −20 °C.
12. We suggest using fluorophores dissolved in PBS 1× as cargo in the beginning due to the easy and sensitive readout.
13. The highly concentrated resulting liposome suspension can be stored at 4 °C until usage.
14. The final concentration of the fluorophores can be adjusted after determination of the click efficiency for the first experiments to yield equal degrees of functionalization for comparison of different nanocarriers.
15. Longer incubation times did not yield substantially higher click efficiencies.
16. For a faster filtration, you can add the lid of the column and attach a PBS-filled syringe via the Luer connection and press the PBS through the column. If aseptic conditions are necessary, you can use D-PBS (#14190, Thermo-Fisher Scientific, Waltham, MA, USA) and perform the purification steps under a laminar flow bench.

17. Typical values using a 2.5 mL column are 500 μL for the nanocarrier fraction and 1.8 mL for the free cargo/ligand fraction with at least 150 μL in between.
18. For absolute quantification, spectrometrical changes due to the conjugation reaction are neglected. To overcome these effects, free alkyne or norbornene residues can be added to the corresponding fluorophores in the same amount as they occur in the nanocarrier suspension and incubated with the reaction conditions described in Subheading 3.2. However, the effect that not all terminal functionalities are accessible when the amphiphilic polymers are formulated in a nanocarrier has to be neglected.

Acknowledgements

The Rotanta 400 dual centrifuge prototype was kindly provided by Andreas Hettich GmbH, Tuttlingen, Germany. The authors would like to thank the collaborative research center SFB 1066 (Project A7) by the German Research Foundation (DFG).

References

1. Strebhardt K, Ullrich A et al (2008) Paul Ehrlich's magic bullet concept: 100 years of progress. Nat Rev Cancer 8:473–480
2. Holmberg E et al (1989) Highly efficient immunoliposomes prepared with a method which is compatible with various lipid compositions. Biochem Biophys Res Commun 175:1272–1278
3. Uster PS et al (1996) Insertion of poly(ethylene glycol) derivatized phospholipid into pre-formed liposomes results in prolonged in vivo circulation time. FEBS Lett 386:243–246
4. Gantert M et al (2009) Receptor-specific targeting with liposomes in vitro based on sterol-PEG(1300) anchors. Int J Pharm 469:168–178
5. Allen TM, Sapra P, Moase E (2002) Use of the post-insertion method for the formation of ligand-coupled liposomes. Cell Mol Biol Lett 7:889–894
6. Fritz T et al (2014) Click modification of multifunctional liposomes bearing hyperbranched polyether chains. Biomacromolecules 15:3114–3118
7. Bangham AD (1983) The liposome letters. Academic Press, New York
8. Bangham AD et al (1965) Diffusion of univalent ions across the lamellae of swollen phospholipids. J Mol Biol 13:238–252
9. Hofmann AM et al (2010) Hyperbranched polyglycerol-based lipids via oxyanionic polymerization: toward multifunctional stealth liposomes. Biomacromolecules 11:568–574
10. Hofmann AM, Wurm F, Frey H (2011) Rapid access to polyfunctional lipids with complex architecture via oxyanionic ring-opening polymerization. Macromolecules 44:4648–4657
11. Müller SS et al (2013) Polyether-based lipids synthesized with an epoxide construction kit: multivalent architectures for functional liposomes. In: Scholz C, Kressler J (eds) Tailored polymer architectures for pharmaceutical and biomedical applications, chap. 2. American Chemical Society, Washington DC, pp 11–25
12. Papahadjopoulos D et al (1991) Sterically stabilized liposomes: improvements in pharmacokinetics and antitumor therapeutic efficacy. Proc Natl Acad Sci U S A 88 (24):11460–11464
13. Kronberg B et al (1990) Preparation and evaluation of sterically stabilized liposomes: colloidal stability, serum stability, macrophage uptake, and toxicity. J Pharm Sci 79 (8):667–671

14. Allen TM et al (1995) Pharmacokinetics of long-circulating liposomes. Adv Drug Deliv Rev 16(2–3):257–284
15. Allen TM et al (1992) Stealth liposomes: an improved sustained release system for 1-β-D-arabinofuranosylcytosine. Cancer Res 52:2431–2439
16. Blume G, Cevs G (1990) Liposomes for the sustained drug release in vivo. Biochim Biophys Acta 1029(1):91–97
17. Fritz T, Voigt M et al (2016) Orthogonal click conjugation to the liposomal surface reveals the stability of the lipid anchorage as crucial for targeting. Chem Eur J 22(33):11578–11582
18. Himo F et al (2005) Copper(I)-catalyzed synthesis of azoles. DFT study predicts unprecedented reactivity and intermediates. J Am Chem Soc 127:210–216
19. Rostovstev VV et al (2002) A stepwise huisgen cycloaddition process: copper(I)-catalyzed regioselective "ligation" of azides and terminal alkynes. Angew Chem Int Ed 41 (14):2596–2599
20. Turnoe CW, Christensen C, Meldal M (2002) Peptidotriazoles on solid phase: [1,2,3]-triazoles by regiospecific copper(I)-catalyzed 1,3-dipolar cycloadditions of terminal alkynes to azides. J Org Chem 67(9):3057–3064
21. Agard NJ, Prescher J, Bertozzi CR (2004) A strain-promoted [3+2] azide-alkyne cycloaddition for covalent modification of biomolecules in living systems. J Am Chem Soc 126 (46):15046–15047
22. Blackman ML, Royzen M, Fox JM (2008) The Tetrazine ligation: fast bioconjugation on inverse-electron-demand Diels-Alder reactivity. J Am Chem Soc 130:13518
23. Devaraj NK, Weissleder R, Hilderbrand SA (2008) Tetrazine-based cycloadditions: application to pretargeted live cell imaging. Bioconjugate Chem 19(12):2297–2299
24. Han H et al (2010) Development of a bioorthogonal and highly efficient conjugation method for quantum dots using tetrazine-norbornene cycloaddition. J Am Chem Soc 123(23):7838–7839
25. Massing U, Cicko S, Ziroli V (2008) Dual asymmetric centrifugation (DAC)—a new technique for liposome preparation. J Control Release 125(1):16–24
26. Hirsch M et al (2009) Preparation of small amounts of sterile siRNA-liposomes with high entrapping efficiency by dual asymmetric centrifugation (DAC). J Control Release 135 (1):80–88

Chapter 17

Polymersomes: Preparation and Characterization

Yumiao Hu and Liyan Qiu

Abstract

Polymersomes, also called polymeric vesicles, are self-assembled by amphiphilic copolymers. Due to their unique characters, polymersomes are attracting more and more interest as an important class of vehicles for nanopharmaceuticals. In this chapter, various methods to prepare and characterize polymersomes are introduced systematically with several applicable examples. In addition, the advantages and disadvantages of each method were compared and analyzed with the aim to help readers choose the appropriate method in the process of experiments. Although some methods we introduced here are effective in preparing and characterizing polymersomes, the remaining challenge in this filed is to develop new tools. The reason is that polymersome is a kind of complex nanostructure, and some minor factors can affect the formation of polymersome. Meanwhile, more advanced technology should be developed to precisely determine the structure of some complex polymersomes such as multilayer polymersomes.

Key words Polymersome, Self-assembly, Amphiphilic copolymers, Light scattering, Microscopy, Morphology

1 Introduction

Polymersomes, also called polymeric vesicles, are self-assembled by amphiphilic copolymers. Compared with liposomes, polymersomes have similar structure with an aqueous lumen surrounded by hydrophobic membrane but own some superior characters such as higher membrane stability and lower membrane permeability, which makes them fascinating on encapsulating and carrying both water-soluble and water-insoluble guest molecules [1]. In addition, the thickness of hydrophobic membrane and the decoration of tumor-targeting groups on polymersome surface can be regulated to some extent [2]. Therefore, polymersomes are considered as an important cluster of vehicles in the field of nanopharmaceuticals.

The polymersome formation of amphiphilic block copolymers has been reported by many groups. It is primarily the packing parameter, $p = v/a_0 l_c$, that determines the morphology of aggregates, where v is the volume of the hydrophobic segment, a_0 is the contact area of the head group, and l_c is the length of the

Volkmar Weissig and Tamer Elbayoumi (eds.), *Pharmaceutical Nanotechnology: Basic Protocols*, Methods in Molecular Biology, vol. 2000, https://doi.org/10.1007/978-1-4939-9516-5_17,

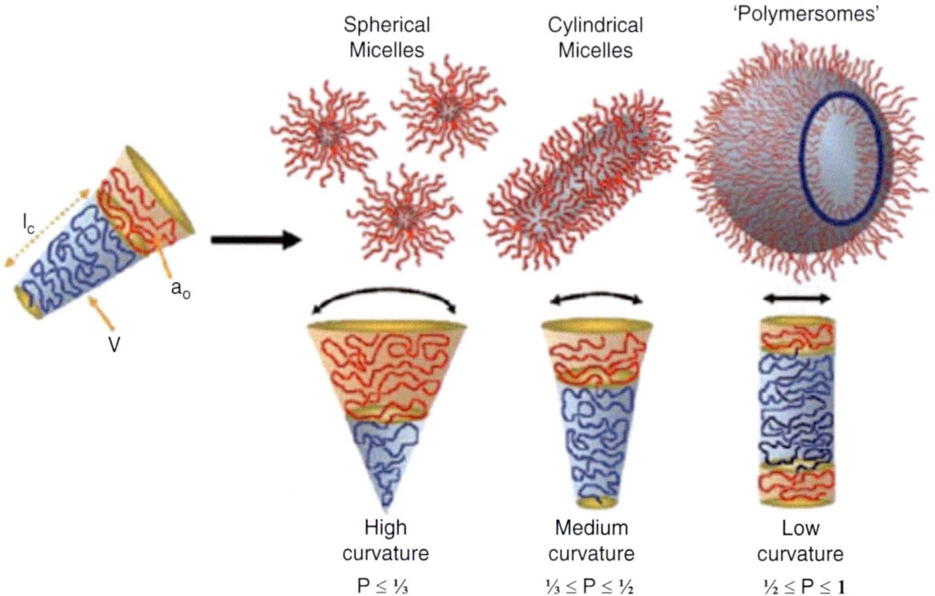

Fig. 1 Several structures formed by the self-assembly of amphiphilic block copolymers as determined by the geometry of the amphiphile. The geometry is captured by the dimensionless packing parameter $p = v/(a_0 l_c)$ (from [1], with permission from The Royal Society of Chemistry)

hydrophobic segment. When $p < 1/3$, spheres are formed; when $1/3 < p < 1/2$, cylinders are formed; when $1/2 < p < 1$, flexible lamellae or vesicles are formed; finally, when $p = 1$, planar lamellae are obtained. If $p > 1$, inverted structures can be observed, as it is depicted in Fig. 1. Generally speaking, the hydrophilic fraction (f) is better suited to predict the expected morphology. Block copolymers with a hydrophilic fraction of more than 45–50% will mostly yield micelles. In the region where f is around 35 ± 10% in many cases, polymersomes are observed, and in case the hydrophilic fraction is less than 25%, inverted structures can be expected. Finally, there is a small region where wormlike micelles have been reported, when the hydrophilic fraction is around 50% [1, 3]. Although this basic empirical rule holds quite well in most cases, the exact aggregation behavior has to depend strongly on the type of block copolymer and the conditions applied. Therefore, the synthesis and control of amphiphilic copolymers with suitable hydrophilic fraction are vital.

In this chapter, methods of preparation and characterization of polymersomes are introduced, respectively. According to different preparation methods, the disadvantage and advantage of each preparation method can be easily comprehended. It is convenient for the readers to decide which kind of method can be used for preparing polymersome under different conditions. Besides, the common characterization methods are mentioned here including theory, sample preparation, characterization examples, advantages, and disadvantages. Sometimes several characterization methods are combined to use in order to confirm the structure of polymersome, while details for the process can be obtained here.

2 Methods of Preparation

2.1 Solvent Switch Method (See Note 1)

Solvent switch method is the most common method to prepare polymersomes. By this way, the most important step is to find an appropriate organic solvent or solvent mixture to dissolve amphiphilic copolymers. The common organic solvents include *N*,*N*-dimethylformamide (DMF), tetrahydrofuran (THF), and dioxane.

1. Amphiphilic copolymers are first dissolved in DMF, which is a kind of common solvent for both hydrophilic and hydrophobic components.
2. Subsequently, deionized water, a solvent only for hydrophilic components, is added to the polymer/DMF solutions (ca. 10 mL) at a rate of 1 drop every 5–10 s with vigorous stirring.
3. As the water is added, the quantity proportion of the solvent for the hydrophobic block decreases gradually, and the hydrophobic block of the amphiphilic copolymers begins to aggregate, as indicated with the solution becoming turbid, especially occurring when the water content reaches 3–6 wt% in most cases, depending on the composition of the block copolymers. The addition of water is continued until water content reaches 25 wt%.
4. After that, the obtained colloidal solution is placed in dialysis bags in deionized water to remove DMF [4].

2.2 Film Rehydration (See Note 2)

Film rehydration is another common method to prepare polymersomes. Unlike solvent switch method, low-boiling organic solvent is often used to dissolve amphiphilic copolymers, such as chloroform, which can be removed completely by evaporation.

1. At first, an amphiphile film should be prepared on a solid surface. It can be achieved by dissolving amphiphilic copolymers in an appropriate organic solvent or solvent mixture, and then evaporating organic solvent by means of a rotary evaporator, assisted by high-vacuum pump or nitrogen stream. The solid surfaces often used are glass vials or roughened Teflon. After evaporating the organic solvent, a thin and fine film can be obtained. This is the first step of film rehydration method.
2. After an amphiphile film is obtained, addition of aqueous buffer leads to spontaneous swelling of vesicles off solid surfaces and dispersing into the solution. This is called film

hydration. The mechanism of swelling procedure is proposed to be as follows: water infiltrates into polymer layers, which is driven by hydration force.

3. Thus, the layers are successively inflated to form bulges, which yield vesicles and separate from the surface. The swelling process can be influenced by the agitation, so effective methods such as turraxing and sonication are commonly used. Using film rehydration method, small multilamellar vesicles with a broad size distribution can be obtained [5].

2.3 Solid Rehydration (Bulk Rehydration) (See Note 3)

The solid rehydration method is similar to film rehydration. The only difference is that the amphiphilic copolymer is not hydrated as a film on a solid surface but hydrated as bulk power. That means solid rehydration needs stronger and longer agitation to let power hydrate completely [4]. The general process is as follows:

1. Firstly, the copolymer power should be dissolved in an aqueous solution of tetrahydrofuran (THF) at 1 wt% to achieve a final polymer concentration of 0.5–1%.
2. Secondly, a certain amount of each solution is pipetted into a glass vial and the solvent is evaporated.
3. At last, rehydrate with deionized water under bubbling nitrogen for 1 h.

2.4 Electroformation (See Note 4)

Closely related to film rehydration, electroformation is another common method to obtain giant lipid vesicles, and also can be used for giant polymersome formation. The general process is as follows:

1. Firstly, the amphiphilic copolymers should be dissolved in chloroform with the concentration of 20 mg/mL. The obtained solution can be applied to two indium tin oxide (ITO) slides and allowed to dry under nitrogen or vacuum. A silicone gasket is affixed in one of the two ITO slides to create a solution reservoir.
2. Secondly, electrodes are immersed in buffer using a Teflon spacer, and the gasket is then sandwiched between the two electrodes. The buffer is often prepared by NaCl (145 mM) and HEPES (2.5 mM) at pH 7.4.
3. Thirdly, the ITO electrodes are connected to an Agilent function generator with AC voltage 5 V at a frequency of 10 Hz for 3 h, and 5 V at a frequency of 0.5 Hz for 30 min. In general, increasing the voltage may lead to the inhibition of the polymersome formation. So the optimal voltage was found to about 5 V [6].

3 Characterization

In this section, the frequently used characterization methods including light scattering methods and microscopy to confirm the structure of polymersome are introduced in detail. These techniques are often combined since single method has its own advantages and drawbacks [7–9].

3.1 Light Scattering Methods (See Note 5)

Turbidity measurements have become a main tool to study aggregation morphology of the nanoparticles. Laser light scattering is able to probe aggregates in the size range of 1–1000 nm. After the beam of the laser light passes the polymer solution, most of the light could pass through the solution, but some of them would be scattered. The intensity of the scattered light will be measured by instruments [10]. In dynamic light scattering (DLS), fluctuations of the intensity of scattered light in the microsecond timescale appear because of diffusive motions of particles in solution. The angle-dependent apparent diffusion coefficient (D_{app}) can be converted to the hydrodynamic radius (R_h) using the Stokes-Einstein equation by extrapolating it to zero concentration and zero momentum transition transfer [11]. For static light scattering (SLS), structural properties are available, such as weight-averaged molecular weight (Mw), z-average radius of gyration (R_g), and second viral coefficient (A_2). Particle-particle as well as particle-solvent interactions can be obtained via A_2 using berry plots. By this way, both the hydrodynamic radius and the radius of gyration could be measured. More importantly, the characteristic radius ratio (R_g/R_h) is an indication to investigate the self-assembly structure of the amphiphilic copolymers. The values of parameter (R_g/R_h) are 1.0 for vesicles and 0.799 for homogeneous hard balls. And the values more than 1.0 demonstrate that polymers are in extended conformations [12]. The advantage of light scattering is that it is a fast and precise method, but it needs to analyze data in complex surfactant systems. High-throughput scattering methods such as combinatorial small-angel X-ray scattering (SAXS)/wide-angle X-ray scattering (WAXS) can provide information about structural features of colloidal size to analyze both concentrated and dilute samples. The small-angle neutron scattering (SANS) technique is a very important method for studying chain conformation and interaction parameters in the one-phase region. It can be applied for investigating the morphology and thermodynamics of polymer blends and copolymers. In addition, structure and self-assembly of block copolymers and control of drug encapsulation by multilamellar vesicles can be investigated with SANS. Morphological and spatial segmental distribution in block copolymer-homopolymer mixture during vesicle formation was also studied with SAXS and SANS [12].

3.2 Microscopy

Microscopy is a very powerful method to observe the morphology of polymersome, because most microscopy techniques are fast and easy and provide relatively straightforward specimen visualization. Many important parameters like size and homogeneity can instantly be revealed. The microscopy techniques we discuss here include optical microscopy, fluorescence microscopy, atomic force microscopy, and transmission electron microscopy, which will be introduced one by one as follows.

3.2.1 Optical Microscopy

The advantage of optical microscopy is the possibility to directly visualize polymeric vesicles under "physiological" conditions. It is not necessary to dry or stain specimens; instead, they can be kept in aqueous buffer. However, compared to electron microscopy, the major drawback of light microscopy is the limited resolution. So optical microscopy is often used to observe polymersomes with large size: giant vesicles with diameters above one micron are best suited for such studies. Optical microscopy includes transmission light microscopy (*see* **Note 6**), phase-contrast microscopy (*see* **Note 7**), and differential interference contrast microscopy (*see* **Note 8**).

3.2.2 Fluorescence Microscopy

In fluorescence microscopy, the excitation light irradiates a specimen and then the red-shifted emitted fluorescent light is separated from the brighter excitation light. There are several important advantages of fluorescence microscopy over transmission microscopy techniques: (1) specific labeling with fluorophores enables distinguishing between nonfluorescent regions in one specimen; (2) multiple staining with different fluorophores allows visualization of individual target molecules; and (3) the presence of fluorescent material is revealed with exquisite sensitivity (at 50 fluorescent molecules/mL). The presence of fluorescent molecules becomes visible even below the diffraction limit [10]. As shown in Fig. 2, polymersomes from oligoanhydride-PEG block copolymer can be observed under fluorescence microscopy. Fluorescent labeling of polymeric vesicles can be achieved through different approaches. In most cases, the amphiphilic polymers do not exhibit intrinsic fluorescence and therefore a dye needs to be encapsulated, or the vesicle membrane has to be stained [13].

Fluorescence microscopy includes wide-field microscopy, total internal reflection fluorescence microscopy, and confocal fluorescence microscopy. A wide-field fluorescence microscope uses a lamp, for example, a mercury arc lamp, to illuminate and excite the specimen. This is a fast and economical way to obtain fluorescent images that can be viewed directly with the eyes through the ocular or captured with a camera. For total internal reflection fluorescence microscopy, the background fluorescence is dramatically reduced by the passage of the refractive light from an optical denser to an optically less dense medium. Image contrast is thus improved and resolution significantly increased to 200 nm or less.

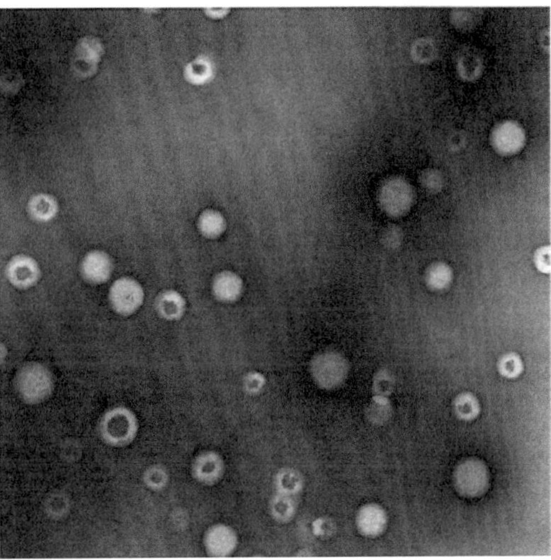

Fig. 2 Polymersomes from oligoanhydride–PEG block copolymer under fluorescence microscope (from [13], with permission from ACS Publications Division)

In laser scanning confocal microscopy (LSCM), both the high lateral resolution and the contrast are dramatically improved due to the reduction of background fluorescence and improved signal-to-noise [11].

3.2.3 Atomic Force Microscopy

Scanning force microscopy (SFM) methods, such as atomic force microscopy (AFM) and scanning tunneling microscopy (STM), are the most frequently applied techniques for determining the surface structures on solid substrates. Both allow obtaining high-resolution (a few Å) images. The AFM principle stems from measuring the force of interaction between the most exterior molecular layer of the sample and the tip. The tip undergoes vertical deflection when it interacts with the surface, which is proportional to the interaction force with the surface molecules (10^{-9}–10^{-10} N). The deflection is recorded by using a very sensitive tip-integrated spring or the deflection of a laser beam. In this way, a surface in homogeneity profile is obtained at the atomic scale [10]. The AFM is a basic tool in laboratories to investigate the properties of thin films on solid substrates, but it has also been proven useful in studies of polymer vesicles (*see* **Note 9**). The sample preparation is described as follows:

1. Firstly, the sample slides used for AFM were thermally treated at 430 °C in the course of the vapor deposition of the aluminum layer.
2. Secondly, this treatment was followed by 1 h of annealing at 450 °C.
3. Thirdly, in order to minimize contamination of samples and tips, the solutions made with ultrapure water were carefully filtered with 0.2 μm standard filters (Acrodisc) just before use.

3.2.4 Transmission Electron Microscopy

Transmission electron microscopy (TEM) was first developed in the 1930s. Compared to optical microscopes, the theoretical resolution of TEM is about hundred thousand times higher. Additionally, the electron microscope offers about a 1000-fold increase in resolution and a 100-fold increase in depth of field. These advantages make TEM an important choice to study surface and subsurface properties, particularly for vesicles. TEM was also used to monitor the morphologies of the aggregates and to determine the sizes of the vesicles (*see* **Note 10**). The TEM sample is prepared as follows:

1. Copper TEM grids are pre-coated with a thin film of Formvar.
2. Then copper TEM grids are coated with carbon.
3. 0.01 mL of the diluted colloid solutions are deposited on the resulting grids.
4. After drying in air overnight, the samples are used for TEM studies.

Although TEM is a convenient method to monitor the morphologies of polymersomes, there are several disadvantages of electron optics. Electrons are high-energy particles, which induce strong interactions with the sample and therefore do not penetrate deeply into a specimen. In addition, for the same reason, an electron microscope has to be kept under high vacuum. The TEM is ideal for studies in "synthetic" systems. Sometimes, the dry specimen of TEM shows collapsed polymersomes. For example, the TEM of the amphiphilic peptides shown in Fig. 3 is a collapsed vesicle. It was verified that the objects own a hollow structure, and the vesicles were collapsed due to the drying effect on the TEM grid. These disadvantages exist especially for biological samples, where the specimen is always dead [14].

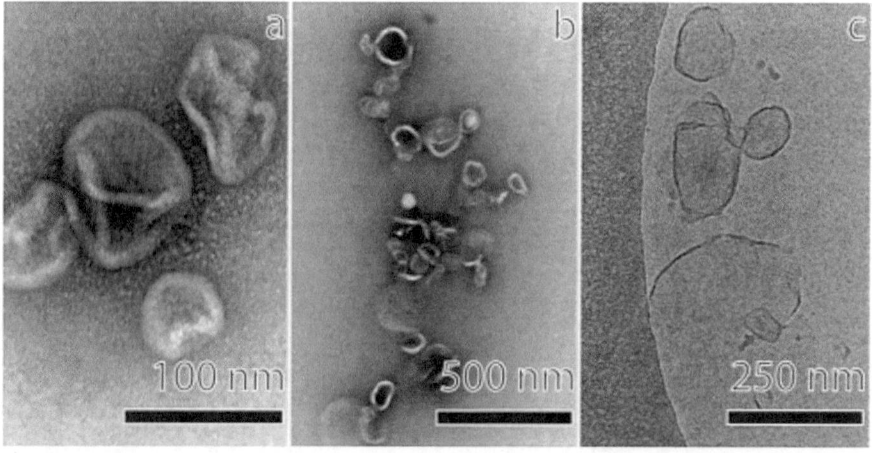

Fig. 3 TEM of self-assembled structures of (**a**) Ac-X8-gA-OEt and (**b**) Ac-E6-gA, and (**c**) cryo-TEM of Ac-E6-gA (from [14], with permission from Wiley-VCH Verlag GmbH & KGaA)

Owing to this fact, cryo-TEMs are rather used in studying biological systems, which allow specimens that are "wet," uncoated. Cryo-TEM offers another advantage: since specimens are frozen and viewed in vitreous ice, they are seen in a natural hydrated state, which is as close to their natural state as possible. Indeed, when a sample is perfectly frozen, the osmotic effects due to chemical fixation are almost suppressed and dehydration is avoided, which are responsible for aggregation and loss of biological materials that almost inevitably happened during the sample preparation for TEM. In addition, cryo-TEM allows investigating the self-assembly behavior of amphiphilic copolymers in water: micellar polymorphism, spontaneous formation of vesicles, and their transition to lamellar structures. The procedure for the cryo-TEM sample preparation is described as follows:

1. A drop of the dispersion is deposited on an electron microscopy copper grid coated by a perforated polymer film.
2. The excessive liquid is blotted by a filter paper, leaving a thin film of the dispersion on the grid.
3. The film on the grid is vitrified by plunging the grid into liquid ethane.
4. The vitrified sample is then transferred to the microscope for observation at liquid nitrogen temperature.

Freeze-fracture reveals intra-vesicular details in three dimensions as shown in Fig. 4. Samples are frozen rapidly in liquid nitrogen and fractured to reveal internal structure. The procedure for the FF-TEM sample preparation is described as follows.

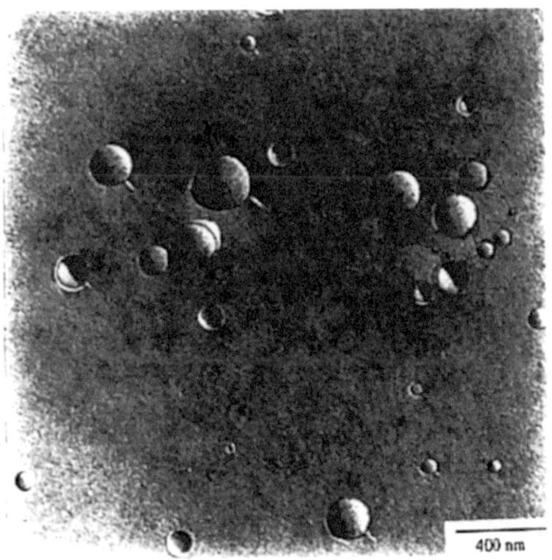

Fig. 4 A freeze-fracture electron micrograph of triblock copolymer vesicles; scale bar: 400 nm (from [16], with permission from ACS Publications Division)

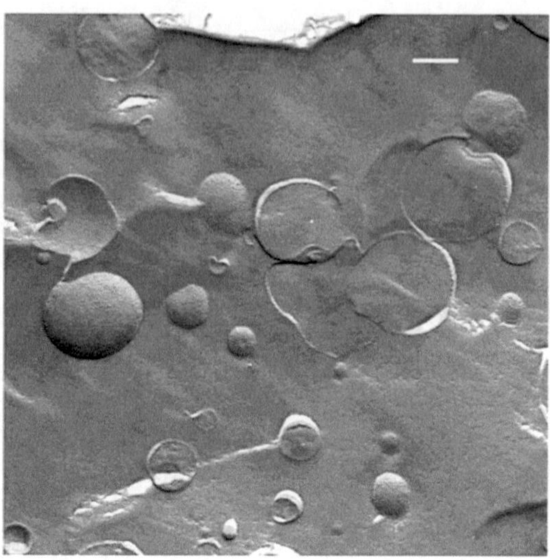

Fig. 5 FF-TEM image of a $EO_{16}PS_{25}EO_8$ vesicle suspension (1 wt%) in water stored at room temperature for 3 months (bar represents 250 nm) (from [15], with permission from ACS Publications Division)

1. Block copolymer aqueous dispersions were placed between two copper platelets using a 25 μm thick gold grid as spacer and frozen from room temperature using the propane-jet method.
2. The cryo-fixed samples were fractured at 123 K and 10^{-6} Pa in a Balzer BAF 300 freeze-etching apparatus and shadowed by 20 D platinum/carbon at an angle of 40°.
3. The replicas were cleaned with chloroform/ethanol to remove sample residuals and examined with a Philips EM 301 electron microscope [15].

The fracture surface is etched under vacuum and rotary shadowed with metal. The resulting replicas contain fine morphological details and have proven to be particularly useful for studies of lipid bilayers. Lyotropic behavior of amphiphilic ABA triblock copolymers in water has been investigated by freeze-fracture TEM [5]. The presence of single uni- or multilamellar vesicles was noticed, as shown in Fig. 5.

4 Notes

1. Although solvent switch method is a fast and convenient method to yield homogenous polymersomes, the drawback is that organic solvent cannot be removed completely and remains in polymersome solution. This is unfavorable for polymersomes to apply in drug delivery, as the residual organic solvent may have toxic effect. Besides, polymersome obtained

by this method often owns a rather broad size distribution [16]. In order to prepare drug-loaded polymersome, the drug is dissolved in distilled water, and then the drug solution was added dropwise under magnetic stirring. After that, the drug-loaded polymersome is obtained by dialyzing against distilled water to remove DMF and unencapsulated drug.

An instance of practical polymersome is as follows. The amphiphilic graft copolymer PEP was first dissolved at 10 mg/mL concentration in DMF and distilled water was added dropwise at 1:1 volume ratio under magnetic stirring. Then, this solution was put into dialysis bag (MWCO = 14,000) to dialyze against distilled water. As DMF was gradually removed within 12 h, the blank PEP polymersome was obtained. The preparation of drug-loaded PEP polymersome is the similar process to blank PEP polymersome. PEP copolymer was dissolved at 10 mg/mL concentration in DMF, the drug was dissolved at 2 mg/mL concentration in distilled water, and then the drug solution was added dropwise at 1:1 volume ratio under magnetic stirring. The drug-loaded PEP nanoparticles were obtained by dialyzing against distilled water to remove DMF and unencapsulated drug. The polymersome can load both the hydrophobic and hydrophilic drugs due to the aqueous cavity [17].

The amphiphilic block copolymer poly(lactic acid)-b-pluronic-b-poly(lactic acid) (PLA-F127-PLA) vesicles were prepared by solvent switch method. The procedure is as follows: PLA-F127-PLA block copolymer (60 mg) was dissolved in THF (3 mL) and the polymer solution was added dropwise to water solution (3 mL) under gentle stirring. The solution was put into a cellulose membrane tubing (MWCO = 12,000–14,000) to dialyze against distilled water to remove DMF. After that, the solution was lyophilized using a freeze dryer (Labconco, USA) to obtain the blank vesicles. In order to prepare insulin-loaded vesicles, the protocol is as follows: PLA-F127-PLA block copolymer (60 mg) was dissolved in THF (3 mL) and the polymer solution was added dropwise to insulin solution (3 mL) under gentle stirring. The insulin-loaded polymer aggregates were dialyzed against ultrapure water for 4 h using a cellulose membrane tubing (MWCO = 12,000–14,000) to remove THF and free insulin outside PLA-F127-PLA aggregates. The water was exchanged at 1-h intervals. After dialysis, the solution remaining in the dialysis tubing was frozen and lyophilized using a freeze dryer for 2 days to obtain dried insulin-loaded PLA-F127-PLA vesicles. The loading capacity of insulin in the PLA-F127-PLA vesicles was determined by deducting the amount of free insulin outside the dialysis membrane from the initial amount of insulin added [18].

2. For example, the diblock polybutadiene-b-polyethylene oxide (PB-PEO) vesicles were prepared by film rehydration procedure. Briefly, a few drops (about 30 μL) of the polymer solution (40 mg/mL in chloroform) was spread on a roughened Teflon disk and dried under vacuum overnight. The sample was prehydrated and swollen in 100 mM sucrose solution at a temperature of 38 °C for a few hours. The vesicles were harvested and incubated in 110 mM excess glucose solution to allow for modest deflation and enhancement of contrast [19].

   PEG-PLA and PEG-PCL vesicles are often prepared by film rehydration. First, PEG-PLA or PEG-PCL polymer was dissolved at desired molar ratios in chloroform. The organic solvent was then evaporated under nitrogen, followed by vacuum drying for 7 h to remove trace amounts of chloroform as the polymer film dried onto the glass wall of a dram vial. The film was subsequently hydrated with solutions of hydrophilic encapsulants such as sucrose. Upon hydration, vesicle self-assembly was further promoted in a 60 °C oven for 12 h. To prepare drug-loading vesicles, a variation of the ammonium sulfate-driven permeation method was applied by Barenholz et al. Unencapsulated ammonium sulfate was removed by dialysis into isotonic PBS. After that, the drug was added to the vesicle suspension by membrane permeation and accumulation due to the drug gradients between inside and out of vesicles. A 10-h incubation at 37 °C was to promote the drug encapsulation sufficiently. Then, another 10-h dialysis is done to remove the unencapsulated drug. The drug-loaded vesicles were obtained by lyophilization using a freeze dryer (Labconco, USA) [4, 20].

   Poly(l-Asp-g-DEAP)-b-PEG vesicles were prepared via the film rehydration method. Poly(l-Asp-g-DEAP)-b-PEG was dissolved in dichloromethane (10 mL), which was added to a round-bottomed flask. After removing dichloromethane using a rotary evaporator (EYELA, N-1000), a thin film was formed on the surface of a round-bottomed flask. By adding $Na_2B_4O_7$ buffer solution (5 mM), the vesicle was easily fabricated via a self-assembly of poly(l-Asp-g-DEAP)-b-PEG film. To prepare DOX-loaded poly(l-Asp-g-DEAP)-b-PEG vesicle by film rehydration, the drug was added to the vesicle suspension by membrane permeation. Non-encapsulated DOX was removed by dialysis against HCl (or NaOH)–$Na_2B_4O_7$ buffer solution (5 mM) in a dialysis membrane (MWCO = 2000) [10].

3. For example, the diblock copolymer poly(butyl acrylate)-b-poly(acrylic acid) (PBA-PAA) vesicles were prepared by solid rehydration. The technique was to disperse a powder in an aqueous solution of THF at 1 wt% to achieve a final polymer concentration of 0.5–1%. Then 25 μL of each solution was

pipetted into a glass vial and the solvent was evaporated. At last, rehydrate with 250 μL deionized water under bubbling nitrogen for 1 h. The preparation procedure of drug-loading vesicle by solid rehydration is also similar to film rehydration. After obtaining the blank vesicles, ammonium sulfate was added into the vesicle solution. The next step is to dissolve drug into the vesicle solution. A 10-h incubation at 37 °C was performed after another 10-h dialysis. The drug-loading vesicle was achieved by freeze-drying [21].

The solid rehydration method was also adopted to prepare polyethylene oxide-block-polybutadiene (PEO-PBA) vesicles. In brief, the protocol consisted of four steps: (1) addition of 10 mg of PEO-PBA to 10 mg of PEG500 in a 1.5 mL centrifuge tube followed by heating for 20 min at 95 °C; (2) mixing by vortexing and cooling to room temperature followed by the addition of 10 μL of methylene blue (21 mg/mL) or myoglobin solution (150 mg/mL) in PBS (10 mM, pH 7.4); (3) dilution with 20, 70, and 900 μL of PBS with mixing (via vortexing); and (4) dialysis for 30 h at room temperature or at 4 °C (MW = 1000 k) to remove unencapsulated methylene blue or myoglobin, respectively. The dialysis buffer was exchanged with 1 L of fresh buffer every 7–8 h. At last, the vesicle was collected by freeze-drying [22].

4. Electroformation was used to obtain lipids, PMOXA-PDMS-PMOXA vesicle, and mixed vesicles. The PMOXA-PDMS-PMOXA copolymer is the ABA block copolymer with two water-soluble poly(2-methyloxazoline) (PMOXA) side blocks and a hydrophobic and flexible poly(dimethylsiloxane) (PDMS) middle block. Briefly, 500 μL chloroform solution containing 5 mg of ABA/egg-phosphatidylcholine/egg-phosphatidylethanolamine/Lα-phosphatidyl-ethanol amine-*N*-(lissamine-rhodamine-B-sulfonyl (79/12/8/1, mol/mol/mol) was sprayed on two indium tin oxide-coated glass electrodes (Merck LCD division). Chloroform was removed under nitrogen and vacuum. Electrodes were immersed in buffer D (145 mM NaCl, 2.5 mM HEPES, pH 7.4) using a Teflon spacer. Vesicles were formed by applying a simple AC voltage of 5 V at a frequency of 10 Hz for 3 h and 5 V at 0.5 Hz for 30 min. Vesicles were observed under a fluorescence microscope equipped with a 100× Zeiss plan apochromat objective [23].

5. For example, the nanostructure of mPEG-b-(polyHis)$_2$ was characterized by LS technology. After monitoring the normalized relaxation time distribution (G(tR)) using DLS mode (Fig. 6), a relationship between the angular-dependent diffusion coefficient (D_{app}) and the magnitude of scattering (q) of the polymersome was plotted to estimate the intrinsic

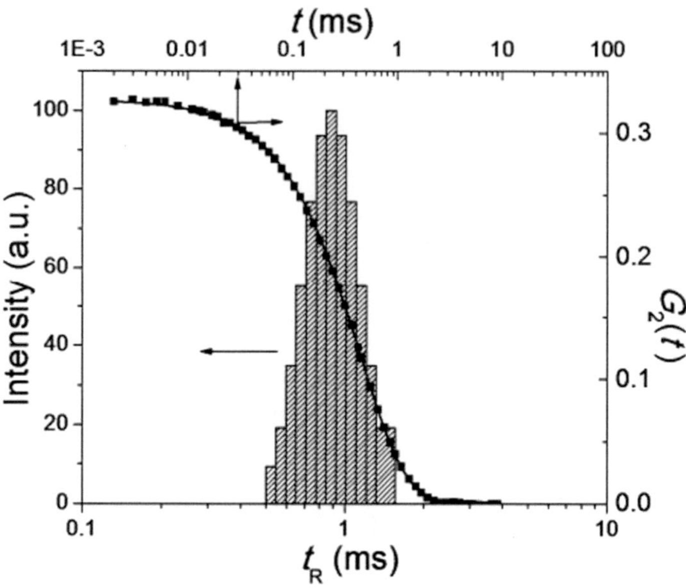

Fig. 6 DLS autocorrelation function and normalized time relaxation distribution at 90° for the polymer solution (pH 9.0, 0.3 mg/mL) (from [24], with permission from The Royal Society of Chemistry)

translational diffusion coefficient ($D_0 = 3.4 \times 10^{-8}$ cm^2/s) (Fig. 6). According to the Stokes-Einstein equation and the estimated D_0, the Rh can be calculated as 70.2 nm. The size polydispersity index (PDI) was 0.05 by the measurement of DLS, indicating a narrow size distribution of the nanostructure. In order to calculate R_g, SLS was conducted from 30° to 120°, and a partial Zimm plot was generated from the excess intensity of scattering (q) (Fig. 7). The value of R_g was calculated as 71.9 nm. Based on the results from DLS and SLS, the ratio of R_g and R_h (R_g/R_h) was determined to be 1.02. This value suggests that the nanostructure of mPEG-b-(polyHis)$_2$ is vesicle [24].

6. Transmission light microscopy is worthless without sufficient contrast in the image. Contrast is not an inherent property of the specimen, but is dependent upon interaction of the specimen with light and efficiency of the optical system used. Polymersomes neither absorb light nor seem to stain with chemical dyes achievable. Therefore, contrast is so weak that the specimen remains essentially invisible and contrast has to be enhanced using other techniques [11].
7. This technique provides an excellent method of improving contrast in unstained biological specimens without significant loss in resolution [21]. An example of phase-contrast imaging in vesicular systems is provided in Fig. 8. Direct visualization of polymeric aggregates, providing information on structural

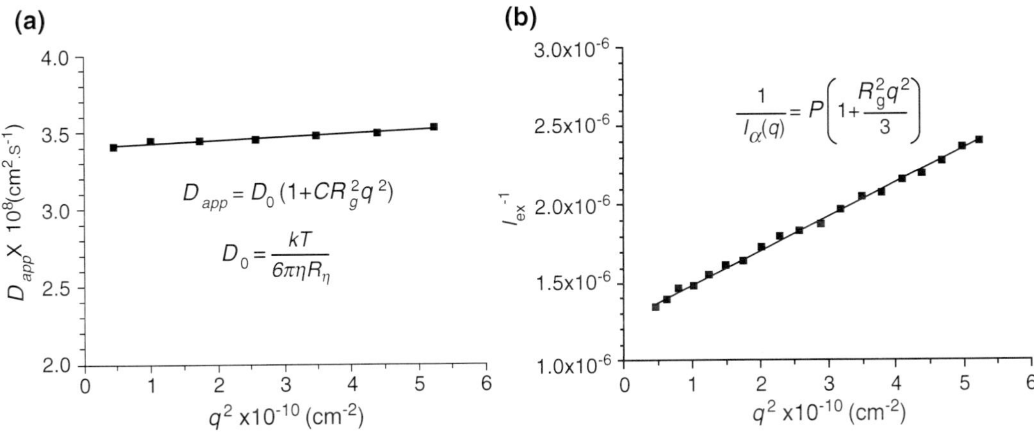

Fig. 7 (**a**) Angle dependence (DLS mode) and (**b**) a partial Zimm plot (SLS mode) of the mPEG-b-(polyHis)$_2$ nanostructure in borate buffer (0.3 mg/mL, pH 9.0). Solid lines represent linear fits to the data points (from [24], with permission from The Royal Society of Chemistry)

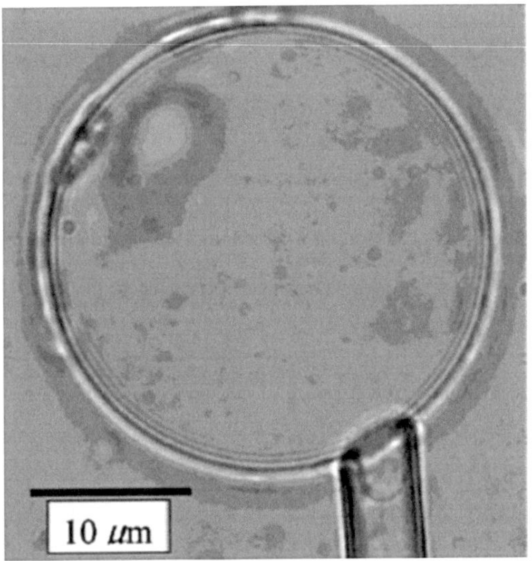

Fig. 8 Vesicles from PBA–PAA (70–30) in 1 wt% aqueous solution in THF; scale bar: 10 mm (from [21], with permission from ACS Publications Division)

details and the kinetics of transition between different aggregate morphologies, is possible in the micrometer regime.

8. Unlike phase-contrast microscopy, differential interference contrast converts gradients in specimen optical path length into amplitude differences, which can be visualized as improved contrast in the resulting image. The method is excellently suited for thick, non-stained specimens [25], as presented in Fig. 9. In addition, it is often employed in combination with fluorescence microscopy to reveal the cellular morphologies associated with fluorescent regions.

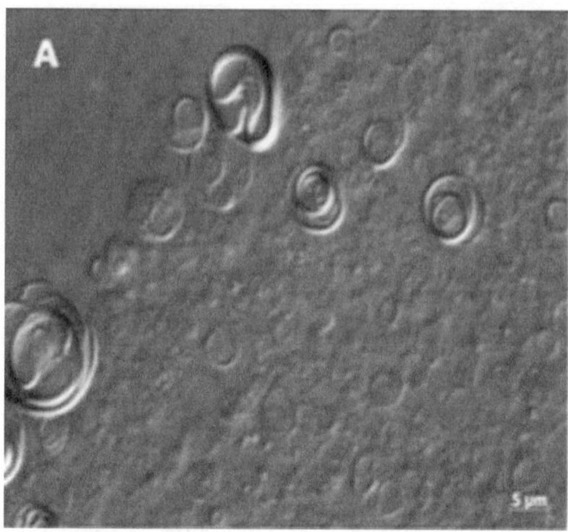

Fig. 9 A differential interference contrast micrograph of giant vesicles formed by PMOXA–PDMS–PMOXA triblock copolymer (from [21], with permission from The Royal Society of Chemistry)

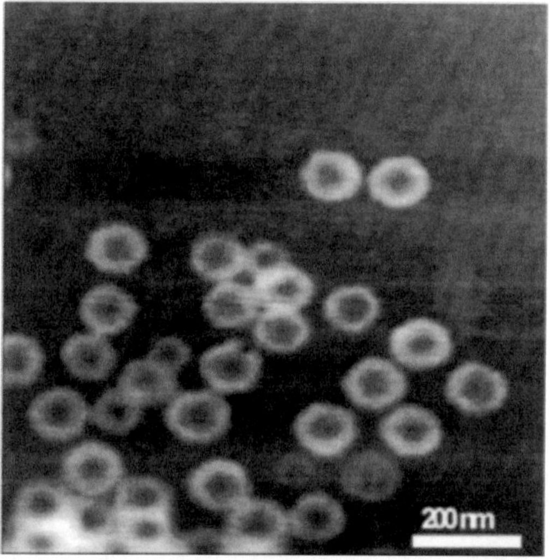

Fig. 10 Scanning force microscopy (SFM) image of BPD_{210}-$P2VPM_{99}$ vesicles on mica surface; scale bar: 200 nm (from [26], with permission from J. Wiley and Sons, Ltd.)

9. For example, on the hydrophilic mica surface, a preferred adsorption should occur due to the positively charged vinylpyridine block in the ambient solution [26]. The self-assemble behavior of poly(butadiene–block-2-vinylpyridene) can be observed by scanning force microscopy. As shown in Fig. 10, the structures of PB_{210}-$P2VPM_{99}$ detected on mica surfaces can be interpreted as the large spherical vesicles. The observed shape can be explained by the fact that the inner core of the vesicles was dried during the drying process of the sample [27].

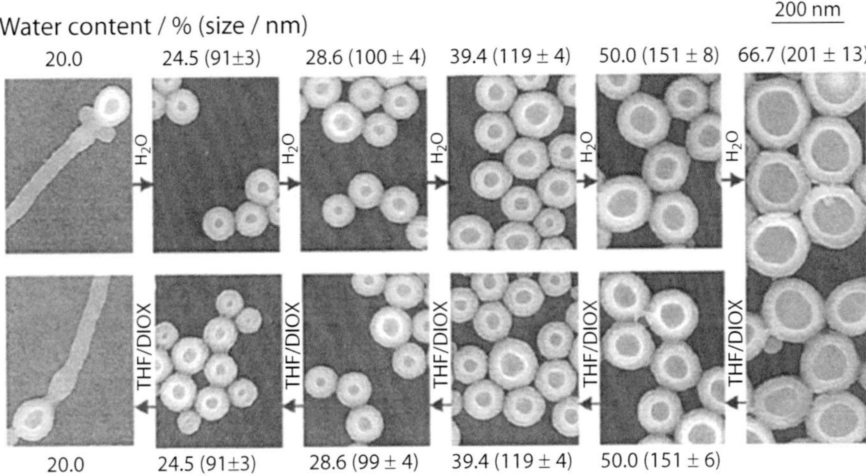

Fig. 11 Reversibility of vesicle sizes in response to changes of water contents for a PS_{300}-b-PAA_{44} system in THF/dioxane; scale bar: 200 nm (from [28], with permission from ACS Publications Division)

10. As shown in Fig. 11, PS_{300}-b-PAA_{44} diblock copolymers self-assemble into a coexisting mixture of rods and vesicles at a water content of 20.0 wt%. When the water content is increased to 24.5 wt%, vesicles with a diameter of 91 ± 3 nm and wall thickness of 35 ± 3 nm are observed. The sizes and distributions are based on an analysis of 50 vesicles in each case. When the water content is increased progressively to 28.6 wt%, 39.4 wt%, 50.0 wt%, and 66.7 wt%, the vesicle sizes are increased to 100 ± 4, 119 ± 4151 ± 8, and 201 ± 12 nm, respectively. The wall thicknesses of vesicles are 35 ± 3, 35 ± 3, 36 ± 3, and 36 ± 3 nm; obviously, no significant changes in the vesicle wall thicknesses are found. The THF/dioxane mixture is then added to decrease the water content progressively to 50.0 wt%, 39.4 wt%, 28.6 wt%, and 24.5 wt%; in the process, the vesicle sizes decreased to 151 ± 8, 119 ± 4, 99 ± 4, and 91 ± 3 nm, respectively. The wall thicknesses of vesicles remained at 36 ± 3, 35 ± 3, 36 ± 3, and 35 ± 3 nm, respectively. When the water content was decreased further to 20.0 wt%, a mixture of rods and vesicles reappeared in the solution. These results show that vesicle sizes are dependent on the water content in solution, and can be changed reversibly by changing the water content [28].

Acknowledgments

The authors would like to thank the National Natural Science Foundation of China (81222047, 81673384).

References

1. Brinkhuis RP, Rutjes FPJT, van Hest JCM (1973) Polymeric vesicles in biomedical applications. Poly Chem 2:1449–1462
2. Oltra NS, Nair P, Discher DE (2014) From stealthy polymersomes and filomicelles to "self" Peptide-nanoparticles for cancer therapy. Annu Rev Chem Biomol Eng 5:281–299
3. Mai Y, Eisenberg A (2012) Self-assembly of block copolymers. Chem Soc Rev 41:5969–5985
4. Dionzou M, Morere A, Roux C et al (2016) Comparison of methods for the fabrication and the characterization of polymer self-assemblies: what are the important parameters? Soft Matter 12:2166–2176
5. Lee JC, Bermudez H, Discher BM et al (2001) Preparation, stability, and in vitro performance of vesicles made with diblock copolymers. Biotechnol Bioeng 73:135–145
6. Theogarajan L, Desai S, Baldo M et al (2008) Versatile synthesis of self-assembling ABA triblock copolymers with polymethyloxazoline A-blocks and a polysiloxane B-block decorated with supramolecular receptors. Polym Int 57:660–667
7. Mertins O, da Silveira NP, Pohlmann AR (2009) Electroformation of giant vesicles from an inverse phase precursor. Biophys J 96:2719–2726
8. Eissa AM, Smith MJP, Kubilis A et al (2013) Polymersome-forming amphiphilic glycosylated polymers: synthesis and characterization. J Polym Sci Part A 51:5184–5193
9. Patil YP, Kumbhalkar MD, Jadhav S (2012) Extrusion of electroformed giant unilamellar vesicles through track-etched membranes. Chem Phys Lipids 165:475–481
10. Oh NM, Oh KT, Youn YS et al (2012) Poly (L-aspartic acid) derivative soluble in a volatile organic solvent for biomedical application. Colloids Surf B Biointerfaces 97:190–195
11. Kita-Tokarczyk K, Grumelard J, Haefele T et al (2005) Block copolymer vesicles-using concepts from polymer chemistry to mimic biomembranes. Polymer 46:3540–3563
12. Matter Y, Enea R, Casse O et al (2011) Amphiphilic PEG-b-PMCL-b-PDMAEMA triblock copolymers: from synthesis to physicochemistry of self-assembled structures. Macromol Chem Phys 212:937–949
13. Kabanov AV, Bronich TK, Kabanov VA et al (1998) Spontaneous formation of vesicles from complexes of block ionomers and surfactants. J Am Chem Soc 120:9941–9942
14. Schuster TB, de Bruyn Ouboter D, Bruns N et al (2001) Exploiting dimerization of purely peptidic amphiphiles to form vesicles. Small 7:2158–2162
15. Napoli A, Tirelli N, Wehrli E et al (2002) Lyotropic behavior in water of amphiphilic ABA triblock copolymers based on poly(propylene sulfide) and poly(ethylene glycol). Langmuir 18:8324–8329
16. Nardin C, Thomas H, Jörg L et al (1999) Polymerized ABA triblock copolymer vesicles. Langmuir 16:1035–1041
17. Xu J, Zhao Q, Jin Y et al (2014) High loading of hydrophilic/hydrophobic doxorubicin into polyphosphazene polymersome for breast cancer therapy. Nanomedicine 10:348–349
18. Xiong XY, Li YP, Li ZL et al (2007) Vesicles from Pluronic/poly(lactic acid) block copolymers as new carriers for oral insulin delivery. J Control Release 120:11–17
19. Yewle J, Wattamwar P, Tao Z et al (2016) Progressive saturation improves the encapsulation of functional proteins in nanoscale polymer vesicles. Pharm Res 33:1–17
20. Ahmed F, Discher DE (2004) Self-porating polymersomes of PEG-PLA and PEG-PCL: hydrolysis-triggered controlled release vesicles. J Control Release 96:37–53
21. Cohen R, Steiner A et al (2013) Chemical and physical characterization of remotely loaded bupivacaine liposomes: comparison between large multivesicular vesicles and small unilamellar vesicles. J Mater Chem B 1:4619–4627
22. Parnell AJ, Tzokova N, Topham PD et al (2009) The efficiency of encapsulation within surface rehydrated polymersomes. Faraday Discuss 143:29–46
23. Ruysschaert T, Sonnen AFP, Haefele T et al (2005) Hybrid nanocapsules: interactions of ABA block copolymers with liposomes. J Am Chem Soc 127:6242–6247
24. Yin H, Kang HC, Huh KM et al (2012) Biocompatible, pH-sensitive AB(2) miktoarm polymer-based polymersomes: preparation, characterization, and acidic pH-activated nanostructural transformation. J Mater Chem 22:91968–19178

25. Nardin C, Meier W (2002) Hybrid materials from amphiphilic block copolymers and membrane proteins. J Biotechnol 90:17–26
26. Ahmed F, Hategan A, And DED et al (2003) Block copolymer assemblies with cross-link stabilization: from single-component monolayers to bilayer blends with PEO-PLA. Langmuir 19:6505–6511
27. Regenbrecht M, Akari S, Förster S et al (1999) Fusion of micelles of poly(butadiene-block-2-vinylpyridene) and derivatives on different substrates. Surf Interface Anal 27:418–421
28. And LL, Eisenberg A (2001) Thermodynamic size control of block copolymer vesicles in solution. Langmuir 18:6804–6811

Chapter 18

Partially Polymerized Phospholipid Vesicles for Efficient Delivery of Macromolecules

Megha Goshi, Nicholas Pytel, and Tamer Elbayoumi

Abstract

Lipid-based vesicles, namely cationic liposomal nanocarriers have been recognized early on as one of the most attractive delivery systems for RNA, protein, and oligonucleotides. Despite several advantages of conventional liposomal carriers for therapeutic macromolecules, their flexible and unsupported bilayered membranes can pose some limitations for efficient intracellular delivery of their sensitive cargos. Hence, polymerized liposomes, a concept conceived about 20 years ago, might offer structural solution to current in vivo efficiency concerns affecting traditional cationic phospholipid vectors, especially when adapted to enable superior loading and stability, typically required for effective intracellular delivery of proteins and polynucleotides.

Our recent approach attempted to remodel polymerized liposomal vesicles—specifically their semi-rigid membrane structure—to create block-polymerized bilayered vesicles (generally composed of DOTAP: DOPE: Diyne PC in 0.1:1:1 molar ratio). Adopting a modified freeze-dry-rehydration technique allowed modular reassembly of such partially polymerized lipidic vesicles (PPL). Different prototype cationic partially polymerized liposomal preparations (PPLs) were successfully developed (mean particle size range 150–300 nm), demonstrating enhanced physicochemical stability and loading capacity, thus promoting improved intracellular delivery of model RNAi and protein cargos.

Key words Cationic phospholipids, 1,2-Dioleoyl-3-trimethylammonium-propane (DOTAP), 1,2-Dioleoyl-*sn*-glycero-3-phosphoethanolamine (18:1 DOPE), 1,2-Bis(10,12-tricosadiynoyl)-*sn*-glycero-3-phosphocholine (23:2 Diyne PC), Freeze-dry-rehydration vesicles (FRV)

1 Introduction

Liposomes are one of the most widely studied nanoscale carrier systems. They have been extensively investigated for both controlled release and targeted delivery of many therapeutic macromolecules, thanks to their favorable pharmaceutical characteristics, such as biocompatibility and the ease of large-scale production. One of the major limitations facing the therapeutic applications of proteins, and polynucleotides, is the inefficiency of delivering these molecules to target cell populations in vivo. This is mostly due to the instability of these molecules as well as their poor cellular uptake

Volkmar Weissig and Tamer Elbayoumi (eds.), *Pharmaceutical Nanotechnology: Basic Protocols*, Methods in Molecular Biology, vol. 2000, https://doi.org/10.1007/978-1-4939-9516-5_18,

and pharmacokinetic profiles in vivo [1]. Much research efforts have been devoted towards addressing such concerns of long-term physicochemical stability, both in vitro and in vivo, to optimize liposomal system designs for macromolecule delivery. Interest in polymeric lipids arose a few years ago as an option to combine in one system both liposomes and polymer characteristics. Among these lipidic polymer strategies, Lipid polymerization attracted quite an early yet renewable attention, specifically for large molecular weight drug cargos. Controlled polymerization reactions leads to stable covalent bonding between lipids chains, thus further enhancing the noncovalent interaction that keeps lipid lamellar phase formed. Henceforth, polymerized liposomes carry an important effect on the stability of the whole system, especially in circulation upon systemic administration [2, 3]. Here, we are presenting our different strategies for formulating siRNA-loaded polymerized lipid particles. In addition, preliminary exploratory studies about the stability and the in vitro traceability and efficacy of the constructs will be discussed in some detail.

The utilization of polymerizable lipid vesicles for DNA, RNA, and oligonucleotides delivery has been repeatedly studied for both vaccine and therapeutic applications [4]. Most of these lipid chain polymerization reactions relied on incorporation of diacetylininc lipids in the liposomal matrix, with the process initiated mostly via ultraviolet (UV) light irradiation or reactive oxygen species. Nevertheless, most chain-polymerization reactions within liposomes would adversely affect the loading efficiency and stability of the drug cargo, especially in case of sensitive molecules, such as polypeptides and proteins. Another recent approach focused on lipid-based formulation relying on combinatorial branched synthetic lipidic molecules [5–7]. These complexed lipidic vesicles, namely lipidoids, have demonstrated marked cell transfection efficiency as siRNA delivery vehicle, owing to their distinctive structural stability after systemic administration [8–10]. Yet, despite their therapeutic potential, lipidoids suffered from a few technical drawbacks that may significantly hinder furthering their broad in vivo clinical application for proteins and polynucleotides. The lipidoids, as a general drug delivery platform, can only be produced via complex custom synthesis schemes, with very low yield of the proper configuration of the monomers to be used to constitute the final vector particles [11].

Based on analogous structural analysis, we proposed simple and modular polymerized diacetylene liposomes as vehicles for intracellular delivery of macromolecules, which would potentially carry similar advantages to the polymeric lipid bilayered vesicle system. Unlike fully polymerized liposomes described before [15, 16], our modular design of the polymeric-phospholipid-cholesterol would allow for separate and sequential processes for vesicle assembly and protein/polynucleotide loading, which can facilitate any further upscaling in production. Our modular polymerizable liposomal membrane design would offer not only enhanced stability and protection of the encapsulated "big and sensitive" cargo but also gradual breakdown of these membrane modules, by various digestive organelles inside the cell. This cationic polymeric liposome system may not only facilitate delayed intracellular release of loaded macro-therapeutics but also can possibly utilize the lipid transport and metabolism machinery inside the cell, to process the phospholipid component of this delivery system.

The current work details our attempt to remodel polymerized liposomal carrier formulation and membrane structure, in order to create a block-polymerized backbone that would allow for enhanced loading, stability, and intracellular delivery of model RNA and protein cargos. Our developed partially/block-polymerized liposome (PPL) formulations (composed of 5 mol% DOTAP, 40 mol% DOPE, and 50 mol% of diyne PC, with or without 5 mol% cholesterol) were prepared using a step-wise approach: reactive oxygen species (ROS)-mediated polymerization, followed by modified freeze-dry-rehydration method (modified FRVs) for macromolecule loading. Prototype PPLs, with unique membrane/surface chemistry, were successfully prepared, as confirmed by both UV and FTIR-spectroscopy, and physicochemically characterized (Table 1) with a broad size range (150–300 nm) [12, 13]. The block-remodeling of cationic polymerized liposomal carriers (mean $\zeta = +17 \pm 2.3$ mv) produced a reinforced liposomal membrane backbone that would allow for enhanced loading, stability (shelf-life and biological) and intracellular delivery of the macromolecular cargos [14, 15]. Moreover, this re-enforced polysome membrane can possibly permit steering intracellular release or trafficking of DNA/ RNA and proteins [15, 16].

It was possible to further control the size via repeated extrusion into different populations, depending primarily on the encapsulated material (e.g. approx. 150–200 nm for empty and siRNA-loaded, and up to 300 nm for model protein-loaded PPLs). Our developed PPLs also demonstrated good physical storage stability (<15% particle size change over 60 days) and superior in vitro stability (<10 leakage of FITC-labeled BSA cargo) following a 7-day incubation in simulated biological media, in comparison to nonpolymerized diyne PC-containing liposome controls. Fluorescence microscopy analysis revealed marked and rapid intracellular localization of labeled cationic PPL, only within 4 h of co-incubation with murine colon carcinoma (C26) and breast cancer (4T1) cells.

The newly developed sequential design for structured cationic diyne PC-liposomal carriers, exhibiting reinforced membrane structure via block/partial polymerization, carries a considerable potential for macromolecule delivery. Such modular polymeric liposome system would allow for enhanced physical and biological

Table 1
Physical characterization and stability of polymerizable Diyne liposomes

Formulation (wt%)	Particle size D1 (nm)	Particle size D14 (nm)	Size change (%)
50:50 DC9,8 PC: DOPE:	1198 ± 530 **P:382 ± 105**	1756 ± 490 **P:427 ± 130**	+46.6% **+11.8%**
50:45:5 DC9,8 PC: DOPE: Oleylamine	697 ± 280 **P:184 ± 39**	962 ± 374 **P:216 ± 35**	+38% **+17.4**

stability and enhanced intracellular delivery, which would ultimately result in improved active protein concentration/target gene knockdown inside target cancer cells.

2 Materials

2.1 Preparation of Partially Polymerized Liposomes

1. Chloroform (100%, dry).
2. 1,2-Bis(10,12-tricosadiynoyl)-*sn*-glycero-3-phosphocholine (23:2 Diyne PC or [DC(8,9)PC]) (Avanti Polar Lipids, Inc., Alabaster, AL). Dissolve 0.219 M 23:2 Diyne PC in 10 mL chloroform to make 20 mg/mL 23:2 Diyne PC stock solution. Store at −20 °C.
3. 1,2-Dioleoyl-*sn*-glycero-3-phosphoethanolamine (18:1 (Δ9-Cis) PE or DOPE) (Avanti Polar Lipids, Inc., Alabaster, AL). Dissolve 0.269 M DOPE in 10 mL chloroform to make 20 mg/mL DOPE stock solution. Store at −20 °C.
4. 1,2-Dioleoyl-3-trimethylammonium-propane (chloride salt) (18:1 TAP or DOTAP) (Avanti Polar Lipids, Inc., Alabaster, AL). Dissolve 0.288 M DOTAP in 10 mL chloroform to make 20 mg/mL DOTAP stock solution. Store at −20 °C.
5. Ethanol (200% proof, denatured).
6. 25 mL pear-shaped and round-bottom glass flasks that fit rotary evaporator spout for organic/co-solvent evaporation.
7. Rotary evaporator with vertical coiled condenser, RE100-Pro (Scilogix, LLC, Rocky Hill, CT) with rotation speed control, connected to a dry-vacuum pump capable of providing at least 100 mtorr of vacuum.
8. Stock HEPES (2×) buffered saline, pH 7.05. Dissolve 280 mM Sodium Chloride (16.4 g), 50 mM HEPES, free acid (11.9 g), 1.5 mM Na_2HPO_4 (0.21 g) in 100 mL of MQ water. Titrate to pH 7.1 with 5 M NaOH, adjust final volume to 1 L. Store at 4 °C.

9. Stock 0.1 M Citrate (1×) buffer, pH 5.5. Dissolve 400 mM sodium citrate dihydrate (12.04 g), 60 mM citric acid (11.34 g), 800 mL of MQ water. Titrate to pH 5.5 with 0.1 N HCl, adjust final volume to 1 L using MQ water. Store at 4 °C.
10. Bovine serum albumin (BSA, 66,400 Da Molecular weight).
11. 0.45 μm pore size, Thermo Scientific™ Nalgene™ 25 mm Nylon Syringe Filter (Thermo Fisher Scientific, Hampton, NH).
12. Benchtop lyophilyzer, FreeZone 4.5 (Labconco, Kansas City, MO).
13. Nitrogen gas source, with flow meter regulator, adjustable from 10 to 60 CF/H.
14. Nitrogen gas-operated LIPEX™ extruder (Northern Lipids Inc., Burnaby, BC, CA).
15. LIPEX™—compatible Polycarbonate filter disks—size 100 nm and 200 nm (Northern Lipids Inc., Burnaby, BC, CA).
16. Noncontact Infrared thermometer with laser sight (Temperature range: −50 °C to 38 °C, and minimum accuracy ±1.5 °C), for accurate measurement of external temperature of the LIPEX™ extruder barrel.
17. Weigh balance (up to 0.001 mg in precision for accuracy).
18. Pipette(s) capable of dispensing at 10 μL, 500 μL, and 1 mL.
19. 5 mL glass vials.
20. Milli-Q (MQ) water.

2.2 Spectroscopic Analysis of Prepared Partially Polymerized Liposomes

1. Multichannel (12 or 8 channel) pipette(s) capable of dispensing at 50 μL, and 300 μL.
2. Speed adjustable/digital orbital mixer.
3. Stock HEPES (2×) buffered saline, pH 7.05. Dissolve 280 mM Sodium Chloride (16.4 g), 50 mM HEPES, free acid (11.9 g), and 1.5 mM Na_2HPO_4 (0.21 g) in 100 mL of MQ water. Titrate to pH 7.1 with 5 M NaOH, adjust final volume to 1 L. Store at 4 °C.
4. 96-well microplates: clear polystyrene plates with clear flat-bottom (300 μL well capacity), compatible with UV/visible spectrophotometer (Corning Inc., Corning, NY).
5. Fluorescence plate reader with excitation 380–410 nm and emission 495–525 nm filter pair, Victor X3 Multilabel microplate reader (PerkinElmer, Santa Clara, CA).

2.3 Physicochemical Characterization of Prepared Partially Polymerized Liposomes

1. Pipette(s) capable of dispensing at 10 μL, 500 μL, and 1 mL.
2. 5 mL glass test tubes.
3. Milli-Q (MQ) water.
4. Malvern Zetasizer Nano ZS (Malvern Instruments, Westborough, MA).
5. Disposable folded capillary (electrophoretic) cells for zeta potential measurements (Malvern Instruments, Westborough, MA).
6. Disposable low volume (1.5 mL capacity) 12 mm square polystyrene cuvettes, for particle size analysis.

3 Methods

3.1 Preparation of Empty and Protein-Loaded Partially Polymerized Liposomal Nanocarriers

Prepare all PPL formulation using only clean glassware. Thoroughly clean the glassware and spatulas with concentrated nitric acid followed by ethanol. Make sure no residue of whitish phospholipids or drug remains in the glassware. Furthermore, use MQ water during the entire formulation processes to guarantee purified grade final product.

1. Turn on the hot plate and adjust to 30 °C. Warm clean 25 mL beaker on the hot plate for 5 min, filled with 1× HEPES buffered saline adjusted to pH 7.4.
2. In a 25 mL pear-shaped glass flask, add phospholipid matrix components, 90.72 mMol of DOTAP: DOPE: Diyne PC in 1.0:1.0 M (as 1.52 mL from 20 mg/mL of 18:1 DOPE stock solution in chloroform, and 2.27 mL from 20 mg/mL of 23:2 Diyne PC stock solution in chloroform) (*see* **Note 1**).
3. Connect pear-shaped glass flask to the rotary evaporator, and slowly evaporate organic solvent under 100 mtorr (26 Hg) vacuum set at 50–60 rpm rotation and 40 °C water bath temperature for approx. 60 min (*see* **Notes 2** and **3**).
4. Release vacuum pressure and carefully disconnect flask (*see* **Note 4**).
5. Using 1 mL pipette, gradually add 2 mL of warm citrate buffer, pH 5.5, onto the warm mixture inside the 25 mL pear-shaped glass flask, mixing thoroughly but slowly, using vortex mixer at about 800 rpm for 10 min, or until the entire lipid film on the glass has been dispersed in buffered solution (*see* **Note 5**).
6. Add 9.07 mM of $Na_2S_2O_7/K_2S_2O_8$ ROS mixture in 1:1 M ratio (as 1.08 mg of $Na_2S_2O_7$; and 1.22 mg of $K_2S_2O_8$) directly to the lipid dispersion inside the pear-shaped flask.

7. Mix the dispersed contents of the flask thoroughly, seal under nitrogen gas, and store in the dark at TR for at least 12 h or overnight (*see* **Note 6**).
8. Using 1 mL pipette, transfer the entire PPL dispersion carefully onto the top of a 8.3 mL Sephadex G-25 (PD-10) desalting column, pre-rinsed in 10 mM ascorbic acid (as 44.0 mg) dissolved in 25 mL of 1× HEPES buffered saline, pH 7.1, employed here as equilibration buffer (*see* **Note 7**).
9. Place the sample-loaded PD-10 desalting column into a new 50 mL collection tube, using the provided adapter, and elute by centrifugation at 700 × g for 2 min, to remove nonreacted species.
10. Collect the PPL dispersion eluate and transfer all volume into a new round-bottom glass flask.
11. Using 1 mL pipette, dilute PPL dispersion 1:1 vol:vol using 10 mM ascorbic acid (as 17.6 mg) dissolved in 10 mL of 1× HEPES buffer, pH 7.1.
12. For macromolecule loading of PPL (e.g. using bovine serum albumin, BSA, as model protein), dilute PPL dispersion 1:1 vol:vol using 0.3 M BSA (0.2 g) in 10 mM ascorbic acid (17.6 mg) dissolved in 10 mL of 1× HEPES buffer solution, pH 7.1.
13. Freeze the diluted PPL dispersion by immersing the flask in liquid nitrogen and vacuum-freeze drying with FreeZone 4.5 Lyophilizer with $p < 200 \times 10^{-3}$ mbar, condenser temperature ≤ -50 °C), overnight.
14. Using 1 mL pipette, carefully and gradually hydrate the lyophilized PPL film in the round-bottom flask, with continuous vortex mixing, utilizing a total of 2.0 mL of 1× HEPES buffered saline, pH 7.1. Leave aside to equilibrate for 1 h at RT.
15. Transfer the final PPL samples (empty or drug-loaded) to the LIPEX™ extruder, pass twice under 300 PSI nitrogen gas pressure, using first 0.4 mm filter disk. Run another two passes using the 200 nm filter (*see* **Note 8**).
16. Store the PPL formulations at 4 °C for later use.

3.2 Spectroscopic Analysis of Empty Partially Polymerized Liposomal Nanocarriers

The efficiency of the ROS-mediated 23:2 diyne-polymerization reaction of PPL is estimated, in comparison with ROS-unreacted control diyne-liposomes, by measuring the absorbance of the liposomal suspensions in the UV/visible region (520 nm), utilizing the Victor X3 fluorescence microplate reader (PerkinElmer, Santa Clara, CA) (Fig. 1).

1. Measure 200 μL vol. from each liposomal preparation, now dispersed in 1× HEPES buffered saline adjusted to pH 7.1,

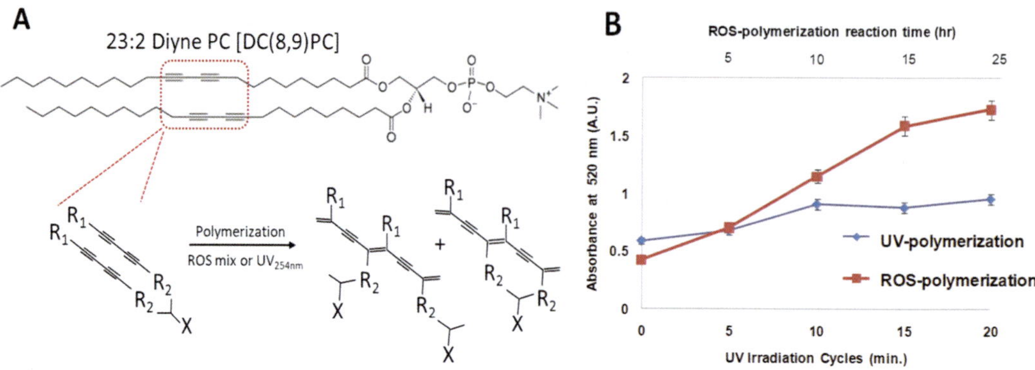

Fig. 1 Polymerizable diacetylene phospholipids. Schematics (**a**); and UV-spectroscopic characterization of Diyne PC-liposomes (**b**)

after filtration using 0.45 μm pore size polycarbonate membrane syringe filters, and then transfer into first row of corresponding wells on the clear polystyrene plates with clear flat-bottoms (allow for at least three replicates of each sample).

2. In all subsequent wells, transfer 100 μL vol. of 1× HEPES buffered saline, pH 7.1, using a multichannel pipette capable of dispensing 300 μL.
3. Perform twofold serial vol./vol. dilution for all samples, using the multichannel pipette and repeat this twofold dilution process four times (*see* **Note 9**).
4. Transfer the sample micro-plate into temperature-controlled orbital shaker, and incubate at room temperature (RT), for 30 min, with continuous shaking at 100 rpm.
5. Finally, measure the absorbance of all serially diluted samples in the microplate ($\lambda = 520$ nm) using Victor X3 Multilabel microplate reader (PerkinElmer, Santa Clara, CA).

3.3 Physical Characterization of Empty and Drug-Loaded Partially Polymerized Liposomal Formulations

Produced PPL formulations are characterized for particle size and size distribution using the dynamic light scattering (DLS) technique with a Malvern Zetasizer Nano ZS (Malvern Instruments, Westborough, MA) at 273° fixed angle and at 23 °C temperature.

1. Dilute each PPL formulation, for particle size analysis, using MQ water at about 50-folds vol/vol, in disposable polystyrene cuvettes. The numbered average particle hydrodynamic diameter and the polydispersity index (DPI) will be determined (*see* **Note 10**).
2. For the zeta potential, dilute PLL samples in MQ water, pH 6.8, at 100- to 200-folds vol/vol, then employ a 1 mL syringe, horizontally, to inject the almost transparent solution carefully inside the folded capillary electrophoretic cell of the Malvern Zetasizer Nano ZS, while making sure to avoid

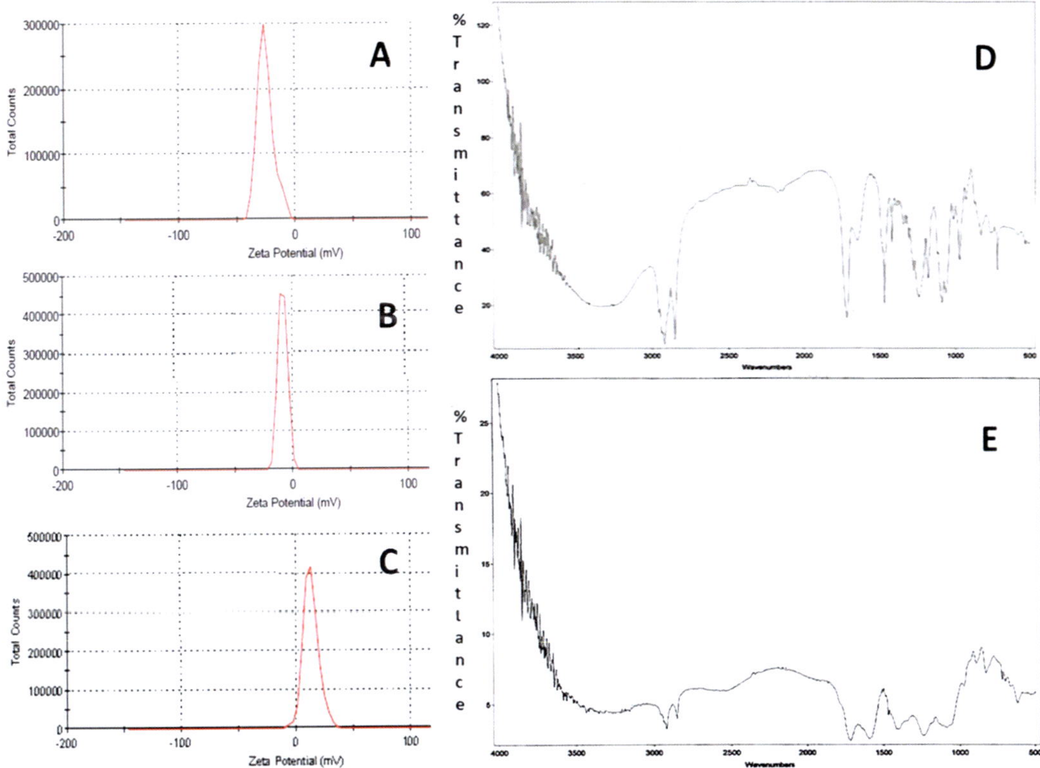

Fig. 2 Characterization of polymerizable diyne liposomes. Zeta-potential measurements (**a–c**) of anionic (**a**), cationic (**b**), and heat-extruded/cationic (**c**) polymerized liposomes; and FTIR analysis of Diyne liposomes (**d**, **e**), showing marked reduction in C≡C bands before (**d**) and after (**e**) overnight ROS-mediated polymerization of DC8,9 PC liposomes

forming any air bubbles. The average surface charge will be measured (Fig. 2a–c).

4 Notes

1. For cationic PPL containing 5 M% of DOTAP in formulations, add 4.54 mM of OA (as 0.23 mL from 20 mg/mL of DOTAP stock solution in chloroform) directly as a part of the phospholipid mixture, all dissolved in organic solvent.
2. While the vacuum is best adjusted based on each solvent, for optimized solvent removal via the rotary evaporator system, apply the 20/40/60 "technical" rule, which correlates to at least 20 °C difference in temperature between the system's main components. Use operating bath temperature of at least 40 °C (glass transition temperature of main phospholipid, diyne PC), to yield a solvent vapor temperature of 20 °C, which is subsequently condensed at about 0 °C (using ice-water to cool the condenser).

3. For efficient and complete evaporation of organic solvent, make sure to keep the connected pear-shaped flask, is tilted at about 60–45° to the plane of the surrounding warm water bath. For improved ethanol evaporation, also increase the flask's rotation speed to about 90 rpm.
4. Complete removal of organic solvent is confirmed when a translucent off-white dry film residue remains in the flask, which may get somewhat opaque as the flask temperature cools down. The dried lipid film must be clear from any suspending white or yellowish lipid clumps or precipitates. Otherwise, redissolve in another 4 mL of chloroform, and repeat the evaporation process, using slightly lower water bath temperature (−1.0 °C), and vacuum settings.
5. Optional: briefly put pear-shaped flask in bath sonicator (only for 2–3 min) to dislodge resistant dried lipid film remains, present on the inside glass wall of flask.
6. A minor color change of the lipid dispersion into purplish-white, may be observed, which is indicative of the progression of the polymerization process.
7. Preparation and equilibration of PD-10 desalting column need to be performed prior to adding the liposome sample (not to exceed 2.5 mL). Briefly after pouring the column storage solution, remove top filter using forceps, followed by cutting the sealed end of the column, prior to placing the PD-10 column into a 50 mL collection using the column adapter. Fill up the column with equilibration buffer (10 mM ascorbic acid, dissolved as 44.0 mg in 25 mL of 1× HEPES buffered saline, pH 7.1), allow it to enter the packed bed completely, and repeat three times before discarding the flow-through. Finally, fill up the column with equilibration buffer and spin down at 1000 × *g* for 2 min, before discard the flow-through.
8. Prewarm the thermobarrel LIPEX™ extruder to about 48 °C (measured externally using noncontact IR thermometer), before running the PPL sample, to guarantee smooth flow-pass through the filter. Make sure to run sample through the larger pore-size filter disk first, before the smaller pore size one, to avoid clogging of the filter disk.
9. Perform the twofold serial dilution step by sequentially transferring 100 μL vol. of first row original liposomal dispersion into the following corresponding row of wells, then mixing via 3× repeated pipetting, followed by transferring the same volume (100 μL) into the corresponding subsequent row of wells.
10. Optional: for enhanced numerical accuracy, compare measured average particle hydrodynamic diameter and the polydispersity index (DPI) for un-reacted and partially polymerized liposomes, consistently using the volume analysis/volume statistics function, available in the output analysis parameters.

References

1. Santel A, Aleku M, Keil O, Endruschat J, Esche V, Durieux B, Loffler K, Fechtner M, Rohl T, Fisch G, Dames S, Arnold W, Giese K, Klippel A, Kaufmann J (2006) RNA interference in the mouse vascular endothelium by systemic administration of siRNA-lipoplexes for cancer therapy. Gene Ther 13:1360–1370
2. Nabel GJ, Nabel EG, Yang ZY, Fox BA, Plautz GE, Gao X, Huang L, Shu S, Gordon D, Chang AE (1993) Direct gene transfer with DNA-liposome complexes in melanoma: expression, biologic activity, and lack of toxicity in humans. Proc Natl Acad Sci U S A 90:11307–11311
3. Gill DR, Southern KW, Mofford KA, Seddon T, Huang L, Sorgi F, Thomson A, MacVinish LJ, Ratcliff R, Bilton D, Lane DJ, Littlewood JM, Webb AK, Middleton PG, Colledge WH, Cuthbert AW, Evans MJ, Higgins CF, Hyde SC (1997) A placebo-controlled study of liposome-mediated gene transfer to the nasal epithelium of patients with cystic fibrosis. Gene Ther 4:199–209
4. Auguste DT, Furman K, Wong A, Fuller J, Armes SP, Deming TJ, Langer R (2008) Triggered release of siRNA from poly(ethylene glycol)-protected, pH-dependent liposomes. J Control Release 130:266–274
5. Akinc A, Goldberg M, Qin J, Dorkin JR, Gamba-Vitalo C, Maier M, Jayaprakash KN, Jayaraman M, Rajeev KG, Manoharan M, Koteliansky V, Rohl I, Leshchiner ES, Langer R, Anderson DG (2009) Development of lipidoid siRNA formulations for systemic delivery to the liver. Mol Ther 17:872–879
6. Frank-Kamenetsky M, Grefhorst A, Anderson NN, Racie TS, Bramlage B, Akinc A, Butler D, Charisse K, Dorkin R, Fan Y, Gamba-Vitalo C, Hadwiger P, Jayaraman M, John M, Jayaprakash KN, Maier M, Nechev L, Rajeev KG, Read T, Rohl I, Soutschek J, Tan P, Wong J, Wang G, Zimmermann T, de Fougerolles A, Vornlocher HP, Langer R, Anderson DG, Manoharan M, Koteliansky V, Horton JD, Fitzgerald K (2008) Therapeutic RNAi targeting PCSK9 acutely lowers plasma cholesterol in rodents and LDL cholesterol in nonhuman primates. Proc Natl Acad Sci U S A 105:11915–11920
7. Nguyen DN, Chen SC, Lu J, Goldberg M, Kim P, Sprague A, Novobrantseva T, Sherman J, Shulga-Morskaya S, de Fougerolles A, Chen J, Langer R, Anderson DG (2009) Drug delivery-mediated control of RNA immunostimulation. Mol Ther 17:1555–1562
8. Wu SY, McMillan NA (2009) Lipidic systems for in vivo siRNA delivery. AAPS J 11 (4):639–652
9. Whitehead KA, Langer R, Anderson DG (2009) Knocking down barriers: advances in siRNA delivery. Nat Rev Drug Discov 8:129–138
10. Huang YH, Bao Y, Peng W, Goldberg M, Love K, Bumcrot DA, Cole G, Langer R, Anderson DG, Sawicki JA (2009) Claudin-3 gene silencing with siRNA suppresses ovarian tumor growth and metastasis. Proc Natl Acad Sci U S A 106:3426–3430
11. Akinc A, Zumbuehl A, Goldberg M, Leshchiner ES, Busini V, Hossain N, Bacallado SA, Nguyen DN, Fuller J, Alvarez R, Borodovsky A, Borland T, Constien R, de Fougerolles A, Dorkin JR, Narayanannair Jayaprakash K, Jayaraman M, John M, Koteliansky V, Manoharan M, Nechev L, Qin J, Racie T, Raitcheva D, Rajeev KG, Sah DW, Soutschek J, Toudjarska I, Vornlocher HP, Zimmermann TS, Langer R, Anderson DG (2008) A combinatorial library of lipid-like materials for delivery of RNAi therapeutics. Nat Biotechnol 26:561–569
12. Weissmann G, Finkelstein M (1980) Uptake of enzyme-bearing liposomes by cells in vivo and in vitro. In: Gregoriadis G, Allison AC (eds) Liposomes in biological systems. Wiley, New York, NY, pp 153–162
13. Kirby C, Clarke J, Gregoriadis G (1980) Effect of the cholesterol content of small unilamellar liposomes on their stability in vivo and in vitro. Biochem J 186:591–598
14. Torchilin VP, Weissig V (2003) Liposomes: a practical approach, 2nd edn. Oxford University Press, Oxford
15. Chen H, Torchilin V, Langer R (1996) Lectin-bearing polymerized liposomes as potential oral vaccine carriers. Pharm Res 13:1378–1383
16. Alonso-Romanowski S, Chiaramoni NS, Lioy VS, Gargini RA, Viera LI, Taira MC (2003) Characterization of diacetylenic liposomes as carriers for oral vaccines. Chem Phys Lipids 122:191–203

Chapter 19

Fabrication of Nanostructured Lipid Carriers (NLC)-Based Gels from Microemulsion Template for Delivery Through Skin

Medha D. Joshi, Rashmi H. Prabhu, and Vandana B. Patravale

Abstract

Nanostructured lipid carriers (NLC) represent the novel and widely explored generation of lipid nanoparticles. These are the second-generation solid lipid nanoparticles (SLN) developed with the aim to overcome limitations of SLN mainly with respect to limited drug loading and drug leakage during its storage. NLC are fabricated by mixing solid lipids with spatially incompatible (liquid) lipids leading to nanoparticulate structures with improved drug loading and controllable release properties. Out of the numerous methods reported to prepare NLC, microemulsion template (ME) technique is the most simple and preferred method. This methodology of preparation of lipid nanoparticles obviates the need for specialized equipment and energy to generate NLC, enables achieving desirable particle size of nanoparticles by modulating the size of the emulsion droplet, and is also feasible for easy scale-up. This chapter describes microemulsion template technique for fabrication of NLC based gel for topical delivery, particularly with respect to its method of preparation and product analysis.

Key words Nanostructured lipid carriers, Microemulsion, Gels, Topical delivery

Abbreviations

GRAS	Generally recognized as safe
HPLC	High-performance liquid chromatography
ME	Microemulsion
MWCO	Molecular weight cut-off
NLC	Nanostructured lipid carriers
o/w	Oil-in-water
SLN	Solid lipid nanoparticles

Volkmar Weissig and Tamer Elbayoumi (eds.), *Pharmaceutical Nanotechnology: Basic Protocols*, Methods in Molecular Biology, vol. 2000, https://doi.org/10.1007/978-1-4939-9516-5_19, © Springer Science+Business Media, LLC, part of Springer Nature 2019

1 Introduction

Nanostructured lipid carriers (NLC) have emerged as potential drug delivery systems for treatment of numerous disorders [1]. NLC are the second generation of lipid nanoparticles obtained by modification of solid lipid nanoparticles (SLN) [2]. SLN are similar to an oil-in-water (o/w) emulsion wherein the liquid lipid (oil) of the emulsion has been replaced by a solid lipid having a liquid-to-solid phase transition well above the body temperature (37 °C) [3]. SLN are lipid nanoparticles produced by one of the following techniques namely high-pressure homogenization, high shear homogenization, microemulsion (ME) template technique, solvent emulsification evaporation technique, solvent displacement technique, solvent emulsification diffusion method, phase inversion and membrane contractor technique [4, 5]. The major disadvantages associated with SLN are limited drug loading, the risk of gelation, physical instability, and drug leakage during storage caused due to lipid polymorphism [6]. NLC were developed to overcome these shortcomings of SLN. NLC consist of a solid lipid (s) matrix with a high content of spatially incompatible liquid lipid (s). The resultant lipid particulate matrices show depression in melting point as compared to the original solid lipid and eventually, they remain solid at body temperature [2, 7]. NLC have been considered as an efficient alternative to liposomes and nanoemulsions due to their various advantageous features such as the use of low-cost excipients, ease of manufacturing and scale-up, particulate nature of nanocarrier, high drug loading, good physical stability and ability to sustain release of the drug [7].

Generally, NLC are composed of lipids which are physiological and biodegradable in nature, having low systemic toxicity, and low cytotoxicity [8]. Lipids having an approved status for topical application or those which are employed as excipients in commercially available pharmaceutical topical or cosmeceutical preparations are preferred. The topical application of drug-loaded NLC causes close contact of these nanocarriers with stratum corneum due to its nanosized lipid particles and also increases drug penetration into mucosa or skin. Controlled release of drug from these nanocarriers is achievable due to their solid lipid matrix. Control of drug release is essential to prolong the release of drug from carrier system, reduce systemic absorption, and minimize direct exposure to a drug which is irritant or causes allergic reactions. NLC affords occlusive properties to skin as a result of film formation on topical application. These properties ascertain the successful applicability of NLC as potential topical drug delivery system [9, 10].

ME template method is the simplest and commercially feasible technique to generate NLC. The preliminary solubility of the drug in different solid lipids, oils, surfactants, and solubilizers is screened. Based on the solubility data, selected components are used to determine the boundaries of the microemulsion domains by employing pseudoternary phase diagrams. Briefly, the NLC preparation by ME template method involves melting of solid lipid, addition of liquid lipid, followed by drug solubilization in this mixture to constitute the lipid phase. The aqueous phase is comprised of surfactant, solubilizer, and water. Both the phases are maintained at a temperature above the melting point of the solid lipid and then mixed together to obtain hot microemulsion. This hot microemulsion is then diluted into an excess of cold water (2–3 °C) under stirring. This dilution step leads to the breaking of the microemulsion, converting it into an ultrafine nanoemulsion, which then gradually recrystallizes the internal lipid phase; thereby forming the so-called nanostructures of lipid particles, i.e., NLC (Fig. 1). Except for the heating and cooling to form the microemulsion, ME template method does not necessitate energy during production of NLC nor specialized equipment as required for high-pressure homogenizer. The particle size of NLC is easily controllable by this method as it depends on the size of the emulsion droplet. Since this ME template method is based on the simple procedure of heating, mixing, and stirring; scale-up of this technology is easy and feasible [9–12].

In this chapter, we have described ME template technique for fabrication of NLC loaded with an anti-inflammatory agent, valdecoxib. The NLC suspension is then converted into a gel intended for topical delivery, with emphasis on formulation methodology and characterization techniques [9].

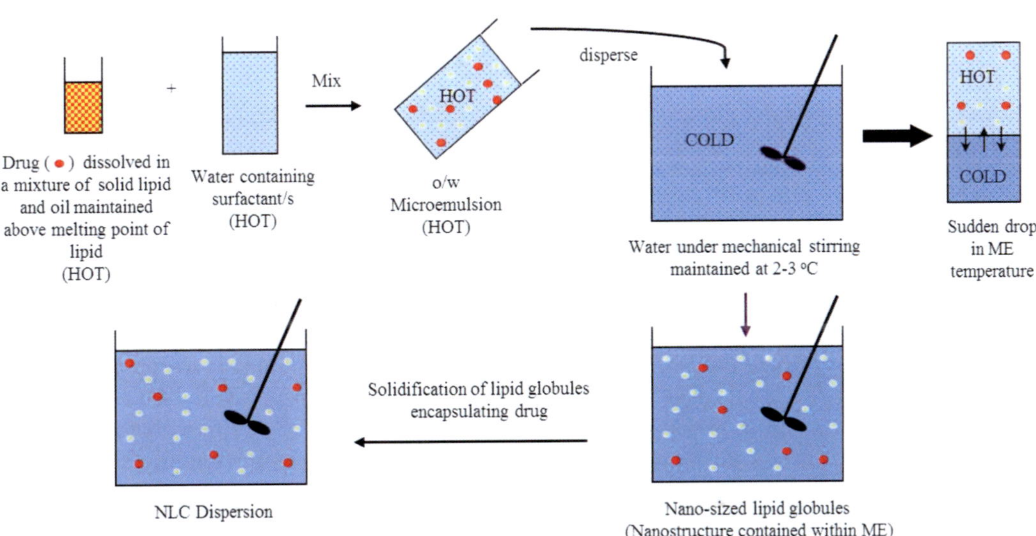

Fig. 1 Schematic representation of NLC production from Microemulsion template

2 Materials

1. Lipophilic drug: Valdecoxib (Cipla Ltd., Mumbai, India).
2. Oils, surfactants and solubilizers: Capmul MCM (glyceryl mono-dicaprylate) Caproyl 90 (propylene glycol monocaprylate, containing 90% monoesters), Labrasol (caprylocaproyl macrogol-8 glycerides) and Transcutol P (purified diethylene glycol monoethyl ether; Solutol HS 15 (macrogol 15 hydroxystearate), Miglyol 812 (caprylic/capric triglyceride), and Softigen 767 (PEG-6 caprylic/capric glycerides); Cremophor EL (PEG 35 castor oil); PEG 400, Tween 20, and Tween 80.
3. Solid lipids: Glyceryl dilaurate; Gelucires [glycerol esters of saturated fatty esters] and Apifil pastills [PEG-8 Beeswax] (Gattefosse, France); Glyceryl monostearate from.
4. Gelling agent: Carbopol Ultrez 10.
5. Nanosep® centrifuge tubes (Pall Life sciences, Mumbai, India).
6. Equipment required for formulation: Water shaker bath, controlled temperature water bath, cyclomixer, and cross-polarizers.
7. Equipment required for characterization: Particle size analyzer (Zetasizer Nano, Malvern Instruments, UK), Centrifuge (Eltek TC4100D Research Centrifuge), UV-VIS spectrophotometer (Shimadzu), Equip-tronic Digital pH meter (Model EQ. 610), Brookfield Synchro-Lectic Viscometer (Model RVT), Dissolution (Modified USP Type II) apparatus, Diffusion cell (Erweka), and High performance liquid chromatography (HPLC).
8. Distilled water or Milli-Q reagent grade water.
9. All other chemicals were of analytical grade.

3 Methods

3.1 Solubility Studies for Screening Components

The solubility of drug needs to be determined in different solid lipids, oils, surfactants, and solubilizers that are preferably approved or generally recognized as safe (GRAS)-listed.

1. Add an excess amount of valdecoxib individually to oils, surfactants, and solubilizers (5 mL each) viz. Miglyol 812, Capmul MCM, Caproyl 90, Tween 20, Tween 80, Labrasol, Cremophor EL, Softigen 767, Solutol HS 15, PEG 400, and Transcutol P in screw-capped tubes. Place these tubes in water shaker bath maintained at 37 °C.
2. After 24 h, centrifuge each sample. Suitably dilute 0.5 mL clear supernatant layer and then analyze the valdecoxib content by a

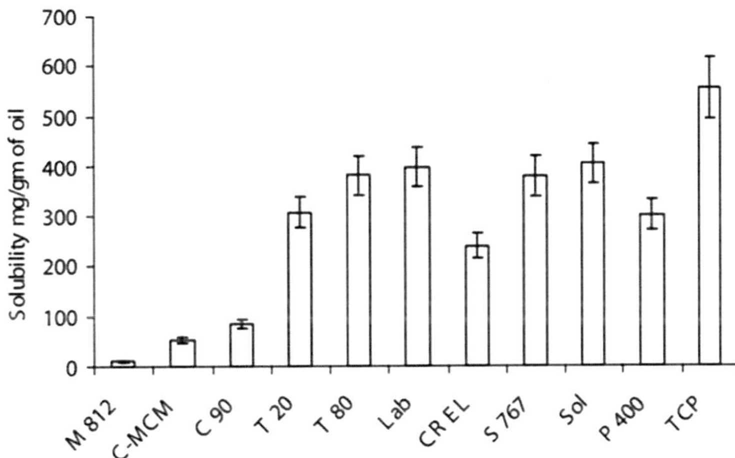

Fig. 2 Solubility of Valdecoxib in different oils, surfactants, and solubilizers, M 812 Miglyol 812, C-MCM Capmul MCM, C 90 Caproyl 90, T 20 Tween 20, T 80 Tween 80, Lab Labrasol, CR EL Cremophor EL, S 767 Softigen 767, Sol Solutol HS 15, P 400 PEG 400, TCP Transcutol P. (Reproduced from ref. 9 with permission from the publisher, Taylor & Francis Ltd.)

validated HPLC method. Determine the oil and solubilizer that solubilizes the highest amount of drug and surfactant that has least drug solubility (*see* Fig. 2, **Notes 1** and **2**).

3. For determining the solubility of the drug in solid lipids individually, take 100 mg valdecoxib in a tube. Add solid lipids viz. Glyceryl Dilaurate, Gelucire 62/05, Gelucire 50/13, Gelucire 53/10, Apifil Pastills, and Glyceryl Monostearate in increments of 0.5 g and heat the test tube in a controlled temperature water bath kept at 80 °C. Note the amount of lipid required to solubilize the drug in a molten state and identify visually the solid lipid that solubilizes the highest amount of drug (*see* Fig. 3, **Note 3**).

3.2 Fabrication of Drug-Loaded NLC-Based Gels Using Microemulsion Template Technique

3.2.1 System Selection

The feasibility to form microemulsion system is checked by using varying combinations of selected lipid and oil, surfactant and solubilizers by titration method. The simplified method is to screen varying surfactant–to-solubilizer ratio (Km ratio) keeping the ratios of the solid lipid and oil fixed in order to determine the emulsifying ability of surfactant-solubilizer mixture.

1. Melt the solid lipid followed by addition of oil and maintain it above the melting point of the lipid.
2. Take different ratios of surfactant and solubilizers (also maintained above the melting point of lipid) in another test tube.
3. Gently mix the contents of both the test tubes to form a monophasic mixture and then titrate against aliquots of preheated distilled water.

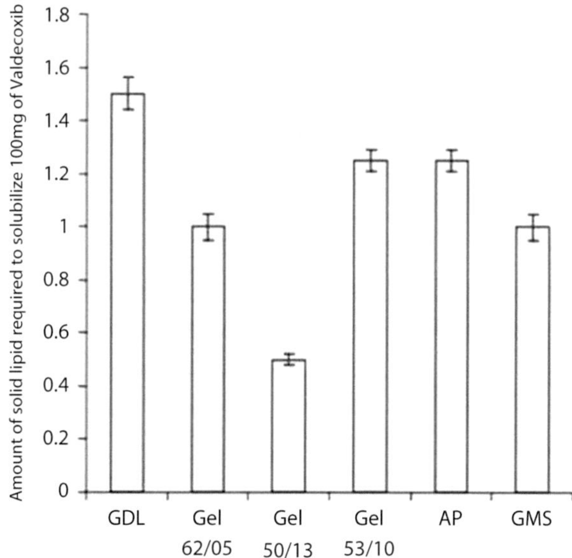

Fig. 3 Solubility of Valdecoxib in different solid lipids, GDL Glyceryl Dilaurate, Gel 62/05 Gelucire 62/05, Gel 50/13 Gelucire 50/13, Gel 53/10 Gelucire 53/10, AP Apifil Pastills, GMS Glyceryl Monostearate. (reproduced from ref. 9 with permission from the publisher, Taylor & Francis Ltd.)

4. Add water gradually to these mixtures till first sign of turbidity is observed.
5. Evaluate the physical state of the various microemulsion based on these parameters consistency (liquid or semisolid), phase separation tendency and stability on dilution and finalize the Km ratio (*see* **Note 4**).

3.2.2 Feasibility of Microemulsion Formation Using Pseudoternary Phase Diagram

Selection of an ME system is based on the drug-solubilizing capacity of the excipient. The boundaries of the microemulsion domains are determined with the help of pseudoternary phase diagrams with the selected excipients (solid lipid, oil, surfactant, solubilizer, and water) as the constituents of the microemulsion.

Components of microemulsion are:

1. The lipid phase consisting of a solid lipid and liquid lipid (oil)—mixture of Glyceryl Dilaurate and Caproyl 90.
2. The surfactant phase consisting of a surfactant and solubilizer—mixture of Cremophor RH 40, Solutol HS 15, and Transcutol.
3. The aqueous phase is double-distilled water:
   (a) Briefly, prepare the mixtures of the lipid phase and surfactant phase at ratios (w/w) of 10:0, 9:1, 8:2, 7:3, 6:4, 5:5, 4:6, 3:7, 2:8, 1:9, 0:10 in pre-weighed test tubes as follows:

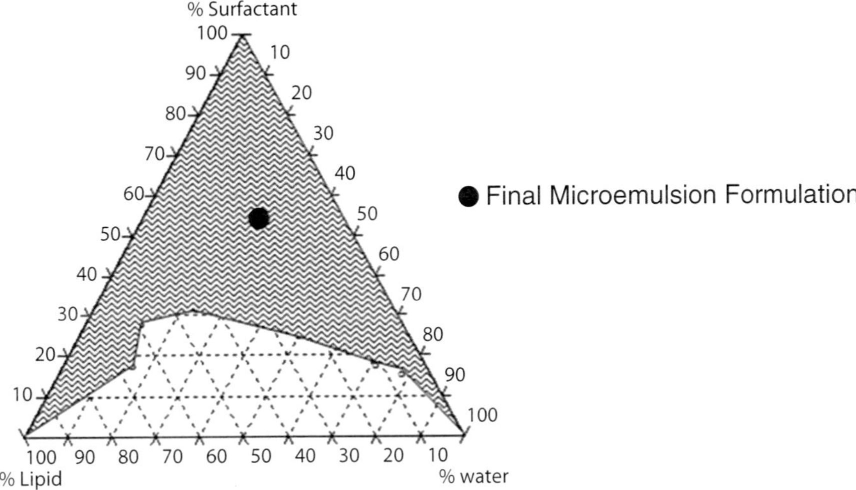

Fig. 4 Microemulsion phase diagram. (Reproduced from ref. 9 with permission from the publisher, Taylor & Francis Ltd.)

- Heat the required quantity of lipid phase on a water bath (at a temperature above melting point of lipid) to melt solid lipid.
- Heat the required quantity of surfactant phase (as per the Km ratio optimized) and gently mix it to lipid phase to form a monophasic mixture.
- Slowly titrate the mixture with aliquots of distilled water and stir at 60 °C for a sufficiently long time to attain equilibrium.

(b) After equilibrium is achieved, check the mixtures visually for transparency and through crossed polarizers for optical isotropy (*see* **Note 5**).

(c) Plot the pseudoternary phase diagram using the TriDraw software and identify the microemulsion region from the plot (*see* Fig. 4, **Note 6**).

3.2.3 Characterization of Microemulsion

1. Freeze-Thaw Cycling: Apply freeze-thaw cycles (−4 to 40 °C) of 24 h for a period of 1 week to the series of selected microemulsion and examine for physical instabilities such as phase separation and precipitation (*see* **Note 6**).
2. Optical Birefringence: Check the microemulsions both visually and using cross-polarizers for optical isotropy to confirm the absence of other phases (*see* **Note 6**).

3.2.4 Formulation of NLC from ME Templates

1. Prepare the oil phase by melting lipid and then mix with the oil.
2. Dissolve valdecoxib in this mixture at 60 °C.

3. Prepare the aqueous phase by mixing surfactant, solubilizer, and water.
4. Maintain the temperature for both phases above the melting point of the lipid, i.e., 60 °C.
5. At this temperature, mix the oil phase with the aqueous phase using a vortex to form a microemulsion.
6. Dilute this warm microemulsion in cold water (2–3 °C) under mechanical stirring to form NLC dispersion such that the concentration of valdecoxib in the final dispersion remains 1% w/w.

3.2.5 Formulation of NLC-Based Gel

Selection of suitable gelling agent to formulate an NLC-based gel is based on the compatibility with the nanoparticulate dispersion, ease of preparation, and esthetic appeal.

1. Screening of carbopol concentration: Disperse carbopol in water using an overhead stirrer at the speed of 600 rpm for 3 h. Screen various concentrations (ranging from 0.5–1% w/w) of carbopol for gelling and select the concentration that yields optimum viscosity (*see* **Notes** 7 and **8**).
2. Add carbopol (0.6% w/w) to the NLC dispersion under overhead stirring at 800 rpm.
3. Continue stirring until carbopol disperses uniformly in NLC dispersion.
4. Neutralize the carbopol dispersion using 50% w/w triethanolamine and check pH.
5. Transfer to an appropriate container (either laminated tube or wide mouth glass container) and label.

3.3 Characterization of NLC

1. Particle size and size distribution determination: Measure the particle size and size distribution of NLC dispersion obtained after suitable dilution of NLC using a particle size analyzer, e.g., Zetasizer Nano ZS (Malvern, UK).
2. Drug encapsulation efficiency determination:
   (a) Prepare known dilution of NLC dispersion and transfer to the upper chamber of Nanosep® centrifuge tubes equipped with an ultrafilter (MWCO 100 KD).
   (b) Centrifuge Nanosep® tubes at 11573 × *g* for 40 min.
   (c) Dilute supernatant and the filtrate appropriately and inject in HPLC. Determine the amount of valdecoxib in both phases using validated HPLC method.
   (d) Calculate entrapment efficiency using the following equation:

$$\text{Entrapment efficiency } (\%) = \left[\left(W_{\text{initial drug}} - W_{\text{free drug}}\right)/W_{\text{initial drug}}\right] \times 100.$$

wherein $W_{\text{initial drug}}$ is the mass of initial drug in the known dilution of NLC utilized for this determination, which is deduced from the mass of the drug incorporated into NLC, and $W_{\text{free drug}}$ is the mass of free drug obtained in the filtrate of the lower chamber of the Nanosep® on centrifugation of the aqueous dispersion.

3.4 Characterization of NLC-Based Gel

3.4.1 Determination of Drug Content, Spreadability, and pH

1. Dug content: Weigh one gram of gel in a 100 mL volumetric flask and dissolve it in methanol. Dilute appropriately and analyze the solution on a UV-VIS Spectrophotometer at a λ-max of 240 nm.
2. Spreadability: Place 0.5 g gel on a glass plate within a circle of 1 cm diameter marked on it. Place a second glass plate on the first plate. Allow a weight of 500 g to rest on the upper glass plate. Measure the increase in diameter due to the spreading of gels using a ruler.
3. pH: Determine the pH of 10% w/w gel using pH meter previously standardized using pH 4.0 and 7.0 standard buffers.

3.4.2 Study of Rheological Properties of NLC-Based Gel

1. Use Brookfield viscometer with helipath stand for rheological studies.
2. Place 30 g of gel in a beaker and allow to equilibrate for 5 min prior to recording the dial reading using T-C spindle at various speeds viz. 0.5, 1, 2.5 and 5 rpm.
3. Note the corresponding dial reading in triplicate at ambient temperature on the viscometer by successively lowering the speed.
4. Compute the viscosity in centipoises by direct multiplication of the dial readings with factors mentioned in the Brookfield viscometer catalog (*see* **Notes 9** and **10**).

3.4.3 In Vitro Release Study

1. Use the paddle and vessel assembly from USP type II apparatus along with the addition of a small stainless steel disk assembly designed for holding the gel at the bottom of the vessel. Maintain the temperature of the assembly at 32 °C ± 0.5 °C.
2. Place phosphate buffer pH 7.4 (900 mL) solution in the vessel and equilibrate at 32 °C ± 0.5 °C.
3. Apply 1 g of the gel on the disk assembly, taking care that the release surface was as flat as possible.
4. Gently insert the disk assembly at the bottom of the dissolution vessel.
5. Maintain the speed of rotation of the paddle at 25 rpm.

6. Withdraw aliquots every hour and analyze for drug content spectrophotometrically.
7. Plot a graph of % cumulative release against time in hours (*see* **Note 11**).

3.4.4 Skin Irritation Test

This test indicates the skin acceptability of the developed formulation for topical application. Test the developed gel formulations for primary skin irritation using following Draize patch test in rabbits:

1. Use three rabbits weighing 2.5–3 kg.
2. Clip the back and sides of the rabbits free of hair 24 h prior to the application of formulations.
3. Apply 0.5 g of each gel formulation groups (viz. NLC-based valdecoxib gel, Marketed valdecoxib gel and Placebo gel) on the hair-free skin of rabbits after 24 h by uniform spreading within the area of 4 cm^2.
4. Observe the skin for any visible change such as erythema (redness) or edema (swelling) at the end of 24, 48, 72 h and evaluate using the scale given by Draize to compute irritation score [13, 14] (*see* **Note 12**).

3.4.5 In Vitro Skin Permeation Studies [10]

1. Place the rat skin specimens (obtained from the abdominal area of male albino rat) individually placed on Franz-type diffusion cells having a surface area 3.14 cm^2, set at 32 °C using a thermostated water bath.
2. Receptor phase constitutes of 10 mL of phosphate buffer saline (PBS) (pH 7.4).
3. Place carefully the excised sections of rat skin between the donor and receptor compartments of the diffusion cells (*see* **Note 13**).
4. Place the stopper on the withdrawal port for removing the receptor phase.
5. Place 4.5 mg/cm^2 of the gels uniformly in the donor phase, in contact with the excised section of rat skin.
6. Stir the receptor phase constantly throughout the experiment and maintain the temperature maintained at 32 °C.
7. Remove 2.5 mL of the receptor phase at set intervals of 0, 1, 2, 5, 7, 9, 12, and 24 h and immediately replace by the equal volume of PBS (pH 7.4) solution.
8. Determine the amount of drug released into the receptor phase from the formulations by HPLC and plot the graph of cumulative % of drug permeated through rat skin versus time (h). Also, plot the graph of the amount of drug diffused per unit area (Q/A) versus time (h) to compute the flux values for the gel formulations (*see* **Note 14**).

9. Also, calculate the skin deposition potential of the gels at the end of 24 h after application of gel formulation by analyzing the amount of drug deposited within the skin (*see* **Note 15**).

3.4.6 Pharmacodynamic Efficacy Study of NLC-Based Gel

The pharmacodynamic efficacy of the gel was determined by aerosil-induced rat paw edema method using aerosol as the phlogistic agent. The study protocol is as follows:

1. Male wistar rats weighing 150–180 g were randomly divided into three groups of six rats, each group receiving different topical treatments.
2. Prepare 0.1 mL of 2.5% w/v aerosil suspension in distilled water and inject in the right hind foot of each rat under the planter aponeurosis [15].
3. Apply the topical treatments to the rats of the standard group with the marketed formulation of valdecoxib while the experimental group with a developed NLC-based formulation of valdecoxib and the control group with no topical treatment.
4. Measure the foot volume by the displacement technique using plethysmometer immediately before and after the injection of aerosil at fixed time intervals (1, 2, 3, 6, 9, 12 and 24 h).
5. Calculate the edema rate and percentage inhibition of each group as follows:

$$\text{Edema rate } (E) = V_t - V_o / V_o$$

$$\text{Inhibition rate } (I)\% = (E_c - E_t / E_c) \times 100$$

where V_o is the mean paw volume before aerosil injection; V_t is the mean paw volume after aerosil injection. E_c is the edema rate of the control group, and E_t is the edema rate of the treated group.

6. Plot a graph of % inhibition versus time in hours.

4 Notes

1. Saturation solubility studies enable to streamline the choice of excipients for the formation of the microemulsion. Lipids exhibiting maximum drug solubility are chosen with the aim of achieving good drug–lipid association and thereby enabling high entrapment efficiency in lipid core-surfactant system. Generally, surfactant showing least drug solubility is preferred for the following reasons: to entrap the drug moieties in the lipid core and not in the aqueous fraction of the dispersion. It also retards the burst release of the drug from NLC and ensures firm association of the drug with lipid matrix.

2. Among the oils screened, maximum solubility of valdecoxib was found in Caproyl 90; while among the surfactants Cremophor EL showed least solubilizing potential for valdecoxib; Transcutol P and Solutol HS 15 proved to be best solubilizers (Fig. 2).
3. The criteria for screening of solid lipids are as follows:
   (a) Safety (absence of any adverse reaction upon application to the skin).
   (b) Physical state (solid state at room temperature having a melting point above body temperature).
   (c) Solubilization (ability to solubilize the drug).
   (d) Compatibility with other excipients and packaging material.
   (e) Ease of availability.

   Equilibrium (or saturation) solubility studies cannot be carried out in the case of solid lipids. Hence, a modified method is used to identify the solid lipid having better solubilization potential for valdecoxib wherein amount of solid lipid required to solubilize fixed amount of drug in molten state is determined [9]. Owing to the dispersible nature of Gelucires, Glyceryl monostearate and Glyceryl dilaurate are preferred over Gelucires to fabricate NLC. The dispersible nature of Gelucires has a tendency to prevent the formation of a solid structure in the ultimate dispersion, which is a prerequisite for fabrication of an NLC. Among solid lipids, the highest solubility of valdecoxib was found in Glyceryl dilaurate (Fig. 3).
4. System composition at a given Km ratio that has a liquid consistency, no phase separation and is stable on dilution is considered for further study. System composition forming liquid microemulsion is preferred than semisolid consistency as it has to be poured in cold aqueous phase under stirring for NLC formation.
5. Only those systems which appear black when visualized through the crossed polarizers are considered to be within the microemulsion region. In pseudoternary phase diagram, the shaded area represents the microemulsion existence region. The area outside this frame indicates a turbid region with multiphase systems.
6. The microemulsion represented as a dot in Fig. 4 prepared using the optimized quantity of selected components could survive the freeze-thaw cycling. This microemulsion is considered to be thermodynamically stable. At temperatures below freezing, ice crystals are formed in an oil/water (o/w) type of microemulsion that may cause oil particles to elongate and flatten. Also, the lipophilic portion of the surfactant molecule

could lose its mobility while the hydrophilic portions get simultaneously "dehydrated" due to the freezing effect of water. During thawing of the sample, water is released and travels rapidly through the microemulsion. If the system can restore itself before coalescence occurs, then the microemulsion survives the freeze-thaw cycling. However, if the coalescence occurs, instability sets in case of the microemulsion, which is not related to normal temperature processes. In the optical birefringence study, the microemulsion was found to be isotropically clear [9, 10].

7. Some gelling agent has tendency to clump when added to the dispersing medium in a haphazard manner. In such cases, the outer molecules of the gelling agent contact the medium first and get hydrated, forming a surface layer that is more difficult for the medium to penetrate. The clumps will gradually hydrate, but it needs more time. Carbopol type gelling agents require a "neutralizer" or a pH adjusting chemical to create the gel after the gelling agent has been wetted and dispersed in the dispersing medium, thus avoiding clump formation.
8. The viscosity of the NLC gel was about 85×10^5 cps at 5 rpm.
9. The NLC-based gel showed a flow index of 0.386, indicating pseudoplastic flow behavior. Flow index gives an idea of the flowability of the gel from the container. Generally, thicker the base, lower is the flow index.
10. Alternatively, use digital Brookfield viscometer which directly displays viscosity in cps.
11. The NLC-based gel show burst release in initial hours followed by a steady release. This pattern of drug release could be due to the following reasons: diffusion of the unencapsulated drug, solubilization by the surfactant micelle, diffusion of the drug from oil nanodroplets in first 2 h followed by diffusion from the solid lipid surface and from the core thereafter. Both burst release as well as sustained release is of interest for dermal application. Burst release is useful for faster onset of action and improves penetration of drug whereas sustained release supplies the drug over a prolonged period of time [9, 10].
12. The irritation score (or primary skin irritation index) enables to assess the effect of cumulative application of the formulations on skin irritation at the end of 48 h and 72 h. The developed formulation should exhibit no skin irritation on intact rabbit skin compared to the marketed formulation in skin irritation test.
13. Make sure that the skin is completely in contact with the receptor phase to eliminate any air bubbles.

14. The flux value for NLC type system is generally lower in comparison to other systems which can be attributed to the slow permeation of drug from the gel owing to its encapsulation in the lipid core [10].
15. The deposition potential of NLC gels is usually higher as these tend to remain in the skin, thus acting as a depot to give sustained release of the drug.

Acknowledgements

Authors are thankful to Department of Science and Technology (INSPIRE scheme) and Technical Education Quality Improvement Programme (TEQIP), Government of India for providing research fellowship.

References

1. Beloqui A, Solinis MA, Rodriguez-Gascon A et al (2016) Nanostructured lipid carriers: promising drug delivery systems for future clinics. Nanomedicine 12(1):143–161
2. Muller RH, Radtke M, Wissing SA (2002) Nanostructured lipid matrices for improved microencapsulation of drugs. Int J Pharm 242 (1–2):121–128
3. Muller RH, Mader K, Gohla S (2000) Solid lipid nanoparticles (SLN) for controlled drug delivery–a review of the state of the art. Eur J Pharm Biopharm 50(1):161–177
4. Muller RH, Mehnert W, Lucks JS et al (1995) Solid lipid nanoparticles (SLN): an alternative colloidal carrier system for controlled drug delivery. Eur J Pharm Biopharm 41(1):62–69
5. Patravale VB, Date AA, Kulkarni RM (2004) Nanosuspensions: a promising drug delivery strategy. J Pharm Pharmacol 56(7):827–840
6. Muller RH, Mader K, Lippacher A et al (2000) Solid-liquid (semi-solid) lipid particles and method of producing highly concentrated lipid particle dispersions. German Patent Application 199,45,203.2
7. Wissing SA, Kayser O, Muller RH (2004) Solid lipid nanoparticles for parenteral drug delivery. Adv Drug Deliv Rev 56(9):1257–1272
8. Muller RH, Ruhl D, Runge S et al (1997) Cytotoxicity of solid lipid nanoparticles as a function of the lipid matrix and the surfactant. Pharm Res 14(4):458–462
9. Joshi M, Patravale V (2006) Formulation and evaluation of nanostructured lipid carrier (NLC)-based gel of valdecoxib. Dru Dev Ind Pharm 32(8):911–918
10. Joshi M, Patravale V (2008) Nanostructured lipid carrier (NLC) based gel of celecoxib. Int J Pharm 346(1–2):124–132
11. Joshi M, Pathak S, Sharma S et al (2008) Design and in vivo pharmacodynamic evaluation of nanostructured lipid carriers for parenteral delivery of artemether: Nanoject. Int J Pharm 364(1):119–126
12. Gasco MR (1997) Solid lipid nanospheres from warm microemulsions. Pharm Technol Eur 9:52–58
13. Kligman AM, Leyden JJ (eds) (1982) Safety and efficacy of topical drugs and cosmetics. Grune & Stratton, New York
14. Vermeer BJ (1991) Skin irritation and sensitization. J Control Release 15(3):261–265
15. Vogel GH, Vogel HW (eds) (1997) Drug discovery and evaluation: pharmacological assays. Springer, New York

Chapter 20

Preparation and Characterization of Solid Lipid Nanoparticles-Based Gel for Topical Delivery

Vandana B. Patravale and Amit G. Mirani

Abstract

Solid lipid nanoparticles (SLNs) have been extensively investigated for effective delivery of both hydrophilic and lipophilic drugs by topical route. There are several scalable techniques for the preparation of SLNs such as homogenization, microemulsion template, and solvent emulsification diffusion. This chapter describes step-wise methodology for the preparation and characterization of SLNs using solvent emulsification diffusion method. Tretinoin, a lipophilic entity, was chosen as a model drug. The critical aspects and the important interpretations with respect to the preparation and characterization of SLNs are reported in "Notes" section.

Key words Solvent-emulsification diffusion, Solid lipid nanoparticles, Tretinoin, Topical route

1 Introduction

Solid lipid nanoparticles (SLNs) are submicron colloidal nanocarriers (50–1000 nm) comprising of the drug either encapsulated or in matrix form with lipid particles. SLNs have witnessed a global regulatory acceptance due to their high safety profile and are hence preferred as an alternative to polymeric nanoparticles for the delivery of the lipophilic and hydrophilic drugs. SLNs have potential applications in drug delivery as they exhibit several advantages such as drug loading of both lipophilic and hydrophilic moieties, improved stability, biodegradability, ease in scale-up, and cost-effectiveness. Till date, several methods are described in the literature for SLN preparation, including high-pressure homogenization, i.e., hot homogenization [1, 2] and cold homogenization [3], microemulsion template technique [4, 5], melt dispersion technique [6], ultrasonication technique [7, 8], double emulsion technique [9, 10], solvent emulsification-evaporation technique [11], solvent emulsification-diffusion technique [12, 13]. Herein, we are discussing, solvent emulsification-diffusion technique for preparation of solid lipid nanoparticles [14]. Solvent

Volkmar Weissig and Tamer Elbayoumi (eds.), *Pharmaceutical Nanotechnology: Basic Protocols*, Methods in Molecular Biology, vol. 2000, https://doi.org/10.1007/978-1-4939-9516-5_20,

emulsification-diffusion technique was mainly used for polymeric nanocarriers, however, its use for fabrication of solid lipid nanoparticle was first explored by Trotta et al. [15]. Solvent emulsification-diffusion technique is an easily scalable technique, requires less physical stress and ensures loading of both hydrophilic and lipophilic drugs. The preparation of SLNs using the solvent emulsification-diffusion technique involves preparation of a solvent-in-water emulsion using a "partially" water-miscible solvent containing the lipid in rational amounts. Upon transferring the transient oil-in-water emulsion into water, lipophilic material dissolved in the organic solvent solidifies instantaneously due to diffusion of the organic solvent from the droplets to the continuous phase [15]. A schematic representation of the solvent emulsification-diffusion technique is depicted in Fig. 1 and detailed preparation and its characterization protocol are discussed in the following section.

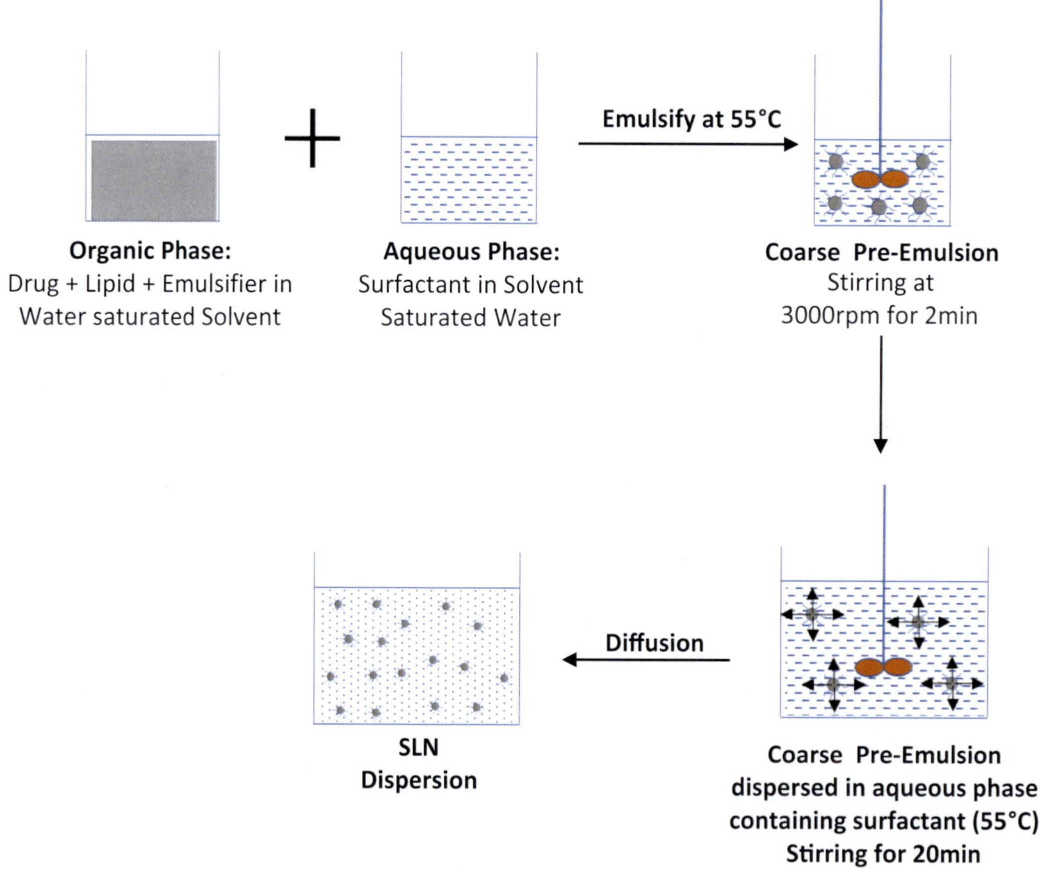

Fig. 1 Preparation of SLN using Solvent Emulsification Diffusion Technique

2 Material

2.1 Preparation of Solid Lipid Nanoparticles

1. Drug: Tretinoin (*see* **Note 1**).
2. Solid lipid: Glyceryl monostearate; Compritol 888 ATO®; Dynasan 116® and Cutina CBS® (*see* **Note 2**).
3. Surfactants/Stabilizer: Epikuron 200®; Tween 20 and Tween 80 (*see* **Note 3**).
4. Partial water miscible solvents: Benzyl alcohol (*see* **Note 4**).
5. Preservative: Methyl paraben, Propyl paraben.
6. Chelating agent: Disodium EDTA.
7. Antioxidant: Butylated hydroxytoluene.
8. Distilled water or Milli-Q reagent grade water.
9. Water shaker bath with controlled temperature.
10. Controlled temperature water bath.
11. Cyclomixer.

2.2 Characterization of Solid Lipid Nanoparticle

2.2.1 Particle Size Analysis

1. Particle size analyzer (Photon Correlation Spectroscopy; Beckman Coulter N4 plus, Wipro, India) (*see* **Note 5**).
2. To dilute the sample: water (Milli Q®).

2.2.2 Entrapment Efficiency

1. Ultrafilter: Nanosep®, (MWCO100KD) (*see* **Note 6**).
2. Centrifuge (Eltek TC 4100, Mumbai, India) for separation of encapsulated and unencapsulated drug.
3. UV-spectrophotometric method (Shimadzu UV-1650, Shimadzu Analytical Pvt. Ltd. India) for analysis of filtrate.

2.2.3 Morphological Studies Using Scanning Electron Microscopy

1. Cameca SU-SEM probe (resolution: upto 40°A; magnification: upto 40,000×; accelerating voltage: upto 30 kV; fully integrated EDS/WDS system).
2. SLN dispersion for analysis.

2.3 Preparation of Gel

1. Gelling polymer: Carbopol Ultrez 10®, Carbopol® 940, and Carbopol® ETD 2020.
2. pH modifier: Triethanolamine.

2.4 Evaluation of Gel

2.4.1 Drug Content

1. Methanol to dissolve formulation component (*see* **Note 7**).
2. UV -spectrophotometric method (Shimadzu UV-1650, Shimadzu Analytical Pvt. Ltd. India) for analysis of dissolved formulation.

2.4.2 Spreadability

1. TA-XT Texture Analyzer (Stable Micro Systems, New Delhi, India) comprising spreadability study related component such as heavy duty base plate, male cone, female cone, load cell 5 kg.
2. SLN dispersion gel for analysis.

2.4.3 Rheological Studies

1. Brookefield Synchro-Lectric Viscometer (Model RVT) with helipath stand.
2. SLN dispersion gel for analysis.

2.4.4 In Vitro Permeation Studies

1. Franz diffusion cell cells with a surface of 3.14 cm^2 and a receptor volume of 10 ml.
2. Permeation medium: pH 7.4 buffer (*see* **Note 8**).
3. HPLC system (Jasco PU-2080 Plus Intelligent (Jasco, Japan) equipped with a Jasco UV-2075 Intelligent UV/vis detector (Jasco, Japan), a Rheodyne 7725 injector (Rheodyne, USA), a Jasco Borwin Chromatography Software (version 1.50) integrator software and a Hi-Q-Sil C (4.6 mm × 250 mm and 10 μm particle size) column. Mobile phase: methanol: acetonitrile: pH 6.8 phosphate buffer (65:20:15, v/v) at a flow rate of 1.2 ml/min.

3 Method

3.1 Screening of Lipid

The lipids are selected on the basis of their drug solubilizing capacity. Solubility of drug in lipid is of paramount importance as it governs loading of drug in the formulation. The commonly employed method for solubility determination is equilibrium solubility studies. However, the same cannot be employed for lipids. Thus, the method proposed by our research group [16, 17] was utilized.

1. Weigh and add 10 mg of drug individually in screw-capped tubes (*see* **Note 9**).
2. Heat the solid lipid (glyceryl monostearate/compritol 888 ATO/dynasan 116/cutina CBS) above its melting point.
3. Add the molten lipid gradually in a tube containing drug under continuous stirring with aid of a cyclomixer.
4. Inspect visually the amount of lipid required to solubilize the drug in a molten state and identify the solid lipid that solubilizes the highest amount of drug.

3.2 Screening of Surfactant

The surfactants are screened for their emulsification capacity. The screening method is as below;

1. Weigh and add drug and screened lipid in a capped tube containing partially miscible solvent (water-saturated benzyl

alcohol maintained at 55 ± 0.5 °C) (*see* **Notes 10** and **11**). Vortex it for 3 s to achieve a homogenous organic phase (*see* **Note 9**).

2. Weigh and add surfactant (Tween 80) in a tube containing Milli Q® water (benzyl alcohol-saturated water maintained at 55 ± 0.5 °C) (*see* **Note 12**). Vortex it for 3 s to achieve a homogeneous aqueous phase (*see* **Note 9**).
3. Gradually add an aqueous phase into an organic phase using an overhead stirrer at 850 × *g* for 2 min to form the primary emulsion.
4. Precipitate the drug-loaded solid lipid nanoparticles by adding the preformed emulsion into an aqueous phase containing a mixture of surfactant (Tween 80 and Tween 20) maintained at 55 ± 0.5 °C and continuously stir for 20 min using an overhead stirrer to extract the benzyl alcohol into the continuous phase.
5. The surfactant which gives stable emulsion with smaller particle size will be considered for preparation of SLN.

3.3 Preparation of Solid Lipid Nanoparticle (SLN) System Using Solvent Emulsification-Diffusion (SED) Method

1. Weigh and add lipid (glyceryl monostearate), surfactant (Epikuron 200), drug, and anti-oxidant (butylated hydroxy toluene) in a capped tube containing 2 g of partially miscible solvent (water saturated benzyl alcohol maintained at 55 ± 0.5 °C) (*see* **Notes 10** and **11**). Vortex it for 3 s to achieve a homogenous organic phase (*see* **Note 9**).
2. Weigh and add 200 mg of surfactant (Tween 80), preservative (methyl paraben, propyl paraben) in a tube containing Milli Q® water (benzyl alcohol-saturated water maintained at 55 ± 0.5 °C) (*see* **Note 12**). Vortex it for 3 s to achieve a homogeneous aqueous phase (*see* **Note 9**).
3. Gradually add aqueous phase into an organic phase using an overhead stirrer at 3000 rpm for 2 min to form the primary emulsion.
4. Precipitate the drug-loaded solid lipid nanoparticles by adding the preformed emulsion into an aqueous phase containing a mixture of surfactant (Tween 80 and Tween 20) maintained at 55 ± 0.5 °C and continuously stir for 20 min using an overhead stirrer to extract the benzyl alcohol into the continuous phase.
5. No further purification will be required for removal of benzyl alcohol and surfactant from SLN dispersion if the amount present in the dispersion is within the acceptable limits for topical formulation (*see* **Note 13**).

3.4 Characterization of SLN Dispersion

3.4.1 Particle Size Analysis

1. To study the particle size of SLN dispersion, dilute the SLN dispersion with water (Milli Q®) (*see* **Note 14**).
2. Add the diluted sample in the cuvette and measure the particle size and polydispersion by photon correlation spectroscopy (PCS) using Zetasizer Nano Series or Brookhaven Instruments standard setup (*see* **Note 15**).

3.4.2 Entrapment Efficiency

The entrapment efficiency (EE), which corresponds to the percentage of drug encapsulated within and adsorbed on to the nanoparticles can be determined by measuring the concentration of free drug in the dispersion medium.

1. To determine total drug content, dissolve SLN dispersion by adding 1 ml of methanol and filter the solution using 0.22 μm membrane. Measure the concentration of drug using the validated UV-spectroscopic method.
2. To determine the encapsulation efficiency, add SLN dispersion (approximately 500 μl) to the Nanosep® centrifuge tube fitted with an ultrafilter (MW cut-off 100KD) and centrifuge it at 18,650 × g for 40 min at 25 ± 0.5 °C.
3. Determine the free drug (i.e., drug non-associated with the solid lipid nanoparticles) by measuring the concentration of drug in the supernatant using the validated UV-spectroscopic method.
4. The encapsulated drug in solid lipid nanoparticles (encapsulation efficiency, EE) can be calculated from the ratio between the difference of the total and the free drug concentrations (TD and FD, respectively) divided by the total concentration, multiplied by 100 (Eq. 1).

$$\mathrm{EE} = \frac{\mathrm{TD} - \mathrm{FD}}{\mathrm{TD}} \times 100 \tag{1}$$

3.4.3 Morphological Studies

To study the morphology of the SLN dispersion, place drops of dispersion over the aluminium grid and completely dry them at ambient temperature, thereby leaving only a thin layer of particles on the grid. Observe the developed gird under scanning electron microscope (magnification: 20,000×; accelerating voltage:20.0 kV) at 25 ± 2 °C.

3.5 Preparation of SLN Dispersion Gel

1. Slowly disperse gelling polymers (viz., Carbopol Ultrez 10®, Carbopol® 940 and Carbopol® ETD 2020) in SLN dispersion under rapid overhead stirring and allow to hydrate for 20–30 min (*see* **Note 16**).
2. After complete hydration of gelling polymer, add dropwise triethanolamine, a pH modifier till the pH of 7 ± 0.2 is achieved to obtain the SLN-based gel formulation (*see* **Note 17**).

3. The gelling polymer which offers compatibility with nanoparticulate dispersion and superior texture (feel and spreadability) is finalized for gel formulation.

3.6 Evaluation of Gel

3.6.1 Drug Content

1. Dissolve approximately 1 g of SLN dispersion gel by adding 1 ml of methanol and filter the solution using a 0.22 μm membrane.
2. Analyze the filtrate using Shimadzu UV-1650 PC UV-vis spectrophotometer managed by Shimadzu UV probe version 2.10 at a wavelength of 344 nm.

3.6.2 Spreadability

1. Perform spreadability study using TA-XT texture analyzer and express the results as firmness (positive force) and stickiness (negative force).
2. Calibrate the instrument for height. The height calibration is required for spreadability study as it ensures the fix-distance between the male cone and female cone, such that a perfect fit can be ensured.
3. To perform calibration, fix female cone in the base plate and male cone to load cell holder. Allow the male cone to move downwards and touch the base of the female cone. Once the base is touched, the male cone reverts to a distance of 25 mm.
4. After calibration of the instrument, place 2.0 g of test compound (SLN dispersion gel) in female cone using a curved spatula to ensure no air bubble is entrapped within it.
5. Allow male cone to move downwards with a distance of 23 mm.
6. The force required for the male cone to travel test compound till the base of the female cone is considered as firmness and the force required for a male cone to detach from the gel is considered as stickiness.

3.6.3 Rheological Studies

1. Perform rheological evaluation of SLN dispersion gel using Brookfield Synchro-Lectric Viscometer (Model RVT) using T-C spindle.
2. Place the sample in a beaker and allow to equilibrate for 5 min.
3. Place the spindle in a beaker containing test sample and measure the dial reading at different spindle speed (0.5, 1, 2.5 and 5 rpm).
4. Note the dial reading by lowering the spindle speed from 5 to 0.5 rpm.
5. Viscosity in centipoise can be calculated by multiplication of dial reading with the factor as specified in Brookfield viscometer catalogue. The factor value varies with the spindle type.

3.6.4 In Vitro Permeation Studies

1. In vitro skin permeability is best measured using Franz diffusion cell comprising of two compartments, i.e., receptor compartment and donor compartment separated by the permeation barrier (skin) (*see* **Note 18**).
2. Perform in vitro skin permeation studies on excised abdominal skin obtained from Wistar rat (Age: 3 months; Weight range: 200–250 g) with prior animal ethical committee permission (*see* **Note 19**).
3. Insert magnetic needle in receptor compartment and mount the skin in between receptor and donor compartment.
4. Fill the receptor compartment with 10 ml of modified permeation medium (i.e., pH 7.4 phosphate buffer containing albumin) which simulates the physiological condition and allow to equilibrate for 2 h.
5. Add test compound (SLN dispersion gel) in donor compartment (typically 0.1–0.5 g) with curved spatula enabling the gel film to cover the entire skin surface evenly.
6. Cover the diffusion cells with aluminium foil to prevent light exposure as the drug is light sensitive.
7. Remove sample (0.3 ml) of fluid from receptor compartment at 1, 4, 6, 8, and 12 h interval and replace the equivalent amount with fresh permeation medium.
8. Determine the concentration of drug withdrawn from receptor compartment using validated HPLC method (*see* **Note 20**).
9. The total quantity of drug that diffuses through to the receptor compartment in time "t" during the steady state and the *flux* at steady state, Js [μg/(cm^2 h)] can be calculated using linear portion of the correlation between the accumulated quantity of drug that diffused through the skin by unit area and time.

4 Notes

1. Both lipophilic and hydrophilic drugs can be delivered using SLN system. Tretinoin is a model drug which exhibits high lipophilicity and poor water solubility [14]. Thus, Tretinoin is an ideal candidate for incorporation into a lipid nanoparticles.
2. The lipid which exhibits highest drug solubility is considered for preparation of SLN. The high solubility will ensure high drug encapsulation within the system [14].
3. The surfactants act as emulsifying agents which slow down the inevitable separation of two phases as well as lower down the particle size. In some cases, one surfactant is not sufficient and may result in coalescence upon standing. Thus, the combination of surfactants is preferred which acts by forming film at the

interface with sufficient viscosity and thereby prevent aggregation upon standing [14, 15].

4. The solvent which exhibits partial miscibility in water is considered for preparation of SLN by the solvent-emulsification diffusion method [14, 15].
5. Other nanoparticle analyzers such as Malvern Instruments/ Horiba Scientific should also work, provided they can measure the size range less than 100 nm.
6. Nanoseps® with different cut off are available and can be selected based on particle size of nanoparticles and molecular weight of the drug entity. For example, The nanoparticle with particle size of 30–60 nm and molecular weight in range of 300–600 kDa requires Nanoseps® 100 kDa which exhibits pore size of 10 nm [18].
7. The solvent which completely dissolves the solid lipid nanoparticulate system is selected for the preparation of sample for drug content analysis.
8. The pH at which drug exhibits maximum solubility is considered for flux study and the volume of media should be atleast three times to the drug solubilizing capacity. This ensures maintenance of sink condition [19].
9. The tube should be capped tightly to avoid the loss of material during vortexing, and to prevent the evaporation of solvent afterwards.
10. Mutual saturation of water miscible solvent, i.e., benzyl alcohol and water is a very critical step for preparation of stable SLN. Missing this step may result in the generation of microparticles [14].
11. Temperature (55 °C) of water saturated benzyl alcohol is ver-y critical, as the solubility of lipid varies with solvent temperature [14].
12. Temperature (55 °C) of benzyl alcohol-saturated water is very critical, so as to achieve the equilibrium with the organic phase [14].
13. Purification of SLN dispersion using dialysis is performed if the solvent concentration is not within the acceptable limits [14].
14. The usual dilution will be decided on the basis of instrument parameters. For example, (in Zetasizer the count rate should be in between 150 and 450 kcps). In general, the dilution ranges between 10 and 1000 folds.
15. The diameters usually observed for nanoparticles prepared using preformed polymers range between 10 and 1000 nm. Polydispersity indexes lower than 0.2 indicate homogeneous systems presenting a satisfactory narrow particle distribution.

16. Hydration time may vary with the type of gelling polymer (15 min to 4 h). The complete hydration of polymer is necessary to get gel with optimum viscosity.
17. Excess of pH modifier may result in gritty appearance or reduced viscosity of final gel formulation.
18. A variety of other diffusion cells is also available including automated cells with flow-through receptor chambers.
19. Some laboratories use cadaver skin. Skin from at least two donors should be used, with skin from each donor used in equal numbers of cells.
20. For HPLC test, all samples are required to be filtered through a filter membrane (0.2 μM) to avoid blocking of the column.

References

1. Liu J, Hu W, Chen H et al (2007) Isotretinoin-loaded solid lipid nanoparticles with skin targeting for topical delivery. Int J Pharm 328:191–195
2. Kumar VV, Chandrasekar D, Ramakrishna S et al (2007) Development and evaluation of nitrendipine loaded solid lipid nanoparticles: influence of wax and glyceride lipids on plasma pharmacokinetics. Int J Pharm 335:167–175
3. You J, Wan F, Cui FD et al (2007) Preparation and characteristic of vinorelbine bitartrate-loaded solid lipid nanoparticles. Int J Pharm 343:270–276
4. Ma QH, Xia Q, Lu YY et al (2007) Preparation of tea polyphenols loaded solid lipid nanoparticles based on the phase behaviors of hot microemulsions. Solid State Phenom 121:705–708
5. Ugazio E, Cavalli R, Gasco MR (2002) Incorporation of cyclosporin a in solid lipid nanoparticles (SLN). Int J Pharm 241:341–344
6. Zhang D, Tan T, Gao L (2006) Preparation of oridonin-loaded solid lipid nanoparticles and studies of them in vitro and in vivo. Nanotechnology 17:5821–5828
7. Hou DZ, Xie CS, Huang KJ (2003) The production and characteristics of solid lipid nanoparticles (SLNs). Biomaterials 24:1781–1785
8. Luo YF, Chen DW, Ren LX et al (2006) Solid lipid nanoparticles for enhancing vinpocetine's oral bioavailability. J Control Release 114:53–59
9. Utada AS, Lorenceau E, Link DR et al (2005) Microcapillary device monodisperse double emulsions generated from a microcapillary device. Science 308:537–541
10. Liu J, Gong T, Wang C et al (2007) Solid lipid nanoparticles loaded with insulin by sodium cholate-phosphatidylcholine- based mixed micelles: preparation and characterization. Int J Pharm 340:153–162
11. Zhang N, Ping Q, Huang G et al (2006) Lectin modified solid lipid nanoparticles as carriers for oral administration of insulin. Int J Pharm 327:153–159
12. Pandey R, Sharma S, Khuller GK (2005) Oral solid lipid nanoparticle based antitubercular chemotherapy. Tuberculosis 85:415–420
13. Hu FQ, Hong Y, Yuan H (2004) Preparation and characterization of solid lipid nanoparticles containing peptide. Int J Pharm 273:29–35
14. Shah KA, Date AA, Joshi MD, Patravale VB (2007) Solid lipid nanoparticles (SLN) of tretinoin: potential in topical delivery. Int J Pharm 345(1–2):163–171
15. Trotta M, Debernardi F, Caputo O (2003) Preparation of solid lipid nanoparticles by a solvent emulsification–diffusion technique. Int J Pharm 257:153–160
16. Joshi MD, Patravale VB (2006) Formulation and evaluation of nanostructured lipid carrier (NLC) based gel of valdecoxib. Drug Dev Ind Pharm 32:911–918
17. Pattani AS, Mandawgade SD, Patravale VB (2006) Development and comparative antimicrobial evaluation of lipid nanoparticles and nanoemulsion of Polymyxin B. J Nanosci Nanotechnol 6:1–5
18. Nanosep MF & Nanosep® Centrifugal devices (2017). https://shop.pall.com/INTERSHOP/web/WFS/PALL-PALLUS-Site/en_US/-/USD/ViewProduct-Start?SKU=gri78m16&CatalogID=Laboratory. Accessed 25 Feb 2017
19. Rohrs BR (2001) Dissolution method development for poorly soluble compounds. Dissolut Technol 8(3):1–5

Chapter 21

Molecular-Level "Observations" of the Behavior of Gold Nanoparticles in Aqueous Solution and Interacting with a Lipid Bilayer Membrane

Priyanka A. Oroskar, Cynthia J. Jameson, and Sohail Murad

Abstract

We use coarse-grained molecular dynamics simulations to "observe" details of interactions between ligand-covered gold nanoparticles and a lipid bilayer model membrane. In molecular dynamics simulations, one puts the individual atoms and groups of atoms of the physical system to be "observed" into a simulation box, specifies the forms of the potential energies of interactions between them (ultimately quantum based), and lets them individually move classically according to Newton's equations of motion, based on the forces arising from the assumed potential energy forms. The atoms that are chemically bonded to each other stay chemically bonded, following known potentials (force fields) that permit internal degrees of freedom (internal rotation, torsion, vibrations), and the interactions between nonbonded atoms are simplified to Lennard-Jones forms (in our case) and coulombic (where electrical charges are present) in which the parameters are previously optimized to reproduce thermodynamic properties or are based on quantum electronic calculations. The system is started out at a reasonable set of coordinates for all atoms or groups of atoms, and then permitted to develop according to the equations of motion, one small step (usually 10 fs time step) at a time, for millions of steps until the system is at a quasi-equilibrium (usually reached after hundreds of nanoseconds). We then let the system play out its motions further for many nanoseconds to observe the behavior, periodically taking snapshots (saving all positions and energies), and post-processing the snapshots to obtain various average descriptions of the system. Alkanethiols of various lengths serve as examples of hydrophobic ligands and methyl-terminated PEG with various numbers of monomer units serve as examples of hydrophilic ligands. Spherical gold particles of various diameters as well as gold nanorods form the core to which ligands are attached. The nanoparticles are characterized at the molecular level, especially the distributions of ligand configurations and their dependence on ligand length, and surface coverage. Self-assembly of the bilayer from an isotropic solution and observation of membrane properties that correspond well to experimental values validate the simulations. The mechanism of permeation of a gold NP coated with either a hydrophobic or a hydrophilic ligand, and its dependence on surface coverage, ligand length, core diameter, and core shape, is investigated. Lipid response such as lipid flip-flops, lipid extraction, and changes in order parameter of the lipid tails are examined in detail. The mechanism of permeation of a PEGylated nanorod is shown to occur by tilting, lying down, rotating, and straightening up. The nature of the information provided by molecular dynamics simulations permits understanding of the detailed behavior of gold nanoparticles interacting with lipid membranes which in turn helps to understand why some known systems work better than others and aids the design of new particles and improvement of methods for preparing existing ones.

Volkmar Weissig and Tamer Elbayoumi (eds.), *Pharmaceutical Nanotechnology: Basic Protocols*, Methods in Molecular Biology, vol. 2000, https://doi.org/10.1007/978-1-4939-9516-5_21,

Key words Gold nanoparticles, Molecular dynamics simulations, Gold nanorod, Membrane permeation, PEGylated

1 Introduction

Contributions to this volume about Basic Protocols in Pharmaceutical Nanotechnology include many types of nanomedicine systems (lipid and surfactant based, plant virus based, dendrimer stabilized, in polymeric matrices, sustained release, pH operated, light operated) for various purposes such as targeted drug delivery, gene delivery, imaging, photothermal treatment, antitumor/cancer treatment, and biosensor. Preparation and characterization methods used in these examples are at the laboratory and clinical levels, involving physical, chemical, and biological aspects. Like the other contributions to this volume, we consider only specific nanosystems that we have studied, rather than taking a global perspective. Our contribution is different from the others in this volume in that ours is a theoretical-computational approach at the atomic and functional group interaction level and at the femtosecond-to-hundreds of nanosecond timescale. Nevertheless, some results and insights gained from our studies could have more general applications.

We use molecular dynamic simulations to characterize the behavior of ligand-coated nanoparticles in aqueous solution with and without electrolytes, examining at the dynamic molecular level the configurations of the ligands, the distributions of configurations, and the interaction of the ligands with the solvent. Then, since nanopharmaceuticals have to interact with and sometimes permeate the cell membrane to be able to carry out their function, we use molecular dynamics simulations of the interactions of nanoparticles, bare, or coated with hydrophobic ligands (alkanethiols), or coated with hydrophilic ligands (PEGylated), with model lipid bilayer membranes of uniform composition. We examine the details of such interactions and the dynamics of the permeation process from the point of view of all participants, the ligands and the lipid molecules constituting the membranes and the water and ions. We first study nanoparticles with various sizes of spherical gold cores and various ligand lengths, and then we also consider gold cores with aspect ratio different from unity, to investigate the permeation mechanism of a PEGylated nanorod. Thus, we provide a detailed picture of the behavior of a typical functionalized nanoparticle with mobile functional groups interacting with a simplified model of a cell membrane, albeit one that reproduces the physical and electronic characteristics observed for planar lipid bilayers. In molecular dynamics simulations, one puts the individual atoms and groups of atoms of the physical system to be "observed" into a simulation box, specifies the forms of the potential energies of interactions between them (ultimately quantum based), and lets them individually move classically according to Newton's equations of motion, based on the forces arising from the assumed potential energy forms. The atoms that are chemically bonded to each other stay chemically bonded, following known potentials (force fields) that permit internal degrees of freedom (internal rotation, torsion, vibrations), and the interactions between nonbonded atoms are simplified to Lennard-Jones forms (in our case) and coulombic (where electrical charges are present) in which the parameters are previously optimized to reproduce thermodynamic properties or are based on quantum electronic calculations. The system is started out at a reasonable set of coordinates for all atoms or groups of atoms, and then permitted to develop according to the equations of motion, one small step (usually 10 femtosecond time step) at a time, for millions of steps until the system is at a quasi-equilibrium (usually reached after hundreds of nanoseconds). We then let the system play out its motions further for many nanoseconds to observe the behavior, periodically taking snapshots (saving all positions and energies), and post-processing the snapshots to obtain various average descriptions of the system. We obtain such average descriptions as end-to-end distances, tilt angles of bonds relative to a fixed axis, etc., and distributions of configurations of ligands and of lipids, and distributions of groups of atoms with respect to three-dimensional space, specifically along an axis, within a plane, or among various regions or compartments.

Molecular dynamics simulations can only probe events that occur over a relatively short timescale, typically under a microsecond; thus, we necessarily choose model systems that do not include all the parts that are present in vivo, not even for in vitro experiments. Yet, we expect to visualize molecular level events that would typically occur in such experiments, for example on supported lipid bilayers [1], provided that we use parameters for the model systems in our simulations that are validated by the experimental physical properties of the model systems. The types of information we obtain are, to some extent, testable, but we also provide very detailed information that is not otherwise available by experiments, which, if used judiciously, sheds some mechanistic light on the behavior of nanoparticles interacting with and permeating lipid membranes.

We structure our contribution in the following form: (a) a question that we pose, (b) the simulation systems that we construct to help answer the posed question, (c) the observed behavior of our simulation system, (d) the interpretation of the behavior in quantitative terms, and (e) the conclusions and the caveats and limitations associated with our answers to the posed question.

2 Simulation Methods and Methods of Interpretation of "Observations" in the Simulations

2.1 Coarse-Grained Force Field

Three saturated lipid bilayer systems with different hydrocarbon chain lengths, DCPC (C_8), DMPC (C_{14}), and DPPC (C_{16}), have been widely studied both experimentally and in simulations. While we have studied all three, [2, 3], in this report we focus on DPPC, which has been more extensively studied and is in general of greatest interest. The molecular structures of these three lipids are shown in Fig. 1.

Our simulation studies have been carried out using the MARTINI coarse-grained force field. It is based on a four-to-one mapping strategy, with four heavy atoms represented by one active interaction site [4]. Details of this model have been previously published [5]. In summary, the model has four main categories of interaction sites: polar (P), nonpolar (N), apolar (C), and charged (Q). Within each category, subcategories are denoted by a letter indicating the hydrogen-bonding characteristics (d = donor, a = acceptor, da = both, o = neither) or a number denoting the level of polarity (from 1 = lowest polarity to 5 = highest polarity).

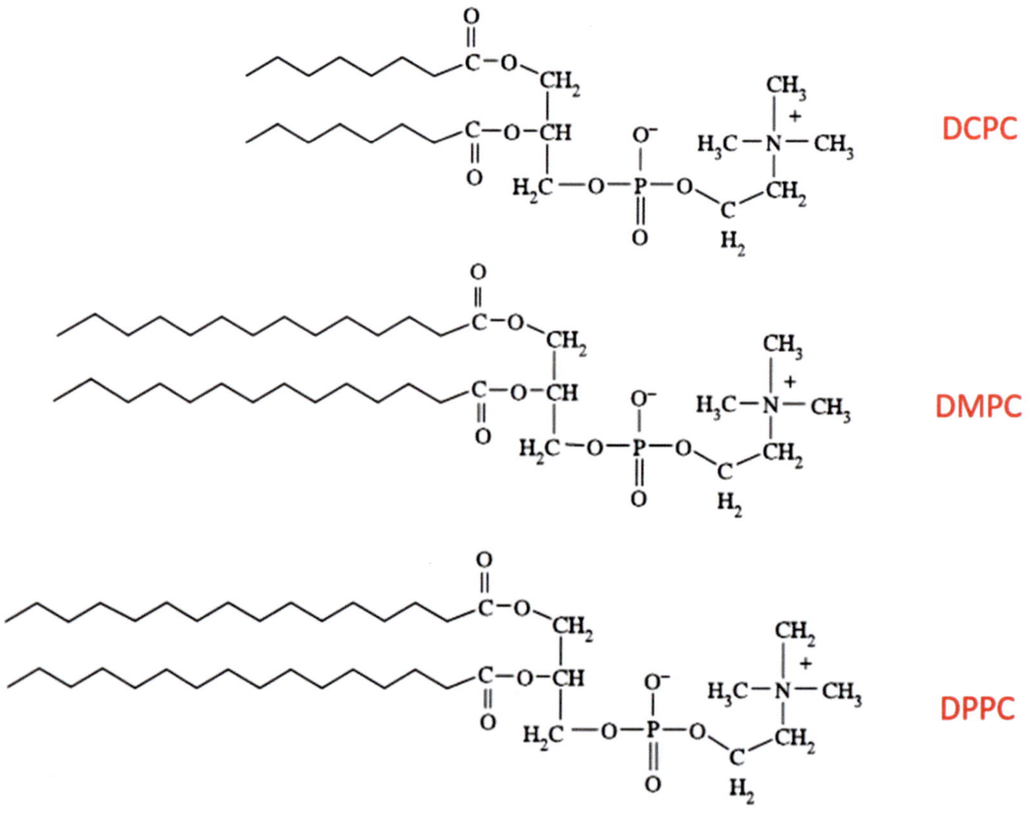

Fig. 1 Molecular structures of DCPC (C_8), DMPC (C_{14}), and DPPC (C_{16}) reproduced from ref. 2

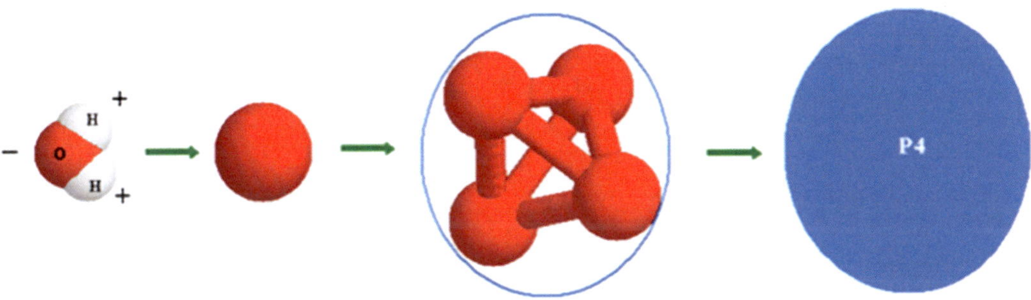

Fig. 2 Coarse-grained mapping strategy for water. Figure reproduced from ref. 2

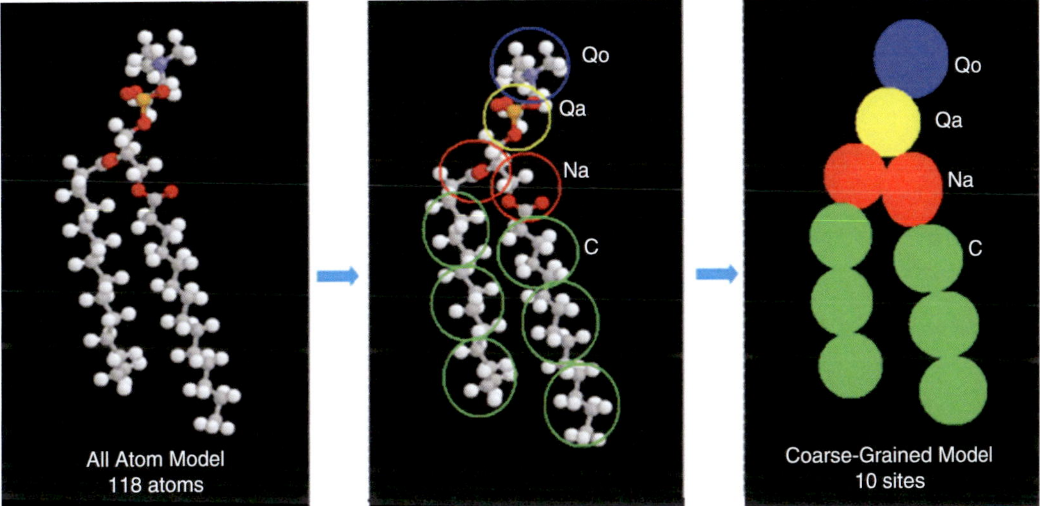

Fig. 3 Coarse-grained mapping strategy for a DMPC molecule (blue = cholinc group, yellow = phosphate group, red = glycol group, green = acyl chain). Figure reproduced from ref. 3

As an example, the mapping of a coarse-grained water site (P_4) is shown in Fig. 2. For the lipids, we used a similar mapping strategy; the phospholipid DPPC is modeled with 12 CG sites; the mapping of a DPPC molecule is shown in Fig. 3.

All site-site interactions between sites i and j at distance r_{ij} are modeled via a Lennard-Jones (LJ) potential:

$$V_{LJ}(r_{ij}) = 4\varepsilon_{ij}\left[\left(\frac{\sigma_{ij}}{r_{ij}}\right)^{12} - \left(\frac{\sigma_{ij}}{r_{ij}}\right)^{6}\right] \quad (1)$$

The LJ energy parameter ε_{ij} values range in value from ε_{ij}=5.6 kJ/mol to ε_{ij}=2.0 kJ/mol for strong polar groups (higher end) and between polar and apolar groups to capture the hydrophobic effect (lower end). The LJ size parameter is set at $\sigma = 0.47$ nm for all interaction types. An exception is made for interactions between charged (Q type) and most apolar types

(C1 and C2) for which it is set at $\sigma = 0.62$ nm to increase the range of repulsion. A shifted coulombic potential function is added in addition for charged groups to the LJ interaction:

$$U_{\text{elec}} = \frac{q_i q_j}{4\pi\varepsilon_0\varepsilon_r r} \tag{2}$$

The nonbonded interaction in our simulations has a cutoff distance of $r_{\text{cut}}= 1.2$ nm. To ensure smooth potentials, the LJ potential is shifted at $r_{\text{shift}} = 0.9$–1.2 nm while the electrostatic potential is shifted from $r_{\text{shift}}= 0.0$ nm to 1.2 nm using the usual standard shift function [6].

Bond vibrations are modeled by a simple harmonic potential, V_{bond} (R):

$$V_{\text{bond}}(R) = \frac{1}{2} K_{\text{bond}}(R - R_{\text{bond}})^2 \tag{3}$$

while for bond angles a cosine-type harmonic potential, $V_{\text{angle}}(\theta)$, is used:

$$V_{\text{angle}}(\theta) = \frac{1}{2} K_{\text{angle}}\{\cos(\theta) - \cos(\theta_0)\}^2 \tag{4}$$

In our simulations, we followed the following atomistic to coarse-grained mapping strategy: the gold and sulfur atoms are mapped 1:1 and assumed rigid/fixed. The residues of alkyl chains are 4:1 mapped and flexible. The interaction between the gold nanoparticle and lipid molecules is modeled by an L–J potential model. A wide range of potential parameters have been used in previous studies for gold atoms; these include all-atom [7–9] or coarse grain (these include atomistic structures but gold atoms modeled as either C-class [10] or P-class [11] using MARTINI force fields). In our studies we have used MARTINI C5-type interaction sites for gold atoms; N0 type has been used for sulfur atoms and C1 type for alkyl chains. These classes of interaction sites and the corresponding potential parameters have been tested and verified against atomistic simulations by Marrink et al. [5] and we have, in addition, tested them in our simulation studies by comparing with experimental data for lipid membranes [12]. The parameters for nonbonded and bonded interactions for gold nanoparticles are shown in Table 1. As was done previously, for the cross-interactions we used the standard Lorentz-Berthelot mixing rules.

2.2 Cycled Annealing

Cycled annealing is often useful in simulations when a model system may have multiple local minima and where a global minimum needs to be reached. When the temperature of a model system is increased, its parameter space is more accessible for the many degrees of freedom of the system. At lower temperatures, the system molecules can get stuck at local minimum structures that are

Table 1
Nanoparticle force field

Nonbond		Bond			Angle		
interaction site	Type[a]	connecting block	R_0	K_{bond}	connecting block	Θ_0	K_{angle}
Au	C5	Au-S	0.445	1250	Au-S-ligand	180	25
S	N0	S-ligand	0.445	1250	S-ligand-ligand	180	25
Ligand	C1	Ligand-ligand	0.47	1250	Ligand-ligand-ligand	180	25

(R_0 in nm, K_{bond} in kJ mol^{-1} nm^{-2}, Θ_0 in deg., K_{angle} in kJ mol^{-1} rad^{-2})
[a]MARTINI classification

unlikely to replicate experimental observations. In cycled (or simulated) annealing, a model system is heated to temperatures high enough to allow the entire parameter space for molecules to be more accessible. This is followed by stepwise cooling to the final desired temperature. If cooling is carried out at a slower rate, the global minimum is more likely attainable. Simulated annealing is used in many computational applications including obtaining the structure of functionalized nanoparticles with various ligand types [13] or determining the correct folded structure of a protein [14]. We employed cycled annealing in our work to obtain equilibrium structures for gold nanoparticles with various lengths of attached alkanethiol or PEG ligands.

2.3 Construction of Nanoparticles

Gold nanoparticles (AuNP) functionalized for biological and biomedical applications are of interest in a wide range of applications; some examples include bio-imaging, single-molecule tracking, drug delivery, and related diagnostic applications [15–18]. Gold nanoparticles can often be engineered to target tumor cells preferentially, using appropriately functionalized ligands. This could then be an effective tool for cancer diagnosis and therapy [19]. Such applications have motivated our work to employ gold as our model nanoparticle. The structure of the gold nanocrystals (nanoparticle) without ligands can be simply obtained by cutting a nearly spherical nanocrystal out of a bulk face-centered cubic (FCC) structure gold lattice, with a diameter of 2.1 nm. Ligands can then be attached to the surface of such a 2.1 nm gold nanocrystal using the following methodology for nonpolar ligands: The nanocrystal is placed in the center of a 12.0 × 12.0 × 12.0 nm^3 simulation cube to which are added butanethiol (ligands) in excess of what would be required to form a compact monolayer. We then carried out cycled annealing (*see* Subheading 2.2) simulations to condense the ligands onto the surface of the nanocrystal, as reported by Luedtke et al. [20] in a similar atomistic investigation. The temperature was cycled between 200 and 500 K to allow adsorption, stable binding, and finally desorption of excess ligand molecules. We obtained the final number of the equilibrated butanethiol chains on the gold core to be 87, resulting in a thiolate surface gold atom coverage of 48.3% (a surface density of 6.28 ligand nm^{-2}), which is within the range of experimental coverage measurements, up to 52–57% for 2.1 nm core diameter alkanethiolate gold nanoparticles [21]. As can be seen from Fig. 4a, following the annealing process, the surface

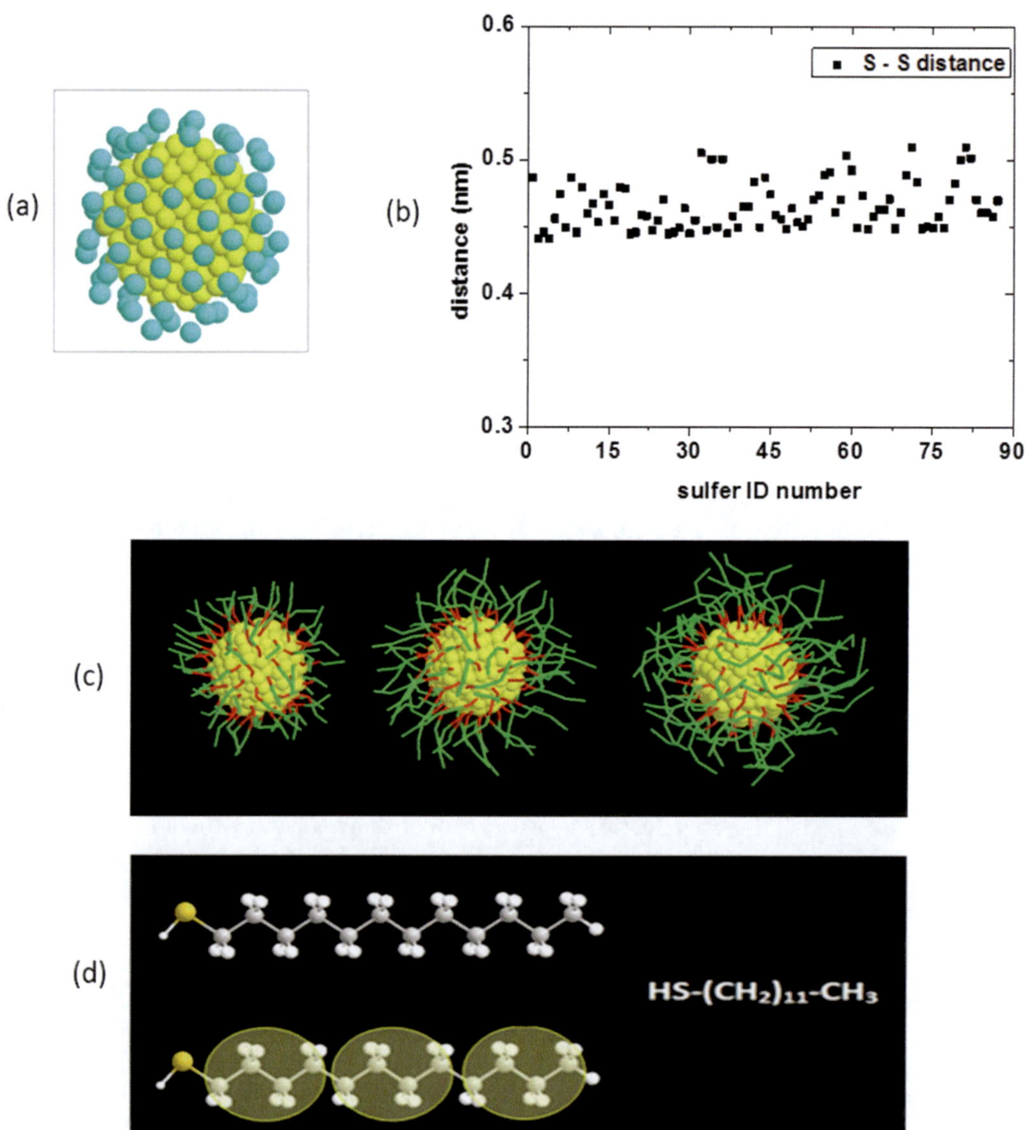

Fig. 4 A schematic illustration of the structures of nanoparticles used in simulations. (**a**) Distribution of surface atoms on the gold core (2.1 nm); (**b**) the distance between sulfur-sulfur atoms on the surface of the gold core; (**c**) the structure of the nanoparticles. The residues are replaced by R = $(CH_2)_n$ to form nanoparticles; shown here are structures for $n = 4$, 8, and 12 from left to right; (**d**) the structure of one alkyl residue and the coarse-grained mapping strategy from atomic sites to coarse-grained sites. Figure reproduced from ref. 12

sulfur atoms are uniformly distributed on the nanoparticles and the distances between sulfur atoms range from 0.44 to 0.51 nm (Fig. 4b). To investigate longer ligand lengths, the butanethiol ligands are replaced by R = $(CH_2)_8$ and R = $(CH_2)_{12}$ to form gold nanoparticles with neutral hydrophobic ligands of medium length and longer length, as shown in Fig. 4c. We used these alkanethiol-coated gold nanoparticles as examples of membrane permeation by nanoparticles with hydrophobic ligands.

The gold nanoparticles used here (2.1 nm diameter) are smaller than those typically used in many biomedical applications. In experimental studies, a wide range of sizes of Au nanoparticles have been used for applications such as drug delivery and as imaging agents. Nanoparticles in most studies range from 1 to 100 nm [22]. PEG-coated AuNPs (4 and 100 nm) have been reported to be administered intravenously to mice [23]. Pan et al. have studied the size dependence of cell toxicity of gold nanoparticles that are water soluble with sizes between 0.8 and 15 nm in diameter and concluded that all were most sensitive to gold particles 1.4 nm in size in all four cell lines investigated [24]. Hainfeld used 1.9 nm diameter AuNPs for imaging in mice [25]. We believe that our 2.1 nm nanoparticles, while small, are appropriate for many applications. We do note that gold nanoparticles used for biomedical applications such as gene and drug delivery are usually larger (20–100 nm). This is because they are often conjugated with other biomolecules or drugs and thus these larger sizes are known to permeate cell membranes efficiently using mechanisms such as endocytosis, which is not studied in our work.

To prepare the AuNP with hydrophilic ligands, we used PEG which has been methyl terminated in order to have a neutral particle. The PEG ligands are coarse-grained as in ref. 26. The original Marrink CG description of PEG was developed in 2009 [27], and the PEG-lipid parameters were developed in ref. 26. A new MARTINI CG PEG model, suitable for the longer molecular dynamics time steps typically used with MARTINI, has recently been developed and validated against the radius of gyration data from atomistic simulations and experiment [28]. The main difference between the latter [28] and the original [26] is using a new bead type P0, leading to slightly smaller equilibrium values for bonds and angles to increase stabilities. This is the model we used.

Gold nanoparticles are functionalized by thiol-terminated methoxy-poly(ethyleneglycol) (mPEG-SH) for colloidal stability. Thiol-terminated polyethylene glycol (PEG) is commonly used to functionalize the surface of gold nanoparticles (AuNPs) in order to improve their in vivo stability and to avoid uptake by the reticular endothelial system (*see*, for example, ref. 29). For PEGylated nanoparticles we equilibrated a $10.0 \times 10.0 \times 10.0$ nm^3 simulation box of PEG3-SH which resulted in a density of 1053.6 kg/m^3 using the NPT ensemble at 400 K (above the glass transition temperature of

PEG) [30, 31] for 5 ns. This agrees well with the value of 1048 kg/m^3 from experimental measurement and appropriate extrapolation [32]. We then inserted a 3.1 nm diameter gold nanoparticle in this system but now in an NVE ensemble at 400 K for another 5 ns. Once we approached equilibration, we switched the simulation to an NPT ensemble once again for cycled annealing. Many groups have simulated polymer nanocomposites, nanoparticles immersed in polymer melts, where polystyrene or polyethylene oxide polymer melts were doped with spherical nanoparticles or nanorods of various aspect ratios [30, 32–35]. Generally, simulations with polymer melts must be carried out with temperatures higher than the glass transition temperature to properly capture the structural properties of the polymer melt, which would otherwise begin to order at lower temperatures [36].

In all of the cycled annealing simulations, the thiol group is the attachment site of the PEGn-SH ligand. We began by heating the system to a high temperature of 1200 K to allow the ligands to explore a variety of stable binding sites on the nanoparticle surface and to favor desorption of excess ligands from the nanoparticle surface. In previous simulations of silica nanoparticles with PEG melts, temperatures of up to 1200 K [37] have been used. During the system cooling stages, we permitted the temperature to drop slowly by 20 K/ns for 5 ns. Once the system had been cooled to the final temperature (323 K for our simulations), we equilibrated the system for over 50 ns.

When constructing PEGn-SH-functionalized gold nanoparticles with various ligand lengths, it is not reasonable to assume that coverage is independent of ligand length, since experimentally synthesized PEGylated nanoparticles show that the coverage density of PEG on the nanoparticle surface decreases as the chain length/molecular weight increases [38]. The cycled annealing simulation with short PEG ligands resulted in a PEGylated AuNP with high coverage density. In the case of PEG3-SH ligands, we found that 75 ligands condensed on the nanoparticle surface which translates to a coverage of 2.49 ligands/nm^2. To investigate PEG6-SH AuNP and PEG12-SH AuNP systems, we used our equilibrated PEG3-SH AuNP nanoparticle in solution of PEG3-SH ligands as a starting point for the next simulations to construct PEGylated AuNPs with longer PEG ligands; to construct PEGylated gold nanoparticles with longer ligands, we attached additional beads to the former and attached additional beads to the ligands in the melt. After we equilibrated the new nanoparticle with longer ligands in the solution of its respective melt, we subjected the system to the same cycled annealing procedure. From this, we obtained a PEGylated AuNP with longer PEG ligands that had, in comparison, a lower coverage density. For example, these simulations resulted in a PEG6-SH AuNP with 50 ligands condensed and a PEG12-SH AuNP with 32 ligands condensed, corresponding to a coverage of

1.66 ligands/nm^2 and 1.06 ligands/nm^2, respectively. In separate simulations, we also completed a cycled annealing simulation for a bare gold nanoparticle in an isotropic melt of longer PEG ligands, directly. Using this alternate method, we obtained the same surface coverages for longer length PEGylated AuNP as we obtained using the replacement method. Depending on the initial ligand, coverages of 0.41–1.63 chains/nm^2 have been measured for Au nanospheres [39]. Other groups who have synthesized PEGylated nanoparticles have also reported a range in PEG coverage of 0.2–-2.0 ligands/nm^2 [29].

2.4 Nanoparticle Permeation Method

In nature, nanoparticles, especially ligand-coated nanoparticles, permeate lipid membranes spontaneously. This is observed with often hundreds of nanoparticles over several seconds/minutes. In our simulations, due to computational constraints we only used one nanoparticle, and our simulations were of the order of 100 ns. To observe permeation events within our constraints, we used an external force in the range of 50–1000 pN to aid the permeation of the nanoparticles in the membrane with velocity in the range of 0.35–1.4 m/s [3, 32]. The nanoparticle permeation velocities we examined here are 0.35, 0.525, 0.7, and 1.4 m/s, respectively. The external forces we applied are significantly smaller than the forces for example between two nanoparticles—0–12 nN [40]—or nanoparticles and cell membranes—50–1200 pN [41]. The nanoparticle velocities investigated (resulting from the forces applied) are larger than some experimental studies; these are however still several orders of magnitude smaller than the thermal velocities of water, ions, and lipid molecules (96.6–334.5 m/s) at the temperature investigated, and an order of magnitude smaller than thermal velocities of nanoparticles (7.5–51.0 m/s). Velocities explored in our studies are smaller than typical flow velocities of particles carried in the bloodstream. Therefore, we believe that our simulations still represent the permeation process realistically, although the process has been facilitated to shorten the permeation time significantly due to computational constraints (we can refer to these studies as "directed" simulations). This is also demonstrated by the recovery of the lipid layer between two permeation cycles indicating no permanent damage to the membrane at these velocities [12]. Other simulations [42, 43] have used similar velocities and their results also appear in reasonable agreement with experiments.

To determine the conditions under which our method of directed simulations may be considered realistic, we carried out some preliminary tests. We monitored the minimum driving forces needed for bare gold nanoparticles to permeate the first and second layers of the lipid as a function of size which are shown in Fig. 5a. In our simulations, we have defined the minimum force as that required to permeate the membrane in 160 ns or less. The minimum force for permeation across the first layer is in the range of

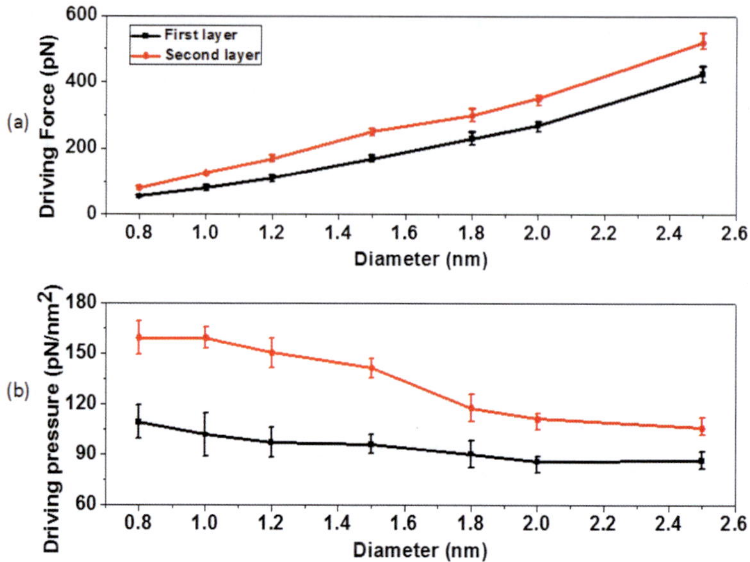

Fig. 5 Minimum driving force (**a**) and pressure (**b**) needed for various sizes of nanocrystals to permeate the first and second layers of the lipid membrane. Figure reproduced from ref. 3

55–425 pN, and for permeating both layers between 80 and 520 pN. The larger the nanocrystals, the larger the force needed. We also examined the minimum driving pressure (external force/ cross-sectional area of nanocrystals), which is shown in Fig. 5b. We found that the pressure needed for the nanocrystal to permeate the first layer is almost independent of the size of the nanocrystals. For the permeation of both the first and second layers, the required pressure decreased with increase of nanocrystal size. A larger nanocrystal introduces more disruption in the bilayer as it permeates across the first layer. At the same time, the larger nanocrystal is closer to the second layer of the lipid membrane after it gets past the first layer, so it more easily penetrates the second layer provided that it can get across the first layer. This is consistent with the observed minimum pressure for larger nanocrystals being smaller than that for the smaller crystals.

For ligand-coated nanoparticles we carried out similar tests. The minimum driving force for crossing the first and second layers (for a 2 nm nanoparticles with ligands) of the lipid membrane is shown in Fig. 6. The minimum force for nanoparticles permeating across the first layer is in the range of 175–225 pN, and 350–550 pN for permeating both layers. Typical forces applied to single cells for AFM imaging, which are not large enough to cause cell rupture, are in the range of 50–1200 pN [41]. For example, in the Vakarelski experiments AFM 20–25 nm tips applied loads of only 100–200 pN [44]. The external forces used in our simulations are of the same order of magnitude. To permeate the first layer, the

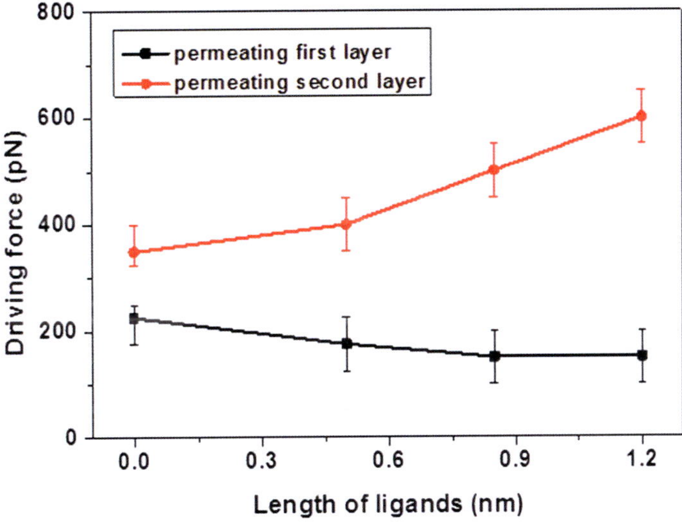

Fig. 6 Minimum driving force for nanoparticles permeating the first and second layers of the lipid membrane. Reproduced from ref. 12

force needed for a nanoparticle with ligands is smaller than for a nanoparticle without ligands. Compared to the bare nanoparticle, the ligands introduce more disruption in the first layer as the ligand-coated nanoparticles get close to the surface of the first layer under the same external forces, making it easier to open up the lipids to accept the nanoparticles into the bilayer.

To compare the dynamic characteristics of nanoparticles with and without ligands during the permeation process, we also monitored the velocity and force profile along the direction of motion of nanoparticles (*z* direction). We consider AuNP_bare, AuNP_SL (4 CH_2 groups), AuNP_ML (8 CH_2 groups), and AuNP_LL (12 CH_2 groups) nanoparticles permeating under the same driving force (600 pN) to study how the ligands change the permeation dynamics. We obtained the force profile by letting the nanoparticles permeate at a constant velocity (0.41 m/s) through the lipid membrane. Typical velocity profiles and force profiles for the permeation are shown in Fig. 7a, b.

Our results show that the method used here has the ability to describe the mechanism of nanoparticle permeation with and without ligands. For example, it correctly shows that for all cases, the velocity is reduced when nanoparticles approach the headgroups of the lipid membrane due to increased resistance in this region. This also indicates that an external force would be needed to permeate the membrane. In the entry region (2.5–4 nm in the *z*-axis in Fig. 7), to permeate the first layer of the lipid membrane, the nanoparticle compresses the first layer and pushes the headgroups apart to make room for the permeation. Thus the velocity of the nanoparticle decreases in this region while the resistance from the

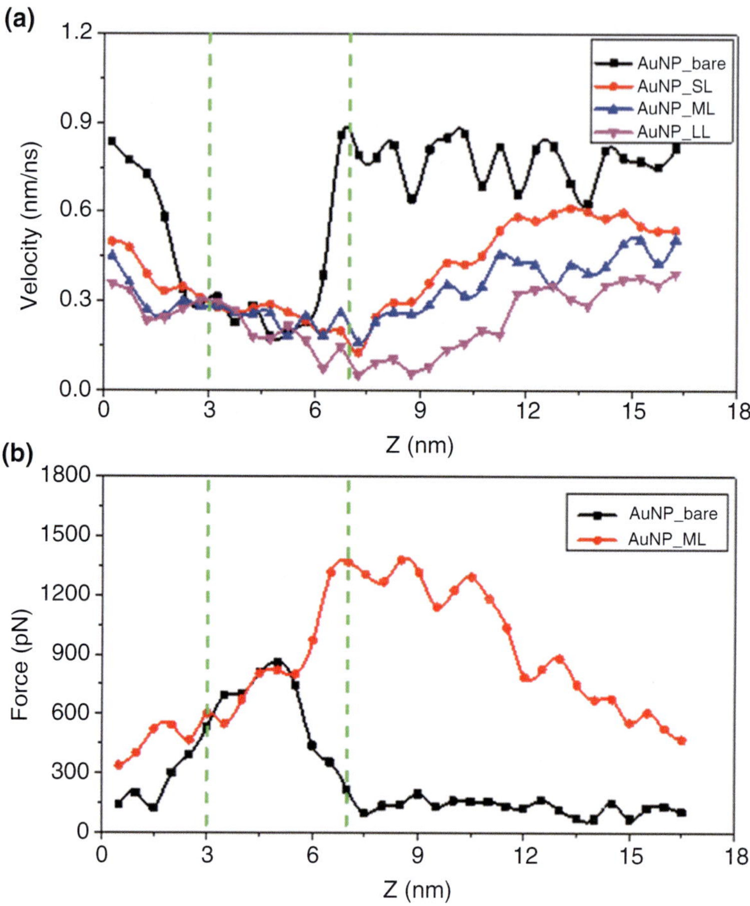

Fig. 7 Velocity and force profiles for permeation of an Au core of 2.1 nm. The green dashed line indicates the equilibrium position of the phosphate headgroups. (**a**) Using the same driving force (600 pN), we obtain velocity profiles of gold nanoparticles (*z* component). (**b**) Using a fixed velocity (0.41 m/s), we obtain the force profile (AuNP_ML nanoparticle). Figure reproduced from ref. 12

headgroup increases accordingly. When the nanoparticles move closer to the membrane center (4–5.5 nm in the *z*-axis in Fig. 7), the velocity of nanoparticles goes down further. The deformation of the first layer leads to the gradual deformation of the second layer, maintaining a pore in the direction of motion of the nanoparticle. Similar velocity and force curves for AuNP_bare, AuNP_SL, AuNP_ML, and AuNP_LL nanoparticles are obtained in this region. When the bare nanoparticles have completely crossed the first layer, this layer begins to recover while the second layer is compressed which leads the tails of second layer to separate from each other. This drags the headgroups apart to form a pore in the second layer even before the nanoparticle arrives there, which leads to a significant speedup in the velocity profile and a significant decrease in the force profile. All ligand-coated gold nanoparticles, however, move more slowly in this exit region and exhibit a minimum in the velocity profile. This is due to the attractive interactions between lipid tails and ligands.

3 The Lipid Bilayer Membrane, Properties, and Dynamics

A real cell membrane is an asymmetric phospholipid bilayer which is heterogeneous and is constituted by more than just lipids and typically has embedded proteins; membrane lipids are highly diverse and include various phospholipids (phosphoglycerides and sphingolipids), glycolipids that have a carbohydrate group, and sterols (e.g., cholesterol). However, we use a phospholipid bilayer model system for a cell membrane, using only one type of lipid so that the compositions of both leaflets are identical and remain so throughout the simulation. It is just as easy to set up a model system constituting a combination of lipids and including cholesterol, or including an embedded transmembrane protein, but for the simulations involving nanoparticles permeating a lipid membrane described herein we started with a homogeneous system with lipids having only saturated tails, because then it is possible to validate our simulation results on the lipid bilayer itself by comparing with experimental properties of such symmetric homogeneous bilayers. Also, the interpretation of our simulation results would be more general and not be specific to the particular chosen composition of the bilayer.

Does a lipid bilayer self-assemble from an isotropic solution in water to a bilayer with indefinitely long lifetimes, maintaining an average thickness and certain mechanical properties? This is an important validation requirement for the simulation results on the model system to be meaningful.

We started out with an isotropic mixture (random initial orientations and positions) of water and the lipid molecules and put the system through 100 ns steps. An example of the results at various simulation times is shown in Fig. 8 for 128 DPPC lipid molecules and 2000 CG water molecules. This is a rather stringent test of both the model and the simulation algorithm. We display snapshots of the configurations at suitable intervals for a 100 ns simulation in Fig. 8, which clearly shows a lipid bilayer membrane being formed spontaneously.

The same mechanism for self-assembly is found in the coarse-grained MD as in atomistic MD simulations, for example, in bilayer self-assembly of eight different types of phospholipids in unbiased molecular dynamics (MD) simulations using three widely used all-atom lipid force fields. Irrespective of the underlying force field, the lipids are shown to spontaneously form stable lamellar bilayer structures within 1 μs [45], the majority of which display properties in satisfactory agreement with the experimental data. In

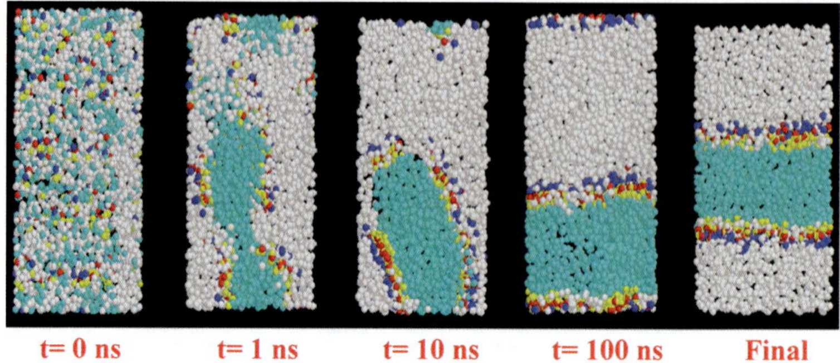

Fig. 8 Self-assembly observed for DPPC lipid molecules from an isotropic solution in water to form a lipid bilayer membrane (blue = choline group, red = phosphate group, yellow = glycol group, cyan = acyl chain, white = water). Figure reproduced from ref. 47

our coarse-grained simulations, the equilibrated lipid bilayer that has formed is the starting point for collecting the properties and comparing with experiments. The bilayer persists indefinitely and the thickness can be measured over a 1 ns period after equilibration for 10 ns once the bilayer has fully formed. We then examined a range of properties of this membrane. Important quantities characterizing a lipid bilayer membrane are (a) the surface area per lipid; (b) the thickness of the membrane which may be characterized by providing the average distance between two extreme points such as phosphate groups on opposite sides of the bilayer; and (c) the thickness of the interior, for example the projected distance along the bilayer normal for carbons on opposite sides of the bilayer. In a CG simulation, we can provide (a) and (b). The area per lipid measured for the self-assembled DPPC at 323 K (50 °C) was found to be 0.68 nm^2, which agrees well with the experimental measurements of 0.63 nm^2 from simultaneous analysis of neutron and X-ray scattering data by Kucerka et al. [46]. The density profile of each component of the lipid was obtained during the simulation, from which we obtained the distance between phosphate groups as 3.7 nm which is in close agreement with the experimental value of 3.80 nm from Kucerka et al. [46]. The distribution of the mass of the lipid along the direction of the axis normal to the bilayer surface can be measured over the same period. The results are shown in Fig. 9.

What is the dynamic nature of a lipid bilayer membrane, i.e., in terms of the extent of lateral movement of the lipids, order parameter of the lipid tails, and cross-sectional distribution of the lipid molecule parts along the normal to the membrane surface? How do these simulation "observations" compare to the experimental properties of bilayer membranes? Again, this is an important validation requirement for the simulation results using the model system for the membrane to be meaningful.

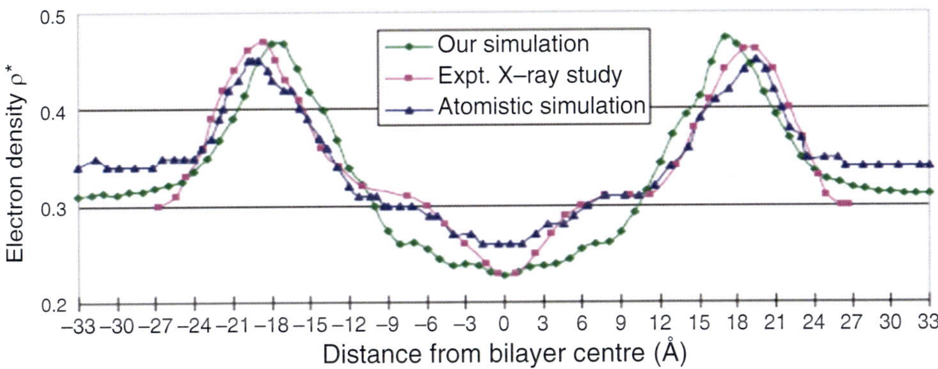

Fig. 9 Comparison of simulated electron density (ρ* in electrons/A° [3]) with X-ray experimental measurements and atomistic simulation from the compilation of Nagle et al. [48]. Figure reproduced from ref. 47

The structure of a lipid bilayer membrane, as obtained from X-ray measurements, is deduced from measured electron densities as a function of the position across the bilayer. By associating the positions of our CG sites with the electron densities accompanying the group of atoms included in that CG site, we may directly compare the results of our average positions from simulations with the electron densities from experiments [47]. This comparison is shown in Fig. 9 We note that our simulations not only compare well with atomistic simulations of the same system, but also agree well with the experimental electron density profile from the compilation of Nagle et al. [48].

The dynamic structure of the middle section (the lipid tails) of the membrane has been probed experimentally by deuteron nuclear magnetic resonance (NMR). Generally, the conformation of the hydrocarbon tails of the lipid in a bilayer membrane is disordered. The conformational and orientational order/disorder can be quantified by various quantities in an atomistic simulation, including the very useful order parameter of C–H bond directions, but in a CG simulation only a limited number of indicators of internal order may be obtained from the simulation. One we have considered above, in Fig. 9, is the probability distribution of different groups along the membrane normal axis, which gives some indication of the average alignment of the various parts of the lipid molecule. Another measure of the internal order of our lipid bilayer is the order parameter,

$$P_2 = \left\langle \frac{3\cos^2\theta - 1}{2} \right\rangle \tag{5}$$

where θ is the angle between the bond and normal to the bilayer. The value $P_2 = 1$ denotes perfect alignment, $P_2 = -0.5$ anti-alignment, and $P_2 = 0$ a random orientation.

Because we are using a coarse-grained scheme, the order parameter of our sites cannot be compared directly with C_{n-1}–C_n

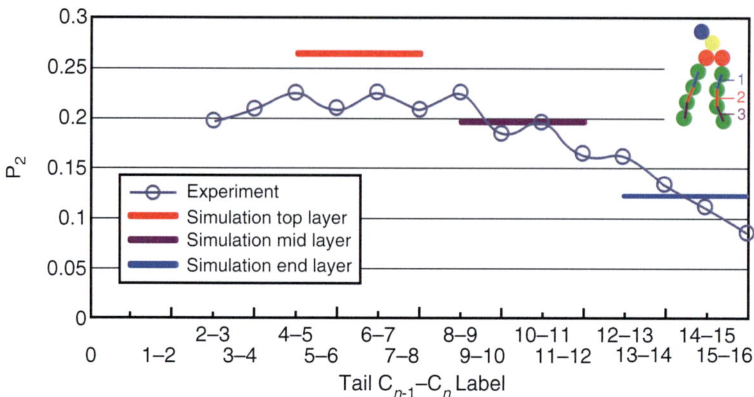

Fig. 10 Comparison of tail segment order parameter with experimental measurements from deuteron NMR [49]. Figure reproduced from ref. 47

order parameters derived from experimental NMR C–D bond order parameters [49]. The four sites that constitute the tail form three "bonds" between them, and the order parameters for these can be effectively compared with experimental values. From the results shown in Fig. 10, we can see that the coarse-grained model qualitatively reproduces the correct trends, with our lipid model being somewhat more highly aligned near the head than the actual DPPC lipid layer.

3.1 How Does a Lipid Bilayer Membrane Behave Under Compression?

There have been atomistic MD simulations of the DPPC bilayer under tension. The results were obtained by starting with various increasing values of area per lipid ranging from 0.635 to 0.750 nm^2, corresponding to tensions ranging from −2.6 to 15.9 mN m^{-1} [50]. The resulting structural and dynamical properties are entirely as expected for increasing area per lipid. Increasing the surface area resulted in a decrease of the lipid density at the headgroup region and a concurrent increase in the local density at the midplane of the bilayer. This indicates increased interdigitation of the acyl chains of the opposing leaflets due to extension of the chains beyond the bilayer midplane; spreading of the acyl chains also takes place. There is a corresponding expected increase in lipid lateral diffusion, and a decrease of order parameters for the tails. Other MD simulations start with zero stress in all directions and impose cyclic expansion or contraction of the bilayer area [51]. We instead consider the compression across the bilayer while keeping the surface area of the membrane the same.

For the membrane compressibility study, we prepared a simulation box with two walls thermally fluctuating and impermeable to water. We then increase the pressure by moving both walls toward the lipid bilayer at a rate of 0.27 m/s for 1 ns, and then the system is allowed to relax for 2 ns. Our studies indicate that after 2 ns the system did not change significantly and appeared to be close to

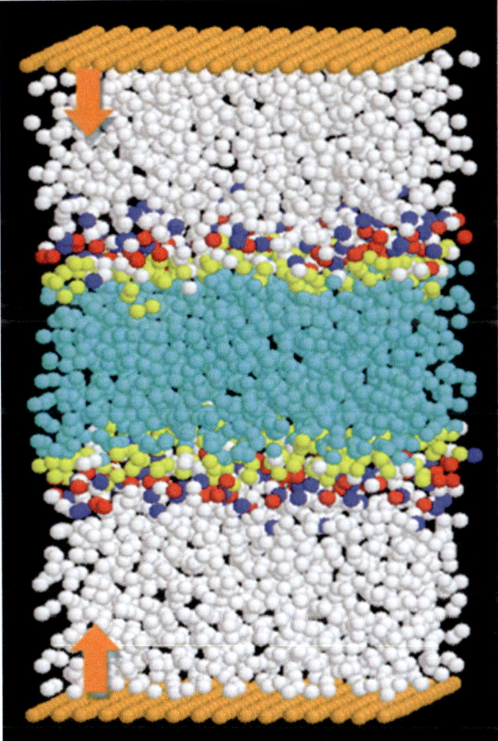

Fig. 11 Schematic of the simulation system to examine the compressibility of DPPC lipid membranes under pressure. Blue dots represent the choline group, red the phosphate group, yellow the glycerol group, and cyan the acyl chain tail

equilibrium. The pressure on the membrane was then calculated from the net force on the wall, that is, the net force exerted by water molecules on the lower side of the upper wall and on the upper side of the lower wall. This pressure should closely match the pressure on the lipid membrane since it is a connected continuous system at equilibrium or very close to it. Membrane thickness was calculated from the density profiles collected during the relaxation period. This procedure was repeated and resulted in developing pressures of up to 68.6 kbar. In order to keep the CG water from "freezing" during compression, we added 0.1 mole percent of antifreeze particles into the simulation system as suggested for the MARTINI force field [5]. The setup is shown in Fig. 11.

The results using a moving velocity of 0.27 m/s for 1 ns are shown in Fig. 12.

Simulations allow us to extrapolate experimental studies to much higher pressures that are not possible in experiments (experiments were at pressures up to 2 kbar). Thus, our simulations can be used to determine membrane behavior at pressures not easily accessible experimentally. Atomistic molecular dynamics simulations of the effects of high pressure up to 3 kbar on fully hydrated DPPC bilayers have been carried out up to 1 μs [52]. It was found that

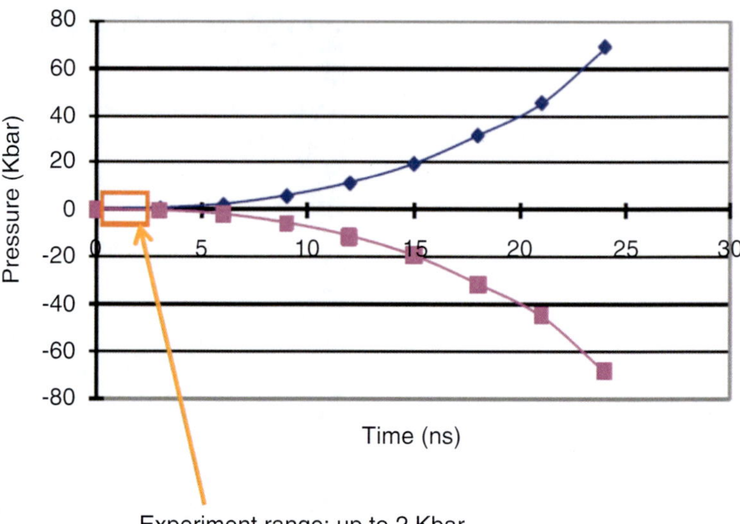

Fig. 12 Pressure variations during the compression process, in blue is the pressure on upper wall, in red is the pressure on lower wall, and range box denotes the experimental range

DPPC bilayers formed a rippled gel-like phase comprising a minor disordered fluidlike region and a major ordered gel-like region at 1 kbar, a partially interdigitated gel-like phase at 2 kbar, and a gel-like phase with most of the lipid acyl chains tilted with respect to the plane of the bilayer at 3 kbar. This was different from our work in that their pressure was applied isotropically. In our coarse-grained simulations, the results are as follows. During the compression, changes in overall atom positions can be expressed as density profile changes. The density profile of each group of atoms of the lipid at the beginning and at the end of the compression cycle is shown in Fig. 13.

The density profiles in general become more nonuniform during compression to a more solid-like state, as a result of the compression of the lipid bilayer. We noticed that significant density profile changes happen at a pressure range up to 10 kbar, which suggests that the lipid membrane is more compressible in this range. Also, the most significant component profile change is observed for the acyl tail, which shows the largest change in the number density, and exhibits the formation of a very distinct solid-like structure. In addition, from the density profile it is evident that most of the compression is resulting from the compaction of the tail section, which is not surprising considering the structure characteristics of bilayers. The orientation order parameter of the tail section also confirms such behavior. We characterized the tail orientations of the lipid molecules during the compression process by calculating order parameters during the relaxation period after each compression step. Bond 1–3 in the CG molecule schematic shows the

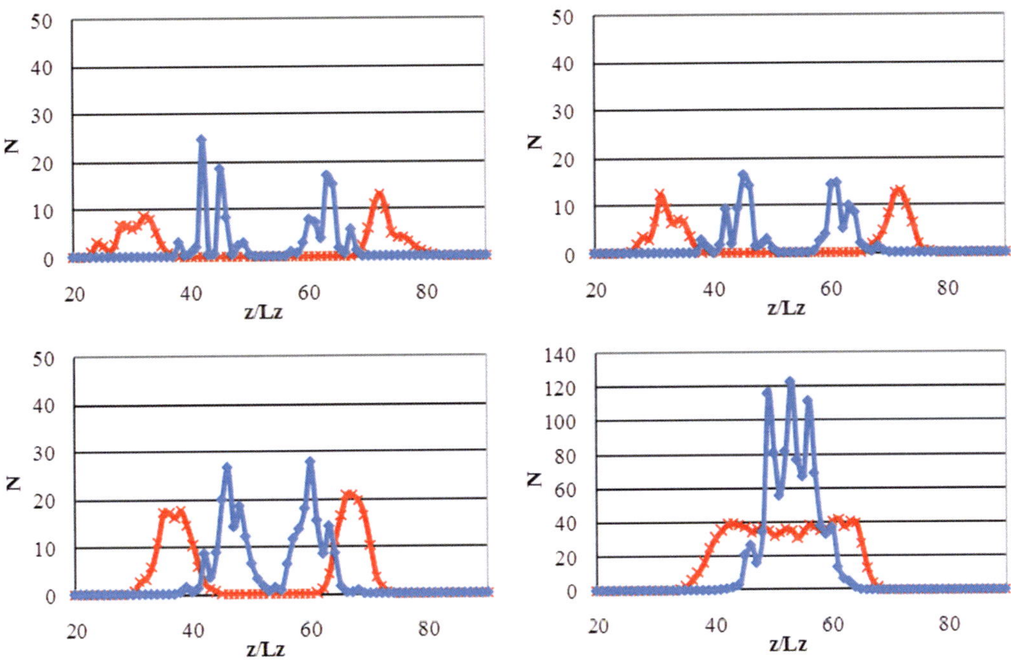

Fig. 13 Lipid density profiles at the beginning and the end of compression (in red from beginning of compression, in blue from end of compression; NC3 = choline group, P04 = phosphate group, GL = glycerol group, Tail = acyl chain group)

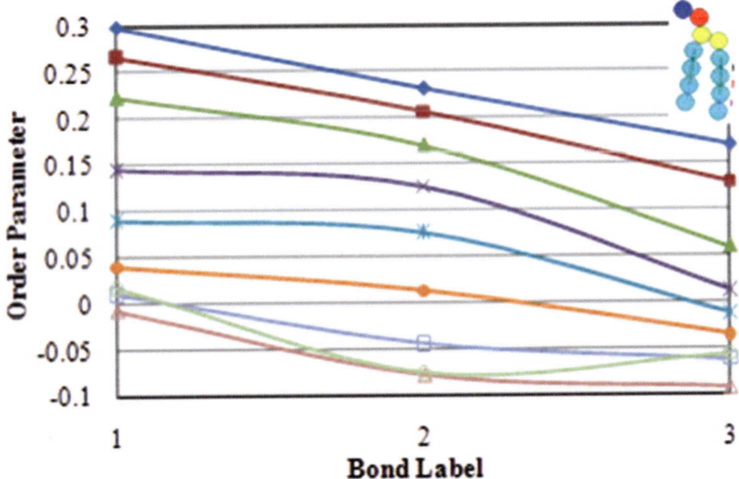

Fig. 14 Lipid tail segment order parameter change during the compression; red arrow shows direction of changes along the compression process. The bond labels are the same as in Fig. 10

connections between the four tail beads on each tail, the same designation as in Fig. 10. As shown in Fig. 14, the order parameter of the lipid tail decreased during the process, which indicates that the lipid tails went from largely aligned along the normal axis to a

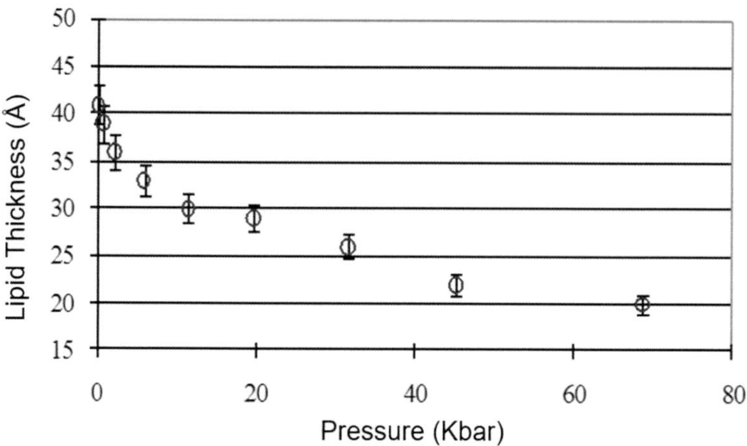

Fig. 15 The thickness of the lipid bilayer membrane changes with applied pressure

more random arrangement, with interdigitation of the tails of the two leaflets. At the beginning (no compression) the lipid is at its most aligned state with order parameters in agreement with previous simulation studies. As the pressure along the normal to the bilayer increases, the free space available for the tail segments decreases and their movement becomes constrained at different orientations. At the highest compression state the tail segments start to show anti-alignment behavior and the order parameter even becomes negative, which means the tails bend toward the lipid bilayer surface because of compression.

The thickness of the membrane (defined as the distance between the phosphate headgroups of the upper and lower layers of the lipid membrane) changes with the pressure as shown in Fig. 15. In atomistic simulations in which the pressure is applied isotropically up to a maximum of 3 kbar, the thickness of the DPPC bilayer decreased to the same extent [52] when compared to our data at the same pressure.

We calculated the compressibility of the lipid membrane from the pressure exerted on it and the change in thickness of the lipid membrane. The compressibility of the lipid membrane K is defined as

$$K(P) = \frac{1}{V_{\text{atm}}} \left(\frac{\partial V}{\partial P} \right)_{\text{T}} \tag{6}$$

Here, P is the pressure, V is the lipid volume, and V_{atm} is the lipid volume at atmospheric pressure. Since the membrane area is fixed in our studies, the volume of the lipid is proportional to the thickness of the lipid membrane. This leads to a more simplified definition of the compressibility in our case to

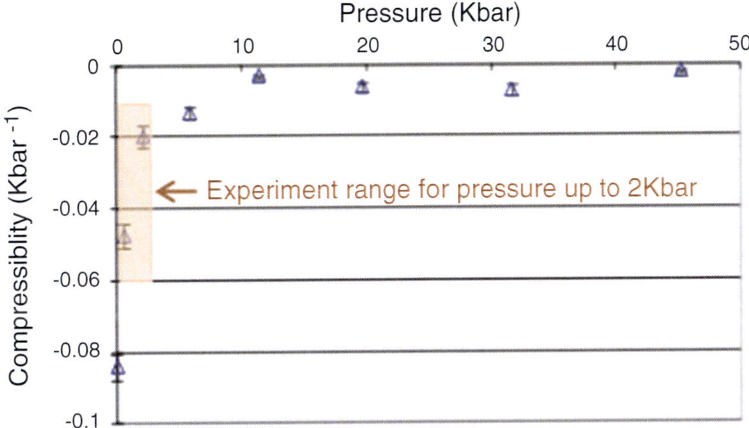

Fig. 16 Compressibility of the DPPC lipid bilayer as a function of pressure, calculated using Eqs. 6 and 7

$$K(P) = \frac{1}{h_{\mathrm{atm}}}\left(\frac{\partial h}{\partial P}\right)_{\mathrm{T}} \tag{7}$$

By measuring the thickness changes during the compression, we can calculate the compressibility K using Eq. 7 and the results are shown in Fig. 16.

As seen in Fig. 16, we can calculate the compressibility for pressures up to 45 kbar (about ten times the experimental range). The estimated values of compressibility from our simulations in the experimental pressure region (0–2 kbar) are between −0.088 and −0.019 kbar^{-1}, which agree well with experimental measurements [48]. Scarlata [53] measured the compressibility of DMPC at 40 °C, and reported values in the range of −0.088 to −0.057 kbar^{-1} from different probe positions. Our results are in agreement with measurements that are in the range of −0.01 to −0.06 kbar^{-1} obtained for oriented DPPC and DPPC/cholesterol multilayers by neutron diffraction [54]. Atomistic simulations also provide area compressibilities for DPPC that are in good agreement with experiment [55].

Our results show that there is no significant change in compressibility beyond 2 kbar. It had been previously suggested that high hydrostatic pressures may cause the elimination of some integral membrane proteins due to increased lipid packing [54, 56]. In a previous study, we had studied the gas permeation through a lipid bilayer membrane with an embedded OmpA channel, a common protein channel in lipid membranes [47]. Proteins have intrinsic compressibility of about 0.01 kbar^{-1}, which is lower than the typical lipid bilayer compressibility of −0.01 to −0.06 kbar^{-1} [56]. It will therefore be of interest to simulate the compression of a DPPC lipid membrane with an embedded OmpA (to use a simple barrel protein as an example), to observe whether the protein becomes disengaged from the bilayer as the system undergoes compression.

4 A Neutral Nanoparticle in Aqueous Solution, Properties, and Dynamics

There are a wide range of nanoparticles in current use or under investigation for medical applications. We focus our computational work on gold nanoparticles because they are simpler than most others, they have been widely studied experimentally, and also because their applications include not only delivery of pharmaceuticals but also photothermal therapy [57–60], as contrast agents for imaging [61, 62] and for targeted cancer therapy [63]. Gold nanoparticles are taken up by human cells but do not cause acute cytotoxicity [18]. A recent review considers the issues of size, shape, and surface chemistry of nanogold in these applications, *vis-à-vis* its cellular interactions, uptake, and toxicity [64]. An interesting possibility is suggested by the studies that have shown that PEGylated gold nanoparticles permeate through brain microvasculature [65].

What is the dynamic nature in solution of a nanoparticle with hydrophobic ligands bonded to its surface, i.e., can we say something about whether the ligands are coiled or extended, and how does this nature change with ligand length and surface coverage? How do these simulation observations compare with experimental properties of gold nanoparticles?

It is an important validation requirement for the simulation results using the model system for the alkane thiol gold nanoparticles in solution, despite the fact that observations on these types of nanoparticles in solution generally provide only average properties, and any details provided by SEM and TEM microscopy images are of dry particles after solvent removal. The ligands are mobile, but we can have a measure of their extension by using the radius of gyration. We can also characterize the AuNP diffusion through a liquid solvent. In order to be able to compare with experimental observations, we prepared a simulation box of chloroform solvent molecules at the appropriate density and equilibrated the alkanethiol-coated AuNP in it. We did this for several ligand lengths. The comparison of our CGMD simulation results with experiment and with atomistic simulations available for two types of ligands is shown in Table 2, where we compare our results with those from atomistic simulations from ref. 10 and with experiments from ref. 66.

We see that our simulations compare very well with atomistic simulations; both our coarse-grained simulations and the atomistic simulations reproduce the experimental radius of gyration reasonably well; and both come up with diffusion coefficients that are of

Table 2
Properties of AuNP with two types of ligands, 8 and 12 carbons long

	Radius of gyration (nm)			Diffusion coefficient (10^{-6} cm^2/s)		
Ligands	**Simulation**	**Simulation**[a]	**Expt**[b]	**Simulation**	**Simulation**[a]	**Expt**[b]
$HS–(CH_2)_7–CH_3$	0.991			1.310	1.333 ± 0.25	2.6 ± 0.3
$HS–(CH_2)_{11}–CH_3$	1.138	1.145 ± 0.004	0.924 ± 0.005	1.073	0.918 ± 0.04	2.3 ± 0.2

Our CGMD simulations of these nanoparticles in $CHCl_3$ solvent are compared with atomistic simulations and experimental observations
[a]From ref. 10
[b]From ref. 66

the same order of magnitude but somewhat smaller than found experimentally. Figure 4 shows typical configurations of the ligands for three lengths of ligands at a coverage of 48%, i.e., 0.48 S atoms per surface Au atom. First, we note that not all the ligands are extended as they would be in a "brush" model. We can see already in the snapshots shown in Fig. 4 that different ligand lengths have different configurations; in particular, the longer ligands show a larger fraction of bent configurations, whereas the shorter ligands do not. The internal degrees of freedom (torsional rotations around "bonds" between CG units and changes of angle between "bonds") permit the ligands to take various configurations over time, and the configurations taken by any one ligand can correspond to some average end-to-end distance (R_{ee}), one end being the S attached to the surface of the Au core. Therefore, one way to characterize the configurations of the ligands for a nanoparticle in solution is to display the distribution of R_{ee} values found for a given NP (for a given diameter of Au core and a given ligand type and length), with a given coverage, in a given solvent, at a given temperature; all of the mentioned quantities affect the distribution of ligand configurations. This would be interesting to show for alkanethiol-covered AuNP, but since $CHCl_3$ is not the usual solvent where the NP would meet a lipid bilayer we do not show this. Instead, we wait until the next section, to carry out this characterization of ligand configuration distributions for an AuNP that is actually used in aqueous solution.

What is the dynamic nature in solution of a nanoparticle with hydrophilic ligands bonded to its surface, i.e., can we say something about whether the ligands are coiled or extended, and how does this nature change with ligand length and surface coverage?

It is an important validation requirement for the simulation results using the model system for the PEGylated gold nanoparticles in solution, despite the fact that observations on these types of nanoparticles in solution generally provide only average properties. Typical snapshots of two PEGylated AuNPs are shown in Fig. 17. In these snapshots, we can see that on any one AuNP, there is a distribution of ligand configurations: some are extended more than others, and there are some that are coiled or in loops and at these relatively low surface coverages some are even lying down on the gold surface [67].

One way to quantify the configuration of a particular ligand is to measure its end-to-end distance; for our coarse-grained model this corresponds to the distance between the S attached to a surface Au atom and the end bead of a ligand. This end-to-end distance for each ligand is averaged every 100 ps over a simulation (after equilibration) run of 1 ns after 200 ns of equilibration is complete. We examined the average end-to-end distance of each of the ligands in PEG12 AuNP at the low coverage of 1.06 ligands/nm^2 (like the right snapshot shown in Fig. 17) and also of PEG12 AuNP at the

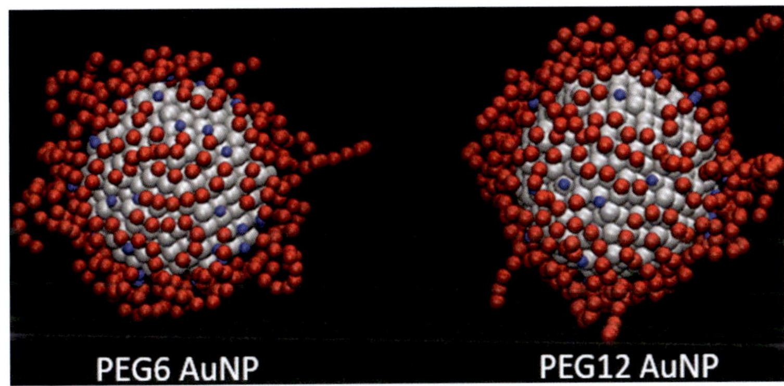

Fig. 17 Snapshots of PEG6 AuNP (left) and PEG12 AuNP (right) at low coverage, respectively, 1.66 and 1.06 ligands/nm^2, in water. The gold nanoparticle core is pictured in white with the sulfur beads (blue) and PEGn ligands (red). For clarity, the solvent molecules are not shown. Figure reproduced from ref. 67

high coverage of 2.49 ligands/nm^2, in water. These averages were 1.56 ± 0.66 nm and 1.89 ± 0.67 nm, respectively. For the same ligand length, the average end-to-end distance is longer at the higher coverage. The distribution of end-to-end distances that lead to these averages can also be examined; the large dispersions associated with the averages hint at broad distributions of values contributing to the averages. We binned the average end-to-end distance of each of the ligands into narrow ranges, in order to represent quantitatively the answer to the question: can we say something about whether the ligands are coiled or extended, and how does this nature change with ligand length and surface coverage? The distribution of the average end-to-end distances, given in Fig. 18, shows that for the same ligand length, the distribution of configurations of the ligands depends on the surface coverage.

First, we note that on any one nanoparticle not all ligands have the same configuration; rather a distribution of configurations is found on a single NP. Next, we note that the distribution of configurations is broader for high coverage than for low coverage. This is not unexpected because the greater number of ligands gives rise to a larger number of configurations. We also note that coiled configurations are found at low coverage as well as high coverage, and extended configurations too are found at low and high coverages. The difference between the two cases is that the proportion of extended configurations is greater at high coverage, and the most extended configurations, with end-to-end distances of 3.0–3.5 nm, are found only at high coverage. Although nanoparticles have been described in the literature as having ligands that are "mushroom" (coiled) style or else ligands that are "brush" (extended) style [68], we see that there are no all-mushroom or all-brush arrangements; that is, the simulations actually provide a broad distribution of configurations for each NP, and this distribution shifts, as the

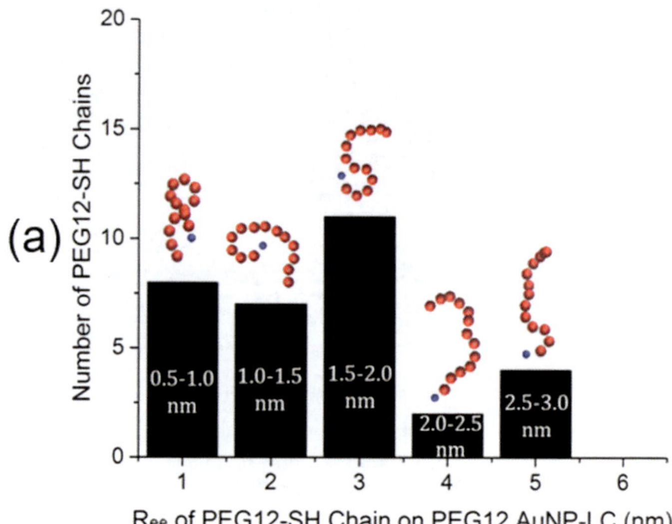

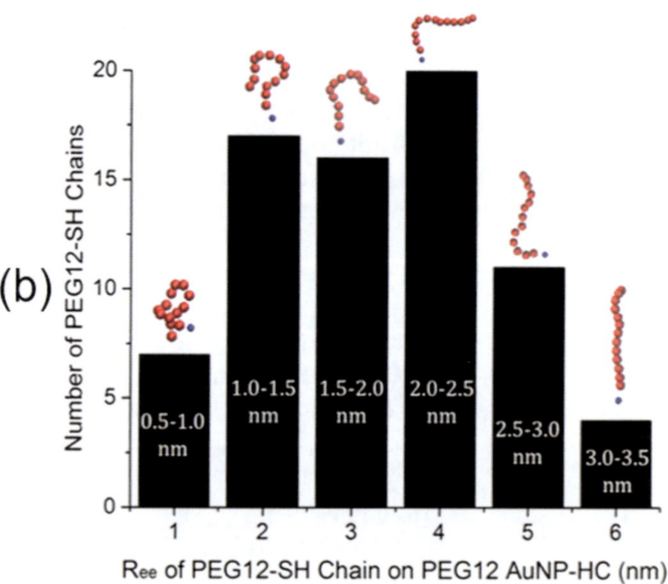

Fig. 18 Distribution of R_{ee} of ligands on (**a**) PEG12-AuNP (1.06 ligands/nm^2), $\langle R_{ee} \rangle = 1.56 \pm 0.66$ nm, and (**b**) PEG12-AuNP (2.49 ligands/nm^2), $\langle R_{ee} \rangle = 1.89 \pm 0.67$ nm, equilibrated in water. Snapshots of typical PEG12-SH from each distribution are included with blue representing the thiol atom and red the beads of PEG12. These are averages over 1 ns after equilibration is complete. Figure reproduced from ref. 67

coverage increases, to a greater weighting of the extended configurations. The icon of a ligand snapshot that is used to represent each of the binned sets is only a typical one; obviously, many types of coiling, for example, can give rise to R_{ee} average values within the range of 0.5–1.0 nm. From our results, we find that the transition from mushroom to brush behavior of individual PEG ligands would occur gradually as the surface coverage density increases, as the distributions depicted in Fig. 18 would change from panel a to panel b. Also worth noting is that another often used quantity, the radius of gyration, R_g, does not provide any indication of the breadth of the distribution of ligand configurations; and R_g is very similar (0.75 ± 0.12 and 0.76 ± 0.12 nm) for the two coverage cases shown in Fig. 18.

Next, we can ask how the distributions of ligand configurations change with ligand length. We found that $\langle R_{ee} \rangle$ does not change significantly as the coverage density increases on the nanoparticle with shorter ligands, PEG6-AuNP. Fewer configurations for each ligand are available for the shorter ligand length, leading to a narrower distribution of configurations (not shown here).

5 Interaction of a Neutral Spherical Nanoparticle with a Lipid Bilayer Membrane

We have examined the nanoparticles, each in solution by itself, and the membrane (with aqueous solution on both sides) by itself; now we can consider putting them together and observing the nature of the dynamic interactions between them.

As a ligand-coated nanoparticle approaches a lipid bilayer membrane, what happens, i.e., how does the lipid bilayer respond to the nanoparticle at close approach? At the same time, how do the ligands respond to the surface of the lipid bilayer? How does the nature of the ligands (hydrophilic or hydrophobic) affect the nature of these responses?

Our simulations provide the answers [69]. In the snapshots shown in Fig. 19, the first thing to note is that the PEG ligands are interacting with the phosphate and choline groups, but the positions of the headgroups of the top leaflet look relatively undisturbed in the plane of the membrane, as seen in the bottom image; all the disturbance is occurring along the direction normal to the membrane, with a slight curvature observed for the top leaflet, and less so for the bottom leaflet. On the other hand, the interactions with the membrane of the NP coated with alkane ligands are very different. There is a barrier to be overcome, but in exactly the same position of the center of the NP with respect to the membrane, already the alkane ligands are beginning to reach out to the lipid tails in the inner region, pushing the headgroups apart in the process, as is clearly seen in terms of the hole created in the top membrane leaflet, seen in the bottom image of Fig. 19 on the right.

When the first leaflet is breached by the nanoparticle, what is going on?

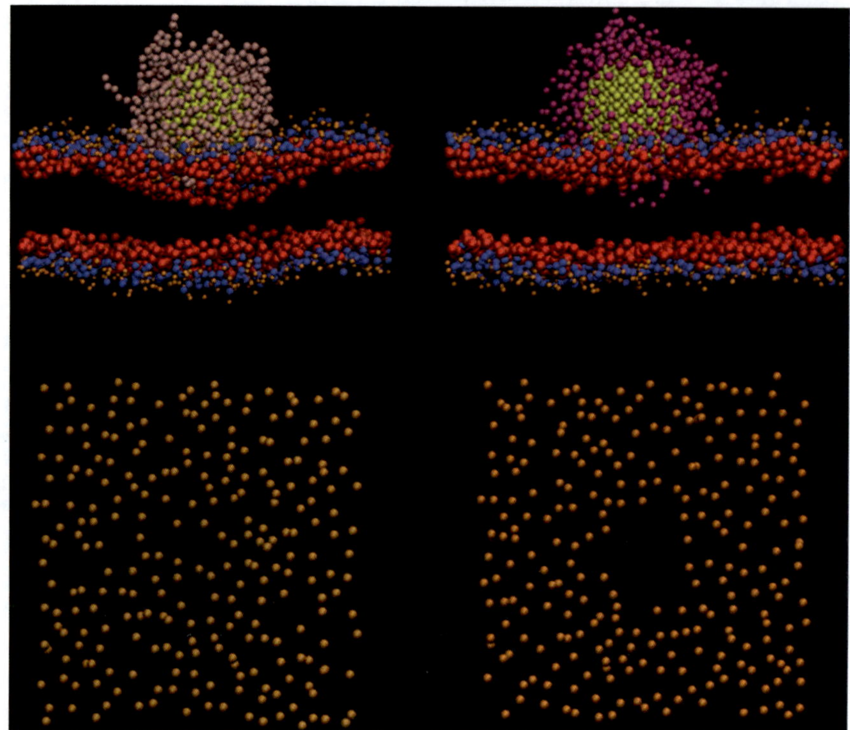

Fig. 19 Snapshot of the gold nanoparticle on approaching the top leaflet of the membrane, (top) and the top view of the choline headgroups of the top membrane leaflet (bottom). On the left is the PEGylated AuNP and on the right the alkanethiol-coated NP. Here, orange = choline, blue = phosphate, red = glycerol, yellow = gold nanoparticle core, pink = PEG ligands, and magenta = alkane ligands. Other atoms/groups are omitted for clarity. Figure reproduced from ref. 69

As seen in the snapshot in Fig. 20, the PEG ligands of the PEGylated AuNP are interacting with the headgroups of the lipid, moving them along with the AuNP as the latter moves through the top leaflet. Even at this position, the PEG is beginning to perturb the headgroups of the *bottom* leaflet. On the other hand, at exactly the same position of the NP center with respect to the membrane, the alkane ligands of the hydrophobic-coated AuNP have pushed away the lipid headgroups in order to reach and interact with the alkyl tails of the lipids, not interacting with the headgroups of the bottom leaflet at all, leaving them relatively unperturbed.

When the nanoparticle reaches the center of the membrane, does it tend to stay there? How do the lipids react to the presence of the nanoparticle between the leaflets? How does the nature of the ligands (hydrophilic or hydrophobic) affect what happens?

With the center of the AuNP in the center of the membrane, we see in the snapshot in Fig. 21 that the PEG ligands behave quite differently from the alkyl ligands; the PEGs at the top and bottom portion of the AuNP are engaged in interacting with the headgroups of the top and bottom leaflets, but those PEGs in the

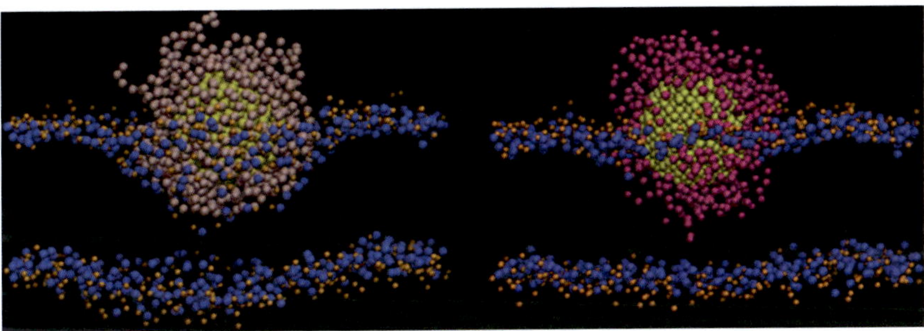

Fig. 20 Snapshots of the gold nanoparticle within the top leaflet of the lipid bilayer membrane. On the left is the PEGylated AuNP, and on the right is the alkanethiol-coated AuNP. As in Fig. 19, many atoms/groups are omitted for clarity. Figure reproduced from ref. 69

middle are tucked in and avoid the lipid tails. At the same time, the lipid heads in *both* the top and bottom leaflets become disordered by the interaction with the PEG ligands when the PEGylated AuNP lies within the center of the membrane. Atomistic MD simulations of a hydrophilic coated AuNP (with mercapto-undecane sulfonate) with a gold core of 2 and 3 nm embedded in a DOPC bilayer show results that are similar to our PEGylated AuNP in DPPC, in terms of the number density distribution of lipid tail groups, lipid headgroups, and ligands around the NP [70]. Accompanying the permeation process in our simulations are lipid flip-flop events, in which the head of a lipid molecule from one leaflet is caused to be dragged toward the interior and this leads to a reproducible tumbling event in which the two tail parts open away from each other and close again as the lipid completes its rotation and joins the opposite leaflet. We have examined and characterized lipid flip-flops in detail elsewhere [67, 71]. Our mechanism for lipid flip-flop events agrees with that from atomistic simulations in the details [72–74]. The number of flip-flop events increases with increasing length of the PEG and with increasing surface coverage. Lipid flip-flop events are known to occur naturally [75]. In our case, we have used a uniform composition (DPPC) and symmetric situation for both leaflets, so there is no consequence in the composition of each leaflet when flip-flops occur. Flip-flops change the composition of each leaflet in natural asymmetric lipid membranes, with important consequences in function [76, 77].

On the other hand, the alkyl ligands of the hydrophobic coated Au NP are extended out into the lipid tail regions, as is obvious from the snapshot in Fig. 21 (right). The entanglement of the alkyl ligands with the lipid tails leads to order parameter changes in the lipid tails (considered in detail elsewhere [71]). Only infrequent flip-flops are observed since the alkyl ligands do not engage the lipid heads.

Potential of mean force calculations (Fig. 22) shows that the hydrophobic coated AuNP experiences a barrier at the water-lipid

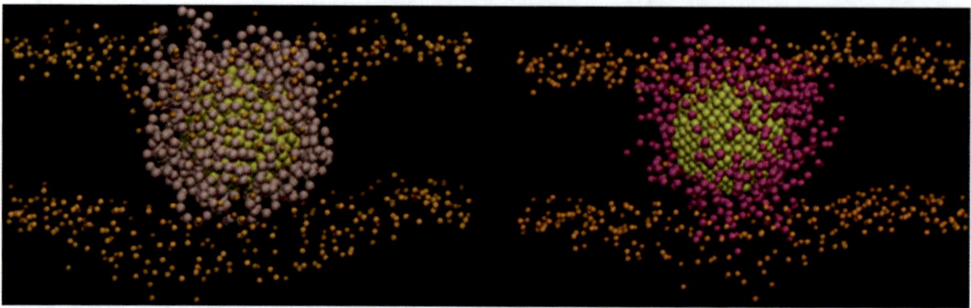

Fig. 21 Snapshots of the gold nanoparticle within the center of the membrane. On the left is the PEGylated AuNP, and on the right is the alkanethiol-coated Au NP. Here the phosphate groups have also been omitted for clarity. Figure reproduced from ref. 69

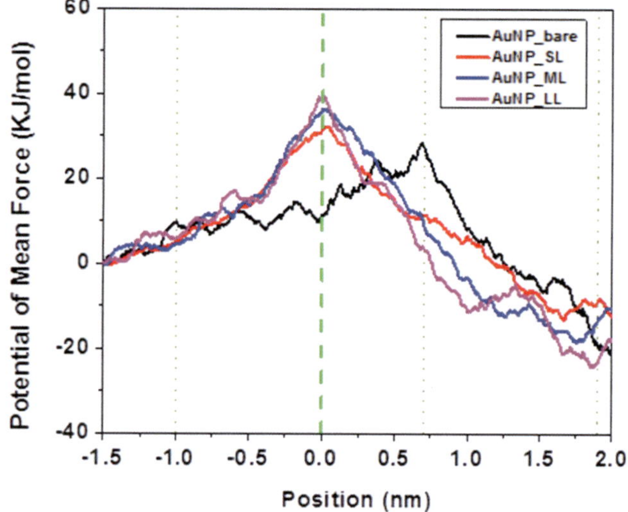

Fig. 22 Free energy profile for the AuNP with alkylthiol ligands permeating across the lipid bilayer membrane. The water-lipid interface is at 0 nm and the center of the membrane is at 2.0 nm. Figure reproduced from ref. 12

interface; the longer the tail, the larger the barrier. This NP has a strong preference to stay in the center of the membrane. This is not surprising since here the hydrophobic alkane ligands of the NP can associate with the hydrophobic alkyl tails of the lipid; as seen in the snapshot in Fig. 21 (right), the alkyl ligands extend out. Clearly, the AuNP with hydrophobic ligands would tend to stay in the center of the membrane, as indicated by the minimum in the potential of mean force profile in Fig. 22.

We should emphasize here that the mobility and articulation of the alkyl (or PEG) ligands permit the NP to experience a very different interaction with the membrane than the bare hydrophobic Au crystal, as we have shown in detail elsewhere [12, 69]. This is significant to note, *a propos* the large number of simulations in which bare NPs with special receptor-seeking surface atoms represent ligand-coated nanoparticles [78–87]. Clearly, the mobile, articulated hydrophobic or hydrophilic sites on the ligand create a distinctly different response from the lipid membrane than a bare NP, and these effects on the dynamics of both the NP and the membrane are missing from those simulations which use a bare NP with "active" sites on immobile surface atoms.

How does the nanoparticle exit the membrane, i.e., what is going on with the ligands and the lipids as the exit process is occurring?

As the PEGylated AuNP leaves, the top leaflet has largely recovered, and as the particle exits the PEG ligands continue to interact with the headgroups of the bottom leaflet, but leave the lipid tails undisturbed, thereby exiting cleanly, as seen in the snapshot in Fig. 23. No lipids at all are seen to cling to the leaving PEGylated AuNP. On the other hand, the hydrophobic coated AuNP displaces and drags many lipid molecules away with it as the particle leaves (note the green tails of the lipids wrapping the exiting AuNP in the right image in Fig. 23), resulting in a net lipid displacement from the membrane from both the *upper and lower leaflets, which we have quantified in detail elsewhere* [88]. *This* net lipid displacement from the membrane occurs because the alkane ligands continue to interact with the tails of the lipid molecules as it moves past the lower leaflet; the alkyl lipid tails get entangled with the alkyl ligands as the AuNP is exiting, thus resulting in lipids becoming displaced from both leaflets of the membrane. This entanglement is more severe with longer ligand lengths and with larger diameter AuNPs; and significantly more lipid molecules are lost from the lower membrane leaflet, where the exit process is occurring, compared with the top leaflet.

Atomistic simulations of an AuNP coated with a monolayer of 1:1 octane thiol and 11-mercapto-1-undecanesulfonate (MUS) that is embedded in a DOPC bilayer membrane provide similar results to our case; they show a significant probability of finding

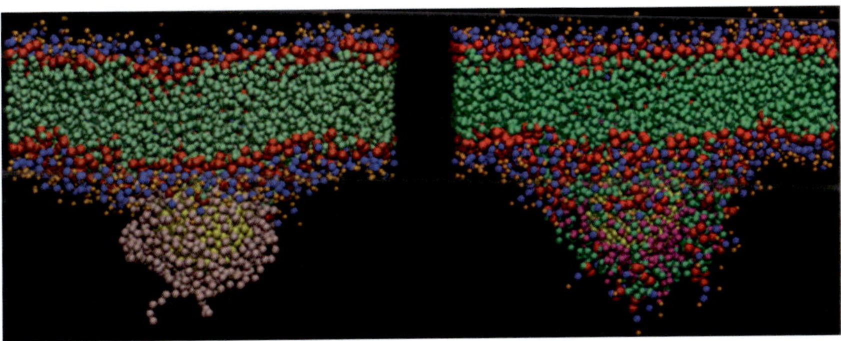

Fig. 23 Snapshots of the gold nanoparticle exiting the membrane. On the left is the PEGylated AuNP, and on the right is the alkanethiol-coated Au NP. All components are shown; only the water molecules have been omitted. Figure reproduced from ref. 69

lipid tails in a spatial region that overlaps with the density of NP ligands [70]. This observation implies that lipids can be "extracted" from the bilayer entirely due to the chemical similarity between the hydrophobic lipid tails and the hydrophobic octane ligands in the NP monolayer. Examination of their simulation trajectories led to the identification of transient lipid extraction events, in which lipids intercalated within the NP monolayer, for all of the simulations, just as we had reported for our coarse-grained MD simulations with alkanethiol ligands and DPPC [88]. It is encouraging that coarse-grained MD simulations provide the same kind of detailed mechanisms (of lipid flip-flop, lipid extraction, etc.) that agree with results from atomistic simulations.

Can water molecules enter into the membrane? If there are ions in the "outside" of the membrane can they accompany the water? What then happens to the water that entered?

We have examined in detail the water column formation for both types of ligands [67, 88]. In the course of a simulation, we can monitor the number of water molecules in the membrane at each position of the AuNP; we display this in Fig. 24 for the bare AuNP, and for the ligand-coated AuNPs. We can see in Fig. 20 that, at the outset, the hydrophobic coated AuNP causes a hole to be formed in the top leaflet because the headgroups are pushed aside by the NP as it enters. Water molecules could enter through this hole in the top leaflet, but they do not go into the alien environment of hydrophobic lipid tails, preferring to stay close to the phosphate heads. As the AuNP permeates the membrane, the pore persists, but is blocked to a large extent by the NP itself while it is still in the membrane. As the NP exits, however, water can come into the membrane interior; the number of waters in the interior reaches a maximum after the NP has left because, once a water chain or bridge completes across the pore, more water molecules are prompted to go in. Experiments show that water permeability correlates most strongly with the area/lipid and is poorly correlated with bilayer thickness and other previously determined structural and mechanical properties of single-component bilayers such as DPPC [89].

We observe in Fig. 24 that the numbers of water molecules present in the hydrophobic interior of the membrane at various positions of the permeating nanoparticle vary according to the nature of the ligands. Although the hole created by the NP with hydrophobic ligands has the potential to let water molecules in, the latter do not enter, whereas the PEGylated AuNP actually lets some water into the interior as it is passing the first leaflet, since water molecules associate with the PEG ligands and move along with the NP as it permeates, giving rise to the first maximum in Fig. 24 that occurs for the PEGylated AuNP but not for the alkanethiol-coated NP. Water content in the membrane interior drops when the PEGylated AuNP is in the middle, and reaches a maximum as the

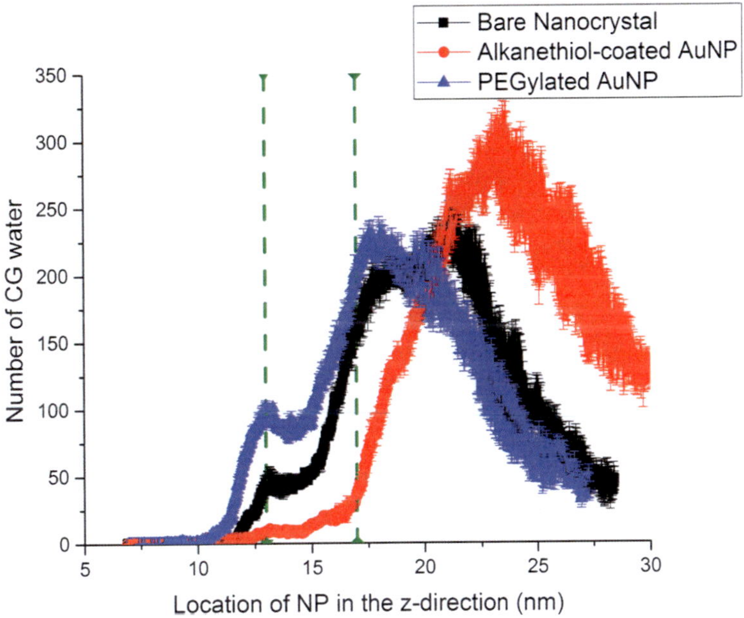

Fig. 24 Number of coarse-grained water molecules present in the hydrophobic membrane interior during permeation of a bare nanocrystal, alkanethiol-coated gold nanoparticle, and PEGylated gold nanoparticle with the same permeation velocity. The green dashed lines represent the equilibrated positions of the phosphate headgroups at the top and bottom membrane leaflets. Each data point has error bars included based on three independent simulations. Figure reproduced from ref. 69

NP is passing through the second leaflet. However, as the PEGylated AuNP exits, the water molecules return to their own compartments soon after the NP leaves. The amount of water that resides temporarily in the interior is less for the AuNP with shorter PEG ligands. On the other hand, water molecules inside the membrane interior come to a maximum *after* the hydrophobic coated AuNP has completely exited the membrane, as seen in Fig. 24. In Fig. 23 we had observed that the alkanethiol-coated gold nanoparticle drags many lipid molecules along with it during the exit stage of permeation, inducing a net displacement of lipids from the membrane; thus, we would expect that with the alkanethiol-coated case, in comparison with the PEGylated NP, expulsion of water occurs later, and membrane recovery is slower. Indeed, this is the case, as seen in Fig. 24. Eventually water molecules do return to their respective compartments, later for the hydrophobic coated NP than for the PEGylated AuNP. At 200 ns all the water has been expelled from the membrane in our simulations.

Both CG and atomistic simulations have demonstrated that ion leakage through transient water pores in lipid membranes can be driven by ionic charge imbalance across the membrane [71, 90]. In our simulations where ions are present in only one of the water compartments to start with, ions may enter the interior of the membrane accompanying water pore formation, with a slight bias to transport of Cl^- ions compared to Na^+ ions. This is in agreement with results of atomistic simulations [90]. CGMD simulations of ion permeation in preformed tension-stabilized pores in DPPC by Leontiadou et al. also show that for larger pores (3.6 nm diameter), the chloride ion flux was an order of magnitude larger than that of the sodium ions [91]. However, the incidence of ion transport through to the other side, although increasing with the size of Au core, was observed to be very low for alkanethiol-coated AuNP [71]. Ions placed in the aqueous compartment to a large extent tended to stay close to the phosphate heads and do not enter the interior during permeation by AuNP coated with hydrophilic or hydrophobic ligands, except rarely, and only when a continuous water column persists [67, 71, 88]. Any transient water column that forms is surrounded only by hydrophobic lipid tails, and thus begins to collapse immediately, in the case of permeation with a AuNP coated with hydrophilic ligands. Post-permeation with a AuNP coated with hydrophobic ligands, the water column collapses eventually although requiring up to 40 ns longer to completely collapse. This occurs in our simulations, as well as in CG and atomistic simulations by others [92–95].

After the nanoparticle has made its exit to the other side of the membrane, what is the membrane's condition, i.e., does the membrane recover its original unbreached condition in every aspect, or not?

Some of the events that we have observed in the simulations include a small number of lipid flip-flops accompanying the permeation of the PEGylated AuNP. On the other hand, the number of lipid flip-flops is negligible in the case of the hydrophobic coated AuNP; instead the lipid molecules do start to flip but the tails become entangled with the alkane ligands, so that the lipid molecule does not complete the transfer to the other leaflet; instead they are displaced from the membrane entirely. After the permeating AuNP has exited and the water molecules have returned to their own compartments (approximately 50–100 ns), the membrane eventually returns to its original equilibrium condition. Lipid molecules temporarily displaced from the membrane by the passage of the hydrophobic coated AuNP do eventually return to the membrane after a long enough time, since after all this same DPPC membrane self-assembles from an isotropic solution. In the end, by monitoring the identities of the lipid molecules, we find that some lipids have indeed exchanged between leaflets as a consequence of the permeation process. Less than 3% of the total lipid molecules displaced during permeation completed the flip-flop in the case of PEGylated AuNP permeation. For alkanethiol-coated nanoparticle permeation, less than 1% of displaced lipid molecules completed the flip-flop since the majority of displaced lipids are dragged into the bulk solution entangled in the alkanethiol ligands of the nanoparticle. Most water molecules do return to their original compartments (by tracking the numbered CG waters we found that no more than 2% of the total water molecules ended up in the other compartment at lower nanoparticle velocities and no more than 6% translocated to the other compartment at high permeation velocities, for both the spherical and rod-shaped AuNP) and the very few ions that had entered do too.

6 Interaction of a Neutral Rod Nanoparticle with a Lipid Bilayer Membrane

The above observations in molecular dynamics simulations used only nanoparticles with spherical gold cores. Gold nanoparticles with aspect ratios different from 1.0 have properties quite different from those that are spherical. Other nanoparticle shapes have been used in nanomedicine, nanorods, in particular. Gold nanorods (NR) modified with polyethylene glycol (PEG) were prepared for the first time by Nidome et al. [96]; they then evaluated its cytotoxicity in vitro and its bio-distribution after intravenous injection into mice. Gold nanorod (NR) medical applications include tumor imaging, photothermal therapy, gene delivery, and drug delivery [57, 58, 97–110]. Although, experimentally, the highest grafting density was achieved with spherical gold nanoparticles, photothermal therapy and drug delivery performance of the nanorods and nanostars were found to be far superior [111]. Therefore, we consider separately the permeation of PEGylated nanoparticles based on gold nanorod cores. We have investigated only a nanocylinder core with a particular aspect ratio (length/diameter = 2.2), and our detailed studies are in ref. 112. It would be straightforward to extend the studies to even greater aspect ratios; however, larger simulation boxes containing larger numbers of solvent and greater numbers of lipids would have to be used in order to avoid interactions with images created by periodic boundary conditions. We cut a cylindrical rod from the same FCC lattice of gold atoms to construct an AuNR. For comparison with spherical NP results, we choose the volume of the NR core to be identical to the spherical core. Thus, we use an AuNR which has the same volume as that of the 3.0 nm diameter spherical AuNP, and we chose an aspect ratio of 2.2, in the middle of the range of aspect ratios (1–5) of Au nanorods synthesized experimentally for biomedical applications [113, 114]. The cellular uptake of rodlike bionanoparticles with different aspect ratios is size dependent and has been investigated recently [115].

As a hydrophilic ligand-coated rod-shaped nanoparticle approaches a lipid bilayer membrane, what happens, i.e., how does the lipid bilayer respond to the approaching nanoparticle? At the same time, how do the hydrophilic ligands respond to the surface of the lipid bilayer? How does this depend on the angle of entry of the rod? How does the rod proceed, i.e., what happens to its orientation, as it permeates the membrane?

Unlike a spherical AuNP, the NR has a long axis, presenting an anisotropic collection of ligands to the lipid membrane. Thus, it is of interest to monitor the changes in orientation of this long axis during the process of permeation. We show in Fig. 25 a series of snapshots for the NR entering with its long axis at an angle of 0°, i.e., normal to the membrane surface. In our simulations, we permit the NR free rotation in all directions; we only constrain the center of mass of the core gold rod to move directly across the membrane

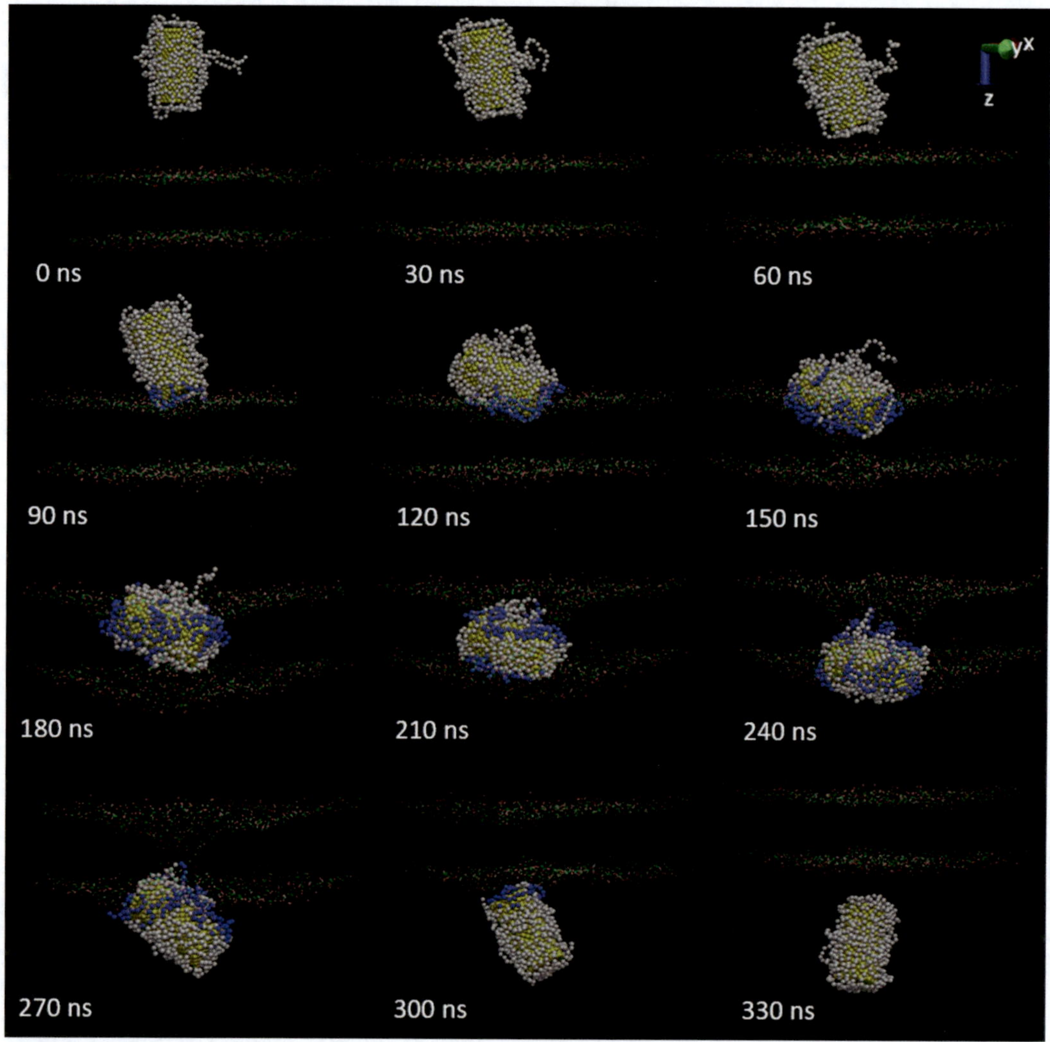

Fig. 25 Molecular snapshots of permeation of a PEG18 nanorod in the lipid bilayer membrane where pink = choline, green = phosphate, yellow = AuNR core, white = PEG18-SH ligands, and blue = PEG18 beads within 1.0 nm of choline and phosphate groups. Figure reproduced from ref. 112

along a line normal to the membrane surface. Although each of the trajectories, each lasting for 320 ns, differs in small details, for the same entry angle, or for any one of the entry angles we used (0°, 10°, 45°, 90°), the same characteristic mechanism for permeation occurs. First the NR tilts toward a lying-down position on the plane of the membrane, and then rotates toward a standing-up position (i.e., with the NR long axis along the normal to the membrane). We see in the series of snapshots in Fig. 25 the NR entering, tilting, lying down, rotating, straightening up, and exiting the membrane.

A cartoon representation of the NR in Fig. 25 is shown in Fig. 26. All MD trajectories provide a consistent scheme that is similar to that in Fig. 26 for the permeation pathway of a PEGylated AuNR through a lipid bilayer membrane, irrespective of entry angle, independent of surface coverage. In each and every case, the NR goes through the following stages: a tilting toward a direction that is in the plane of the membrane, lying down on its side in the membrane plane, and tilting further and straightening up as it is leaving the membrane. It is interesting that if we instead pull on the center of mass of the AuNR along the same direction as the angle of entry, the NR goes through the same trajectory as well, except that the tilting toward a lying-down position occurs much earlier in the trajectory.

We can see, in the snapshots of Fig. 25, the details of the cooperative response of ligand and lipid heads (for clarity water and the lipid tails are omitted in the images) that drives this permeation pathway. To call attention to the favorable interactions that are occurring, in Fig. 25 we have colored in blue those PEG beads that are within 1.0 nm of a choline or phosphate head of a lipid. Regardless of its initial angle of entry at time zero, PEG ligands get

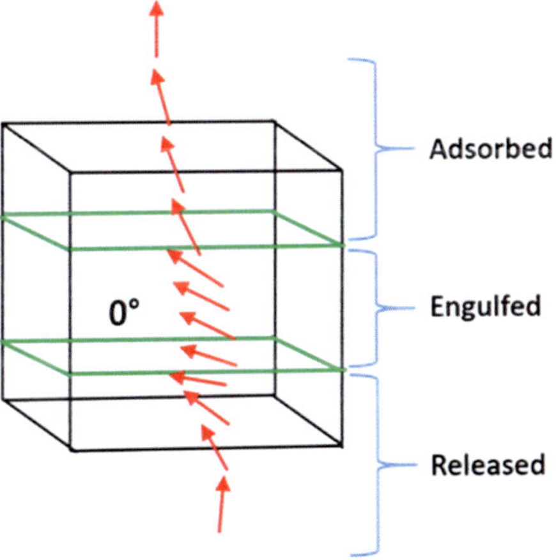

Fig. 26 Cartoon representation of permeation of PEG18-NR in the lipid bilayer membrane corresponding to the molecular snapshots of Fig. 25. Figure reproduced from ref. 112

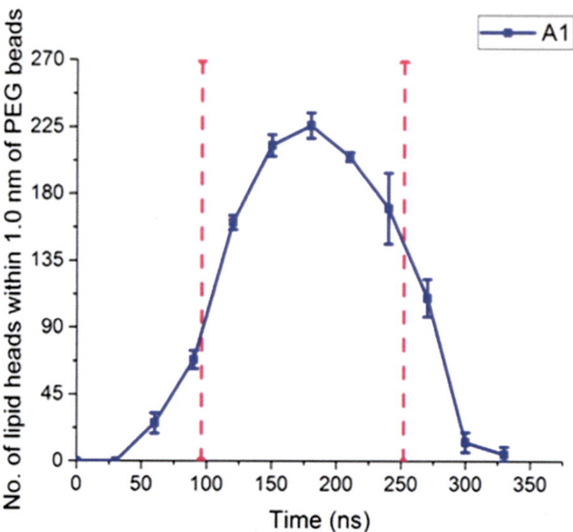

Fig. 27 Total interaction energy between all PEG beads and all choline and phosphate heads of lipids along the course of permeation of a nanorod starting at an entry angle of 10°. Figure reproduced from ref. 112. All trajectories give very similar plots, regardless of the initial angle

attracted to and reach out to lipid heads, causing the NR to tilt toward a lying-down position (at 120 ns) that maximizes the attractive interactions. We can see the PEG beads stretching out to lipid heads in both leaflets as the NR continues to permeate the membrane (at 180 ns); straightening up permits PEGS at both the top and bottom parts of the NR to reach out to the top leaflet heads and the bottom leaflet heads (at 240 ns). Finally, the NR exits; even as it is leaving (300 ns) the PEGs at the top of the NR hang on to the lipid heads of the bottom leaflet. For one example of a particular trajectory, the total interaction energy between all PEG beads and phosphate and choline lipid headgroups along the course of nanorod permeation is shown in Fig. 27. All trajectories give very similar plots. Here we can see that favorable interactions between PEG and lipid heads drive the changes in orientation of the NR axis as it permeates the membrane.

When the first leaflet is breached by the nanorod, can water molecules enter into the membrane? How does this depend on the angle of entry of the rod?

Just as in the case of the PEGylated spherical AuNP, water molecules do enter the interior of the membrane during the permeation, and later leave. We find that the number of water molecules leaking into the membrane interior increases with increasing NR entry angle beyond 45°. At initial angles of 45° and larger, a larger water pore is formed which allows more water molecules to enter the membrane interior, compared to entry angles of 0°–10°. Despite differences of these profiles compared to spherical

PEGylated nanoparticles, water molecules will enter the hydrophobic region, but ultimately do exit the membrane due to recovery of the bilayer leaflets (for favorable total interaction energies, lipid tails become ordered in each leaflet and in the process squeeze out the water molecules). In the end, (after 50–100 ns) all water molecules will have been expelled from the membrane. No permanent change to the membrane is sustained, especially at lower permeation velocities. And then, only a tiny fraction (less than 2%) of the waters end up in a compartment different from the original. That is, of the 300 or so CG waters that entered the middle region of the membrane, all but 5 or so did return back to their original compartment. In other words, hardly any waters transported across the membrane to the other side during PEGylated NR permeation.

When the nanorod reaches the center of the membrane, does it tend to stay there? How do the lipids react to the presence of the nanoparticle between the leaflets?

Although there are attractive interactions that cause the NR to change its orientation as it permeates through the membrane, the PEGylated NR, like the PEGylated spherical NP, does not prefer to stay in the interior of the membrane, unlike the AuNP with alkyl ligands. We can see in Fig. 25 that while the center of mass of the NR is in the center of the membrane, there is considerable disturbance of the lipid heads from their original equilibrated positions, the membrane does bulge out somewhat in the region where the NR is in its interior, and the lipid tails are pushed away (not shown). We expect a considerable change in the order parameters of the lipid tails in the immediate vicinity of the NR in the interior (not shown here), just as we had found for those lipids closest to the Au nanocrystal or the AuNP coated with alkyl ligands [3, 12]. For the size of NR we used in our simulations, the NR is already beginning to straighten up when the center of mass is in the center of the membrane. This permits the PEG to reach the headgroups of both upper and lower leaflets, thereby maximizing the attractive interactions between the NR's PEG ligands and the lipid's phosphate and choline groups (as seen in Fig. 27).

How does the nanoparticle exit the membrane, i.e., what is going on with the ligands and the lipids as the exit process is occurring? How does this depend on the original angle of entry of the rod?

As the NR is leaving the bottom leaflet, PEG ligands on the end surface of the NR continue to stretch out to the lipid heads on the bottom leaflet, until the distance becomes longer than our cutoff distance for interactions. The NR was observed exiting at an orientation with the NR long axis nearly perpendicular to the plane of the membrane, regardless of the initial angle of entry, in all trajectories. However, unlike in Figs. 25 and 26, due to the rotational behavior of the rod, the NR sometimes exits upside down from its original orientation when it entered the membrane, i.e., the top of the Au nanocylinder at entry sometimes ends up at the bottom as the NR leaves the membrane.

After the nanorod has made its exit to the other side, what is the membrane's condition, i.e., does the membrane recover its original unbreached condition in every aspect or not? How does this depend on the original angle of entry of the rod?

Absolute time to recovery is not accurate because we are using a coarse-grained rather than an atomistic model; however, we can compare relative values using the same potential parameters and similar coarse graining. Recovery time for PEGylated nanorods is slightly longer compared to PEGylated AuNP with spherical core of the same volume; it is shorter for those NR with shorter ligands. Recovery time is very nearly the same regardless of the entry angle. All but a tiny fraction of the waters end up in the same compartment where they started, and all but a tiny fraction of the lipids end up in the same leaflet where they started. Note, however, that our conclusions about recovery are limited by the fact that we have used a very stable lipid bilayer (DPPC); the stages toward recovery could well be different for other lipids, or for unsymmetrical lipid bilayers that have compositions that are more like those of biological cell membranes.

Is it possible to observe experimentally the rotational behavior of a nanorod as it permeates a membrane to verify the simulation results? What is the implication of the result that, for all angles of entry, there is a common mechanism by which a nanorod permeates a lipid bilayer?

First, it is now possible to observe this rotational behavior with the new techniques for imaging. It is currently technically possible to simultaneously visualize with nanometer precision a single gold nanorod in a live cell, localize its position, and determine its orientation and rotational motion using microscopy to take images in a differential interference contrast mode simultaneously with bright-field mode [116, 117]. Our simulations display the results that would be observed if this method were used on a model planar membrane on a solid support. In another new experimental technique, direct observation of a transmembrane event, namely, (a) negatively charged gold nanorods approaching the plasma membrane from the open solution, (b) being confined rotationally and laterally static at a membrane site (with a narrow distribution of angles close to perpendicular to the membrane surface), and (c) the exact moment of the NR detaching from the inner surface of the membrane, has been captured in a movie. This new method used annular oblique illumination, positioning the focal plane of the microscope objective at the sidewall of the cell, with a birefringent prism to split the AuNR plasmonic scattering into *two channels of polarization*, thereby providing azimuthal and polar angles. This experimental method permits monitoring of the distribution of angles (θ, ϕ) that the gold nanorod axis makes relative to the normal to the membrane surface [118]. Our simulation results report the NR axis (θ, ϕ) as a function of time in the permeation process of one PEGylated gold nanorod through a planar DPPC membrane. Since this new experimental method permits the observation of a single gold nanorod entering a cell membrane, it may be possible to make observations of the rotational behavior that our simulations predict.

Our results obviously apply strictly only to the NR with these dimensions relative to the thickness of the lipid bilayer and probably only to aspect ratios close to 2.2. It is easy to imagine that a nanorod with a very large aspect ratio would enter the lipid bilayer in a different way, with different consequences. Nevertheless, in comparison with its equal-volume counterpart PEGylated nanosphere, the ease of permeation of the PEGylated AuNR, attributable to the observed rotational behavior, is not inconsistent with the finding that the experimental drug delivery performance of the nanorods was found to be far superior compared to spherical AuNPs [111].

Our molecular simulations are of the direct penetration of a membrane by diffusion, permeation, and pore formation, involving only nonspecific interactions. A completely different pathway of internalization into a membrane is the endocytosis process, which includes a wrapping stage. It is interesting to note that a similar rotational behavior of a sphero-cylindrical nanoparticle has been observed in dissipative particle dynamics (DPD) simulations of the sequence of events in the endocytosis process of docking, recognition, and binding of the ligands on the NP to the complementary receptors on the cell membrane surface; wrapping of the NP by the membrane; and completion of internalization [79, 81, 82, 84, 119, 120]. Review articles by these groups continue to affirm this description of the sequence of events for endocytosis [121, 122]. In their DPD simulations (not traditional molecular dynamics), the model is a sphero-cylindrical nanorod with no mobile ligands, rather surface beads that interact specifically with receptors on a simple model of the membrane, namely a two-dimensional fluid surface in 3D space, constituted of beads of two types (50% receptor and 50% non-receptor). Unlike their models, we have no special receptors in the membrane to drive the sequence: rotation, lying down, rotating further, and exiting perpendicular. Our process arises entirely from fully dynamic interactions between the individual hydrophilic PEG moieties as articulated ligands and the fully dynamic individual phosphate heads, choline groups, and hydrophobic articulated tail sections of the lipid molecules constituting the membrane. Other endocytosis simulations use models of PEGylated NPs and lipids similar to ours in numbers of beads per molecule and articulation; however, they placed special receptor-seeking beads at the *end* of each PEG ligand and 50% receptor beads among the lipid heads

[123, 124]. They find that different shapes constructed so as to have equal surface area with the sphere lead to a similar rotational behavior for the rod-shaped PEGylated NP, that rate of internalization depends on the shape, with the entry angle playing an important role in the internalization process for the rod. Again, we only note the similarity in the rotational behavior, since endocytosis is an entirely different process from the direct membrane penetration that we have simulated.

7 What Are the Lessons Learned from These Model System Simulations Vis-à-Vis a Realistic Neutral Nanoparticle for Drug Delivery?

We started out with simulations on gold nanoparticles because a diverse range of gold nanoparticles have been explored for use in therapy and imaging applications [125]. It is well known that nanoparticle biodistribution is influenced by the size and surface characteristics of the nanoparticle. Rapid clearance from the blood hinders nanoparticle delivery to target sites, and, consequently, most nanoparticles developed for medical applications are coated with chains of polyethylene glycol (PEG). The addition of PEG increases the hydrodynamic particle size, which prevents filtration by excretory organs, also provides a hydrating layer that sterically hinders nonspecific binding of proteins to the particle surface, and delays recognition of the particles by the reticuloendothelial system (RES) [126]. This ultimately increases the blood circulation time of the particles. It has been found that *smaller* PEG-coated AuNPs are overall better, i.e., smaller particles had slower clearance, less uptake by RES cells, and higher accumulation in the tumor. Presumably this is due to larger core AuNP having lower surface density of PEG and smaller particles having more success in dispersing throughout the tumor [126–128]. Recent work evaluating the uptake and distribution of gold nanoparticles in preclinical tumor models has been reviewed [129]. Although gold nanoparticles are often designed for prolonged circulation, eventual clearance of the nanoparticles from the body is desired after therapy is complete. Renal excretion is found to be better for smaller (<5 nm) gold nanoparticles [130]. Although charged ligands provide coulomb interactions that facilitate spontaneous permeation, we chose to model neutral AuNPs because, in applications, neutral nanoparticles, as well as those with a slight negative charge, show significantly prolonged circulating half-lives, which translates to improved accumulation in tumors. In contrast, cationic nanoparticles are known to penetrate cell membranes by creating pores that disrupt the membranes [131], and induce cytotoxicity [132, 133]. Simulations of anionic ligand-coated nanoparticles in membranes show quite different behavior than neutral particles: ligand snorkeling, local disordering of lipids, and cholesterol depletion [134].

A large part of the excitement generated by nanoparticles for drug delivery arises from their potential to preferentially accumulate at sites of injury, infection, and inflammation. Tumor blood vessels have irregular diameters; they are fragile and leaky. Rapid cancer cell growth is fed by chaotic and disorganized vasculature. This feature of tumors, by no means unique to cancer, has led to strategies aimed at enhancing site-specific accumulation of nanoparticles for imaging and therapy [135]. The use of a hydrophilic "stealth shell" of PEG has permitted the medical applications that take advantage of this imperfect vasculature and high surface coverage with PEG helping to achieve the desired goal. Although the highest coverage or grafting density was achieved with spherical gold nanoparticles, photothermal therapy and drug delivery performance of the nanorods and nanostars were found to be far superior [111]. Our simulations have indicated the differences in interaction of a nanorod versus a spherical AuNP with a lipid bilayer that appears to explain this, although we had not been able to investigate the much larger aspect ratios that are closer to those used in photothermal heating therapy.

What do the simulation results imply about optimizing the design of useful AuNPs? The simulation results indicate that ease of permeation of lipid bilayer increases with higher PEG coverage for a given Au core diameter. Therefore, methods of synthesis and ligand replacement with PEG should have this goal, especially for Au nanorods. In our cycled annealing using thiol ligands, we have found that surface coverage, for a given ligand length, and given Au core diameter, tends to become sparser with increasing AuNP diameter. The same cycled annealing, for a fixed Au core diameter, leads to surface coverage that is lower for longer PEG. This appears to be true for experimental synthesis as well [136, 137], so that other methods were necessary to obtain the higher coverages from 0.8 to 1.57 chains/nm^2 by Rahme et al. [138]. They measured the number of PEG molecules grafted per nanoparticle and showed that the surface density decreased in a nonlinear way as a function of increasing PEG length which they attributed to the increased conformational entropy of the polymer with chain length. Similarly, the surface density of m(PEG)n -S- was shown to decrease slightly with increasing Au nanoparticle diameter, while the number of polymers grafted to the particle surface was found to increase in a nonlinear way with the size. That renal excretion (in rats) is found to be better for smaller (<5 nm) diameter AuNPs [130] may in fact be attributable to the higher PEG surface coverage attained with smaller Au cores.

The very different details of interactions between the hydrophilic ligand-coated AuNP and the lipid bilayer compared to their hydrophobic counterparts imply that there may be some advantage to having both types of ligands present in two domains on the same nanoparticle. Many drug delivery nanoparticles are not uniformly functionalized; some can have partly hydrophilic and partly hydrophobic ligands; so-called Janus particles are examples of the latter. MD simulations can find out the major differences between permeation by a Janus nanoparticle and a uniformly functionalized nanoparticle. An attempt to do this used DPD simulations of permeation of a membrane by a Janus nanoparticle with no ligands, only surface atoms (half hydrophobic, the other half hydrophilic) on a nanosphere, and a moving spring force acting on the center of mass to guide the penetration through a simple model membrane, compared with all-hydrophobic or all-hydrophilic surface atoms [139]. However, we have already seen the limitations of using no mobile articulated ligands on the NP; the results for NPs with mobile ligands are quite different from bare NPs. The simulations need to be carried out with mobile articulated ligands on the Janus particle to observe how the dynamics of interactions of both types of ligands with the lipid bilayer affect the mechanism of permeation by a Janus particle.

The differences in details of interactions by hydrophilic and hydrophobic ligands that we have observed in our simulations of uniformly coated AuNPs can be explored in simulations for binary-coated AuNP by using the same alkanethiol ligands, distributed over half the gold core and the same PEG ligands distributed over the other half. The permeation characteristics of a Janus nanoparticle could be "observed" in simulations and compared directly with our previous ones with uniform ligands. Similarly, the spontaneous permeation of nanoparticles having two types of ligands, a hydrophobic (C_8) alkane and a hydrophilic slightly longer alkyl (C_{11}) sulfonate (MUS) in a "striped" arrangement on the AuNP surface [140], has been reported, although it proved controversial [141]. It is still disputed whether a striped particle was indeed formed and observed, but the observed ability of a binary-mixed-ligand-coated AuNP to have cell membrane permeation characteristics different from one which has uniform coverage of the hydrophilic ligand has not been disputed. Thermodynamic conditions under which stripes, rather than random distribution of binary ligands of unequal length, could form have been shown computationally to be quite restrictive [142]. For a particular set of ligands, it has been found experimentally that domains of stripes or ripples form only for specific core particle sizes [143]. The characteristics of AuNPs with mixed monolayers of ligands (random, striped, Janus) have been studied computationally [144, 145], but not their membrane permeation characteristics. We have demonstrated that AuNPs with hydrophobic ligands interact very differently with the lipid bilayer than AuNPs with hydrophilic ligands. Therefore, having both types be present on the surface of the AuNP in domains, such as stripes or ripples, could lead to very revealing simulation observations. It would be interesting for us to carry out MD simulations of a binary-ligand AuNP interacting with a lipid bilayer with the specific goal to examine the mechanistic characteristics of permeation of a striped AuNP compared to a uniformly covered AuNP. The effects of varying the distribution (uniformly mixed or striped domains) of the hydrophilic and hydrophobic ligands can be investigated to help in optimizing parameters for the synthesis, such as ratio of ligand lengths, polarity/charge of the hydrophilic ligand, and relative volumes of the bulky end groups, in order to achieve best permeation properties. An atomistic MD simulation using a 1:1 coating of the 11-mercapto-1-undecanesulfonate (MUS) and either octanethiol or heptadecanethiol AuNP of diameter 2 or 3 nm embedded in a DOPC bilayer has been carried out. These binary mixed ligands were permitted to distribute themselves uniformly on the Au surface [70].

The mechanism for permeation observed for ligand-coated Au nanorods (enter, tilt, lay down, rotate, straighten up, and exit) arises from the favorable interactions that PEG ligands have with the headgroups of the lipids. We posit that nanorods could be functionalized so as to take advantage of this permeation mechanism, i.e., to choose ligands for the nanorod that interact favorably with specific groups at the membrane surface. The same mechanism of cell entry could apply to nanorods coated with ligands that recognize specific receptors on a cell membrane. For example, glycolipids have carbohydrate groups exposed on the membrane surface. Specific glycolipids participate in pathogenicity of protozoan parasitic diseases caused by *P. falciparum* (malaria), or *Trypanosoma brucei* (sleeping sickness), *Toxoplasma gondii* (encephalitis), and *Trypanosoma cruzi* (Chagas disease), to name a few, and protozoan glycolipids and glycoproteins have carbohydrate parts distinguishable from mammalian [146]. If Au nanorods coated with ligands that specifically recognize and interact attractively with the specific sugar groups on cell membranes of these protozoa could cause the latter to be targeted, then we would have a possible defense against these parasitic diseases. MD simulations of ligand-coated AuNRs permeating through lipid membranes containing the specific sugar groups in abundance on the membrane-included glycolipids could be a first step. Molecular dynamics simulations of membranes composed of glycolipids and phospholipids are already being carried out [147].

8 Prospects for Future Simulations to Answer Further Questions

Many drug delivery nanoparticles carry a net charge. The behavior of charged nanoparticles in aqueous solution and their interactions with a lipid bilayer membrane will be significantly different from that of a neutral particle. MD simulations can find out the major differences between charged and neutral nanoparticles by using nearly identical model systems, for example terminating the PEG not with CH_3 groups but with either COO^- or CNH_3^+ groups and including the attendant counterions in the aqueous solution. The MD simulations could also answer questions about the optimal nature of the counterion, e.g., surface charge density (size, mono- or multivalent), and monatomic versus a molecular ion. Clearly there will be differences in behavior if the counterion is too bulky to enter the membrane, or tends to stay at the interface, interacting with the polar lipid heads rather than accompanying the NP into the membrane. Many drug delivery nanoparticles not only are functionalized with PEG, but also carry a payload drug molecule via nonbonded interactions with the PEG ligands. MD simulations can investigate the conditions under which (various PEG length, charge) the payload molecule (various size, functional groups, charge) can be carried around by the NP in aqueous solution without being released along the way.

Some drug delivery nanoparticles are functionalized with highly branched structures such as dendrimers and star polymers rather than extended chains. The behavior of these highly branched structures is expected to be substantially different from extended chain ligands. MD simulations can explore the extent of branching that is optimum for carrying various payloads (specific drugs in use could be used in the simulations). Some drug delivery nanoparticles are functionalized with specific chemical functional groups that are "molecular recognition" probes to attach to specific receptors on the surface of the lipid bilayer membrane. It would be interesting to use the simulations to optimize the length and flexibility of the tether with which such probes are attached to the core of the NP to maximize recognition probability.

We have only shown our "observations" for uniform lipid bilayers, i.e., with the outer and inner leaflets having identical composition of a single lipid molecule type. In nature, mixed composition membranes occur, with the inner and outer leaflets having different lipids and different fractions of a mixture of lipids. All these are easily modeled in MD simulations [148–150], and the presence of small molecules in the membrane itself, cholesterol for example, has also been modeled (*see* review in ref. 151), and experiments on supported lipid bilayers including cholesterol and sphingolipids [152] can provide validation tests of such simulations. We have seen that lipid flip-flop occurs in the MD simulations that are described above. MD simulations of nanoparticle permeation of such nonhomogeneous lipid bilayer membranes can be carried out to answer such questions as to what extent does the permeation of a nanoparticle change the outer leaflet composition in a nonhomogeneous lipid membrane. Many naturally occurring lipid membranes have a net charge, for example, those composed of anionic 1,2-dimyristoyl-*sn*-glycero-3-phosphoglycerol (DMPG) lipid. The electrostatic potential within such phospholipid membranes implies an enormous electric field of 10^8–10^9 V m^{-1} [153], which is likely to be important in controlling the transfer of ions and molecules, and nanoparticles across the membrane. There have not yet been any simulations of nanoparticles interacting with charged lipid membranes.

By systematically considering various combinations of model systems to study, MD simulations can answer many mechanistic questions at the molecular level and make contributions toward answering the relevant questions in nanomedicine related to cytotoxicity, etc. When atomic level details are required, atomistic MD simulations are in order; however, the suggested procedure would be to start doing MD simulations with a coarse-grained model until a suitable equilibrated system is attained, and then do the reverse transformation back to the atomic detailed structure. This approach has been suggested and developed by Marrink and Tieleman [154], using a geometric projection and subsequent force field-based relaxation for the back-mapping. The reconstruction of an atomistic backbone from a coarse-grained model is done using a new dedicated algorithm. A similar approach has been suggested by Brocos et al. [155]. Molecular dynamics simulations have much to contribute to nanomedicine.

Acknowledgments

This research has been funded by a grant from the National Science Foundation (Grant No. CBET/1263107/1545560).

References

1. Alessandrini A, Facci P (2012) Nanoscale mechanical properties of lipid bilayers and their relevance in biomembrane organization and function. Micron 43:1212–1223. https://doi.org/10.1016/j.micron.2012.03.013
2. Yuan HJ, Jameson CJ, Murad S (2009) Exploring gas permeability of lipid membranes using coarse-grained molecular dynamics. Mol Simulat 35:953–961. https://doi.org/10.1080/08927020902763839
3. Song B, Yuan HJ, Jameson CJ, Murad S (2011) Permeation of nanocrystals across lipid membranes. Mol Phys 109 (11):1511–1526. https://doi.org/10.1080/00268976.2011.569511
4. Monticelli L, Kandasamy SK, Periole X, Larson RG, Tieleman DP, Marrink SJ (2008) The MARTINI coarse-grained force field: extension to proteins. J Chem Theory Comput 4:819–834. https://doi.org/10.1021/ct700324x
5. Marrink SJ, Risselada HJ, Yefimov S, Tieleman DP, De Vries AH (2007) The MARTINI force field: coarse grained model for biomolecular simulations. J Phys Chem B 111 (27):7812–7824. https://doi.org/10.1021/jp071097f
6. Baron R, de Vries AH, Hunenberger PH, van Gunsteren WF (2006) Configurational entropies of lipids in pure and mixed bilayers from atomic level and coarse-grained molecular dynamics simulations. J Phys Chem B 110:15602–15614. https://doi.org/10.1021/jp061627s
7. Lee OS, Schatz GS (2009) Interaction between DNAs on a gold surface. J Phys Chem C 113(36):15941–15947. https://doi.org/10.1021/jp905469q

8. Lee OS, Schatz GS (2009) Molecular dynamics simulation of DNA-functionalized gold nanoparticles. J Phys Chem C 113 (6):2316–2321. https://doi.org/10.1021/jp8094165
9. Chang CI, Lee WJ, Young TF, Ju SP, Chang CW, Chen HL, Chang JG (2008) Adsorption mechanism of water molecules surrounding Au nanoparticles of different sizes. J Chem Phys 128(15):154703. https://doi.org/10.1063/1.2897931
10. Lin J, Zhang H, Chen Z, Zheng Y (2010) Penetration of lipid membranes by gold nanoparticles: insights into cellular uptake, cytotoxicity, and their relationship. ACS Nano 4 (9):5421–5429. https://doi.org/10.1021/nn1010792
11. Hoefler L, Gyurcsanyi RE (2008) Coarse grained molecular dynamics simulation of electromechanically-gated DNA modified conical nanopores. Electroanalysis 20 (3):301–307. https://doi.org/10.1002/elan.200704058
12. Song B, Yuan HJ, Jameson CJ, Murad S (2012) Role of surface ligands in nanoparticle permeation through a model membrane: a coarse-grained molecular dynamics simulations study. Mol Phys 110(18):2181–2195. https://doi.org/10.1080/00268976.2012.668964
13. Mang X, Zeng X, Tang B, Liu F, Ungar G, Zhang R, Mehl GH (2012) Control of anisotropic self-assembly of gold nanoparticles coated with mesogens. J Mater Chem 22 (22):11101–11106. https://doi.org/10.1039/C2JM16794H
14. Nilges M, Clore GM, Gronenborn AM (1988) Determination of three-dimensional structures of proteins from interproton distance data by dynamical simulated annealing from a random array of atoms Circumventing problems associated with folding. FEBS Lett 239(1):129–136. https://doi.org/10.1016/0014-5793(88)80559-3
15. Boisselier E, Astruc D (2009) Gold nanoparticles in nanomedicine: preparations, imaging, diagnostics, therapies and toxicity. Chem Soc Rev 38(6):1759–1782. https://doi.org/10.1039/b806051g
16. Jain PK, Huang XH, El-Sayed IH, El-Sayed MA (2008) Noble metals on the nanoscale: optical and photothermal properties and some applications in imaging, sensing, biology, and medicine. Acc Chem Res 41 (12):1578–1586. https://doi.org/10.1021/ar7002804
17. Sperling RA, Rivera Gil P, Zhang F, Zanella M, Parak WJ (2008) Biological applications of gold nanoparticles. Chem Soc Rev 37(9):1896–1908. https://doi.org/10.1039/b712170a
18. Connor EE, Mwamuka J, Gole A, Murphy CJ, Wyatt MD (2005) Gold nanoparticles are taken up by human cells but do not cause acute cytotoxicity. Small 1(3):325–327. https://doi.org/10.1002/smll.200400093
19. El-Sayed IH, Huang XH, El-Sayed M (2005) Surface plasmon resonance scattering and absorption of anti-EGFR antibody conjugated gold nanoparticles in cancer diagnostics: applications in oral cancer. Nano Lett 5 (5):829–834. https://doi.org/10.1021/nl050074e
20. Luedtke WD, Landman U (1998) Structure and thermodynamics of self-assembled monolayers on gold nanocrystallities. J Phys Chem B 102(34):6566–6572. https://doi.org/10.1021/jp981745i
21. Hostetler MJ, Wingate JE, Zhong CJ, Harris JE, Vachet RW (1998) Alkanethiolate gold cluster molecules with core diameters from 1.5 to 5.2 nm: core and monolayer properties as a function of core size. Langmuir 14 (1):17–30. https://doi.org/10.1021/la970588w
22. Tiwari PM, Vig K, Dennis VA, Singh SR (2011) Functionalized gold nanoparticles and their biomedical applications. Nano 1 (1):31–63. https://doi.org/10.3390/nano1010031
23. Cho WS, Cho M, Jeong J, Choi M, Han BS, Shin HS, Hong J, Chung BH, Jeong J, Cho MH (2010) Size-dependent tissue kinetics of PEG-coated gold nanoparticles. Toxicol Appl Pharmacol 245(1):116–123. https://doi.org/10.1016/j.taap.2010.02.013
24. Pan Y, Neuss S, Liefert A, Fischler M, Wen F, Simon U, Schmid G, Brandau W, Jahnen-Dachent J (2007) Size-dependent cytotoxicity of gold nanoparticles. Small 3 (11):1941–1949. https://doi.org/10.1002/smll.200700378
25. Hainfeld JF, Slatkin DN, Smilowitz HM (2004) The use of gold nanoparticles to enhance radiotherapy in mice. Phys Med Biol 49(18):N309–N315. PMID: 15509078
26. Lee H, Pastor RW (2011) Coarse-grained model for PEGylated lipids: effect of PEGylation on the size and shape of self-assembled structures. J Phys Chem B 115:7830–7837. https://doi.org/10.1021/jp2020148
27. Lee H, de Vries AH, Marrink SJ, Pastor RW (2009) A coarse-grained model for polyethylene oxide and polyethylene glycol: conformation and hydrodynamics. J Phys Chem B 113

(40):13186–13194. https://doi.org/10.1021/jp9058966

28. Rossi G, Fuchs PFJ, Barnoud J, Monticelli L (2012) A coarse-grained MARTINI model of polyethylene glycol and of polyoxyethylene alkyl ether surfactants. J Phys Chem B 116 (49):14353–14362. https://doi.org/10.1021/jp3095165
29. Manson J, Kumar D, Meenan BJ, Dixon D (2011) Polyethylene glycol functionalized gold nanoparticles: the influence of capping density on stability in various media. Gold Bull 44(2):99–105. https://doi.org/10.1007/s13404-011-0015-8
30. Chevigny C, Dalmas F, Di Cola E, Gigmes D, Bertin D, Boué F, Jestin J (2010) Polymer-grafted-nanoparticles nanocomposites: dispersion, grafted chain conformation, and rheological behavior. Macromolecules 44 (1):122–133. https://doi.org/10.1021/ma101332s
31. Wu C (2011) Simulated glass transition of poly (ethylene oxide) bulk and film: a comparative study. J Phys Chem B 115 (38):11044–11052. https://doi.org/10.1021/jp205205x
32. Barbier D, Brown D, Grillet AC, Neyertz S (2004) Interface between end-functionalized PEG oligomers and a silica nanoparticle studied by molecular dynamics simulations. Macromolecules 37(12):4695–4710. https://doi.org/10.1021/ma0359537
33. Ghanbari A, Rahimi M, Dehghany J (2013) Influence of surface grafted polymers on the polymer dynamics in a silica–polystyrene nanocomposite: a coarse-grained molecular dynamics investigation. J Phys Chem C 117 (47):25069–25076. https://doi.org/10.1021/jp407109r
34. Corbierre MK, Cameron NS, Sutton M, Mochrie SG, Lurio LB, Rühm A, Lennox RB (2001) Polymer-stabilized AuNPs and their incorporation into polymer matrices. J Am Chem Soc 123(42):10411–10412. https://doi.org/10.1021/ja0166287
35. Smith JS, Bedrov D, Smith GD (2003) A molecular dynamics simulation study of nanoparticle interactions in a model polymer-nanoparticle composite. Compos Sci Technol 63(11):1599–1605. https://doi.org/10.1016/S0266-3538(03)00061-7
36. Ndoro TV, Voyiatzis E, Ghanbari A, Theodorou DN, Böhm MC, Müller-Plathe F (2011) Interface of grafted and ungrafted silica nanoparticles with a polystyrene matrix: atomistic molecular dynamics simulations. Macromolecules 44(7):2316–2327. https://doi.org/10.1021/ma102833u
37. Hong B, Panagiotopoulos AZ (2012) Molecular dynamics simulations of silica nanoparticles grafted with poly (ethylene oxide) oligomer chains. J Phys Chem B 116 (8):2385–2395. https://doi.org/10.1021/jp2112582
38. Karakoti AS, Das S, Thevuthasan S, Seal S (2011) PEGylated inorganic nanoparticles. Angew Chem Int Ed Engl 50 (9):1980–1994. https://doi.org/10.1002/anie.201002969
39. Xia X, Yang M, Wang Y, Zheng Y, Li Q, Chen J, Xia Y (2011) Quantifying the coverage density of poly(ethylene glycol) chains on surfaces of gold nanostructures. ACS Nano 6 (1):512–522. https://doi.org/10.1021/nn2038516
40. Zeng Q, Yu A, Lu G (2010) Evaluation of interaction forces between nanoparticles by molecular dynamics simulation. Ind Eng Chem Res 49:12793–12797. https://doi.org/10.1021/ie101751v
41. Vasir JK, Labhasetwar V (2008) Quantification of the force of nanoparticle-cell membrane interactions and its influence on intracellular trafficking of nanoparticles. Biomaterials 29:4244–4252. https://doi.org/10.1016/j.biomaterials.2008.07.020
42. Lee OS, Schatz GC (2011) Computational simulations of the interaction of lipid membranes with DNA-functionalized gold nanoparticles. Methods Mol Biol 726:283–296. https://doi.org/10.1007/978-1-61779-052-2_18
43. Wallace EJ, Sansom MSP (2008) Blocking of carbon nanotube based nanoinjectors by lipids: a simulation study. Nano Lett 8:2751–2756. https://doi.org/10.1021/nl801217f
44. Vakarelski IU, Brown SC, Higashitani K, Moudgil BM (2007) Penetration of living cell membranes with fortified carbon nanotube tips. Langmuir 23(22):10893–10896. https://doi.org/10.1021/la701878n
45. Skjevik AE, Madej BD, Dickson CJ, Lin C, Teigen K, Walker RC, Gould IR (2016) Simulation of lipid bilayer self-assembly using all-atom lipid force fields. Phys Chem Chem Phys 18:10573–10584. https://doi.org/10.1039/c5cp07379k
46. Kučerka N, Nagle JF, Sachs JN, Feller SE, Pencer J, Jackson A, Katsaras J (2008) Lipid bilayer structure determined by the simultaneous analysis of neutron and X-ray scattering data. Biophys J 95:2356–2367. https://doi.org/10.1529/biophysj.108.132662

47. Yuan H, Jameson CJ, Murad S (2010) Diffusion of gases across lipid membranes with OmpA channel: a molecular dynamics study. Mol Phys 108(12):1569–1581. https://doi.org/10.1080/00268976.2010.484396
48. Nagle JF, Tristram-Nagle S (2000) Structure of lipid bilayers. Biochim Biophys Acta 1469 (3):159–195. PMID: 11063882
49. Douliez JP, Leonard A, Dufourc EJ (1995) Restatement of order parameters in biomembranes–calculation of C-C bond order parameters from C-D quadrupolar splittings. Biophys J 68(5):1727–1739. https://doi.org/10.1016/S0006-3495(95)80350-4
50. Muddana HS, Gullapalli RR, Manias E, Butler PJ (2011) Atomistic simulation of lipid and DiI dynamics in membrane bilayers under tension. Phys Chem Chem Phys 13 (4):1368–1378. https://doi.org/10.1039/c0cp00430h
51. Ayton G, Smondyrev AM, Bardenhagen SG, McMurtry P, Voth GA (2002) Calculating the bulk modulus for a lipid bilayer with nonequilibrium molecular dynamics simulation. Biophys J 82(3):1226–1238. https://doi.org/10.1016/S0006-3495(02)75479-9
52. Chen R, Poger D, Mark AE (2011) Effect of high pressure on fully hydrated DPPC and POPC bilayers. J Phys Chem B 115:1038–1044. https://doi.org/10.1021/jp110002q
53. Scarlata SF (1991) Compression of lipid membranes as observed at varying membrane positions. Biophys J 60(2):334–340. https://doi.org/10.1016/S0006-3495(91)82058-6
54. Wong PTT, Mantsch HH (1988) Reorientational and conformational ordering processes at elevated pressures in 1,2-dioleoyl phosphatidylcholine: a Raman and infrared spectroscopic study. Biophys J 54(5):781–790. https://doi.org/10.1016/S0006-3495(88)83016-9
55. Venable RM, Brown FLH, Pastor RW (2015) Mechanical properties of lipid bilayers from molecular dynamics simulation. Chem Phys Lipids 192:60–74. https://doi.org/10.1016/j.chemphyslip.2015.07.014
56. Braganza LF, Worcester DL (1986) Structural changes in lipid bilayers and biological membranes caused by hydrostatic pressure. Biochemistry 25(23):7484–7488. https://doi.org/10.1021/bi00371a034
57. Alkilany AM, Thompson LB, Boulos SP, Sisco PN, Murphy CJ (2012) Gold nanorods: their potential for photothermal therapeutics and drug delivery, tempered by the complexity of their biological interactions. Adv Drug Deliv Rev 64(2):190–199. https://doi.org/10.1016/j.addr.2011.03.005
58. Bagley AF, Hill S, Rogers GS, Bhatia SN (2013) Plasmonic photothermal heating of intraperitoneal tumors through the use of an implanted near-infrared source. ACS Nano 7 (9):8089–8097. https://doi.org/10.1021/nn4033757
59. Huang X, Jain PK, El-Sayed IH, El-Sayed MA (2008) Plasmonic photothermal therapy (PPTT) using gold nanoparticles. Lasers Med Sci 23(3):217–228. https://doi.org/10.1007/s10103-007-0470-x
60. Kennedy LC, Bickford LR, Lewinski NA, Coughlin AJ, Hu Y, Day ES, West JL, Drezek RA (2011) A new era for cancer treatment: gold-nanoparticle-mediated thermal therapies. Small 7(2):169–183. https://doi.org/10.1002/smll.201000134
61. McQuaid HN, Muir MF, Taggart LE, McMahon SJ, Coulter JA, Hyland WB, Jain S, Butterworth KT, Schettino G, Prise KM, Hirst DG, Botchway SW, Currell FJ (2016) Imaging and radiation effects of gold nanoparticles in tumour cells. Sci Rep 6:19442. https://doi.org/10.1038/srep19442
62. Curry T, Kopelman R, Shilo M, Popovtzer R (2014) Multifunctional theranostic gold nanoparticles for targeted CT imaging and photothermal therapy. Contrast Media Mol Imaging 9(1):53–61. https://doi.org/10.1002/cmmi.1563
63. Cai W, Gao T, Hong H, Sun J (2008) Applications of gold nanoparticles in cancer nanotechnology. Nanotechnol Sci Appl 1:17–32. https://doi.org/10.2147/NSA.S3788
64. Carnovale C, Bryant G, Shukla R, Bansal V (2016) Size, shape and surface chemistry of nano-gold dictate its cellular interactions, uptake and toxicity. Prog Mater Sci 83:152–190. https://doi.org/10.1016/j.pmatsci.2016.04.003
65. Etame AB, Smith CA, Chan WC, Rutka JT (2011) Design and potential application of PEGylated gold nanoparticles with size-dependent permeation through brain microvasculature. Nanomedicine 7(6):992–1000. https://doi.org/10.1016/j.nano.2011.04.004
66. Terrill RH, Postlethwaite TA, Chen C, Poon CD, Terzis A, Chen A, Hutchison JE, Clark MR, Wingall G, Londono JD, Superfine R, Falvo M, Johnson CS Jr, Samulski ET, Murray RW (1995) Monolayers in three dimensions: NMR, SAXS, thermal, and electron hopping studies of alkanethiol stabilized gold clusters. J Am Chem Soc 117(50):12537–12548. https://doi.org/10.1021/ja00155a017

67. Oroskar PA, Jameson CJ, Murad S (2016) Simulated permeation and characterization of PEGylated gold nanoparticles in a lipid bilayer system. Langmuir 32 (30):7541–7555. https://doi.org/10.1021/acs.langmuir.6b01740
68. Jokerst JV, Lobovkina T, Zare RN, Gambhir SS (2011) Nanoparticle PEGylation for imaging and therapy. Nanomedicine 6 (4):715–728. https://doi.org/10.2217/nnm.11.19
69. Oroskar PA, Jameson CJ, Murad S (2017) Molecular dynamics simulations reveal how characteristics of surface and permeant affect permeation events at the surface of soft matter. Mol Simulat 43(5):1–28. https://doi.org/10.1080/08927022.2016.1268259
70. Van Lehn RC, Alexander-Katz A (2014) Membrane-embedded nanoparticles induce lipid rearrangements similar to those exhibited by biological membrane proteins. J Phys Chem B 118(44):12586-–12598. https://doi.org/10.1021/jp506239p
71. Song B, Yuan HJ, Pham SV, Jameson CJ, Murad S (2012) Nanoparticle permeation induces water penetration, ion transport, and lipid flip-flop. Langmuir 28 (49):16989–17000. https://doi.org/10.1021/la302879r
72. Gurtovenko AA, Vattulainen I (2007) Molecular mechanism for lipid flip-flops. J PhysChem B 111(48):13554–13559. https://doi.org/10.1021/jp077094k
73. Tieleman DP, Marrink SJ (2006) Lipids out of equilibrium: energetics of desorption and pore mediated flip-flop. J Am Chem Soc 128 (38):12462–12467. https://doi.org/10.1021/ja0624321
74. Sapay N, Bennett WFD, Tieleman DP (2009) Thermodynamics of flip-flop and desorption for a systematic series of phosphatidylcholine lipids. Soft Matter 5:3295–3302. https://doi.org/10.1039/b902376c
75. Contreras FX, Sánchez-Magraner L, Alonso A, Goñi FM (2010) Transbilayer (flip-flop) lipid motion and lipid scrambling in membranes. FEBS Lett 584 (9):1779–1786. https://doi.org/10.1016/j.febslet.2009.12.049
76. Fadeel B, Xue D (2009) The ins and outs of phospholipid asymmetry in the plasma membrane: roles in health and disease. Crit Rev Biochem Mol Biol 44(5):264–277. https://doi.org/10.1080/10409230903193307
77. Devaux PF (1991) Static and dynamic lipid asymmetry in cell membranes. Biochemist 30 (5):1163–1173. https://doi.org/10.1021/bi00219a001
78. Ding HM, Tian WD, Ma YQ (2012) Designing nanoparticle translocation through membranes by computer simulations. ACS Nano 6:1230–1238. https://doi.org/10.1021/nn2038862
79. Vacha R, Martinez-Veracoechea FJ, Frenkel D (2011) Receptor-mediated endocytosis of nanoparticles of various shapes. Nano Lett 11(12):5391–5395. https://doi.org/10.1021/nl2030213
80. Yang K, Ma YQ (2010) Computer simulation of the translocation of nanoparticles with different shapes across a lipid bilayer. Nat Nanotechnol 5(8):579–583. https://doi.org/10.1038/nnano.2010.141
81. Huang CJ, Zhang Y, Yuan HY, Gao HJ, Zhang S (2013) Role of nanoparticle geometry in endocytosis: laying down to stand up. Nano Lett 13:4546–4550. https://doi.org/10.1021/nl402628n
82. Yi X, Shi X, Gao H (2014) A universal law for cell uptake of one-dimensional nanomaterials. Nano Lett 14(2):1049–1055. https://doi.org/10.1021/nl404727m
83. Yang K, Yuan B, Ma YQ (2013) Influence of geometric nanoparticle rotation on cellular internalization process. Nanoscale 5:7998–8006. https://doi.org/10.1039/C3NR01561K
84. Zhang HZ, Wang L, Yuan B, Yang K, Ma YQ (2014) Effect of receptor structure and length on the wrapping of a nanoparticle by a lipid membrane. Materials 7:3855–3866. https://doi.org/10.3390/ma7053855
85. Chen YB, Liu YH, Zeng Y, Mao W, Hu L, Mao ZL, Xu HQ (2015) Optimal aspect ratio of endocytosed spherocylindrical nanoparticle. Front Physiol 10:108702. https://doi.org/10.1007/s11467-014-0444-y
86. Yue T, Zhang X, Huang F (2015) Molecular modeling of membrane responses to the adsorption of rotating nanoparticles: promoted cell uptake and mechanical membrane rupture. Soft Matter 11(3):456–465. https://doi.org/10.1039/c4sm01760a
87. Li Y, Chen X, Gu N (2008) Computational investigation of interaction between nanoparticles and membranes: hydrophobic/hydrophilic effect. J Phys Chem B 112 (51):16647–16653. https://doi.org/10.1021/jp8051906
88. Oroskar PA, Jameson CJ, Murad S (2015) Surface-functionalized nanoparticle permeation triggers lipid displacement and water and ion leakage. Langmuir 31

(3):1074–1085. https://doi.org/10.1021/la503934c

89. Mathai JC, Tristram-Nagle S, Nagle JF, Zeidel ML (2008) Structural determinants of water permeability through the lipid membrane. J Gen Physiol 131(1):69–76. https://doi.org/10.1085/jgp.200709848

90. Gurtovenko AA, Vattulainen I (2007) Ion leakage through transient water pores in protein-free lipid membranes driven by transmembrane ionic charge imbalance. Biophys J 92(6):1878–1890. https://doi.org/10.1529/biophysj.106.094797

91. Leontiadou H, Mark AE, Marrink SJ (2007) Ion transport across transmembrane pores. Biophys J 92:4209–4215. https://doi.org/10.1529/biophysj.106.101295

92. Bennett WFD, Tieleman DP (2011) Water defect and pore formation in atomistic and coarse-grained lipid membranes: pushing the limits of coarse graining. J Chem Theory Comput 7(9):2981–2988. https://doi.org/10.1021/ct200291v

93. Bennett WFD, Sapay N, Tieleman DP (2014) Atomistic simulations of pore formation and closure in lipid bilayers. Biophys J 106(1):210–219. https://doi.org/10.1016/j.bpj.2013.11.4486

94. Wang S, Larson RG (2014) Water channel formation and ion transport in linear and branched lipid bilayers. Phys Chem Chem Phys 16(16):7251–7262. https://doi.org/10.1039/c3cp55116d

95. Koshiyama K, Yano T, Kodama T (2010) Self-organization of a stable pore structure in a phospholipid bilayer. Phys Rev Lett 105(1):018105. https://doi.org/10.1103/PhysRevLett.105.018105

96. Niidome T, Yamagata M, Okamoto Y, Akiyama Y, Takahashi H, Kawano T, Niidome Y (2006) PEG-modified gold nanorods with a stealth character for *in vivo* applications. J Control Release 114(3):343–347. https://doi.org/10.1016/j.jconrel.2006.06.017

97. Alekseeva AV, Bogatyrev VA, Dykman LA, Khlebtsov BN, Trachuk LA, Melnikov AG, Khlebtsov NG (2005) Preparation and optical scattering characterization of gold nanorods and their application to a dot-immunogold assay. Appl Opt 44(29):6285–6295. https://doi.org/10.1364/AO.44.006285

98. El-Sayed MA, Shabaka AA, El-Shabrawy OA, Yassin NA, Mahmoud SS, El-Shenawy SM, Emad AA, Eisa WH, Farag NM, El-Shaer MA, Salah N, Al-Abd AM (2013) Tissue distribution and efficacy of gold nanorods coupled with laser induced photoplasmonic therapy in Ehrlich carcinoma solid tumor model. PLoS One 8(10):e76207. https://doi.org/10.1371/journal.pone.0076207

99. Huang X, El-Sayed IH, Qian W, El-Sayed MA (2006) Cancer cell imaging and photothermal therapy in the near-infrared region by using gold nanorods. J Am Chem Soc 128(6):2115–2120. https://doi.org/10.1021/ja057254a

100. Huff TB, Tong L, Zhao Y, Hansen MN, Cheng JX, Wei A (2007) Hyperthermic effects of gold nanorods on tumor cells. Nanomedicine (Lond) 2(1):125–132. https://doi.org/10.2217/17435889.2.1.125

101. Li CZ, Male KB, Hrapovic S, Luong JHT (2005) Fluorescence properties of gold nanorods and their application for DNA biosensing. Chem Commun 2005(31):3924–3926. https://doi.org/10.1039/B504186D

102. Lin KY, Bagley AF, Zhang AY, Karl DL, Yoon SS, Bhatia SN (2010) Gold nanorod photothermal therapy in a genetically engineered mouse model of soft tissue sarcoma. Nano Life 1(3–4):277–287. https://doi.org/10.1142/S1793984410000262

103. Link S, Mohamed MB, El-Sayed MA (1999) Simulation of the optical absorption spectra of gold nanorods as a function of their aspect ratio and the effect of the medium dielectric constant. J Phys Chem B 103(16):3073–3077. https://doi.org/10.1021/jp990183f

104. Maestro LM, Camarillo E, Sánchez-Gil JA, Rodríguez-Oliveros R, Ramiro-Bargueño J, Caamaño AJ, Jaque D (2014) Gold nanorods for optimized photothermal therapy: the influence of irradiating in the first and second biological windows. RSC Adv 4(96):54122–54129. https://doi.org/10.1039/C4RA08956A

105. Vigderman L, Khanal BP, Zubarev ER (2012) Functional gold nanorods: synthesis, self-assembly, and sensing applications. Adv Mater 24(36):4811–4841. https://doi.org/10.1002/adma.201201690

106. von Maltzahn G, Park JH, Agrawal A, Bandaru NK, Das SK, Sailor MJ, Bhatia SN (2009) Computationally-guided photothermal tumor therapy using long-circulating gold nanorod antennas. Cancer Res 69(9):3892–3900. https://doi.org/10.1158/0008-5472.CAN-08-4242

107. von Maltzahn G, Centrone A, Park JH, Ramanathan R, Sailor MJ, Hatton TA, Bhatia SN (2009) SERS-coded gold nanorods as a multifunctional platform for densely

multiplexed near-infrared imaging and photothermal heating. Adv Mater 21 (31):3175–3180. https://doi.org/10.1002/adma.200803464

108. Wang H, Huff TB, Zweifel DA, He W, Low PS, Wei A, Cheng JX (2005) *In vitro* and *in vivo* two-photon luminescence imaging of single gold nanorods. Proc Natl Acad Sci U S A 102(44):15752–15756. https://doi.org/10.1073/pnas.0504892102

109. Xiao Y, Hong H, Matson VZ, Javadi A, Xu W, Yang Y, Zhang Y, Engle JW, Nickles RJ, Cai W, Steeber DA, Gong S (2012) Gold nanorods conjugated with doxorubicin and cRGD for combined anticancer drug delivery and PET imaging. Theranostics 2 (8):757–768. https://doi.org/10.7150/thno.4756

110. Zhu J, Huang L, Zhao J, Wang Y, Zhao Y, Hao L, Lu Y (2005) Shape dependent resonance light scattering properties of gold nanorods. Mater Sci Eng B 121 (3):199–203. https://doi.org/10.1016/j.mseb.2005.03.022

111. Adrian NNM, Cheng YY, Ong NMN, Kamaruddin TT, Rozlan E, Schmidt TW, Duong HTT, Boyer C (2016) Effect of gold nanoparticle shapes for phototherapy and drug delivery. Polym Chem 7:2888–2903. https://doi.org/10.1039/C6PY00465B

112. Oroskar PA, Jameson CJ, Murad S (2016) Rotational behavior of PEGylated gold nanorods in a lipid bilayer system. Mol Phys 115(9–12):1122–1143. https://doi.org/10.1080/00268976.2016.1248515

113. Nguyen TM, Gigault J, Hackley VA (2014) PEGylated gold nanorod separation based on aspect ratio: characterization by asymmetric-flow field flow fractionation with UV-Vis detection. Anal Bioanal Chem 406 (6):1651–1659. https://doi.org/10.1007/s00216-013-7318-y

114. Qiu Y, Liu Y, Wang L, Xu L, Bai R, Ji Y, Chen C (2010) Surface chemistry and aspect ratio mediated cellular uptake of Au nanorods. Biomaterials 31(30):7606–7619. https://doi.org/10.1016/j.biomaterials.2010.06.051

115. Liu XX, Wu FC, Tian Y, Wu M, Zhou Q, Jiang S, Niu ZW (2016) Size dependent cellular uptake of rod-like bionanoparticles with different aspect ratios. Sci Rep 6:24567. https://doi.org/10.1038/srep24567

116. Gu Y, Di XW, Sun W, Wang GF, Fang N (2012) Three-dimensional super-localization and tracking of single gold nanoparticles in cells. Anal Chem 84:4111–4117. https://doi.org/10.1021/ac300249d

117. Gu Y, Wang GF, Fang N (2013) Simultaneous single-particle superlocalization and rotational tracking. ACS Nano 7:1658–1665. https://doi.org/10.1021/nn305640y

118. Xu D, He Y, Yeung ES (2014) Direct imaging of transmembrane dynamics of single nanoparticles with dark-field microscopy: improved orientation tracking at cell sidewall. Anal Chem 86(7):3397–3404. https://doi.org/10.1021/ac403700u

119. Shi XH, von dem Bussche A, Hurt RH, Kane AB, Gao HJ (2011) Cell entry of one-dimensional nanomaterials occurs by tip recognition and rotation. Nat Nanotechnol 6 (11):714–719. https://doi.org/10.1038/nnano.2011.151

120. Chen L, Xiao S, Zhu H, Wang L, Liang HJ (2016) Shape-dependent internalization kinetics of nanoparticles by membranes. Soft Matter 12:2632–2641. https://doi.org/10.1039/c5sm01869b

121. Ding HM, Ma YQ (2015) Theoretical and computational investigations of nanoparticle–biomembrane interactions in cellular delivery. Small 11:1055–1071. https://doi.org/10.1002/smll.201401943

122. Zhang S, Gao H, Bao G (2015) Physical principles of nanoparticle cellular endocytosis. ACS Nano 9(9):8655–8671. https://doi.org/10.1021/acsnano.5b03184

123. Li Y, Kroger M, Liu WK (2014) Endocytosis of PEGylated nanoparticles accompanied by structural and free energy changes of the grafted polyethylene glycol. Biomaterials 35:8467–8478. https://doi.org/10.1016/j.biomaterials.2014.06.032

124. Li Y, Kröger M, Liu WK (2015) Shape effect in cellular uptake of PEGylated nanoparticles: comparison between sphere, rod, cube and disk. Nanoscale 7:16631–16646. https://doi.org/10.1039/C5NR02970H

125. Ghosh P, Han G, De M, Kim CK, Rotello VM (2008) Gold nanoparticles in delivery applications. Adv Drug Deliv Rev 60 (11):1307–1315. https://doi.org/10.1016/j.addr.2008.03.016

126. Zhang GD, Yang Z, Lu W, Zhang R, Huang Q, Tian M, Li L, Liang D, Li C (2009) Influence of anchoring ligands and particle size on the colloidal stability and *in vivo* biodistribution of polyethylene glycol-coated gold nanoparticles in tumor-xenografted mice. Biomaterials 30 (10):1928–1936. https://doi.org/10.1016/j.biomaterials.2008.12.038

127. Terentyuk GS, Maslyakova GN, Suleymanova LV, Khlebtsov BN, Kogan BY, Akchurin GG, Shantrocha AV, Maksimova IL, Khlebtsov NG, Tuchin VV (2009) Circulation and distribution of gold nanoparticles and induced alterations of tissue morphology at intravenous particle delivery. J Biophotonics 2 (5):292–302. https://doi.org/10.1002/jbio.200910005
128. Perrault SD, Walkey C, Jennings T, Fischer HC, Chan WC (2009) Mediating tumor targeting efficiency of nanoparticles through design. Nano Lett 9(5):1909–1915. https://doi.org/10.1021/nl900031y
129. England CG, Gobin AM, Frieboes HB (2015) Evaluation of uptake and distribution of gold nanoparticles in solid tumors. Eur Phys J Plus 130:231. https://doi.org/10.1140/epjp/i2015-15231-1
130. Balogh L, Nigavekar SS, Nair BM, Lesniak W, Zhang C, Sung LY, Kariapper MST, El-Jawahri A, Llanes M, Bolton B, Mamou F, Tan W, Hutson A, Minc L, Khan MK (2007) Significant effect of size on the *in vivo* biodistribution of gold composite nanodevices in mouse tumor models. Nanomedicine 3:281–296. https://doi.org/10.1016/j.nano.2007.09.001
131. Leroueil PR, Hong SP, Mecke A, Baker JR Jr, Orr BG, Holl MMB (2007) Nanoparticle interaction with biological membranes: Does nanotechnology present a Janus face? Acc Chem Res 40:335–342. https://doi.org/10.1021/ar600012y
132. Goodman CM, McCusker CD, Yilmaz T, Rotello VM (2004) Toxicity of gold nanoparticles functionalized with cationic and anionic side chains. Bioconjug Chem 15:897–900. https://doi.org/10.1021/bc049951i
133. Feng ZV, Gunsolus IL, Qiu TA, Hurley KR, Nyberg LH, Frew H, Johnson KP, Vartanian AM, Jacob LM, Lohse SE, Torelli MD, Hamers RJ, Murphy CJ, Haynes CL (2015) Impacts of gold nanoparticle charge and ligand type on surface binding and toxicity to Gram-negative and Gram-positive bacteria. Chem Sci 6:5186–5196. https://doi.org/10.1039/C5SC00792E
134. Gkeka P, Angelikopoulos P, Sarkisov L, Cournia Z (2014) Membrane partitioning of anionic, ligand-coated nanoparticles is accompanied by ligand snorkeling, local disordering, and cholesterol depletion. PLoS Comput Biol 10(12):e1003917. https://doi.org/10.1371/journal.pcbi.1003917
135. Maeda H, Nakamura H, Fang J (2013) The EPR effect for macromolecular drug delivery to solid tumors: improvement of tumor uptake, lowering of systemic toxicity, and distinct tumor imaging *in vivo*. Adv Drug Deliv Rev 65:71–79. https://doi.org/10.1016/j.addr.2012.10.002
136. Hinterwirth H, Kappel S, Waitz T, Prohaska T, Lindner W, Lämmerhofer M (2013) Quantifying thiol ligand density of self-assembled monolayers on gold nanoparticles by inductively coupled plasma–mass spectrometry. ACS Nano 7(2):1129–1136. https://doi.org/10.1021/nn306024a
137. Liu HY, Doane TL, Cheng Y, Lu F, Srinivasan S, Zhu JJ, Burda C (2015) Control of surface ligand density on PEGylated gold nanoparticles for optimized cancer cell uptake. Part Part Syst Charact 32 (2):197–204. https://doi.org/10.1002/ppsc.201400067
138. Rahme K, Chen L, Hobbs RG, Morris MA, O'Driscoll C, Holmes JD (2013) PEGylated gold nanoparticles: polymer quantification as a function of PEG lengths and nanoparticle dimensions. RSC Adv 3(17):6085–6094. https://doi.org/10.1039/C3RA22739A
139. Zhang HZ, Ji Q, Huang Q, Zhang S, Yuan B, Yang K, Ma YQ (2015) Cooperative transmembrane penetration of nanoparticles. Sci Rep 5:10525. https://doi.org/10.1038/srep10525
140. Verma A, Uzun O, Hu Y, Hu Y, Han HS, Watson N, Chen S, Irvine DJ, Stellacci F (2008) Surface-structure-regulated cell-membrane penetration by monolayer-protected nanoparticles. Nat Mater 7:588–595. https://doi.org/10.1038/nmat2202
141. Stirling J, Lekkas I, Sweetman A, Djuranovic P, Guo Q, Pauw B, Granwehr J, Lévy R, Moriarty P (2014) Critical assessment of the evidence for striped nanoparticles. PLoS One 9(11):e108482. https://doi.org/10.1371/journal.pone.0108482
142. Ge XW, Ke PC, Davis TP, Ding F (2015) A thermodynamics model for the emergence of a stripe-like binary SAM on a nanoparticle surface. Small 11(37):4894–4899. https://doi.org/10.1002/smll.201501049
143. Carney RP, DeVries GA, Dubois C, Kim H, Kim JY, Singh C, Ghorai PK, Tracy JB, Stiles RL, Murray RW, Glotzer SC, Stellacci F (2008) Size limitations for the formation of ordered striped nanoparticles. J Am Chem Soc 130:798–799. https://doi.org/10.1021/ja077383m
144. Velachi V, Bhandary D, Singh JK, Cordeiro MNDS (2015) Structure of mixed self-assembled monolayers on gold nanoparticles at three different arrangements. J Phys Chem

C 119(6):3199–3209. https://doi.org/10.1021/jp512144g

145. Velachi V, Bhandary D, Singh JK, Cordeiro MNDS (2016) Striped gold nanoparticles: new insights from molecular dynamics simulations. J Chem Phys 144(24):244710. https://doi.org/10.1063/1.4954980

146. Debierre-Grockiego F (2010) Glycolipids are potential targets for protozoan parasite diseases. Trends Parasitol 26:404–411. https://doi.org/10.1016/j.pt.2010.04.006

147. Kapla J, Stevensson B, Dahlberg M, Maliniak A (2012) Molecular dynamics simulations of membranes composed of glycolipids and phospholipids. J Phys Chem B 116 (1):244–252. https://doi.org/10.1021/jp209268p

148. Polley A, Vemparala S, Rao M (2012) Atomistic simulations of a multicomponent asymmetric lipid bilayer. J Phys Chem B 116 (45):13403–13410. https://doi.org/10.1021/jp3032868

149. Kindt JT (2011) Atomistic simulation of mixed-lipid bilayers: mixed methods for mixed membranes. Mol Simulat 37 (7):516–524. https://doi.org/10.1080/08927022.2011.561434

150. Hong C, Tieleman DP, Wang Y (2014) Microsecond molecular dynamics simulations of lipid mixing. Langmuir 30 (40):11993–12001. https://doi.org/10.1021/la502363b

151. Grouleff J, Irudayam SJ, Skeby KK, Schiøtt B (2015) The influence of cholesterol on membrane protein structure, function, and dynamics studied by molecular dynamics simulations. Biochim Biophys Acta 1848 (9):1783–1795. https://doi.org/10.1016/j.bbamem.2015.03.029

152. Gumí-Audenis B, Costa L, Carlá F, Comin F, Sanz F, Giannotti MI (2016) Structure and nanomechanics of model membranes by atomic force microscopy and spectroscopy: insights into the role of cholesterol and sphingolipids. Membranes 6:58. https://doi.org/10.3390/membranes6040058

153. Rønnest AK, Peters GH, Hansen FY, Taub H, Miskowiec A (2016) Structure and dynamics of water and lipid molecules in charged anionic DMPG lipid bilayer membranes. J Chem Phys 144:144904. https://doi.org/10.1063/1.4945278

154. Wassenaar TA, Pluhackova K, Boeckmann RA, Marrink SJ, Tieleman DP (2014) Going backward: a flexible geometric approach to reverse transformation from coarse grained to atomistic models. J Chem Theory Comput 10(2):676–690. https://doi.org/10.1021/ct400617g

155. Brocos P, Mendoza-Espinosa P, Castillo R, Mas-Oliva J, Pineiro A (2012) Multiscale molecular dynamics simulations of micelles: coarse-grain for self-assembly and atomic resolution for finer details. Soft Matter 8 (34):9005–9014. https://doi.org/10.1039/c2sm25877c

Chapter 22

Atomic Force Microscopy for Cell Membrane Investigation

Mingjun Cai and Hongda Wang

Abstract

Atomic force microscopy (AFM) is a very versatile tool for studying biological samples at nanometer-scale resolution. The cell membrane plays a key role in compartmentalization, nutrient transportation, and signal transduction, while the structural feature of both sides of the membrane remains elusive. Here we describe our methods for the preparation of the cell membrane from the red blood cells and nucleated cells. High-resolution AFM topographs reveal substructural details of both sides of the cell membrane. The structure composition of cell membrane can be directly observed by time-lapse AFM and the positional information of membrane proteins can be located by molecular recognition.

Key words Atomic force microscopy (AFM), Topography and recognition imaging (TREC), Cell membrane structures

1 Introduction

Cell membranes possess crucial functions in a living cell, such as compartmentalizing the cell from the environment, cell signaling, and solute transporting. The structure of cell membranes at the molecular level is a fundamental question in cell biology. The study of their structure is vital for various applications, including drug screening, cancer treatment, signal transduction control, etc. Hypotheses that include the liquid mosaic model, lipid raft model, and protein domain model have been constructed in the last four decades. However, the structure of cell membranes is still a controversial topic because these models are based on indirect evidences or nonnative conditions [1–3].

Atomic force microscopy [4], a key family member of scanning probe microscopy, may offer a solution to in situ imaging of biological samples without large structural perturbations. AFM takes advantage of a micro-fabricated cantilever with a sharp tip to scan the surface of the samples, and the deflection of the cantilever is utilized to record the information of the surface properties. AFM allows imaging of cellular membranes at a spatial resolution of a few

Volkmar Weissig and Tamer Elbayoumi (eds.), *Pharmaceutical Nanotechnology: Basic Protocols*, Methods in Molecular Biology, vol. 2000, https://doi.org/10.1007/978-1-4939-9516-5_22, © Springer Science+Business Media, LLC, part of Springer Nature 2019

nanometers that gives real three-dimensional imaging. One of the greatest advantages of AFM is that it can image the samples in solutions under the physiological conditions, which makes it very important in biological applications [5–8]. Furthermore, through the functionalization of the AFM tips AFM can recognize specific molecules in heterogeneous samples at the single-molecule level, which is called topography and recognition imaging (TREC) [9].

Our current approach for the recovery of cell membrane in the liquid environment is by utilizing the AAC mode AFM, which minimizes the AFM tip effect on the sample. We directly observed both sides of cell membranes on the APTES-mica by in situ AFM at molecular resolution under quasi-native conditions. By in situ AFM and molecule recognition technique, we find that the location of oligosaccharides and proteins in human red blood cell (hRBC) membranes might be different from the current membrane model. The cytoplasmic side membrane is covered by dense proteins with fewer free lipids, and the Na^+-K^+ATPases were well distributed in the cytoplasmic side of cell membranes with about 10% aggregations in total recognized proteins [10]. In contrast, the ectoplasmic side membrane is quite smooth; oligosaccharides and peptides supposed to protrude out of the ectoplasmic side surface might be actually hidden in the middle of hydrophilic lipid heads; transmembrane proteins might form domains in the membranes revealed by PNGase F and trypsin digestion. Based on the above observation, we drew a proposed model of red blood cell membranes—semi-mosaic model [11]. Our result could be significant to interpret some functions about red blood cell membranes and guide to heal the blood diseases related to cell membranes. Lipid rafts are membrane microdomains enriched with cholesterol, glycosphingolipids, and proteins [12]. The major characteristics of lipid rafts have been originally inferred from detergent-resistant membranes (DRMs). High-resolution and time-lapse in situ atomic force microscopy is used to directly confirm the existence of lipid rafts in native erythrocyte membranes. Cholesterol contributes significantly to the formation and stability of the protein domains, and Band III is an important protein of lipid rafts in the cytoplasmic side of erythrocyte membranes, indicating that lipid rafts are exactly the functional domains in plasma membrane [13, 14].

For the structure of the nucleated cell membranes, we found that proteins at the ectoplasmic side of the cell membrane form a dense protein layer (4 nm) on top of a lipid bilayer; proteins aggregate to form islands evenly dispersed at the cytoplasmic side of the cell membrane with a height of about 10–12 nm; cholesterol-enriched domains exist within the cell membrane; carbohydrates stay in microdomains at the ectoplasmic side; and exposed amino groups are asymmetrically distributed on both sides. Based on these observations, we proposed a protein layer-lipid-protein island

(PLLPI) model, to provide a better understanding of cell membrane structure, membrane trafficking, and viral fusion mechanisms [11].

2 Materials

2.1 Preparation of the Cell Membrane

1. Phosphate buffer solution (PBS) buffer: 136.9 mM NaCl, 2.7 mM KCl, 1.5 mM KH_2PO_4, 8.1 mM Na_2HPO_4, pH 7.4. Sterilized by autoclaving at 121 °C for 15 min. Store at 4 °C.
2. Piperazine-*N*,*N′*-bis(2-ethanesulfonic acid) (PIPES) buffer solution: 20 mM PIPES, 150 mM KCl, pH 6.2 [15].
3. Hypotonic buffer solution: 4 mM PIPES, 30 mM KCl, pH 6.2.
4. High-salt solution: 2 M NaCl, 1.5 mM KH_2PO_4, 2.7 mM KCl, 1 mM Na_2HPO_4, pH 7.2.
5. Detergent solution: 0.1% (v/v) Triton X-100 in PBS solution; 10 mM methyl-β-cyclodextrin (M-β-CD) in PBS solution [14, 16].
6. Digestive solution: 1 mg/mL Trypsin in PBS solution; 1 unit PNGase F in 200 μL PBS solution [16].
7. Cytoskeleton damage reagent: 20 μM Cytochalasin B; 60 μM nocodazole.
8. Ringer's solution: 155 mM NaCl, 3 mM KCl, 2 mM $CaCl_2$, 1 mM $MgCl_2$, 3 mM NaH_2PO_4, 10 mM glucose in 5 mM HEPES, pH 7.4.
9. Hypotonic Ringer's solution: Prepared by mixing one part of Ringer's solution with two parts of distilled water (18 MΩ).
10. Buffer A solution: 30 mM HEPES, pH 7.4, 70 mM KCl, 3 mM $MgCl_2$, 1 mM ethylene glycol tetraacetic acid (EGTA), 1 mM dithiothreitol, 0.1 mM4-(2-aminoethyl) benzenesulfonyl fluoride hydrochloride (AEBSF).
11. Polylysine solution: 0.5 mg/mL Polylysine dissolved in Ca^{2+}-free Ringer's solution.

2.2 Material for Cell Culture

1. Dulbecco's modified Eagle's medium (DMEM) was supplemented with 10% fetal bovine serum (FBS), 2 mM L-glutamine, 100 unit/mL penicillin, 100 mg/mL streptomycin, 0.25% trypsin, 0.05% EDTA, 0.25% trypsin, 1 M HEPES.
2. For cell cultures square cover glass with side length 22 × 22 mm with #1 thickness (0.13–0.17 mm) is used [17]. Place the cover glass in the 35 mm tissue culture dishes for culturing cells for AFM experiments.
3. Cells were cultured in a humidified incubator with 5% CO_2 at 37 °C.

2.3 Material for Modification: The Mica/Cover Glass

1. Shear the mica sheets with a thickness of about 0.5 mm and 20 mm × 20 mm size.
2. Square cover glass 22 × 22 mm with #1 thickness.
3. Detergent for clean cover glass: concentrated cleaning solution micro-90, 1 M potassium hydroxide.
4. Silylation solution [18]: 3-Aminopropyltriethoxysilane (APTES); *N,N*-diisopropylethylamine (DIPEA).

2.4 Material for Modification: The AFM Tips

1. AFM cantilevers (probe model: DNP-10, Bruker), spring constant 0.06 N/m.
2. Piranha solution: H_2SO_4/30% H_2O_2: 7/3 (*see* **Note 1**).
3. Linker [9, 19]: Aldehyde-PEG21-NHS (SensoPath Technologies, Inc.).
4. Reagents: Chloroform, triethylamine (TEA), ethanolamine, 1 M $NaBH_3CN$ (made from 32 mg $NaBH_3CN$, 50 mL 100 mM NaOH, and 450 μL H_2O).

3 Methods

3.1 Preparation of the AP-Mica

We used mica as a substrate for AFM imaging because of its atomic flat surface that has minimum effect on the feature of cell membranes. On the APTES-mica surface, there is a layer of amino groups so that RBC membranes are attached onto the mica substrate tightly. The mica and glass surface without amino group modification was tested and appeared to be not tight enough to attach the cell membranes [18, 20]. The setup for producing AP-mica is shown in Fig. 1.

1. Cleave mica sheet with scotch tape. Make sure that the cleaved mica surface for sample preparation is completely smooth, and then placed in the desiccator.
2. After a desiccator is purged with argon for 5 min, 30 μL of APTES and 10 μL of DIPEA are placed into small containers at the bottom of the desiccator.
3. Purge with argon for additional 5 min and then seal off the desiccator. The mica is exposed to APTES vapor for 4 h.
4. Remove the APTES and DIPEA from the desiccator carefully. Purge the desiccator with argon for 5 min and seal. The AP-mica is stored in the sealed desiccator under argon for use within a week.

3.2 Preparation of the AP-Cover Glass

1. Cover glass can be used for various cultured cells. Prior to cell culture, strict cleaning steps must be performed to obtain a clean surface on the cover glass.

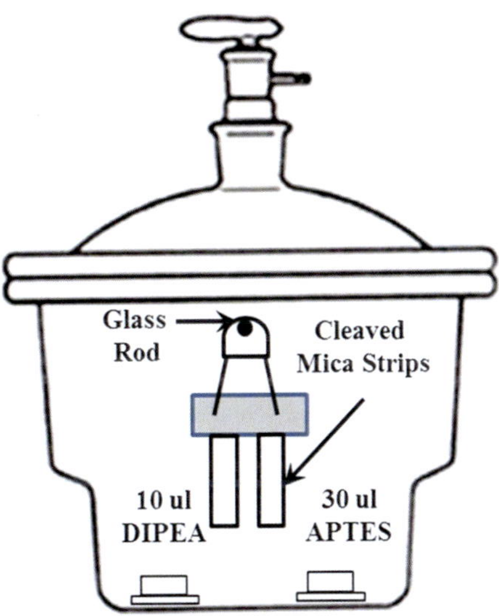

Fig. 1 Schematic diagram of making AP-mica. Mica strips are clamped in metal clips suspended on a glass rod of appropriate length (to nest snugly across the upper part of the desiccator). The reagents DIPEA and APTES are placed in small containers on the bottom of the desiccator

2. Cover glass is cleaned using a detergent (Micro-90) and then sonicated in 1 M potassium hydroxide for 20 min at room temperature.
3. The coverslips are rinsed with sterile distilled water (18 MΩ) and subsequently stored in absolute ethyl alcohol.
4. Prior to use, the coverslips are washed three times with sterile distilled water and dried with pure argon [21, 22].
5. The clean coverslips are placed in a culture dish as the substrate for cell culture.

3.3 Preparation of the RBC Membrane

The shearing open method is appropriate for the preparation of a clean membrane with minimum damage. The schematic of shearing open method is shown in Fig. 2 [23].

1. Two drops of blood are taken from a fingertip and centrifuged in 1 mL of PBS buffer five times ($560 \times g$, for 2 min).
2. A drop of red blood cell in PBS buffer (100 μL) is subsequently deposited on the AP-mica surface for approximately 20 min of absorption.
3. Wash out the non-adsorbed cells with PBS buffer by 1 mL pipette.
4. A syringe is adjusted to obtain a 20° to the sample surface, and 10 mL of 5% (v/v) PBS hypotonic buffer is injected to flush the mica surface and then obtain the flat cell membrane patch.

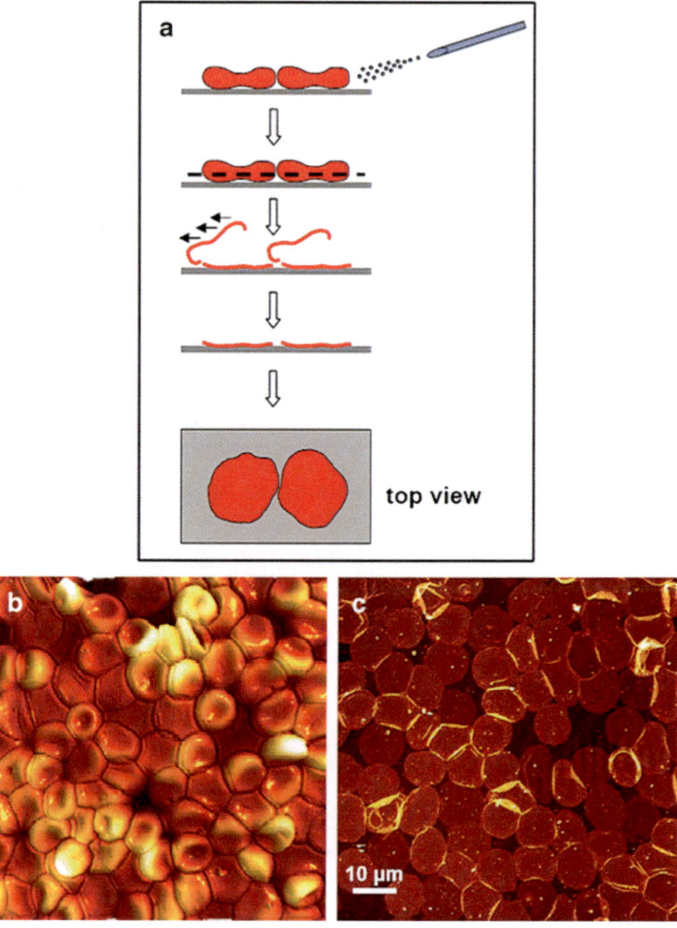

Fig. 2 Shearing open method of membrane preparation scheme. (**a**) RBCs are exposed to fluid flow-imposed shear stress to open the cells. (**b**) AFM images of RBCs attached to AP-mica. (**c**) Cytoplasmic side RBC membranes spread on surface after shear stress

5. Wash out the mica with PBS buffer and then the membrane patch can be imaged by AFM.
6. Digestion of the cytoplasmic side cell membranes: 200 mL 1 mg/mL Trypsin was added onto the membranes for 1 h at 37 °C and the membranes were gently washed by PBS for AFM imaging.
7. Digestion of the ectoplasmic side of cell membranes: Fresh washed RBC in PBS buffer was digested by PNGase F for 2 h at 37 °C, then the RBC was deposited on AP-mica for 15 min, and the cells were gently sheared opened.
8. To digest the outer membrane by both PNGase F and trypsin the fresh washed RBC in PBS buffer was digested by PNGase F for 2 h at 37 °C and trypsin for 1 h at 37 °C in sequence, and then the membranes were imaged by AFM.

3.4 Preparation of Detergent-Resistant Membranes (DRMs)

1. Prepared the RBC membrane as described above.
2. Add 200 μL 0.1% (v/v) Triton X-100 on the sample of the cell membrane for 4 min at 4 °C.
3. The sample was rinsed with PBS buffer three times to remove the remaining Triton X-100. The DRMs of erythrocytes were scanned in PBS buffer.
4. After recording DRM images, 100 μL 10 mM M-β-CD was injected into the AFM sample cell through the flow-through liquid cell. The changes of the sample can be recorded in real time [14, 24].

3.5 Preparation of the Cytoplasmic Side Nucleated Mammalian Cell Membranes

1. Cells were cultivated overnight on glass coverslips. The cell membranes are prepared by the shearing open method.
2. The cells are washed twice with ice-cold PIPES buffer, incubated with ice-cold hypotonic buffer for 3 min, and then sheared opened by a stream of 10 mL of hypotonic buffer through a needle at an angle of 20° [15].
3. The membranes are subsequently treated with high-salt buffer for 30 min at room temperature to remove the cytoskeletons.
4. The prepared membranes are immediately imaged in PBS buffer by AFM.

Ultrasonic stimulation is an alternative method for the preparation of the cytoplasmic side of membranes [25].

1. Cells on coverslips were washed once with HEPES-based Ringer's solution three times, and then with Ca^{2+}-free Ringer's solution three times.
2. The coverslips were soaked for about 10 s in polylysine solution, and then washed three times for a few seconds each in hypotonic Ringer's solution. This induced cell swelling, which enabled the cells to burst easily following ultrasonic stimulation.
3. After immersing in hypotonic solution, the cells were exposed to a small bubble jet by weak ultrasonic vibration in isotonic buffer A. Cells unroofed by the bubble jet were washed briefly in fresh buffer A used for AFM imaging.

3.6 Preparation of the Ectoplasmic Side Nucleated Mammalian Cell Membranes

Two strategies are used to prepare the ectoplasmic side of membranes. First, the ectoplasmic side of membranes can be imaged at the flat edge of a living cell under native conditions. Second, the ectoplasmic side of membranes can be obtained using the hypotonic lysis centrifugation method [17].

1. Cells are incubated with 60 μM nocodazole and 20 μM cytochalasin B for 50 min at 37 °C to destroy the actin filaments and microtubules.

2. The cells are subsequently digested using trypsin (1 mg/mL) and washed with 1 mL of PBS buffer three times.
3. Cells are treated with 1 mg/mL DNase to digest the nuclei/DNA and then centrifuged at 5035 × *g* for 10 min.
4. Cell membrane precipitate is dissolved in PBS buffer and deposited on AP-mica for AFM imaging.

Caution is required when imaging the ectoplasmic side of membrane (*see* **Note 1**).

3.7 AFM Tip Modification

Perform the whole procedure in a well-ventilated hood.

1. AFM tips are cleaned with piranha solution (H_2SO_4/30% H_2O_2: 7/3) and then in a UV cleaner for 15 min to remove any organic contamination on the tips. Please note that piranha is extremely dangerous: It is potentially explosive and extremely corrosive, and it will destroy gloves, clothes, skin, and body tissue within seconds (*see* **Note 2**).
2. The tips are placed in a dish at the bottom of a desiccator and modified with APTES just as in preparing AP-mica. After the treatment process, the APTES is removed and the treated tips (AP-tip) are stored in the sealed desiccator until used. This procedure provides a layer of amino groups on the AFM tips, suitable for TREC.
3. Aldehyde-PEG-NHS (3.3 mg) is dissolved in 0.5 mL chloroform and poured into a small glass reaction chamber (around 5–10 mL glass vial, do not use plastic).
4. Add 30 μL triethylamine to the glass reaction chamber and then the amino-functionalized AFM tips are immediately immersed for 2 h. During this period the reaction chamber is covered with a glass beaker to make sure that chloroform does not evaporate.
5. After 2 h the tips are washed with chloroform (three times) and dried under a gentle stream of nitrogen gas. The tips should be immediately used for further derivatization with antibodies.
6. The antibody solution (about 100 μL of roughly 0.1 mg/mL protein) is put in one drop onto the cantilevers. In addition, 2 μL of 1 M $NaBH_3CN$ is added to the drop and mixed carefully with the pipette. The proteins have to be allowed to react for 1 h to couple via intrinsic amino groups to the aldehyde groups of PEG linkers on the tip.
7. 5 μL of 1 M aqueous ethanolamine is added to the protein solution drop in order to passivate unreacted aldehyde groups. The tips are washed three times with PBS buffer and stored in PBS in well plates or small beakers at 4 °C. They can be used for TREC experiments [26, 27].

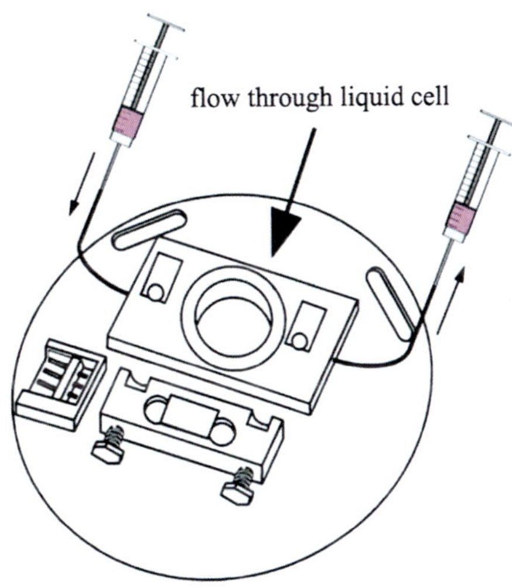

Fig. 3 Schematic of flow-through liquid cell

3.8 In Situ AFM Imaging by a Flow-Through Liquid Cell

AFM imaging was performed by 5500 AFM (Keysight Technologies, Chandler, AZ). Cell membranes were imaged by AAC mode AFM. Oxide-sharpened Si_3N_4 probes (probe model: DNP-10, Bruker) with a spring constant of 0.06 N/m were used for imaging the soft cell membranes. Imaging amplitude was set between 2.0 and 2.5 V. Scanning speed was 1.8 Hz. All the images were recorded with 512 × 512 pixels.

In situ imaging is the greatest advantage of AFM. To maintain biological samples in their native states, in situ imaging of cell membranes is desirable because a dry sample will be deformed and will not represent the physiological features. Environmental control can be realized using a flow-through liquid cell [18, 20]. As shown in Fig. 3, the use of a syringe to inject the appropriate solution from one end and remove the same volume solution from the other end can change the buffer conditions (e.g., the salt concentration, proteinase, and pH) during imaging, which is rather important to observe biomolecules at work (*see* **Note 3**).

3.9 Recognition Imaging

For recognition imaging, cell membrane samples are deposited on the AP-mica and scanned with antibody-modified tips. Imaging amplitude is set between 2.0 and 2.5 V. Recognition imaging was performed by Magnetic AC mode (MAC mode) AFM 5500 with a PicoTREC recognition imaging attachment (Keysight Technologies, Chandler, AZ) (with 6–8 nm amplitude oscillation at 9 kHz, imaging at 70% set point, and scan speed at <1 Hz). For testing specificity of the recognition reaction, 100 μL of a 30 μg/mL solution of a peptide antigenic to the antibody (usually the one that was used to elicit the antibody response) is flowed into the AFM sample cell and the sample is rescanned [9, 28]. If recognition is specific, the presence of the peptide will block the recognition to a very high degree (*see* **Note 4**).

4 Notes

1. Caution is required when imaging the ectoplasmic side of membranes because the sharp AFM tips can destroy the relatively soft ectoplasmic side lipid bilayer membrane and induce imaging artifacts. AAC mode (or MAC mode) in solution and a regular tip radius (approximately 20 nm) are better for imaging a soft lipid bilayer. In addition, the use of glutaraldehyde to fix a cell may cause cross-linking of membrane proteins and produce artifacts in the membrane appearance. It is worth noting that to obtain the native structure of cell membranes without damage, it is not recommended to fix the membrane using chemicals (e.g., glutaraldehyde, paraformaldehyde) because they may cross-link the membrane proteins and result in local destruction of the membranes (e.g., blurred images because of protein cross-linking).
2. Treatment with piranha solution: Caution! Piranha solution can be prepared by adding the peroxide to the acid. Mixing the solution is highly exothermic. Piranha solution reacts violently with organic compounds and should be handled with great care, and using appropriate personal protective measures. Waste should not be stored in closed containers and should be disposed off by approved procedures.
3. It is also important to maintain a constant solution volume and temperature during imaging because a small temperature change may cause drift between the AFM probe and the sample, which thus affects repeated images in the same location.
4. Soft cantilevers with low frequency (6–9 kHz) should be used for recognition imaging. Stiffer cantilevers (higher frequency) do not produce good recognition images. Also, scan speeds >1 Hz should be avoided as they increase the leakage of topography into the recognition image. It is very important to check the specificity of antibodies used for recognition imaging studies. We have noted a number of nonspecific reactions, i.e., antibodies that recognize non-antigens, sometimes quite strongly, in both AFM recognition imaging (including force curves) and standard ELISA assays.

Acknowledgments

This work was supported by National Key R&D Program of China (No.2017YFA0505300), and National Natural Science Foundation of China (No. 21525314, 21503213, 21727816 and 21721003).

References

1. Nicolson GL (2014) The fluid-mosaic model of membrane structure: still relevant to understanding the structure, function and dynamics of biological membranes after more than 40 years. BBA-Biomembranes 1838 (6):1451–1466. https://doi.org/10.1016/j.bbamem.2013.10.019
2. Goni FM (2014) The basic structure and dynamics of cell membranes: an update of the Singer-Nicolson model. BBA-Biomembranes 1838(6):1467–1476. https://doi.org/10.1016/j.bbamem.2014.01.006
3. Engelman DM (2005) Membranes are more mosaic than fluid. Nature 438(7068):578–580. https://doi.org/10.1038/nature04394
4. Binnig G, Quate C, Gerber C (1986) Atomic force microscope. Phys Rev Lett 56:930–933
5. Alessandrini A, Facci P (2005) AFM: a versatile tool in biophysics. Meas Sci Technol 16(6): R65–R92. https://doi.org/10.1088/0957-0233/16/6/r01
6. Kada G, Kienberger F, Hinterdorfer P (2008) Atomic force microscopy in bionanotechnology. Nano Today 3(1–2):12–19
7. Muller DJ, Dufrene YF (2008) Atomic force microscopy as a multifunctional molecular toolbox in nanobiotechnology. Nat Nanotechnol 3(5):261–269. https://doi.org/10.1038/nnano.2008.100
8. Dufrene YF, Ando T, Garcia R, Alsteens D, Martinez-Martin D, Engel A, Gerber C, Muller DJ (2017) Imaging modes of atomic force microscopy for application in molecular and cell biology. Nat Nanotechnol 12 (4):295–307. https://doi.org/10.1038/nnano.2017.45
9. Stroh C, Wang H, Bash R, Ashcroft B, Nelson J, Gruber H, Lohr D, Lindsay SM, Hinterdorfer P (2004) Single-molecule recognition imaging-microscopy. Proc Natl Acad Sci U S A 101(34):12503–12507. https://doi.org/10.1073/pnas.0403538101
10. Jiang JG, Hao X, Cai MJ, Shan YP, Shang X, Tang ZY, Wang HD (2009) Localization of Na^{+}-K^{+} ATPases in quasi-native cell membranes. Nano Lett 9(12):4489–4493. https://doi.org/10.1021/nl902803m
11. Shan YP, Wang HD (2015) The structure and function of cell membranes examined by atomic force microscopy and single-molecule force spectroscopy. Chem Soc Rev 44 (11):3617–3638. https://doi.org/10.1039/c4cs00508b
12. Lingwood D, Simons K (2010) Lipid rafts as a membrane-organizing principle. Science 327 (5961):46–50. https://doi.org/10.1126/science.1174621
13. Shan YP, Wang ZY, Hao XA, Shang X, Cai MJ, Jiang JG, Fang XX, Wang HD, Tang ZY (2010) Locating the Band III protein in quasi-native cell membranes. Anal Methods 2 (7):805–808. https://doi.org/10.1039/c0ay00278j
14. Cai MJ, Zhao WD, Shang X, Jiang JG, Ji HB, Tang ZY, Wang HD (2012) Direct evidence of lipid rafts by in situ atomic force microscopy. Small 8(8):1243–1250. https://doi.org/10.1002/smll.201102183
15. Ziegler U, Vinckier A, Kernen P, Zeisel D, Biber J, Semenza G, Murer H, Groscurth P (1998) Preparation of basal cell membranes for scanning probe microscopy. FEBS Lett 436(2):179–184
16. Wang HD, Hao X, Shan YP, Jiang JG, Cai MJ, Shang X (2010) Preparation of cell membranes for high resolution imaging by AFM. Ultramicroscopy 110(4):305–312. https://doi.org/10.1016/j.ultramic.2009.12.014
17. Zhao WD, Tian YM, Cai MJ, Wang F, Wu JZ, Gao J, Liu SH, Jiang JG, Jiang SB, Wang HD (2014) Studying the nucleated mammalian cell membrane by single molecule approaches. PLoS One 9(5):13. https://doi.org/10.1371/journal.pone.0091595
18. Lohr D, Bash R, Wang H, Yodh J, Lindsay S (2007) Using atomic force microscopy to study chromatin structure and nucleosome remodeling. Methods 41(3):333–341. https://doi.org/10.1016/j.ymeth.2006.08.016
19. Wang HD, Obenauer-Kutner L, Lin M, Huang YP, Grace MJ, Lindsay SM (2008) Imaging glycosylation. J Am Chem Soc 130 (26):8154–815+. https://doi.org/10.1021/ja802535p
20. Wang HD, Bash R, Yodh JG, Hager GL, Lohr D, Lindsay SM (2002) Glutaraldehyde modified mica: a new surface for atomic force microscopy of chromatin. Biophys J 83 (6):3619–3625
21. Wu JZ, Gao J, Qi M, Wang JZ, Cai MJ, Liu SH, Hao X, Jiang JG, Wang HD (2013) High-efficiency localization of Na+-K+ ATPases on the cytoplasmic side by direct stochastic optical reconstruction microscopy. Nanoscale 5 (23):11582–11586. https://doi.org/10.1039/c3nr03665k
22. Wang Y, Gao J, Guo XD, Tong T, Shi XS, Li LY, Qi M, Wang YJ, Cai MJ, Jiang JG, Xu CQ, Ji HB, Wang HD (2014) Regulation of EGFR nanocluster formation by ionic protein-lipid

interaction. Cell Res 24(8):959–976. https://doi.org/10.1038/cr.2014.89

23. Schillers H (2008) Imaging CFTR in its native environment. Pflugers Arch 456(1):163–177. https://doi.org/10.1007/s00424-007-0399-8
24. Lingwood D, Simons K (2007) Detergent resistance as a tool in membrane research. Nat Protoc 2(9):2159–2165. https://doi.org/10.1038/nprot.2007.294
25. Usukura J, Yoshimura A, Minakata S, Youn D, Ahn J, Cho SJ (2012) Use of the unroofing technique for atomic force microscopic imaging of the intra-cellular cytoskeleton under aqueous conditions. J Electron Microsc 61 (5):321–326. https://doi.org/10.1093/jmicro/dfs055
26. Wildling L, Unterauer B, Zhu R, Rupprecht A, Haselgrubler T, Rankl C, Ebner A, Vater D, Pollheimer P, Pohl EE, Hinterdorfer P, Gruber HJ (2011) Linking of sensor molecules with amino groups to amino-functionalized AFM tips. Bioconjug Chem 22(6):1239–1248. https://doi.org/10.1021/bc200099t
27. Ebner A, Wildling L, Kamruzzahan ASM, Rankl C, Wruss J, Hahn CD, Holzl M, Zhu R, Kienberger F, Blaas D, Hinterdorfer P, Gruber HJ (2007) A new, simple method for linking of antibodies to atomic force microscopy tips. Bioconjug Chem 18 (4):1176–1184. https://doi.org/10.1021/bc070030s
28. Hinterdorfer P, Dufrene YF (2006) Detection and localization of single molecular recognition events using atomic force microscopy. Nat Methods 3(5):347–355. https://doi.org/10.1038/nmeth871

Chapter 23

Physicochemical Characterization of Phthalocyanine-Functionalized Quantum Dots by Capillary Electrophoresis Coupled to a LED Fluorescence Detector

Gonzalo Ramírez-García, Fanny d'Orlyé, Tebello Nyokong, Fethi Bedioui, and Anne Varenne

Abstract

Capillary zone electrophoresis (CZE) complemented with Taylor Dispersion Analysis-CE (TDA-CE) was developed to physicochemically characterize phthalocyanine-capped core/shell/shell quantum dots (QDs) at various pH and ionic strengths. An LED-induced fluorescence detector was used to specifically detect the QDs. The electropherograms and taylorgrams allowed calculating the phthalocyanine-QDs (Pc-QDs) ζ-potential and size, respectively, and determining the experimental conditions for colloidal stability. This methodology allowed evidencing either a colloidal stability or an aggregation state according to the background electrolytes nature. The calculated ζ-potential values of Pc-QDs decreased when ionic strength increased, being well correlated with the aggregation of the nanoconjugates at elevated salt concentrations. For the same reason, the hydrodynamic diameter of Pc-QDs increased with increasing background electrolyte ionic strength. The use of electrokinetic methodologies has provided insights into the colloidal stability of the photosensitizer-functionalized QDs in physiologically relevant solutions and, thereby, its usefulness for improving their design and applications for photodynamic therapy.

Key words Colloidal stability, Quantum dots, Capillary electrophoresis, Photodynamic therapy, Phthalocyanines, Photosensitizers

1 Introduction

Quantum dots (QDs) are inorganic semiconductor nanocrystals having unique optical properties such as high luminescent quantum yields, large molar extinction coefficients, tunable excitation and emission spectra, and photostability [1, 2]. Hence they can be used as imaging and analytical probes for biomedical studies and as therapeutic systems in the nanomedicine field [3, 4].

In order to improve their biological applications, surface modification is crucial for providing biocompatible physicochemical properties and aqueous solubility and for preventing aggregation [1, 5]. Various types of nanoparticles (NPs), including QDs have

Volkmar Weissig and Tamer Elbayoumi (eds.), *Pharmaceutical Nanotechnology: Basic Protocols*, Methods in Molecular Biology, vol. 2000, https://doi.org/10.1007/978-1-4939-9516-5_23,

been functionalized and proposed to be applied in photodynamic therapy (PDT) [6]. In this sense, a family of glutathione-capped core/shell/shell QDs has been recently synthesized and covalently functionalized with phthalocyanines (Pc) containing different purposeful groups and metallic centers [7–9]. To facilitate the discussion, they will be denoted below as Pc-conjugated QDs (Pc-QDs). These nanohybrids hold great promise in PDT because they can act as carriers of photosensitizers, energy donors to excite the photosensitizers, and as an imaging contrast agent. Their unique optical and emission properties have been precisely tuned in order to make them able to emit light in the excitation region of their corresponding Pcs, permitting the energy transfer via the Förster resonance energy transfer (FRET) mechanism, giving a theranostic tool for cancer detection and treatment. They have been characterized in terms of photophysical properties in order to be applied in photodynamic therapy, making the synthesized Pc-QDs viable and active photosensitizers.

However, one of the most important current limitations of PDT is that the majority of the photosensitizers have extended delocalized aromatic π electron systems, a characteristic property that allows them to absorb light efficiently. At the same time due to π–π stacking (attractive noncovalent interactions between aromatic rings, since they contain π bonds) and hydrophobic interactions, they easily form aggregates in aqueous media [10]; it is therefore necessary to evaluate their colloidal stability before biological applications [11].

Capillary electrophoresis (CE) has emerged as a powerful tool to separate and characterize different nanoparticles, among which QDs [12–20]. In comparison to the classical methods for NP physicochemical characterization, CE presents several advantages, such as high separation efficiency in aqueous, hydro-organic or organic media, low sample and products consumption, simple sample preparation, accurate quantitative analysis, reduced analysis time, and a high degree of automation.

A family of water-soluble QDs functionalized with various Pc was successfully synthesized and characterized by CE using physiologically relevant buffers as background electrolyte [11]. The colloidal stability, hydrodynamic size, and ζ-potential values of the Pc-QDs were determined by CZE and Taylor dispersion analysis (TDA) performed in the CE apparatus. Thanks to the presence of QDs, a specific detection was obtained by employing an LED-induced fluorescence detector. The Pc-QDs characterization by CE is performed without any extensive purification, and provides in the same run, the identification of impurities in the samples. The results were compared and completed by screening LDE and DLS methods. Due to the photophysical properties of QDs, which cause interferences in the Laser Doppler Electrophoresis (LDE) and Dynamic Light Scattering (DLS) measures, a non-optical method such as CE turn out to be the most attractive technique for the characterization of the colloidal properties of this family of Pc-QDs in buffered solutions at physiological conditions, allowing the determination of the optimal conditions for dispersion and applicability of these nanohybrids in biological systems.

2 Materials

Prepare all solutions using ultrapure water (produced by purifying deionized water to attain a sensitivity of 18 MΩ/cm at 25 °C) and analytical grade reagents. Prepare all reagents at room temperature and store them at 4 °C. Diligently follow all waste disposal regulations when disposing waste materials.

2.1 Background Electrolytes

1. 150 mM ionic strength sodium phosphate (pH 7.4): 45.7 mM Na_2HPO_4/12.8 mM NaH_2PO_4 in water. Add 6.49 g Na_2HPO_4 and 1.53 g NaH_2PO_4 in 1 L water. Degass by sonication for 5 min, and store at 4 °C.
2. Background electrolytes preparation: for 15, 25, 50 and 100 mM ionic strength phosphate (pH 7.4), dilute the 150 mM ionic strength phosphate (pH 7.4) by 10, 6, 3 and 1.5, respectively, with water.
3. Filter the background electrolytes through a 0.20 μm filter before use.

2.2 Phthalocyanine Quantum Dots (See Note 1) and Neutral Marker Solutions

1. Phtalocyanine quantum dots (*see* **Note 2**): dry in oven at 100 °C and store at room temperature in hermetic containers.
2. Stock solutions of phtalocyanine quantum dots: dissolve at a 0.2 $mg.mL^{-1}$ concentration in the background electrolyte. Store at 4 °C (*see* **Note 3**).
3. Neutral marker: 200 μM Rhodamine B. Add 0.01% Rhodamine in the background electrolyte (*see* **Note 4**).

2.3 Capillary Preparation and Storage

1. Bare-fused silica capillary: 45 cm total length, 50 μm internal diameter, 380 μm outer diameter. Condition new capillaries by successive flushes with 1 M and 0.1 M NaOH, and then with water under a pressure of 92.5 kPa for 15 min each. Between runs, wash capillaries with BGE for 3 min. For storage, rinse with BGE and then water for 5 min each. Dry by air when not in use.

3 Methods

Carry out all procedures at room temperature. Analytical protocols, including capillary conditioning, sample introduction, and voltage application, were automatized and thus realized in one step. Each sample was analyzed four times with the average result presented.

3.1 CZE Protocol

Set the temperature in the capillary cartridge at 25 °C. Fill and homogenize the capillary with the BGE by flushing for 15 min after capillary activation or 3 min in between each separation at 92.5 kPa. To realize the successive injection of neutral marker and samples, apply a 2 kPa pressure for 6, 6, and 10 s for neutral marker, BGE, and sample, respectively (*see* **Note 5**). Apply a voltage of 14 kV to achieve the separation of the phthalocyanine quantum dots (*see* **Note 6**).

3.2 Taylor Dispersion Analysis (TDA-CE) Protocol (See Note 7)

Set the temperature in the capillary cartridge at 25 °C. Fill and homogenize the capillary with the BGE by flushing for 15 min after capillary activation or 3 min in between each separation at 92.5 kPa. Continuously introduce the sample via a 30 mbar pressure in the sample inlet vial (*see* **Note 8**).

3.3 Fluorescence Detection

For CZE and TDA-CE, perform detection with an LED-induced fluorescence detection at wavelength of >520 nm, induced by an excitation source of 480 nM (*see* **Note 9**).

3.4 Data Treatment for CZE and TDA-CE

Determine the barycenter of the electrophoretic profiles to calculate the zeta-potential values (*see* **Note 10**). Convert taylorgrams into apparent sphere diameter (r_H) (*see* **Note 11**).

3.5 Influence of Ionic Strength on Electrophoretic Profiles and Pc-QDs Zeta-Potential (See Note 12)

Separate by CZE the Pc-QDs in the phosphate background electrolyte at various ionic strengths (ranging from 15 to 150 mM) (*see* Fig. 1). Calculate the zeta-potential from the electrophoretic barycenter (*see* Table 1) (*see* **Note 13**).

3.6 Influence of Ionic Strength on TDA Fronts and Pc-QDs Hydrodynamic Diameter (See Note 14)

Analyze by TDA-CE the Pc-QDs in the phosphate background electrolyte at various ionic strengths (ranging from 15 to 150 mM) (*see* Fig. 2). Calculate the hydrodynamic diameter via the Stokes Einstein Equation (*see* Table 1) (*see* **Note 15**).

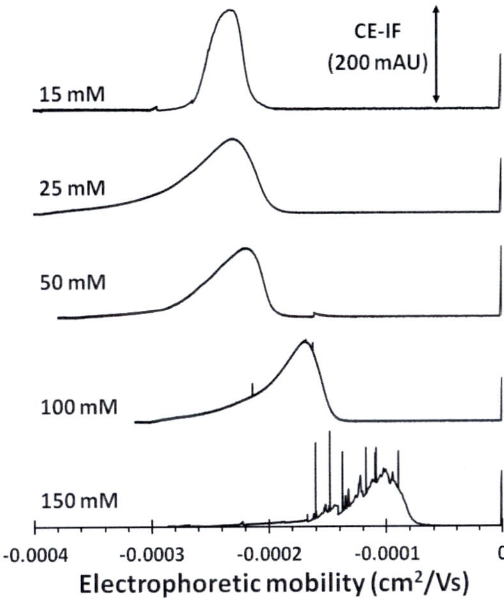

Fig. 1 Ionic strength effect on the electrophoretic mobility of QDs-ZnTCPPc in pH 7.4 phosphate buffer. *IF* intensity of the fluorescence. *Reproduced from* [11] *with permission from Springer*

Table 1
Ionic strength effect on the ζ-potential as obtained by CE, and on the size parameters as obtained by CE-TDA of QDS-ZnTCPPc in pH 7.4 phosphate buffer solution ($n = 4$)

Ionic strength (mM)	ζ-potential by CE (mV)	Diffusion coefficient by TDA ($\times 10^{-12}$ m²/s)	Hydrodynamic diameter by TDA (nm)
15 mM	−40.4 ± 3.4	18.20 ± 5.26	27.0 ± 7.8
25 mM	−41.9 ± 2.5	16.60 ± 6.39	29.6 ± 11.4
50 mM	−31.4 ± 3.6	6.36 ± 0.46	77.0 ± 5.6
100 mM	−25.1 ± 2.1	3.74 ± 0.27	131.2 ± 9.4
150 mM	−12.4 ± 3.2	2.81 ± 0.22	175.4 ± 14.4

4 Notes

1. The glutathione-capped QDs were synthesized and covalently functionalized using previously reported procedures [7–9]. The phthalocyanines are linked to the GSH-QDs via an amide bond using the carboxyl or sulphonate groups of the former and the NH_2 group on the glutathione capping of the latter. The schematic structure of these glutathione-capped

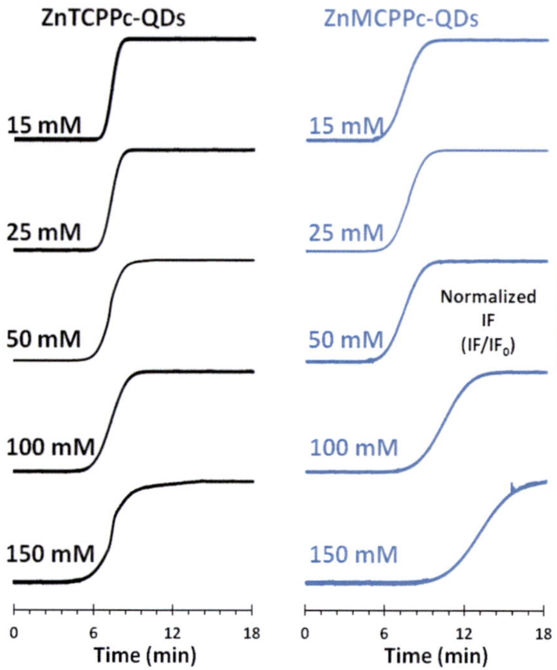

Fig. 2 Taylorgrams for size determination of ZnTCPPc-QDs and ZnMCPPc-QDs at various ionic strengths

QDs conjugates with phthalocyanines is presented in Fig. 3. Results are presented for zinc tetrasubstituted carboxyphenoxy phthalocyanine- GSH-CdTe/CdS/ZnS QDs [6.7 nm] (ZnTCPPc-QDs). The number in square brackets refers to the diameter of the crystalline structure of the nanoconjugate as obtained from X-ray dispersion (XRD) analysis. They were precipitated with excess methanol and purified before to be re-dispersed in the corresponding aqueous buffer solution.

2. The photophysical characterization of the synthesized Pc-QDs has already been described [8]. As a summary of the photophysical properties, the QDs conjugates with ZnTCPPc show maximum emission wavelength (λ_{emi}) at 630 nm, while the maximum absorption wavelength (λ_{abs}) of the corresponding conjugates was at 681 nm. The corresponding FRET efficiency (E_{ff}) was 98% for ZnTCPPc-QDs, with fluorescence lifetimes (τ_F) of 2.9 ns, and fluorescence quantum yields (Φ_F) of 0.12 [7].
3. Before analysis, the synthesized Pc-QDs were initially dispersed in the corresponding buffer and then gently sonicated by using a conventional sonication bath.
4. Rhodamine B is a neutral fluorescent marker, which *p*Ka value is 3.22 [21].

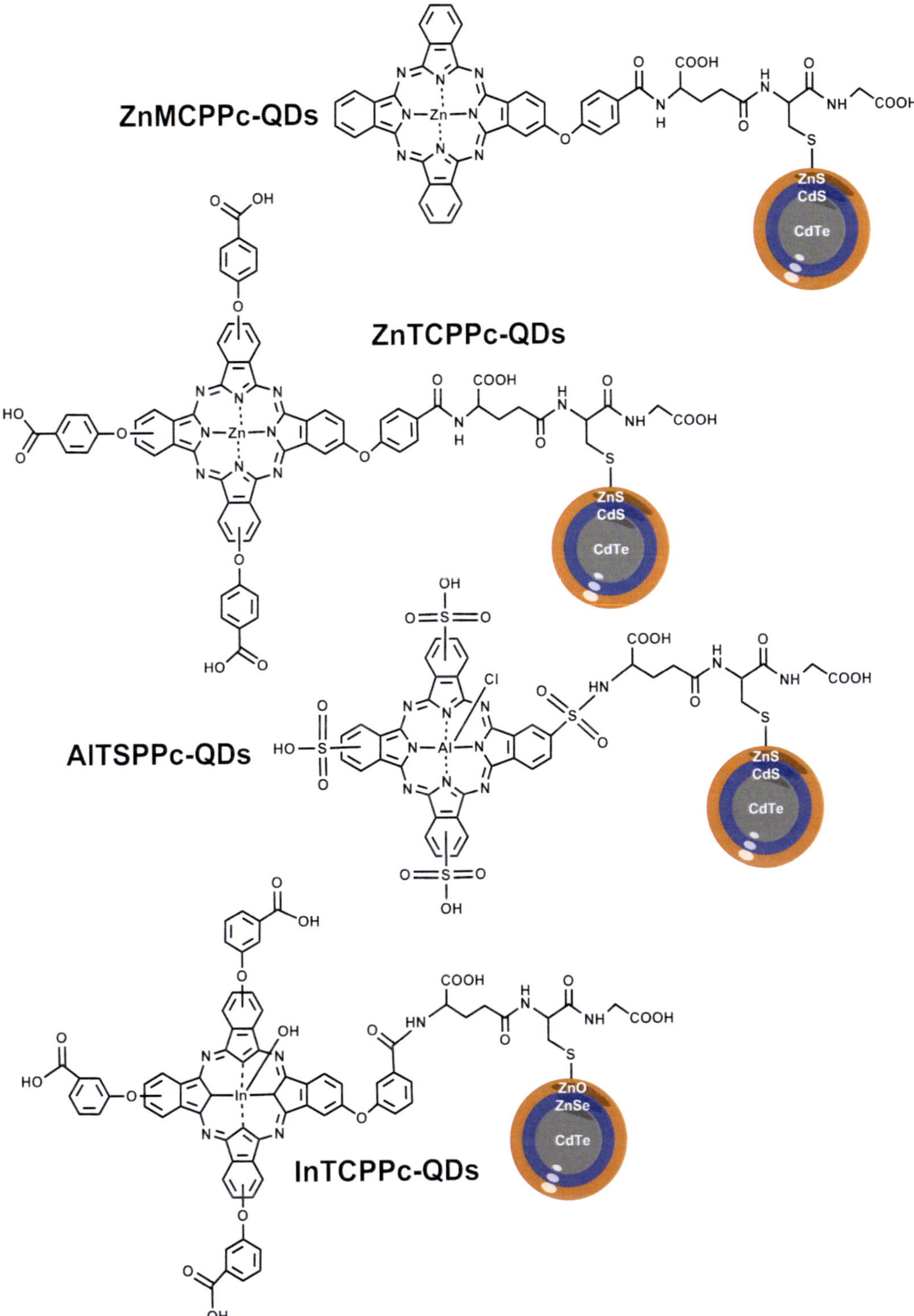

Fig. 3 Schematic representation of the phthalocyanine-capped GSH-QDs nanoconjugates. ZnMCPPc-QDs: zinc monosubstituted carboxyphenoxy phthalocyanine-GSH-CdTe/CdS/ZnS QDs, ZnTCPPc-QDs: zinc tetrasubstituted carboxyphenoxy phthalocyanine- GSH-CdTe/CdS/ZnS QDs, AlTSPPc-QDs: aluminum tetrasulfonated phthalocyanine-GSH-CdTe/CdS/ZnS QDs, InTCPPc-QDs: indium tetracarboxyphenoxy phthalocyanine-GSH-CdTe/ZnSe/ZnO-QDs. *Reproduced from [11] with permission from Springer*

5. The injection of a neutral marker for each sample separation allows controlling the electro-osmotic flow and determining precisely the electrophoretic mobilities.
6. The Pc-QDs have a negative electrophoretic mobility due to their negative surface charge density. However, their electrophoretic mobility is lower in absolute value than the electrosomotic mobility. Therefore the Pc-QDs are detected when a positive polarity is applied.
7. An alternative technique to determine diffusion coefficients of solute molecules was presented long time ago by Taylor [22]. Taylor dispersion analysis (TDA) using a capillary electrophoresis instrument appears to be a fast, simple and reliable technique for performing diffusion coefficient (D) measurements of nanometric particle populations. From diffusion coefficient one can determine the hydrodynamic diameter.
8. The TDA analysis was performed with front concentration profiles (and not pulse profiles) because this approach permits easy signal normalizations especially when the precise NP concentration is unknown and gives more accurate results as it involves less pressure step changes [18].
9. The Pc-QDs can be visualized by the UV-vis detector (diode array at 200 nm) available on each capillary electrophoresis equipment. However, fluorescence detection allows the specific detection of QDs. Therefore a double detection was performed that lead to confirm the presence of QDs.
10. The numeric electrophoretic mobility values were obtained from the barycenter of CE profiles, allowing the calculation of the ζ-potential values thanks to the equation derived by Ohshima et al., and simplified by Pyell et al. [23]:

$$\mu_{\mathrm{E}} = \frac{2}{3}\frac{\varepsilon_{\mathrm{r}}\varepsilon_0\zeta}{\eta}\left[f_1(\kappa a) - \left(\frac{ze\zeta}{kT}\right)^2 f_3(\kappa a) - m_{\mathrm{counter}}\left(\frac{ze\zeta}{kT}\right)^2 f_4(\kappa a)\right] \tag{1}$$

where ε_{r} is the relative electric permittivity, ε_0 the electric permittivity of vacuum, η the vicosity, e the elementary electric charge, k the Boltzmann constant, z the charge number of the counterion, T the absolute temperature, m_{counter} the dimensionless ionic drag coefficient of the counterion, and can be defined as:

$$m_{\mathrm{counter}} = \frac{2\varepsilon_{\mathrm{r}}\varepsilon_0 kT N_{\mathrm{A}}}{3\eta z \Lambda^0_{\mathrm{counter}}} \tag{2}$$

in which N_A is the Avogadro number and $\Lambda^0{}_{counter}$ the limiting equivalent conductance of the counter ion. The functions $f_1(\kappa a)$, $f_3(\kappa a)$ and $f_4(\kappa a)$ are given by the next equations:

$$f_1(\kappa a) = 1 + \frac{1}{2[1 + 2.5/\{\kappa a(1 + 2e^{-\kappa a})\}]^3} \quad (3)$$

$$f_3(\kappa a) = \frac{\kappa a(\kappa a + 1.3e^{-0.18\kappa a} + 2.5)}{2(\kappa a + 1.2e^{-7.4\kappa a} + 4.8)^3} \quad (4)$$

$$f_4(\kappa a) = \frac{9\kappa a(\kappa a + 5.2e^{-3.9\kappa a} + 5.6)}{8(\kappa a + 1.55e^{-0.32\kappa a} + 6.02)^3} \quad (5)$$

where κ is the Debye-Hückel parameter, the reciprocal thickness of the ion cloud, and a the sphere radius, which can be approximated by the number-weighted mean hydrodynamic radius of the investigated nanoparticles population.

11. Adequate mathematical models were also used to convert the taylorgrams into diffusion coefficient (Aris-Taylor equation, Eq. 6) and then into an apparent equivalent sphere radius (r_H) via the Stokes–Einstein equation [18, 22–25]:

$$D = \frac{R_c^2}{24\sigma^2} t_R \quad (6)$$

where R_c is the capillary radius, t_R the mean residence time (the time it takes to the solute moving with the mean velocity of the fluid to reach the detector) and σ^2 the temporal variance of the elution profile related to the dispersion coefficient by the eq. 7:

$$\sigma^2 = 2\frac{k t_R}{u^2} \quad (7)$$

where k is the dispersion coefficient and u is the mean fluid velocity (measured as the effective length of the capillary divided by t_R). The NP hydrodynamic radius (r_H) is then obtained via the Stokes–Einstein equation:

$$D_0 = \frac{k_B T}{6\pi\eta r_H} \quad (8)$$

being D_0 the Brownian diffusion coefficient of a single sphere, k_BT the thermal energy, η the fluid viscocity. By considering an infinite dilution of the NPs suspensions (or in practical approaches, considering very diluted suspensions), the D values can be used for the calculation of the hydrodynamic radius using the Eq. 8.

12. Figure 1 shows the electropherograms obtained by CZE showing the ionic strength effect on the ZnTCPPc-QDs electrophoretic mobility (μ_E) in PB (pH 7.4). As expected, the absolute electrophoretic mobility values decrease with an

increasing ionic strength due to counterions screening effect [26, 27]. The peak width of the CZE profiles increased on going from a relatively fine peak in 15 mM PB to larger profiles at ionic strengths between 25 and 100 mM probably due to electrophoretic dispersion. Indeed in these ζ-potential and ionic strength ranges, peak mobility dispersion should be increasingly sensitive to particle size distribution that may as well increase due to gradually enhanced inter-particle interaction giving rise to aggregate formation at the highest IS values. At 150 mM IS a severe aggregation of ZnTCPPc-QDs was evidenced by the presence of repeatable spiky profiles. Electrophoretic profiles obtained by CE are narrower but comprised within the broad domain of electrophoretic mobilities determined by LDE (results not shown). This observation indicates a clear advantage of CE in terms of precision and sensitivity with respect to LDE.

13. Table 1 presents ζ-potential values for ZnTCPPc-QDs calculated from electrophoretic profiles displayed in Fig. 1, indicating a decrease in absolute ζ-potential values when increasing ionic strength for ZnTCPPc-QDs (from −40.4 mV at 15 mM IS, to −12.4 mV at 150 mM IS). This observation is consistent with the differences in colloidal stability. Indeed, at high ζ-potential values important electrostatic repulsions between nanoparticles occur, thereby preventing their aggregation in aqueous solutions. At 150 mM IS, the ζ-potential value is quite low, dropping below −20 mV which can explain the plain aggregation of the sample, as evidenced by spiky electrophoretic profiles. No significant variations in ζ-potential values could be evidenced by means of LDE across this ionic strength range (15–150 mM) (results not shown). This can be due to the presence of impurities, i.e. some traces of precursors used during Pc-QDs synthesis or functionalization, which is overcome by CE thanks to its resolving power.
14. The mean residence time and the temporal variance σ^2 of each elution profiles were obtained by fitting the front concentration profile to the adequate equation [18]. Experimental profiles were very well fitted to the mathematical models. No significant variation in the retention time was detected in the taylorgrams for ZnTCPPc-QDs, indicating good robustness of the pressure system in addition to the absence of chromatographic retardation derived from interactions between the NPs and the capillary wall.
15. Table 1 presents the hydrodynamic diameter (d_h) value for ZnTCPPc-QDs obtained by using TDA. It is constant for IS under 25 mM at around 28 nm (Tukey test ($p \leq 0.05$), $n = 4$ replicates). It increases with the IS for values higher than 50 mM. This increase can be explained by the lower surface

charge density with higher ionic strength, leading to lower electrostatic repulsions between QDs and therefore to higher aggregation effect.

Similar d_h values were obtained by means of DLS measurements, with a lower precision (higher standard deviation values) than TDA (results not shown), which can be due to the interferences in the light scattering measurements caused by the inherent fluorescence or color presence of the QDs. With DLS, the particle size is not directly measured, but is calculated from the measured Brownian motion (diffusion coefficient) of the sample [28, 29].

Acknowledgments

This work was partially supported by the Department of Science and Technology (DST) and National Research Foundation (NRF) of South Africa, through the DST/NRF South African Research Chairs Initiative for Professor of Medicinal Chemistry and Nanotechnology (UID = 62620) and Rhodes University and by DST/Mintek Nanotechnology Innovation Centre (NIC). GRG is grateful to the Mexican National Council for Science and Technology (CONACYT). The authors acknowledge financial support from PROTEA Project 33885ZJ (France–South Africa).

References

1. Breger J, Delehanty JB, Medintz IL (2015) Continuing progress toward controlled intracellular delivery of semiconductor quantum dots. Wiley Interdiscip Rev Nanomed Nanobiotechnol 7(2):131–151. https://doi.org/10.1002/wnan.1281
2. Mussa Farkhani S, Valizadeh A (2014) Review: three synthesis methods of CdX (X = Se, S or Te) quantum dots. IET Nanobiotechnol 8 (2):59–76. https://doi.org/10.1049/iet-nbt.2012.0028
3. Volkov Y (2015) Quantum dots in nanomedicine: recent trends, advances and unresolved issues. Biochem Biophys Res Commun 468 (3):419–427. https://doi.org/10.1016/j.bbrc.2015.07.039
4. Maysinger D, Ji J, Hutter E, Cooper E (2015) Nanoparticle-based and bioengineered probes and sensors to detect physiological and pathological biomarkers in neural cells. Front Neurosci 9:480. https://doi.org/10.3389/fnins.2015.00480
5. Petryayeva E, Algar WR, Medintz IL (2013) Quantum dots in bioanalysis: a review of applications across various platforms for fluorescence spectroscopy and imaging. Appl Spectrosc 67(3):215–252. https://doi.org/10.1366/12-06948
6. Lucky SS, Soo KC, Zhang Y (2015) Nanoparticles in photodynamic therapy. Chem Rev 115 (4):1990–2042. https://doi.org/10.1021/cr5004198
7. Oluwole DO, Nyokong T (2015) Physicochemical behavior of nanohybrids of mono and tetra substituted carboxyphenoxy phthalocyanine covalently linked to GSH–CdTe/CdS/ZnS quantum dots. Polyhedron 87:8–16. https://doi.org/10.1016/j.poly.2014.10.024
8. Oluwole DO, Nyokong T (2015) Comparative photophysicochemical behavior of nanoconjugates of indium tetracarboxyphenoxy phthalocyanines covalently linked to CdTe/ZnSe/ZnO quantum dots. J Photochem Photobiol

A Chem 312:34–44. https://doi.org/10.1016/j.jphotochem.2015.07.009

9. Oluwole DO, Britton J, Mashazi P, Nyokong T (2015) Synthesis and photophysical properties of nanocomposites of aluminum tetrasulfonated phthalocyanine covalently linked to glutathione capped CdTe/CdS/ZnS quantum dots. Synth Met 205:212–221. https://doi.org/10.1016/j.synthmet.2015.04.015
10. Li L, Huh KM (2014) Polymeric nanocarrier systems for photodynamic therapy. Biomater Res 18:19
11. Ramírez-García G, Oluwole DO, Nxele SR, d'Orlyé F, Nyokong T, Bedioui F, Varenne A (2017) Characterization of phthalocyanine functionalized quantum dots by dynamic light scattering, laser Doppler, and capillary electrophoresis. Anal Bioanal Chem 409 (6):1707–1715. https://doi.org/10.1007/s00216-016-0120-x
12. Li YQ, Wang HQ, Wang JH, Guan LY, Liu BF, Zhao YD, Chen H (2009) A highly efficient capillary electrophoresis-based method for size determination of water-soluble CdSe/ZnS core-shell quantum dots. Anal Chim Acta 647 (2):219–225. https://doi.org/10.1016/j.aca.2009.06.004
13. Stewart DTR, Celiz MD, Vicente G, Colón LA, Aga DS (2011) Potential use of capillary zone electrophoresis in size characterization of quantum dots for environmental studies. TrAC Trends Anal Chem TrAC 30(1):113–122. https://doi.org/10.1016/j.trac.2010.10.005
14. Sang F, Huang X, Ren J (2014) Characterization and separation of semiconductor quantum dots and their conjugates by capillary electrophoresis. Electrophoresis 35(6):793–803. https://doi.org/10.1002/elps.201300528
15. Trapiella-Alfonso L, d'Orlyé F, Varenne A (2016) Recent advances in the development of capillary electrophoresis methodologies for optimizing, controlling, and characterizing the synthesis, functionalization, and physicochemical, properties of nanoparticles. Anal Bioanal Chem 408(11):2669–2675. https://doi.org/10.1007/s00216-015-9236-7
16. Radko SP, Chrambach A (2002) Separation and characterization of sub-mu m- and mu m-sized particles by capillary zone electrophoresis. Electrophoresis 23(13):1957–1972. https://doi.org/10.1002/1522-2683(200207)23:13<1957::aid-elps1957>3.0.co;2-i
17. d'Orlye F, Varenne A, Georgelin T, Siaugue JM, Teste B, Descroix S, Gareil P (2009) Charge-based characterization of nanometric cationic bifunctional maghemite/silica core/shell particles by capillary zone electrophoresis. Electrophoresis 30(14):2572–2582. https://doi.org/10.1002/elps.200800835
18. d'Orlyé F, Varenne A, Gareil P (2008) Determination of nanoparticle diffusion coefficients by Taylor dispersion analysis using a capillary electrophoresis instrument. J Chromatogr A 1204(2):226–232. https://doi.org/10.1016/j.chroma.2008.08.008
19. Wang YH, Wang L (2007) Defect states in Nd3+-doped CaAl2O4 : Eu2+. J Appl Phys 101 (5):Artn 053108. https://doi.org/10.1063/1.2435822
20. Fourest B, Hakem N, Guillaumont R (1994) Characterization of colloids by measurement of their mobilities. Radiochim Acta 66-67. https://doi.org/10.1524/ract.1994.6667.special-issue.173
21. Milanova D, Chambers RD, Bahga SS, Santiago JG (2011) Electrophoretic mobility measurements of fluorescent dyes using on-chip capillary electrophoresis. Electrophoresis 32 (22):3286–3294. https://doi.org/10.1002/elps.201100210
22. Taylor G (1953) Dispersion of soluble matter in solvent flowing slowly through a tube. Proc Royal Soc Lond A Math Phys Eng Sci 219 (1137):186–203. https://doi.org/10.1098/rspa.1953.0139
23. Pyell U, Jalil AH, Pfeiffer C, Pelaz B, Parak WJ (2015) Characterization of gold nanoparticles with different hydrophilic coatings via capillary electrophoresis and Taylor dispersion analysis. Part I: determination of the zeta potential employing a modified analytic approximation. J Colloid Interface Sci 450:288–300. https://doi.org/10.1016/j.jcis.2015.03.006
24. Pyell U, Jalil AH, Urban DA, Pfeiffer C, Pelaz B, Parak WJ (2015) Characterization of hydrophilic coated gold nanoparticles via capillary electrophoresis and Taylor dispersion analysis. Part II: Determination of the hydrodynamic radius distribution - Comparison with asymmetric flow field-flow fractionation. J Colloid Interface Sci 457:131–140. https://doi.org/10.1016/j.jcis.2015.06.042
25. Aris R (1956) On the dispersion of a solute in a fluid flowing through a tube. Proc Royal Soc Lond A Math Phys Eng Sci 235(1200):67–77. https://doi.org/10.1098/rspa.1956.0065
26. Wu L, Zhang J, Watanabe W (2011) Physical and chemical stability of drug nanoparticles. Adv Drug Deliv Rev 63(6):456–469. https://doi.org/10.1016/j.addr.2011.02.001

27. Kuzovkov VN, Kotomin EA (2014) Static and dynamic screening effects in the electrostatic self-assembly of nano-particles. Phys Chem Chem Phys 16(46):25449–25460. https://doi.org/10.1039/C4CP02448F
28. Hoo CM, Starostin N, West P, Mecartney ML (2008) A comparison of atomic force microscopy (AFM) and dynamic light scattering (DLS) methods to characterize nanoparticle size distributions. J Nanopart Res 10 (1):89–96. https://doi.org/10.1007/s11051-008-9435-7
29. Sapsford KE, Tyner KM, Dair BJ, Deschamps JR, Medintz IL (2011) Analyzing nanomaterial bioconjugates: a review of current and emerging purification and characterization techniques. Anal Chem 83(12):4453–4488. https://doi.org/10.1021/ac200853a

Chapter 24

In Vitro Testing of Anticancer Nanotherapeutics Using Tumor Spheroids

Avanti Ganpule, Zishu Gui, Mohammed A. Almuteri, and Gerard G. M. D'Souza

Abstract

Tumor cells grown as spheroids more closely resemble in vivo solid tumors and present physical and physiological barriers to drug action that conventional monolayer cell cultures do not. The physiological relevance of the model can be further increased by incorporating fluid flow and multiple cell types. The protocols described in this chapter use simple and easily available equipment to reproducibly generate spheroid models that can be used for the in vitro testing of a variety of preparations.

Key words Spheroids, 3D cell culture, 3D in vitro model, Coculture spheroids, Fluid models

1 Introduction

The widespread approach of utilizing cancer cell in pure monolayers for testing nanotherapeutics is considered to have low predictability as it lacks the physical barriers present in the in vivo tumors. 3D cell cultures (Spheroids) provide the advantage of simulating these barriers and have been shown to closely mimic the properties of in vivo tumors [1, 2]. Various methods are being used for spheroid production and their analysis [3–5]. Currently, research is focused on obtaining uniform-sized spheroids in a reproducible and high throughput manner and to extend the scope of these models to make them physiologically more relevant [6, 7]. The protocols described herein are based on our efforts that focused on using simple and easily available apparatus to generate large numbers of spheroids for analysis by assays typically used with monolayer cultures [8]. Also described is our recently developed protocol to incorporate fluid flow that mimics blood flow.

Volkmar Weissig and Tamer Elbayoumi (eds.), *Pharmaceutical Nanotechnology: Basic Protocols*, Methods in Molecular Biology, vol. 2000, https://doi.org/10.1007/978-1-4939-9516-5_24, © Springer Science+Business Media, LLC, part of Springer Nature 2019

2 Materials

2.1 Preparation of Spheroids for 96-Well Plate Assays

1. Cell lines (*see* **Note 1**).
2. Complete cell culture medium (*see* **Note 1**).
3. Serum-free cell culture medium (*see* **Note 1**).
4. Cell culture grade Agarose.
5. 96-well polystyrene plates.
6. 0.25% Trypsin-EDTA solution.
7. 10 mM Phosphate Buffered Saline (PBS) pH 7.4.
8. Glass beaker.
9. Aluminum foil.
10. Parafilm.
11. Autoclave.
12. Multichannel pipettor.
13. Cell culture pipettes.
14. Reservoir boats.
15. Plate centrifuge.

2.2 Incorporation of Spheroids in a Simple Fluid Flow Chamber

1. Pasteur pipette.
2. Glass cutter.
3. Plastic or glass tube with 2 mm outer diameter.
4. Flexible plastic connectors with 6 mm internal diameter(ID).
5. Luer lock connectors.
6. Y-shaped connector.
7. Tygon tubing with 2.4 mm ID.
8. Parafilm.
9. Spheroids.
10. Matrigel® matrix (Corning Inc. Life Sciences, Tewksbury, MA).
11. Serum-free culture medium.
12. Complete culture medium.
13. Peristaltic pump.
14. Bead bath.

3 Methods

3.1 Preparation of Spheroids for 96-Well Plate Assays

1. Based on the number of plates desired, combine an appropriate amount of agarose with serum-free medium in a glass beaker that can hold at least three times the volume being prepared.
2. Cover the beaker with aluminum foil and autoclave using an appropriate cycle.
3. Following autoclaving collect the resulting 1% solution of agarose in serum-free medium while still hot and liquid. If necessary, heat in a microwave or water bath to liquefy the agarose solution.
4. Transfer the agarose solution to a reservoir boat and while still liquid transfer 50 μl of the agarose solution to each well of the 96-well plates using a multichannel pipettor. Taking care to avoid trapping air bubbles in the agarose.
5. Allow the agarose to cool and seal plates with parafilm.
6. Parafilm-sealed plates may be stored at room temperature until needed for up to a week.
7. Harvest cells from culture flasks using standard techniques.
8. For spheroids containing single cell type seed appropriate number of cells in each well of previously prepared agarose-coated plates. 5000 or 10,000 cells per well is usually optimal (*see* **Note 1**). For spheroids containing multiple cell types, first prepare a cell suspension containing the desired ratio of the two cell populations and then transfer 5000 or 10,000 cells from the mixture to each well of an agar-coated plate (*see* **Note 1**). Final volume of cell culture medium in the well should not exceed 200 μl.
9. Once seeded, centrifuge the plates for 20–25 min in a plate centrifuge (*see* **Note 2**).
10. Leave the plates undisturbed at 37 °C, 5% CO_2 environment. Allow the spheroids to grow to the desired size while changing media every 2 days. Typically, spheroids take 10 days to reach the optimum size for further testing.
11. To use spheroids for testing they need to be transferred to 96-well plates without agarose. The transfer can be performed using a multichannel pipettor.
12. After transfer of the spheroids, carefully remove medium and replace with fresh medium containing the desired treatments with a final volume of no more than 200 μl in each well. Incubate treated spheroid plates for desired time.
13. After treatment, remove medium and replace with fresh medium or PBS.

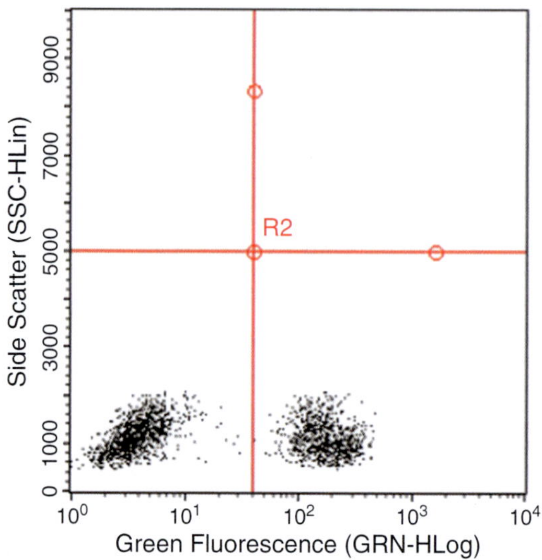

Fig. 1 Flow cytometry analysis of cells recovered from a spheroid containing two cell populations. Lower left quadrant: unstained cells. Lower right quadrant: GFP expressing cells

14. Whole treated spheroids can be directly examined by microscopy.
15. For flow cytometric analysis first dissociate spheroids by replacing cell culture medium with 125 μl of 0.25% Trypsin-EDTA and incubating for 5–10 min. Gentle agitation by pipetting can be used to facilitate dissociation. Dissociated spheroids can then be processed for flow cytometric analysis using similar protocols as for monolayer cultured cells. (*See* Fig. 1 for sample flow cytometric analysis of spheroid containing two cell types.)

3.2 Incorporation of Spheroids in a Simple Fluid Flow Chamber

1. Grow desired spheroids according to previous protocol.
2. Place Matrigel in 4 °C overnight to liquefy it.
3. Cut a glass Pasteur pipette using a glass cutter to obtain a cylindrical glass tube of 24 mm length and 6 mm diameter. Seal one end of the tube with parafilm. This tube will be further referred to as treatment tube.
4. Place a 30 mm long piece of the 2 mm outer diameter plastic or glass tube in the center of the prepared treatment tube from **step 3**.
5. Prepare a 1:1 mixture of chilled serum-free medium and Matrigel.
6. Transfer desired number of spheroids from production plate into a tube containing an appropriate amount of ice-cold 1:1 mixture of Matrigel and serum-free culture medium. Typically,

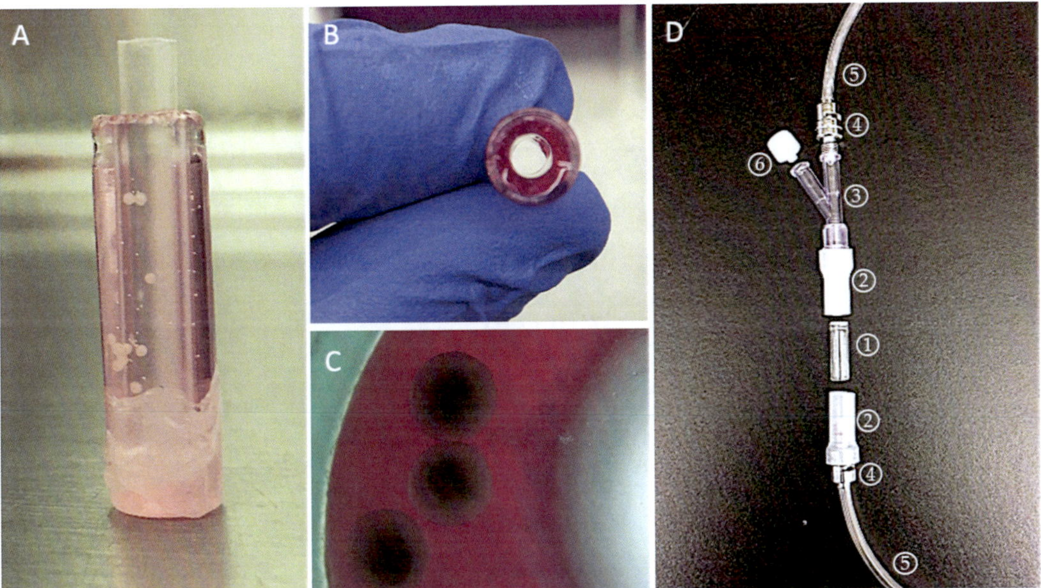

Fig. 2 Spheroid flow chamber setup. (**a**) Spheroid chamber with spheroids suspended in Matrigel before gel formation (**b**) Gel formation with a channel at the center (**c**) Cross-section of spheroid chamber showing spheroids embedded in the Matrigel (**d**) Components of the flow system—(1) Spheroid chamber, (2) 6 mm ID flexible plastic tube, (3) Y-shaped adaptor, (4) Luer lock connectors, (5) 2.4 mm ID plastic Tubing, (6) Luer lock plug

6 spheroids are combined with 400 μl of the Matrigel media mixture for each flow chamber.

7. Carefully pipette 400 μl of spheroid-containing Matrigel mixture into the space around the plastic tube in the treatment tube and allow it to gel at 37 ° C for 1 h. *See* Fig. 2a for how the system should look at this point.
8. Carefully remove the plastic tube from the center of the treatment tube once the media has set. *See* Fig. 2b for how the spheroid chamber system should look at this point.
9. Using a 30–40 mm long piece of 6 mm ID flexible plastic connector connect a Y-shaped adaptor at one end of the spheroid chamber.
10. Using another 30–40 mm long piece of 6 mm ID flexible connector, connect a Luer lock connector to the other end of the spheroid chamber.
11. Using a 300 mm long piece of the 2.4 mm ID tygon tubing connect the Y connector end of the spheroid chamber to the Luer lock connector end to create a closed loop.
12. Insert the loop tubing into an appropriate peristaltic pump.

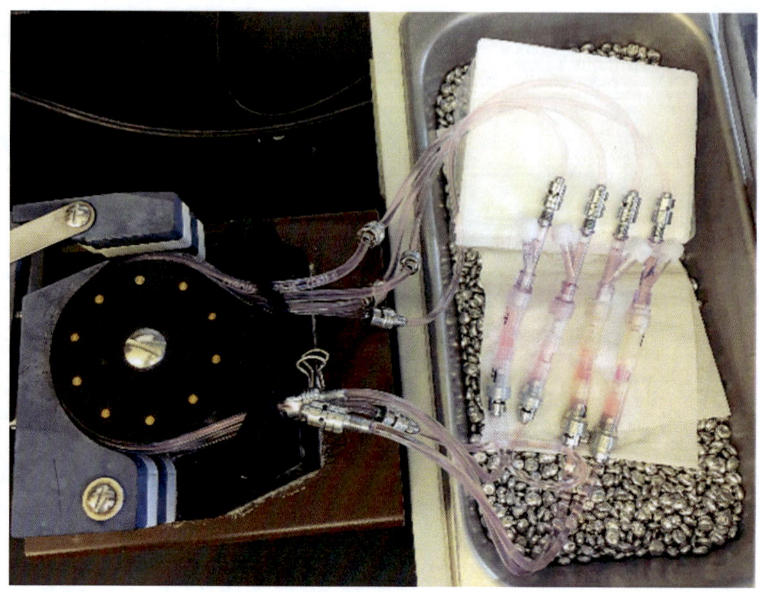

Fig. 3 Complete flow chamber setup with four individual spheroid chambers

13. Add approximately 5 ml of formulation containing complete culture medium through the side arm of Y-shaped adaptor to fill up the loop.
14. Start the peristaltic pump to allow minimum possible velocity of 1 μm/s. However, most peristaltic pumps might not be able to go that low. Instead, a velocity of 1 mm/s might be sufficient; accordingly set the appropriate velocity in μl/s. Check the system for leakage and wrap parafilm around the connections if required.
15. Place the treatment tube in a bead bath and leave it undisturbed for the desired incubation period. Multiple loops can be used depending on the type of peristaltic pump. *See* Fig. 3 for a representative setup.
16. After treatment disconnect the adaptors and recover the spheroid chamber tube. To collect the spheroids, place the treatment tube in 2 ml of PBS and incubate at 4 °C till the Matrigel liquefies.
17. Once the Matrigel has liquefied, collect spheroids using a pipette and process for further analysis.

4 Notes

1. It should be noted that the protocol is written without specific mention of the cell lines and media used. The user is expected to adapt this general protocol to the desired cell type. A

literature search can be used to determine whether a desired cell type can form spheroids and at what minimum seeding density. In case of cell lines that have not yet been tested for spheroid formation, the protocol can be used to empirically determine whether the cell line can form spheroids as well as to optimize the seeding density. If spheroids of multiple cell types are desired, the minimum requirement is that the cell lines form spheroids in pure cultures and can be grown in a common medium. We have in most cases determined such combinations empirically using various ratios of the cell lines to be combined. So far, we have used only two cell lines at a time, but it should be possible to use more than two as long as the facility to resolve the individual populations via flow cytometry or microscopy exists. Our primary approach to the resolution of multiple cell types is to use at least one cell line in the pair that expresses a fluorescent protein marker. Such constitutively labeled cell lines are becoming increasing available from commercial suppliers but could also be generated using standard transfection approaches. Starting cell number, cell ratio, and duration of spheroid growth must be optimized based on the desired size and cell population ratio required at the end of spheroid formation and on the aim of further experiments.

2. Centrifugation promotes cell adhesion. Hence, centrifugation time can vary based on cell line and cell number seeded. At the end of centrifugation, cells should be checked by microscopy for aggregation at the center of the plate.

References

1. Riedl A, Schlederer M, Pudelko K, Stadler M, Walter S, Unterleuthner D et al (2017) Comparison of cancer cells in 2D vs 3D culture reveals differences in AKT-mTOR-S6K signaling and drug responses. J Cell Sci 130(1):203–218
2. Xu X, Farach-Carson MC, Jia X (2014) Three-dimensional in *vitro* tumor models for cancer research and drug evaluation. Biotechnol Adv 32(7):1256–1268
3. Atefi E, Lemmo S, Fyffe D, Luker GD, Tavana H (2014) High throughput, polymeric aqueous two-phase printing of tumor spheroids. Adv Funct Mater 24(41):6509–6515
4. Ivanov DP, Parker TL, Walker DA, Alexander C, Ashford MB, Gellert PR et al (2014) Multiplexing spheroid volume, resazurin and acid phosphatase viability assays for high-throughput screening of tumour spheroids and stem cell neurospheres. PLoS One 9(8):e103817
5. Monjaret F, Fernandes M, Duchemin-Pelletier-E, Argento A, Degot S, Young J (2016) Fully automated one-step production of functional 3D tumor spheroids for high-content screening. J Lab Autom 21(2):268–280
6. Sabhachandani P, Motwani V, Cohen N, Sarkar S, Torchilin V, Konry T (2016) Generation and functional assessment of 3D multicellular spheroids in droplet based microfluidics platform. Lab Chip 16(3):497–505
7. Lao Z, Kelly CJ, Yang XY, Jenkins WT, Toorens E, Ganguly T et al (2015) Improved methods to generate spheroid cultures from tumor cells, tumor cells & fibroblasts or tumor-fragments: microenvironment, microvesicles and MiRNA. PLoS One 10(7):e0133895
8. Solomon MA, Lemera J, D'Souza GG (2016) Development of an in *vitro* tumor spheroid culture model amenable to high-throughput testing of potential anticancer nanotherapeutics. J Liposome Res 26(3):246–260

Index

Volkmar Weissig and Tamer Elbayoumi (eds.), *Pharmaceutical Nanotechnology: Basic Protocols*, Methods in Molecular Biology, vol. 2000, https://doi.org/10.1007/978-1-4939-9516-5,

Zeitfracht Medien GmbH
Ferdinand-Jühlke-Straße 7
99095 Erfurt, Deutschland
produktsicherheit@kolibri360.de